This book describes the origin and evolution of the solar system, with an emphasis on interpretation rather than description. Starting with the Big Bang 15–20 billion years ago, it traces the evolution of the solar system from the separation of a disk of gas and dust, the solar nebula, 4.7 billion years ago. The problems of the formation of the sun and the planets are considered beginning with Jupiter and the other gas giants, and ending with the formation of the earth, the other rocky inner planets and the Moon. All planets, satellites and rings are different and random encounters have played a major role in the evolution of the system. A prologue emphasizes the crucial role of lunar exploration, while an epilogue considers the place of *Homo sapiens* in the solar system. The author concludes that the solar system is probably unique; other planetary systems may be common, but the chances of finding a replica of ours seem remote.

Solar System Evolution

A New Perspective

SOLAR SYSTEM EVOLUTION

A New Perspective

An inquiry into the chemical composition, origin, and evolution of the solar system

Stuart Ross Taylor
M.A., D.Sc. (Oxon), M.Sc. (N.Z.), Ph.D. (Indiana)
F.A.A., Hon. F.G.S., Hon. F.R.S.N.Z.

Lunar and Planetary Institute, Houston, Texas and
The Australian National University, Canberra, Australia

CAMBRIDGE
UNIVERSITY PRESS

Published by the Press Syndicate of the University of Cambridge
The Pitt Building, Trumpington Street, Cambridge CB2 1RP
40 West 20th Street, New York, NY 10011–4211, USA
10 Stamford Road, Oakleigh, Victoria 3166, Australia

First published 1992

Printed in the United States of America

Library of Congress Cataloging-in-Publication Data

Taylor, Stuart Ross, 1925–
Solar system evolution : a new perspective : an inquiry into the chemical composition, origin, and evolution of the solar system / Stuart Ross Taylor.
p. cm.
Includes bibliographical references.
ISBN 0-521-37212-7
1. Solar system. 2. Solar system – Origin. 3. Cosmochemistry.
I. Title.
QB501.T25 1992
523.2–dc20 92-25784
CIP

A catalog record for this book is available from the British Library

ISBN 0-521-37212-7 hardback

To

Noël

Susanna, Judith, and Helen

CONTENTS

Preface

The true dimensions of the terrestrial globe were revealed mostly in the fifteenth and sixteenth centuries, principally through the technical development of truly ocean-going vessels and the magnetic compass, which enabled the exploration of the oceans. Even with these advantages, the real extent of the Great Southern Continent or *terra australis incognita* had to await the voyages of Cook in the eighteenth century. This flood of new geographic knowledge replaced the medieval view of the world (although Flat-Earth societies still persist). Such understanding was gained only very slowly. A detailed knowledge of the topography of the ocean floors and our understanding of their composition and origin has been obtained only a little ahead of our radar pictures of the surface of Venus.

Our exploration of the solar system is at a similarly heroic stage. The distant points of light, barely resolvable in telescopes, have been revealed through the use of space vehicles, the latter-day equivalent of the Portuguese caravels, as separate worlds, with an astonishing amount of diversity.

The information has been rapidly and widely disseminated, electronic media having superseded the printing press of the Renaissance. Everyone is informed of the striking new discoveries. Although no Eldorados have emerged, the pictures reveal a plurality of worlds unimagined by the Elizabethans. Every satellite has turned out to differ in some significant feature from its neighbor: ". . . the sense of novelty would probably not have been greater if we had explored a different solar system" [1].

This comment on the jovian system reveals a fundamental truth. Lurking behind the photographs and radar images is a new observation, uncomfortable for *Homo sapiens*. What combination of circumstances could reproduce in some other planetary system the detail observed in our own solar system or lead to the assembly and evolution of a clone of the Earth? Whether one contemplates the battered face of Mercury, the crumpled crust of Venus, our own water-sheathed planet, the vast deserts and gigantic landscapes of Mars, or the profusion of distinct icy satellites of the giant planets, the uniqueness of the individual planets and satellites informs us of a sober fact: The system is unlikely to be duplicated in detail elsewhere.

Other planetary systems doubtless exist. That they would resemble ours in any but the broadest detail is only a remote possibility. Too many chance events have occurred to bring our system to its present condition. Jaques Monod [2] remarked on the uniqueness of the evolutionary path for life and of the apparently insurmountable difficulties of duplicating our version of intelligent life elsewhere. The planets and satellites are rich in diversity and the difficulty of producing clones of our present solar system makes duplication as unlikely as the possibility of finding an elephant on Mars.

Despite the cornucopia of new information about the solar system, it is curious how little effect this has had on the development of theories for its origin. The Apollo and Luna data from the Moon had almost no impact on theories of the origin of the solar system, or indeed of the Moon, until comparatively recently. The reason for this is partly the unusual circumstances of lunar origin and evolution and the very large amount of new data from separate fields that had to be assimilated.

The complexity of the solar system is not in accord with theories that start from some simple initial condition. Such hypotheses do not predict the diversity of the present system. Thus, while it is tempting to look for grand unified theories to explain our solar system, the basic approach is wrong and the trail is false. Too many chance events occur. Among the many singular events that determined the outcome, a few examples may be cited: the size of the fragment of the molecular cloud that became the solar nebula; the formation of the Sun as a single rather than a double or multiple star; the huge size of Jupiter and its controlling influence on the asteroid belt and Mars; the random accretion events that made the Earth larger than Venus; the unique collisions that produced the bone-dry Moon and the high density of Mercury (a red herring for grand theories); the differences among the satellite systems; and the hundred other curious features of the solar system such as Hyperion tumbling chaotically until the end of time. All are the result of events that might readily have taken a different turn.

Scientists and philosophers have been attempting to find an explanation for the origin of the solar system since the question was first raised by the Greeks, but a universally acceptable scenario has yet to appear. A major reason for this failure to solve one of the oldest scientific questions has lain in the approach adopted. Another planetary system will differ in numbers and sizes of planets and satellites. Thus, the quest for a universally applicable solution to the origin of the system is not fruitful. Attention is better directed to explaining how the various aspects of the system came about.

In this book I have attempted to provide this and to describe our current understanding of the origin and evolution of the solar system. In this respect, I have followed the path of most previous investigators in trying to account for the existence of the planets, satellites, asteroids, and comets. The formation and evolution of the Sun as a quite normal G-type star is well enough understood, so in this account it is mainly an offstage companion to the discussion. The book is biased toward a geochemical point of view because of my field of interest. Cosmochemistry and geochemistry are not, however, without their difficulties, and those readers who wish to find a simple statement of the composition of the Earth will be disappointed to learn that many problems yet remain. I have, nevertheless, been unable to resist treading among other areas, conscious that such a path is fraught with as many perils as those recorded by John Bunyan (1628-1688) in *The Pilgrim's Progress.*

The subject is rendered very difficult by the wide variety of contrary opinions held among investigators and the many alternative explanations offered for the same set of data. Sometimes the data are suspect. It is thus surprising to find that the students of cratering frequently disagree on the numbers of craters involved. One might have hoped that at least the database was agreed on. However, it is a common error to underrate the difficulties of subjects outside one's experience. The diversity of opinion is only exceeded by the strength with which the opposing views are held and defended, supporting the thesis that subjectivity is more common in science than hoped for [3]. This has not made my task easier in trying to reach some conclusions about the status of our understanding of the solar system.

Readers may complain that I have chosen to cover too broad a canvas and that the scope of the book should be restricted to areas of immediate competence. Thus, they may question why topics as disparate as Chiron, comets, crustal development, calcium-aluminum inclusions, chondrules, the Copernican Revolution, and lunar cataclysms are all discussed.

The problem is that detail in one area may provide a crucial constraint in another. If one paints with too broad a brush, error may creep in. Grand theories are useless if they cannot explain the minute details. Thus, the concept of giant gaseous protoplanets condensing from fragments of the nebula was viable until it was demonstrated that Jupiter and Saturn were not of the solar composition required by the theory. The rare-gas isotopic abundances in the terrestrial planets rule out the popular concept of late-accreting veneers of carbonaceous chondrites. Consequently, an apparently minor detail from one topic may negate extensive work from another, just as the presence of the europium depletion in lunar mare basalts removed the model, developed from experimental petrology, that these lavas were derived from a primitive unfractionated lunar interior.

The reverse of this particular coin is that the mere accumulation of the staggering amount of detailed observations in the solar system is of little use unless there is some unifying concept; the discovery of the Periodic Table was needed to make sense out of the bewildering array of properties of the chemical elements.

Although this book treads into far more nebulous territory than, for example, *Lunar Science: A Post-Apollo View* [4], it might be recalled that the state of our understanding of the Moon at that time was also beset with many misconceptions and false trails, a particularly good example being the many ingenious but erroneous explanations advanced to explain the europium anomaly [5].

It therefore behooves workers in this interesting field to consider all the details if they wish to avoid error, unless they want to make all the planets out of CI carbonaceous chondrites, or the Moon from the Earth's mantle. In this context, I had hoped, somewhat naïvely when beginning this task, to conclude with definitive compositions at least for the terrestrial planets. What has emerged in reality is that even the major-element compositions of the inner planets are poorly constrained. A solution will have to wait until the problems of the moments of inertia of Mars and Mercury, the nature of the lower mantle of the Earth, and that of the light element in the core, among many others, are resolved.

However, it is easy to become fascinated with the curious landscapes of Miranda or Triton, the large crater on Mimas, the enormous layered deposits in Valles Marineris, the crumpled terrain of Venus, the rare gas abundances in the atmospheres of the terrestrial planets, the Kirkwood Gaps in the asteroid belt, the relationship of Pluto and Charon, the composition of Comet Halley, or that of interplanetary dust. While contemplating all these marvels, it is not difficult to become lost among the trees, forgetting that the principal purpose of this book is to understand the origin of the system, and to integrate all these wonders into a coherent explanation.

Thus, it is not my intention to provide a detailed travelogue or Cook's Tour of the solar system, tempting though that option may be, since there is already a multitude of good technical books and papers covering both the detail and overviews of the solar system [6]. I assume that most readers are familiar with the spectacular images and I have not attempted to reproduce many of these or provide much mathematical treatment. This is available in the material referenced and it seems pointless to fill up pages with equations of use only to a few specialists.

Rather, I have attempted a commentary on the problems of its origin and evolution. I had originally planned to start the book with a description of the present solar system, and proceed from there into more nebulous regions, moving backwards in time into the unknown. I was wisely persuaded to begin instead with the solar nebula, and to bring the story forward in chronological order, a decision that has proven robust. However, as the one of the editors of *Meteorites and the Early Solar System* remarked, ". . . good reasons could be found for placing every chapter before every other chapter" [7].

In general, I have attempted to reference the most recent reviews and texts; the subject is moving so rapidly that older material quickly becomes of historical interest. In many areas, the subject has come clearly into focus only in the past two or three years as the avalanche of data from the past two decades has been assimilated. Accordingly, much of the older literature is now of historical, rather than scientific, interest [8]. Literature coverage extends to April 1990, with a few later additions.

The deluge of new information and the emergence of some agreement on the general scenario of solar system origin and evolution has turned the subject in a few years from one having the status of a hobby, " . . . pursued by eccentric, elderly gentlemen who fight with one another's theories of the origin of the solar system" [9] to one in which quite detailed scientific questions can be posed. In this context, I have been selective in giving more attention to those studies that fall within the general theme of the planetesimal hypothesis, rather than to give an overview of all competing ideas. Here I have attempted to produce a broad survey pointing to the many questions of interest that can be addressed and the regions in which our understanding is inadequate. Even in those areas such as the surface geology of Mars, where we have much basic knowledge and understanding, many significant questions (for example, the relative importance of sedimentary vs. igneous processes, or the early climate and the role of ice) remain obscure. Only a sample return, followed by a manned mission, will supply the answers to these particular questions.

The wealth of new detail leads to the complaint, common to many scientific disciplines, that there is now too much information for any individual to comprehend. In this view, it was simpler in past ages when the corporate body of knowledge was so much smaller.

However, it is doubtful that this is true; scientists in the Renaissance had to deal with a staggering burden of topics such as astrology, alchemy or numerology, on which Newton wasted so much of his time. In any subject where there are no unifying principles, as in chemistry before the Periodic Table revealed an essential simplicity or biology before evolution cast its revealing light, it is impossible for an individual to be an expert in every nook and cranny. Nevertheless, given some general principles of physics and chemistry as a guide, it is possible to discern the forest, even through the thickest underbrush, just as the architectural unity and splendor of a great building may be glimpsed from a distant prospect.

Among the other intellectual baggage that has been discarded, the differences between the relative ages of the universe and the solar system have clearly distinguished their origins; we no longer have to account for them together. The age of the solar system, obtained from meteorites as 4560 m.y., is only about one-quarter of the age of the visible universe. This fact—a relatively recent discovery—separates the origin of the Earth and the solar system from that of the universe. This knowledge has become so ingrained in thinking that it is surprising to recall that as recently as 1950, the age of the Earth, established by isotopic dating of rocks, approached and sometimes exceeded the astronomical estimates for the age of the universe.

One is comforted on this journey by the steady convergence of scientific ideas toward some kind of consensus, as new facts are acquired. Science is in this way distinct from most other human activities, which display the opposite tendency of divergence with time, a process most clearly revealed by the multitude of religious and philosophical systems.

Stuart Ross Taylor
Houston, Texas
December 1990

NOTES AND REFERENCES

1. Smith B. A. et al. (1979) *Science, 204,* 951.
2. Monod J. (1974) *Chance and Necessity,* Collins, 187 pp.
3. Mitroff I. (1974) *The Subjective Side of Science: A Philosophical Enquiry into the Psychology of the Apollo Moon Scientists,* Elsevier, New York, 329 pp.
4. Taylor S. R. (1975) *Lunar Science: A Post-Apollo View,* Pergamon, New York, 372 pp.
5. Ibid., pp. 154-159.
6. There are so many good texts that it would be invidious to choose among them. A good beginning is Morrison D. and Owen T. (1987) *The Planetary System,* Addison-Wesley, Reading, Massachusetts, 600 pp.
7. Kerridge J. F. (1988) in *Meteorites and the Early Solar System* (J. F. Kerridge and M. S. Matthews, eds.), p. xvi, Univ. of Arizona, Tucson.
8. The fate of most scientific work is to be forgotten. The typical scientific paper has a half-life of five years, forming a curious contrast to the dominant egos of most workers in the field. The correct is incorporated into the general body of knowledge; the erroneous is mostly ignored.
9. Wetherill G. W. (1988) in *LPI Tech. Rpt. 88-04,* p. 81.

Acknowledgments

This book was written over a period extending from October 1987 to December 1990, mainly at the Lunar and Planetary Institute, Houston, during a series of Visiting Scientist appointments. I am grateful to Dr. Kevin Burke, who was Director of the Institute for much of that period, for extending to me the hospitality of the Institute, for useful advice, and stimulating discussions. I also wish to thank Dr. David Black, the current Director, for continuing support and encouragement in this work.

A book that sets out to cover such a broad canvas inevitably requires much advice and information from scientists in fields remote from my own experience. I am grateful to many friends and colleagues for much assistance. In particular, I thank the following for carrying out the necessary but arduous task of reviewing draft material on the various chapters: John Wood for Chapter 1 on planetary accretion; Alan Boss for Chapter 2 on the solar nebula; Jeff Taylor for Chapter 3 on meteorites; Richard Grieve for Chapter 4 on impacts; Jim Head and Jayne Aubele for Chapter 5 on the planets; and Bill McKinnon for Chapter 6 on rings and satellites and for Chapter 7 on the new solar system. Their peer review has proven invaluable and I am greatly indebted to them for contributing their expertise in the many diverse fields into which I have strayed.

Malcolm McCulloch, Tezer Esat, and Larry Nyquist provided insights into early solar system chronology and the intricacies of isotopic anomalies and also read parts of the text. I owe a special debt to Marc Norman, who read the entire text and suggested a better overall arrangement of the first three chapters.

I am grateful to all these people, whose comments have substantially improved the text. The responsibility for interpretations and errors, particularly for a book covering such a broad canvas, remains my own.

I have received reprints, illustrations, and much valuable advice from many of my scientific colleagues, including Ralph Baldwin, Nadine Barlow, Willy Benz, Bruce Bills, Alan Boss, Joyce Brannon, Adrian Brearley, Robin Brett, Al Cameron, Joe Chamberlain, Tezer Esat, Bruce Fegley, Everett Gibson, Bill Hartmann, Roger Hewins, Fred Hörz, Bill Hubbard, Bill Irvine, Gero Kurat, Gunter Lugmair, Melanie Magisos, Mildred Matthews, Malcolm McCulloch, Lucy-Ann McFadden, David McKay, Scott McLennan, Hap McSween, Horton Newsom, Marc Norman, Larry Nyquist, Bob Pepin, Carlé Pieters, Frank Podosek, Ron Prinn, Steve Saunders, Ed Scott, Buck Sharpton, C.-Y. Shih, Brad Smith, Steven Squyres, Tim Swindle, George Tilton, Alan Treiman, Jerry Wasserburg, John Taylor Wasson, Stuart Weidenschilling, George Wetherill (who has always endeavored to point out the correct direction, with results I must leave to others to judge), Paul Weissman, John Wood, and Mike Zolensky.

Once again I wish to acknowledge the excellent support I have received from the staff at the Lunar and Planetary Institute. Stephanie Tindell, Managing Editor, contributed greatly to the efficient production of this book, and Sarah Enticknap was responsible for copyediting. Shirley Brune and Christy Owens skillfully drafted all the figures. Kin Leung provided essential computer support, including a loan at a vital juncture of his personal Macintosh word processor. Fran Waranius provided her customary expertise by supplying reference material in many different fields.

Peter-John Leone, of Cambridge University Press, wisely persuaded me at an early stage to reverse the order in which I had proposed to deal with the topics in this book, a suggestion that proved durable under the stresses of composing this work. Nancy Selzer and Florence Padgett of Cambridge University Press efficiently oversaw the production.

It is a pleasure to acknowledge the considerable help I have received from the prolific sequence of books on planetary matters that have been published by the University of Arizona Press under the general editorial guidance of Tom Gehrels, Melanie Magisos, and Mildred Shapley Matthews. All workers in the field owe a great debt to this successful enterprise.

I owe a great deal to many other people: Brian Mason, Louis Ahrens, and Harold Urey started me on this path long ago; Fred Hörz made the initial suggestion that I write such a book; Robin Brett intervened at a vital juncture of my career and must accept some responsibility for this outcome; Ted and Lil Bence and Arch and Mary Reid provided much support, both logistical and academic. The Department of Nuclear Physics, Australian National University, kindly provided facilities that materially assisted in the completion of this book. Finally, production of this book would not have been possible without the strong support of Dr. William L. Quaide of NASA.

Prologue

The Moon: Rosetta Stones [1] and Large Impacts

Many of the insights gained in this book are based on our study and understanding of the Moon, which, except for the Sun, is the most obvious astronomical object in the sky. It immediately informs us of the existence of other worlds and, accordingly, has had a powerful effect on human development. The other planets and stars are so distant as to be points of light in all but the most powerful telescopes, requiring an extraordinary feat of imagination to understand their true nature. The waxing and waning of the Moon on a monthly cycle provided primitive calendars, and the connection with tides was realized very early and led to concepts of gravitation. Without the Moon, our intellectual development would have taken a different, probably more inward-looking course.

Ever since Galileo demonstrated in 1610 that the Moon was mountainous, fulfilling the prediction of Anaxagoras (*c.* 500-428 B.C.) that the Moon was a stone—an impiety for which he was banished from Athens—it became a possible place to visit. The Greek cosmologists in the third century B.C. had proposed that the heavenly bodies were composed of shining crystal, a fifth element or quintessence (the four elements, earth, air, fire, and water, sufficed for terrestrial geochemistry), but Galileo's observation indicated clearly that the Moon bore some resemblance to the Earth; he was bold enough to distinguish land (*terrae*) and seas (*maria*) and this distinction must mark the birth of comparative planetology.

The Moon has played a central role in the development of theories of the origin and evolution of the solar system. This is not without irony, since it has proven one of the most difficult objects to explain. It is in plain sight, as Harold Urey was accustomed to remind us, accessible even to naked-eye observation, an obvious first object to fit into theories for the origin of the universe. This was the basis for the common belief in pre-Apollo time, and a major justification for the manned lunar missions, that "we could unravel the mysteries of the solar system by going to the Moon" [2].

The Moon eventually provided us with keys for understanding the history of the solar system, but not in the manner imagined by pre-Apollo Earth-based thinkers. One of the principal conclusions of lunar studies was to demonstrate the importance of large collisions. The evidence from the observed wide range of impact crater sizes led to the notion that a hierarchy of objects existed during accretion, and that the planets accreted from these rather than from dust, thus reinforcing the planetesimal hypothesis.

Before the lunar missions, the heroic efforts of cosmologists and a wide variety of experimental scientists all failed to provide an adequate explanation for the existence of the Moon. With the benefit of hindsight, however, it is clear that several key facts, among them the low density of the Moon, the strange orbit, and the high value of the angular momentum of the Earth-Moon system, were available already, awaiting integration into a coherent theory.

The space missions provided crucial additional information on ages, chemistry, and the significance of cratering, in particular of the importance of large basin-forming impacts. The largest structures formed by such processes, such as Mare Orientale, were caused by the impact of objects of the order of 50 km in diameter. A further imaginative leap was required to extrapolate orders of magnitude beyond the scale of Mare Orientale, to propose the single large impact hypothesis for lunar origin, a concept that required an overview of many different fields. Since the hypothesis derived the Moon principally from the silicate mantle of the Mars-sized impacting body, this notion cut the Gordian Knot [3] tying the Moon to the Earth, which itself was eventually recognized as "a strange and beautiful anomaly in our solar system" [4]. Many other features of the solar system such as planetary obliquities could be explained by such large-scale random or stochastic processes, so that a new understanding of the origin of planets and satellites emerged.

Once we properly understood the origin of the Moon [5] with all its implications and connotations, then a clearer view of the rest of the solar system began to emerge, and the somewhat untidy nature of the solar system received a rational explanation. Those who dispute this might ponder the attempts to explain the origin of the Moon and planets before 1969. It is salutary not only to read of the early attempts, involving so many false trails in lunar geology and geochemistry (geophysicists seem to have been in better shape, either because of the paucity of data to deal with, or the dearth of workers in the field), but also to contemplate similar discussions now occurring over the other planets and satellites.

NOTES AND REFERENCES

1. The Rosetta Stone was discovered in 1799 during the French invasion led by Napoleon, near Rosetta (Rashid), Egypt. It is a tablet of black basalt, on which are listed benefactions by Ptolemy V Epiphanes (205-180 B.C.). These were inscribed by the priests at Memphis in two Egyptian scripts (hieroglyphs and demotic, a cursive script related to hieroglyphs) and also in Greek. This discovery enabled the ancient Egyptian picture language (hieroglyphs) to be translated, a task accomplished principally by J. F. Champollion in 1822.
2. Stevenson D. J. et al. (1986) in *Satellites* (J. A. Burns and M. S. Matthews, eds.), p. 88, Univ. of Arizona, Tucson.
3. The Gordian Knot was reputedly tied by King Gordias (father of Midas) in the town of Gordian, capital of Phrygia (in modern west-central Turkey) about 700 B.C. By all accounts, it was a knot of great thickness and complexity, which secured the yoke to the harnessing pole or shaft of a two-horse chariot. King Gordias prophesied that whoever could untie the knot would conquer Asia (which then essentially comprised the known world). All attempts failed until Alexander the Great (356-323 B.C.), on passing through Gordian in 333 B.C. on his expedition against the Persians, solved the problem by cutting through the Gordian Knot with his sword. Alexander's subsequent conquest of Asia was held to have fulfilled the prophesy. The tale has survived to illustrate one manner of solving apparently intractable problems.
4. Lovelock J. E. (1979) *Gaia, A New Look at Life on Earth*, Oxford Univ., New York, 157 pp.
5. Benz W. et al. (1986) *Icarus, 66,* 515; Benz W. et al. (1987) *Icarus, 71,* 30; Newsom H. E. and Taylor S. R. (1989) *Nature, 338,* 29.

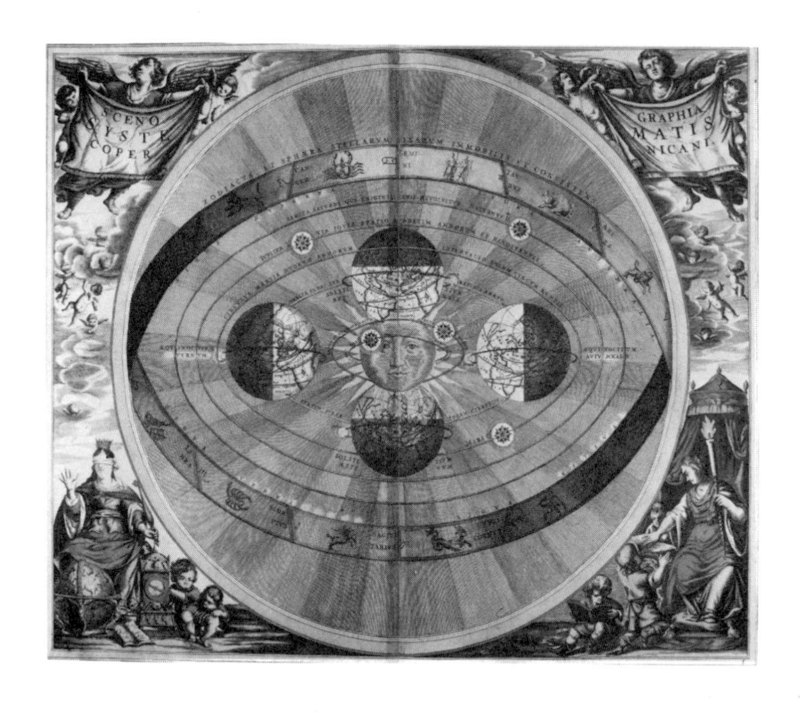

Chapter 1

Planetary Formation: A Historical Perspective

Illustration, preceding page—*The Copernican System, from Macrocosmia seu Atlas Universalis Novus by Andres Cellarius (1661). This illustration by Cellarius demonstrates that the Copernican System had firmly replaced the Ptolemaic System within 100 years of the publication of Copernicus and 40 years after the confirmatory observations by Galileo.*

Planetary Formation: A Historical Perspective

1

1.1. THE NATURE OF THE PROBLEM

The problem of building planets is fundamental to the entire question of the origin of the solar system. Historically, this question has frequently been considered solved, but the wide variety of explanations and solutions that have been offered, from the creation myths of primitive societies to the more recent, but numerous, scientific attempts, have generally collapsed when faced with new information about the system.

There are two principal difficulties. First, we have only one planetary system to study. We do not know if our solar system is unique or if other systems exist that involve many small objects circling a central star. Although we strongly suspect that such systems would be coplanar, it is quite unclear whether a similar mass distribution, with its division into terrestrial and giant planets, is typical or unique. Statistical treatment, so successful in the erection of theories of stellar evolution, cannot yet be applied to a single system, since improbable events can always happen once [1].

Some light can be shed on the problem of uniqueness by considering the satellites. The regular satellite systems of the giant planets are often considered to be miniature solar systems, a philosophical habit traceable back to Galileo's observation of the moons of Jupiter. However, there is not much evidence of the operation of a uniform satellite-forming process. The regular prograde systems of Jupiter, Saturn, Uranus, and Neptune do not resemble one another. They could just as well belong to separate planetary systems, so they provide little help to a search for some general principles underlying the formation of planetary systems. We may deduce from the disparity among the satellites of both gaseous and icy giant planets that other solar systems, if discovered, are likely to show the same sort of variations and not resemble our own, except in the broadest outlines.

The second problem arises because of the unknown initial state. We see only the final product, and have to infer both the evolutionary path by which it came, and the starting point. Both, like the path of terrestrial biological evolution, are nonunique. The wide variety of explanations that have been proposed reflects these uncertainties.

Here I outline the new evidence from astronomical and space exploration, which imposes many new constraints on the origin of the solar system. Despite a general feeling that some sort of scientific consensus is being reached [2], it is salutary to recall in this context the many previous claims to have solved the problem of the origin of the solar system; one must endeavor to avoid this philosophical trap [3].

It is perhaps worth commenting on two obvious truisms. First, the solar system is isolated, with the distance to the nearest stars exceeding the dimensions of the planetary system by factors of more than 50,000 (Fig. 1.1.1). Second, the whole system effectively lies in one plane, with most of the bodies orbiting the Sun and rotating in the same direction. Both these factors constitute evidence for a common origin of the Sun and the planets. Such a configuration is unlikely to result from a random accumulation of a diverse set of objects. The concept of a common origin has been part of the general consensus about the solar system for the past 200 years, ever since it formed the basis for the Laplace hypothesis. A case might otherwise be made for assembling this heterogeneous collection of diverse planets and satellites from some kind of cosmic junkyard.

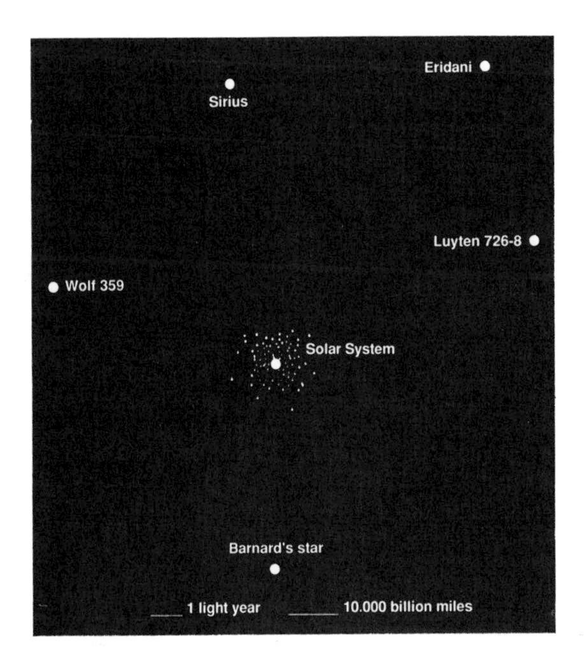

Fig. 1.1.1. *The position of the solar system relative to nearby stars.*

The flatness of the system does seem to be an inherent feature, inherited from the rotating disk form of the solar nebula. It does not appear to have been superimposed on the system by, for example, the early formation of massive Jupiter. Computer studies of the stability of the system have found that Jupiter merely influences small variations in orbital eccentricities of the other planets.

Perhaps the most instructive planet is Neptune, which gravitationally is the most weakly bound of the system. Its orbit is nearly perfectly circular. Any perturbation of the system, for example, by a passing star, would have affected the eccentricity and inclination of the neptunian orbit (see section 5.14). The possibility that large gaseous planets further out were stripped away by passing stars appears to be excluded since such events would have perturbed the orbit of Neptune [4].

Neptune thus appears to be the true outer planet and the planetary system was never more extensive; neglecting Pluto and Charon, only the Kuiper and Oort Clouds of comets occupy the outer reaches. The expectation of Kant that many outer planets existed has not been substantiated. The very diversity of the individual planets and satellites imposes a second-order constraint on hypotheses: clearly much randomness has been superimposed upon the orderly system as it was perceived by Laplace.

In considering the basic question of planetary accretion, our concepts about the primordial conditions are strongly controlled by the evidence from meteorites, which influences much of our thinking about the origin of the inner planets (see Chapter 3). A major uncertainty that cannot be addressed here is the role of cometary material in planetary formation. The composition of Comet Halley, although broadly similar to the solar composition for the volatile gaseous elements, shows discrepancies with solar, CI, and interplanetary dust particle abundances by possessing the lowest Fe/Si and Mg/Si ratios yet discovered in the solar system. Comet Halley does not seem to be the long-sought "primordial" sample. This raises further, as yet unanswered, questions about the initial heterogeneity of and chemical processing in the primitive nebula.

A primary assumption is that the composition for the nongaseous elements (excluding H, C, N, O, and the noble gases) of the primordial solar nebula is given by that of the Type-1 carbonaceous chondrites (CI). The rationale for this is the similarity between the abundances in the solar photosphere and in the CI chondrites.

This implies a broad similarity in the composition of the nebula, on which many second-order heterogeneities (e.g., oxygen isotopes, K/U ratios, asteroidal belt zoning) are superimposed. Although this has been a useful hypothesis, it has been clear for a long time that the compositions of the inner planets differ in detail from those of the CI meteorites, and that substantial chemical fractionation from CI compositions occurred before the final accretion of the terrestrial planets. Models that have used CI compositions to construct the inner planets [5] have encountered numerous problems [6].

A central and philosophical point is whether the planets condense from fragments of the nebula or are built up by accretion of smaller particles. The first model resembles star formation and is somewhat easier to deal with theoretically, but has generally been abandoned. The second alternative raises the question about the size of the accreting material. In such models, did the planets accrete directly from dust and the dispersed material of the nebula, or are they the end product of the accretion of a hierarchical succession of bodies (see section 1.10)?

Several observations point to the latter process. There is ample observational evidence from the battered surfaces of planets and satellites for the impact of innumerable large (<100-km diameter) bodies throughout the solar system. The obliquities of the planets are consistent with the collision of very large objects (>1000-km diameter). The most reasonable hypothesis for lunar origin demands an impacting object of 0.1-0.2 Earth masses at a rather late stage in accretional history, when both it and the Earth had already formed a core. Finally, accretion from a dusty nebula might be expected to lead to rather similar planetary compositions rather than the observed chemical and isotopic diversity, both of meteorites and planets. These comments point to planetary accretion from a hierarchy of massive objects rather than from the infall of dust. This conclusion is reinforced by the evidence from meteorites; individual meteorites do not consist of single lithic components but are always mixtures of several components, or are fragments of larger bodies [7]. A corollary of the concept of planetary accretion from a hierarchy of objects is that internal fractionation may occur in such bodies before their final sweep-up into the planets.

Before continuing this line of inquiry, it is instructive to survey the past history of attempts to account for the planetary system. Historically, theories of solar system origin, although they show much diversity, may be classified into three categories:

1. Tidal theories, in which the formation of the planets occurs from material extracted from the Sun or a passing star after these bodies have formed.
2. Accretion theories, in which material is captured by the Sun from interstellar space.
3. Nebular theories, in which the planets form directly either concurrently or consecutively from the same nebula as the Sun [8].

1.2. PRE-COPERNICAN VIEWS AND THE COPERNICAN REVOLUTION

Pre-Copernican theories in which the Earth was the center of the universe have long lost scientific attention. This is not only because such theories have been superseded since the Copernican Revolution, but also because in such hypotheses the origin of the Earth, Sun, and planets is inextricably bound up with the origin of the universe. In the modern context we are now aware that the solar system is less than one-quarter of the age of the observable universe, so that it is no longer permissible to seek a common origin.

The term "planet" is of course derived from the Greek word meaning "wanderer," but it is curious that although the Greek astronomers devoted much study to planetary motions, not much attention was paid to the origin of the solar system. This topic seems to have been the province of the philosophers rather than the astronomers.

Anaxagoras (c. 500-428 B.C.) considered the Moon to be a stone, and the Sun to be a red-hot mass of iron bigger than the Peloponnesus. The latter opinion was based on meteoritic evidence. An iron meteorite fell near Aegospotamis, a stream in ancient Thrace [9] about 467 B.C. and Anaxagoras concluded that the visitor had come from the Sun. For these views, held to be impious, he was banished from Athens. Little of his work has survived, but apparently he pictured the Earth at the center of a large cosmic vortex, anticipating Descartes in the notion of vortices.

Aristarchus of Samos (about 250 B.C.) placed the Sun at the center of the solar system, including the Earth with the rest of the planets, and apparently was the first to suggest that the Earth both rotates and revolves around the Sun. It is fitting that a prominent and interesting lunar crater is named for him.

The Greek philosophers were mostly concerned with questions of purpose, and distinguished carefully between the Earth, with its obvious imperfections, and the heavens, which they held to be unchanging, composed of the perfect fifth element, or quintessence. The other four elements, earth, air, fire, and water, sufficed for terrestrial geochemistry. Socrates (c. 470-399 B.C.) was concerned with purpose rather than observation, and thus did little to encourage scientific investigation. This doctrine of perfection left no room for evolutionary development, so Plato (c. 428-347 B.C.) concerned himself with the motions of the planets rather than their origin. The heavenly bodies were supposed to move in perfect circles and the wandering of the planets among the fixed stars was disturbing to this viewpoint. Aristotle (384-322 B.C.), the third member of the great trio of Greek philosophers, also subscribed to the view that the heavens were permanent and thus not subject to the laws of terrestrial physics.

Epicurus (341-270 B.C.), in contrast, did not give the heavens any special or separate status, and supposed that the heavenly bodies formed by random collisions of atoms. It is not clear that he distinguished between stars and planets, and his epicurean philosophy, concerned with freedom and happiness, neither encouraged skeptical observation nor quantitative measurements. Lucretius (first century B.C.) in his *De rerum natura* adopted many of the ideas of Epicurus, although he also did not distinguish clearly between the planets and fixed stars. It is refreshing that neither of these workers paid much attention to astrology, unlike Plutarch (46-119? A.D.) who populated the universe with strange inhabitants.

Like Lucretius, very little is known of the life of Ptolemy. He lived in the second century A.D., but his birth and death dates are unknown. Ptolemy was much studied by the Arab astronomers and these sources record that he lived for 78 years. The codification of Greek astronomical thought in the Almagest remained as the definitive work until the end of the Middle Ages, and was a triumph of the geometric approach to the solar system. Nevertheless, Ptolemy remains an enigmatic figure and doubt has been cast on the reliability of his data. Perhaps his chief achievement was to transmit the observations of Hipparchus, the greatest of the ancient astronomers [10]. Like his Greek predecessors, Ptolemy was concerned with purpose, and like them felt that the imperfect Earth could not be accorded a place among the heavenly bodies, composed of shining quintessence. Echoes of this philosophical approach still appear in the very common tendency to ascribe uniformity to essentially unknown or distant regions Examples include the terrestrial mantle, the solar nebula, and the universe, all of which were thought to be uniform; more recent information is rapidly dispelling these myths.

The medieval European thinkers, such as Oresme, Bishop of Lisieux (1325-1382), and Nicholas of Cusa (1401-1464), became concerned with the concept of "clockwork" universes created by a master craftsman. Once created, these could evolve according to fixed laws. These men departed from the Greek philosophers with their division between the Earth and the heavens, by considering the Earth an integral part of the universe.

The discovery of the Galilean satellites circling Jupiter led to the collapse of the Earth-centered Ptolemaic cosmology. The Copernican revolution is customarily dated at 1543 by the publication of the great work of Nicolas Copernicus (1473-1543), *De revolutionibus orbium coelestium, libri VI (On the Revolutions of the Celestial Spheres)*. The acceptance of the Sun, rather than the Earth, at the center of the planetary system caused a profound change, creating the philosophical climate in which we still live [11].

The next significant step was taken by Tycho Brahe (1546-1601), another outstanding figure of Renaissance science, whose chief accomplishment was the precise measurement, before the invention of the telescope, of planetary motions. This work was mostly undertaken in Denmark, before his move to Prague.

Johannes Kepler (1571-1630), the successor to Tycho Brahe at Prague, inherited his monumental calculations, which formed the basis for Kepler's discoveries. Kepler's great contribution was his three laws [12] and his advocacy of the Copernican System. Like many other scientists, he was principally concerned with other matters and "the three major gems in his works on astronomy lay in a vast field of errors, of irrelevant data . . . of mystical fantasies, [and] of useless speculations" [13].

Kepler was able to fit the orbits of the planets into spheres based on the five geometrical solids: octahedron, icosahedron, dodecahedron, tetrahedron, and cube. These are the only solids bounded by identical faces and so were considered "perfect." He first fitted an octahedron around the orbit of Mercury, and a sphere circumscribed about the octahedron corresponded to the orbit of Venus. A sphere circumscribed on an icosahedron around the venusian orbit gave the orbit of the Earth; a sphere on a dodecahedron fitted to the orbit of the Earth gave the orbit of Mars; a sphere on a tetrahedron fitted to the martian orbit gave the orbit of Jupiter; and a sphere on a cube fitted to the jovian orbit gave the orbit of Saturn.

Kepler thus considered that he had answered another fundamental question: why there were only six planets (as known at the time). This cosmic limit was imposed because of the small number of "perfect" solid forms! However, this insight marked the end of the attempts to explain the solar system on the basis of geometrical constructions, for the planetary orbits, on the basis of Kepler's own laws, turned out to be elliptical. The "chief achievement of Kepler's superhuman labours was to discredit the glitter of geometry that turned out to be a false prospect for tracing out . . . the genesis of the system of planets" [14].

Kepler, however, prepared the way for Newton, and the construction of clockwork systems, that, once set in motion by the clockmaker, were unchanging. Plato had remarked that God was a geometer, and this theme of a celestial clockmaker came to dominate thinking about the solar system in the seventeenth and eighteenth centuries. Good examples of this approach are Leibnitz and R. Bentley (chaplain to the Bishop of Worcester) who gave the Boyle lectures (founded by Robert Boyle) in 1692, and was an exponent of the work of Newton. These workers considered that the solar system had been constructed by an omnipotent clockmaker. Once the system was set running, no further attention was needed; it would continue to operate under the laws of physics. Since the Earth was clearly imperfect, some periodic repairs might be needed, at least for this planet. Comets were favored as possible replenishing agents.

1.3. TIDAL THEORIES

Tidal theories are now out of favor, although they were very popular in the past. The initial idea seems to be due to Buffon [15] who proposed that a cometary collision with the Sun ejected a disk of material. The masses of comets, then unknown, were thought to be about 0.1 solar masses. When the true masses of comets became established, the theory languished until the comet was replaced with a collision with a passing star [16]. Other proposals involved a head-on collision [17] or a collision between two nebulae, the true nature of nebulae not being established at that time [18].

The Chamberlin-Moulton hypothesis attempted to overcome the angular momentum difficulties with the Laplacian models, in which the Sun should have an equatorial rotation exceeding 400 km/s, instead of the observed 2 km/s. The Chamberlin-Moulton solution proposed that the approach of another star would cause an increase in solar activity, resulting in the ejection of solar material [19]. This theory is of interest now chiefly because these clouds of ejected material condensed, forming what they termed planetesimals, from which the planets were accreted. The term planetesimals has survived and occurs frequently in current models, although in a different context.

Further attempts to deal with the thorny problem of angular momentum, which had caused the eclipse of the Laplace nebular hypothesis, led to the further development of tidal theories. These used [20] the tidal action of passing stars to draw out a cigar-shaped filament from the Sun. Although these theories were very popular, particularly in the public domain, and explained many features of the solar system, various objections were raised at the time [21, 22].

Although the tidal theories had been erected to account for the angular momentum of the planets, difficulties still emerged in attempting to overcome this particular problem. It was soon shown that the amount of energy transferred by a passing star would be sufficient to cause the filament to escape completely, so it would disperse before condensation into planetary bodies could occur [23]. Another criticism was that in these hypotheses, the planets must form within four solar radii. Mercury, however, lies at 60 solar radii, and even the regular satellite systems of the giant planets lie far beyond this restrictive limit.

Two more recent criticisms appear to raise major difficulties for tidal theories. The first is that planetary compositions differ significantly from the present composition of the Sun. This is particularly marked for the abundances of deuterium and lithium. Both are highly depleted in the Sun due to thermonuclear reactions, but are present at about their cosmic abundance levels in the planets. Accordingly, the material now in the planets could not have not been resident in the Sun for any extended period. Thus, formation of the planets would have to occur by the process of tidal fission more or less immediately following the formation of the Sun.

Since the Sun and the planets indeed appear to have formed within 10^8 years of one another, these tidal hypotheses demand that two unrelated processes, stellar formation and close approach of another star, occur very close in time.

The third problem was long recognized: the present distribution of stars makes such an event unlikely. This criticism has become weaker since it has been realized that stars are likely to form in associations, and so be much closer together at an early stage of stellar formation. Thus, some of these objections fall away, and the coincidence of solar and planetary formation might be explicable. The fatal problem with the tidal theories seems to be that the filament of material would be dispersed rather than condense into planetesimals, let alone planets.

Other attempts to get around the statistical improbability of the near approach of another star led to the proposal that the Sun was originally one of a binary pair, the collision occurring between a third star and the solar companion [24]. Then the stakes were raised by making the Sun one of a triplet [25]. Another attempt to resolve the problem had the solar companion evolve into a nova or supernova [26].

Another version of the tidal hypothesis suggests the event occurred during the formation of a stellar cluster, when the Sun and a protostar of low mass and luminosity interacted to produce a filament from which the planets condensed. This hypothesis goes some way to bridging the gap with more conventional theories. The events all occurred close to the formation time of the Sun, thus avoiding the geochemical problems with the high abundances of deuterium and lithium in the nebula relative to present depleted solar values [27].

This model is essentially another version of the giant gaseous protoplanet hypothesis, and seeks to build the planets by fragmenting the nebula. It suffers from the same problems as that theory. A predictable consequence is that the giant planets, at least, might be uniform in composition, while there appears to be no mechanism to explain their obliquities.

1.4. SOLAR ACCRETION THEORIES

The first modern statement of these concepts is of historical interest as one of the few dissenting views to the then popularly held tidal theories of Jeans and Jeffreys [28]. In these, the Sun captures material from interstellar space. There would thus be no necessary connection between solar and nebular compositions. Material so accreted to the nebula could contain primordial abundances of Li, Be, and B, thus getting over the difficulty that besets the tidal filament hypothesis. In Schmidt's theory, a companion star was present whose task was to distort the accreting material and provide it with angular momentum to account for the observed values. This represents yet another attempt to get around the angular momentum difficulty. A completely distinct hypothesis proposed that the material in the nebula was a plasma so that accretion was dominated by ionization effects [29].

A principal and probably fatal objection to such models is that there is no significant correlation between elemental abundances and the ionization potentials of the elements, which should be expected on the basis of such a theory. Other workers proposed the accretion of a ring of material to the Sun during the final stages of solar formation, and were able to obtain satisfactory agreement for the mass of the individual planets [30].

The general advantage of such hypotheses is to avoid the light-element abundance problem. The disadvantage is that the Sun and the nebula could have different compositions. However, it is clear that the distinction is becoming blurred between hypotheses that propose the accretion of material from the interstellar medium and those that propose the formation of the Sun and the planets from a common nebula that separated as a fragment from a molecular cloud.

1.5. NEBULAR THEORIES

Most currently discussed theories involve a close connection between the formation of the Sun and the rest of the solar system. Such models will be developed in detail in the remainder of this chapter. In this section we are concerned with the historical development of the topic. The earliest theories of Descartes, Kant, and Laplace were of this type, perhaps indicating that such hypotheses are, or were, obvious first choices. In following such intuitive paths, one must beware, of course, of the "flat Earth" syndrome and of geocentric universes.

1.5.1. Descartes

René Descartes (1596-1650) appears to have been unencumbered by convention, if John Aubrey's (1626-1697) biographical sketch is a realistic account [31]. In 1644 Descartes postulated that the universe contained many circular eddies. Matter accumulated in the center of the vortex to form the Sun, while coarser particles (a forerunner of planetesimals?) were captured to form the planets. Satellites formed in secondary vortices surrounding the planets. His work, in contrast to that of Newton a little later, envisaged an evolutionary process, but was distinguished by a lack of quantitative rigor. Moreover, his hypothesis of vortices collapsed before the demonstration by Newton that the apparent complexity of the solar system could be dealt with by exact physical laws. Nevertheless, Descartes was the first to demonstrate that the solar system could be explained on the basis of simple laws of motion, and his cosmological scheme possessed originality. The intellectual successor to Descartes 300 years later was von Weizsacker, who proposed a regular sequence of vortices (Fig. 1.5.1a). In his theory, von Weizsacker was able to account for the Titius-Bode spacing of the planets. Subsequent workers replaced the regular vortices with more random turbulence [32] (Fig. 1.5.1b). A variation on this theme proposed that a turbulent nebula broke up into many small units, termed floccules [33]. A number of subsequent workers have attempted to quantify this model [34].

1.5.2. Kant

Immanuel Kant (1724-1804) provided a correct explanation for the Milky Way, proposing that it was an edge-on view of a disk of stars. He showed remarkable foresight by suggesting that the fuzzy lentil-shaped nebulae were distant island universes similar to the Milky Way, a cosmological leap in understanding that was not substantiated until the second decade of this century, nearly 200 years later. These essentially correct insights perhaps explain why his concepts for planetary origins are usually acclaimed rather than critically examined.

In the words of one critic, "Kant took but a glance and not a thorough look at the staggering problem of the origin of planetary systems. His explanation of it contained more nebulous statements at crucial junctures than there was nebulosity in the rudimentary form of the solar system. ... contrary to the stereotype accounts of Kant's planetary theory, he did

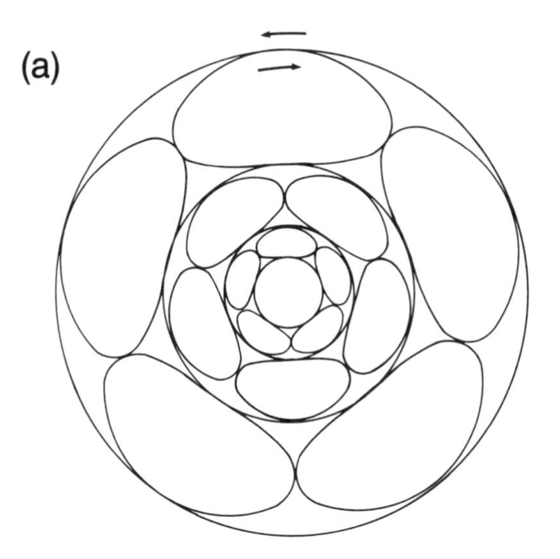

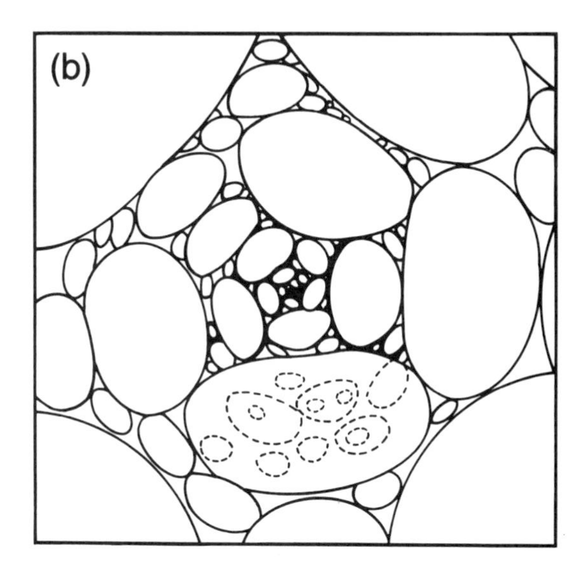

Fig. 1.5.1. ***(a)*** *The portrayal of the nebular disk as sets of concentric rings composed of five eddies. After von Weizsäcker C. F. (1943) Zeit. f. Astrophys., 22, 339.* ***(b)*** *Discontinuities within the von Weizsäcker sets of eddies. After Kuiper G. P. (1951) in Astrophysics: A Topical Symposium (J. A. Hynek, ed.), p. 374, McGraw Hill, New York.*

not compare the rudimentary form of the solar system to a nebulous agglomeration of matter" [35]. He believed that the operations of Newtonian mechanics could not of themselves produce the regularities of the solar system without divine guidance, and that "the material universe . . . has no freedom to deviate from this perfect plan" [36].

His model for the origin of the solar system was based heavily on an analogy with the galaxies. It began with a chaotic distribution of particles with slight density variations, which would accrete material and grow with time. The material was assumed to be rotating and to develop into flattened rotating disks. The Sun formed at the center, and the planets formed at secondary condensations within the disk. He postulated the existence of many additional planets outside the orbit of Saturn, although these possessed large eccentricities, with a gradual transition to the comets. He assumed that density fell off with distance, and accounted for the anomalously low density of the Sun at the center by deriving the material in the Sun from beyond Saturn.

Unlike Laplace, who proposed that the ring of Saturn (no divisions were known at the time) was solid, Kant correctly thought it to be composed of a multitude of small particles, and supposed, with less prescience, that all the planets, including the Earth, were initially surrounded by a ring, meanwhile populating the planets with beings whose intelligence increased with distance from the Sun, so that "our own Newton would not outrank a monkey on Saturn" [37].

It seems clear that the many contradictions in Kant's hypothesis do not accord with the general popular acclaim that it has received. Perhaps this is due to his eminence as a philosopher. It is a commentary on the inherent difficulties of accounting for the solar system that one of the foremost thinkers of the Enlightenment should have failed to produce an internally consistent explanation. His model is often linked with the hypothesis of Laplace, to which we now turn.

1.5.3. Laplace and His Followers

Pierre-Simon, Marquis de Laplace (1749-1827), was impressed, as Newton had been earlier, with the regularities in the solar system as it was known in the late eighteenth century: the planets all lay in a plane, they all moved in the same anticlockwise sense around the Sun, the satellites revolved around their parent planets in the same direction, while the planetary and satellite orbits were nearly circular. This led to his concept that the system had arisen from a primitive rotating cloud, the "solar nebula." Newton had considered such a regular system as evidence for the existence of God; Laplace, an inhabitant of the Age of Enlightenment, had "no need for that hypothesis," in his famous reply to Napoleon's question about who had designed the solar system [38].

The critical feature of the Laplace (1796) model for forming the Sun and the planets from the solar nebula was that as the cloud contracted, rings were successively ejected. These eventually condensed into the planetary sequence. A recent variation on this theme proposed that the gravitational constant, G, was decreasing with time [39]. In this scenario, the Sun is rotating at the edge of instability. As G decreases, an additional ring will be thrown off. Since G decreases only slowly, on timescales of 10^9 years or so, the theory should predict that the outer planets should be measurably older, on timescales of 10^9 years. Evidence for this is not obvious. Transfer of angular momentum outward by magnetic coupling has also been proposed to overcome the angular momentum objections to the Laplacian theory [40].

The failure of tidal and solar accretion theories has led to a rather general return to the nebular hypothesis, and current theories, particularly for the formation of the terrestrial planets, offer several variations on the theme. These hypotheses for planetary origin can be divided into two principal classes: those that form the planets as the result of condensation from an initially hot solar nebula and those in which the planets represent the final stages of the accretion of a hierarchy of smaller bodies.

The latter scenario is most readily addressed with evidence from the meteorites, which are commonly thought to be left-over fragments from a process of hierarchical accretion. However, it is a regrettable fact that the presently available samples of meteorites do not form a viable population from which to construct the Earth or, apparently, the other terrestrial planets.

1.5.4. Equilibrium Condensation Hypothesis

Condensation from an initally hot nebula is usually referred to as the equilibrium condensation model [41]. This theory was a paradigm for the formation of the planets for some 15 years as an explanation for the varying composition of the planets with increasing distance from the Sun. Some readers may be surprised that little more space is given to its discussion than has been accorded to the concepts of Descartes. The theory, although it provoked much research, does not account for many of the features of the solar system. It is a commentary on the rapidity with which our perceptions have changed within the past two or three years that this hypothesis must now take its place along with the host of other elegant but inadequate theories that have attempted to account for the compositions of the planets. Most proponents have been seduced by the apparent zonation from dense Mercury to the gaseous outer planets, which even so eminent a philosopher as Emmanuel Kant believed to be correlated with the intelligence of their imaginary inhabitants.

The equilibrium condensation model postulates that each planet forms within rather narrow concentric zones in a nebula that displays variations in composition with distance from the Sun. As the minerals condense in succession, in accordance with the predictions of the "condensation sequence" (section 3.9), they are accreted to the growing planet. This theory encounters several difficulties. Many meteorites are rather complex mixtures of high- and low-temperature minerals. They are not composed of a simple sequence of minerals that had condensed monotonically from a vaporized nebula. There is much evidence for isotopic heterogeneities and other irregularities that might be homogenized in such a high-temperature nebula [42]. Some revision of nebular models predicts higher temperatures out to 3 AU, which occur, however, at a very early presolar stage in nebular evolution and are not necessarily relevant to the stage of planetary formation [43].

Heating of an inner zone by early intense solar activity (T Tauri and FU Orionis stages involving flares, strong solar winds, enhanced UV flux), seems necessary to account both for the volatile-element depletion in the inner planets and to deplete that zone in the gaseous nebular components (H, He, and the noble gases). However, it is unlikely to have extended much beyond 2-3 AU. Supporting evidence from meteorites for low nebular temperatures in the asteroid belt is the presence of layer-lattice silicates in CI and CM2 chondrites, which would decompose into anhydrous phases at temperatures in excess of 700 K. Accordingly, the concept of an initial hot (~2000 K) nebula that cools monotonically, producing en route the classical condensation sequences of minerals [44], does not seem consistent with several lines of evidence.

In summary, the homogeneous accretion model relies on many dubious assumptions. These include the concept that nebular temperatures decreased monotonically outward from the Sun, and were isothermal at the sites of accretion of the individual planets. The model also assumes that equilibrium was maintained between the gas and dust in the nebula over a large temperature range, and that there was a smooth variation in the composition of the solid phases with distance from the Sun. Furthermore, it is assumed that a constant temperature was maintained while each phase accreted to the planet, and finally, that each planetary accretion zone was isolated.

The theory was anchored at one end by the extreme iron-rich composition of Mercury. The difference in the condensation temperatures between that of silicates and Fe-Ni metal is so narrow that a very delicate relationship had to be preserved during the accretion of that planet. However, other explanations for the anomalous composition are available, with large collisional models removing silicate mantles being preferred (sections 4.8 and 5.3).

1.5.5. Heterogeneous Accretion Model

A related hypothesis of planetary origin is the heterogeneous accretion model [45]. This proposes that as each phase condenses in the nebula, it is accreted, so that the planets grow like a layer cake. The initial condition of the planets is that of a highly refractory interior and a volatile-rich surface. This scenario at least explains the surficial concentration of volatiles on the Earth, but many of the difficulties that beset the homogeneous accretion model apply equally to this hypothesis and it is not at present a serious contender [46].

In contrast to these somewhat theoretical models, the remaining hypotheses rely more heavily on meteoritic analogies. Many scenarios have been based on the two components (fine-grained matrix and coarser metal particles, chondrules, etc.) that are intimately mixed in many meteorites [47]. Sometimes these components have been characterized as volatile-rich and volatile-poor, or refractory [48], and similar scenarios continue to appear [49].

A more complex version, involving seven components, is based on meteorite mineralogy [50]. The philosophy behind this approach is that the variations in the compositions of the inner planets are similar to those observed in chondrites. Inherent in this approach is the concept that varying mixtures of the seven components could account for the complexity of both meteorites and planets, without seeking specific matches between them. The more recent models employing this general scenario have tended to simplify the components to two: one oxidized and volatile-rich, and the other a reduced high-temperature volatile-poor, refractory- and metal-rich component [51].

Such models attempt in this way to account for the obvious presence of reduced metal, oxidized phases, and some volatiles in the inner planets, but do not shed much light on the source or origins of the postulated two components in the nebula. They describe the phenomena rather than explain them, a common intellectual trap. Several properties (e.g., oxygen isotopes, noble gas contents, etc.) indicate that neither the volatile nor the refractory component can be matched with a specific meteorite class.

Although the volatile component is commonly identified with CI chondrites, a specific match is ruled out on many grounds (e.g., oxygen isotopes). The addition of such a component adds large quantities of water and volatile elements. In some models, this induces reactions with metallic iron from the reduced component, and the hydrogen so produced escapes. In general, however, too high a component of volatile elements is added by using CI analogues.

The E chondrites (section 3.6) are the most obvious candidates for the refractory component, but although they are reduced, they are not volatile depleted, and so they compound the problem of adding excess amounts of the volatile elements. They also show extreme variations in major-element compositions such as Mg, Ca, and Al, that make them among the least satisfactory candidates for building the Earth.

Such deviations are usually excused on the grounds that CI and E chondrites are only used as analogues for the volatile and refractory reduced components that went to make up the Earth and the other terrestrial planets. However, their use produces as many problems as it solves, and so many ad hoc adjustments have to be made that the initial advantages of the theory are lost. One might expect that if the inner planets were built out of these two components, they should have been widespread in the inner nebula. Consequently, both planetary and meteoritic compositions should show considerably more homogeneity, rather than exhibiting their considerable diversity in composition.

1.6. TITIUS-BODE RULE

This mathematical regularity in the spacing of the planets has always attracted wide interest, and is frequently cited as one of the significant parameters to be satisfied in any theory for the origin of the solar system. However the Titius-Bode relationship is not really a "rule," and in any case is only approximately satisfied. The best statement of the rule is the Blagg-Richardson formulation, that gives

$$r_n = r_0 A^n$$

where r_n is the distance of the nth planet from the Sun and A is a constant (1.73). The rule was first formulated by Johann Daniel Titius von Wittenberg (1729-1796) and was popularized by Johann Elert Bode (1747-1826).

Apart from the planets, the spacing of the regular satellite systems of Jupiter, Saturn, and Uranus also appear to be consistent with the rule. The position of Pluto does not fit the rule as it applies to the other planets, but this left-over icy planetesimal can, of course, be excluded since it does not justify classification as a planet. Other evidence (section 5.14) points to the orbit of Neptune as the true outer boundary of the planetary system.

Does the Titius-Bode rule have any real significance? It is curious that there is no correlation with other planetary properties such as mass or composition with either the spacing formulated by the rule or heliocentric distance. It seems reasonable to expect that if the rule represents some major physical determinant in the construction of planetary systems, then some other major properties might also correlate with the simple mathematical regularities in spacing of the planets noted by Titius [52]. If the Titius-Bode rule were due to a fundamental property connected with the formation of the system, it is reasonable that some such correlation should be apparent. This deficiency raises the possibility that the rule is a secondary, rather than a primary, property of the solar system.

There is, of course, no real evidence that we are looking at the initial orbital distribution and the spacing may equally well be determined by forces subsequent to planetary formation. This view is supported in the case of satellites, for example, by the prevalence of orbits controlled by stable resonances. It is possible that the early formation of massive Jupiter, followed by Saturn, imposed resonances on the rest of the system, now somewhat altered by the local collisions [53].

Nieto [54], who has made a definitive study of the rule, considered that there are three periods in the history of the solar system. These are (1) the disk period, (2) the period of aggregation, and (3) the planet period. He proposes, after what is the most thorough and objective survey of the entire topic, that "(a) the geometric progression (i.e., the Titius-Bode Rule) is due to a fluid and/or magnetohydrodynamical mechanism that occurred during the disk period and (b) the "evolution" or periodic function of the rule comes from a tidal or point gravitational relaxation that took place during the planet period" [55]. He reserves the position that the rule may have been entirely "caused by gravitational and/or tidal encounter mechanisms" [55, p. 131] that occurred following planetary formation.

This latter view receives support from the interesting fact that the rule is obeyed by the inner satellites of Jupiter, Uranus, and Saturn. It is difficult to imagine that the conditions in the individual planetary nebulae closely reproduced those in the solar nebula. This is apparent from the many morphological and compositional differences among the three regular prograde satellite systems of Jupiter, Saturn, and Uranus.

In summary, it seems likely that the Titius-Bode rule results from gravitational and tidal evolution following planetary and satellite assembly, and is less likely to be due to dynamical conditions within the nebula [56]. Thus the famous Titius-Bode rule appears to be a qualitative relationship of uncertain status. The spacings have probably arisen naturally due to tidal interactions of the planets, perhaps forced by the early growth of Jupiter. These relationships are thus without genetic significance with respect to the origin of the system.

1.7. GIANT GASEOUS PROTOPLANETS

The concept of large protoplanets has a long history. Jeans and Jeffreys in their tidal hypotheses had envisaged gas spheres condensing from the filament of material torn out of the Sun, but the concept achieved its modern form through the work of Kuiper [57]. A similar concept of "gas spheres" was elaborated by Urey [58].

Only two mechanisms apparently exist to produce rapid formation of planets: either through an instability arising in a gaseous disk [59] or through instability in a dust disk [60]. These naturally result in the two contrasting hypotheses for building planets. The first mechanism fragments the gaseous nebula, so that planets form in a manner analogous to the formation of stars. The second mechanism builds them out of smaller units. The giant gaseous protoplanet theory is of the first type and calls for the condensation or fragmentation of gaseous protoplanets directly from the nebula [61]. Since such a process would occur rapidly, it overcomes the long timescale problems inherent in early versions of the planetesimal hypothesis, although at the cost of requiring a massive (2 solar masses) nebula to induce the instability for breakup. Fragmentation should occur on timescales of a few million years at most, and could thus form Jupiter well in advance of the asteroids or the terrestrial planets. This dubious advantage is offset by many other problems.

In another version of the giant protoplanet hypothesis, six standard protoplanets of mass 10^{30} gm form [62]. Two inside Jupiter break up into the four terrestrial planets, Jupiter and Saturn suffer some mass loss, while H and He escape from the outer solar system, accounting for the gas-poor composition of Uranus and Neptune. By combining the Earth, Moon, and Mars into one planet, and Venus and Mercury into another, this theory accounts for "one of the most puzzling things about the solar system, the assortment of compositions of the inner planets without [appealing to] any differentiation within the solar system itself" [63].

This and similar gaseous protoplanet hypotheses have some interesting and predictable consequences. If the nebula breaks up into fragments, all planets should exhibit similar compositions or exhibit some regular and systematic change with heliocentric distance if the initial solar nebula was zoned compositionally. Even if subsequent evolution removes the gases in some manner from the region of the terrestrial planets, the giant planets at least should all be similar, unless the nebula breakup is not coeval for all. However, there is little uniformity; the ratios of H and He to the heavier elements varies among all four giant planets, and none possesses "solar" or "cosmic" ratios of H and He to the heavier elements.

There is no apparent mechanism in the hypothesis to fractionate elements before the breakup of the nebula, so that separation of volatile from refractory elements would be minimal. In the giant gaseous protoplanet model, the inner planets should be essentially of CI composition reflecting the initial abundances in the nebula. Depletion of volatile relative to refractory elements cannot take place from large condensed bodies, but must occur when the material in the nebula is finely dispersed. Perhaps processes connected with the dissipation of the gaseous envelopes might be invoked to accomplish some separation of the more volatile elements. One might expect the rare gas signatures to have more uniformity, rather than the observed diversity [64]. A general deduction from this evidence is that the terrestrial planets accumulated in a gas-free environment and this seems to be a fairly firm constraint.

The principal attraction of the giant gaseous protoplanet hypothesis is that it appears possible to fragment the nebula into Jupiter-sized bodies, thus accounting for the giant planets. However, the moments of inertia of Jupiter and the other giant planets indicate the existence of rocky cores, of about 15 Earth masses. In the giant gaseous protoplanet hypothesis, these would form by gravitational settling of the denser components. A major objection is that at pressures above about 10 mbar (the central pressure in an icy body of about 40 Earth masses), the rocky, icy, and gaseous material will all become completely miscible, so that a central core cannot "rain out" [65]. In the planetesimal hypothesis, these cores are conventionally regarded as composed of rocky planetesimals that act as centers allowing the gravitational collapse of the H and He from the nebula.

Would it be possible to transform a giant gaseous protoplanet into a terrestrial rocky planet? In this hypothesis, grains condensed as the planet cooled, and rained out forming a rocky core. In the inner reaches of the solar system, a strong solar wind then removed the gaseous outer envelope. Mercury is a test case. Two options exist. The first is that Mercury was a giant gaseous protoplanet, most of whose mass was removed by early solar activity during an FU Orionis or T Tauri stage. First, of course, this means that the planet had to form within 10^6 years of the Sun and probably earlier. Even after the gas has gone, some difficulties remain. Studies of the evaporative loss of the silicate mantle of Mercury in sufficient amounts to account for the high density show that about 80% evaporation is needed. The alkali elements are totally lost. This conflicts with the observation that Na and K are present in the tenuous mercurian atmosphere [66]. A more reasonable second hypothesis is that the high density of Mercury is due to collisional removal of part of the silicate mantle.

In summary, the giant gaseous protoplanet hypothesis encounters some serious geochemical problems. It should in general predict uniform compositions for the giant planets as fragments of the original nebula, while the terrestrial planets should likewise be the rocky cores of protoplanets, and uniform in the absence of mechanisms for fractionating the elements. Even if the gases were removed by early intense solar winds, there is no easy way to remove volatile elements, such as K, from condensed large bodies. The hypothesis does not readily account for, for example, the depletions in volatile (e.g., K) relative to refractory (e.g., U) elements observed throughout the inner solar system. For these reasons, new versions of the planetesimal hypothesis have replaced the concept of giant gaseous protoplanets [67].

1.8. DIFFERENCES BETWEEN INNER AND OUTER PLANETS

The most striking dichotomy in the solar system is the distinction between the giant planets and the rocky inner or terrestrial planets. The compositional differences among the planets are discussed in Chapter 5. Although the giant planets are superficially similar, there is really a threefold division (classifying Pluto as a minor body), since Uranus and Neptune are in effect ice giants rather than gas giants. Compared to Jupiter and Saturn, these two large outer planets are depleted in H and He, and contain a much higher proportion of the heavier elements. Jupiter and Saturn are neither equivalent in composition, nor does either match solar abundances. These distinctions, although significant, are minor compared with the gas-free compositions in the inner solar system.

What is the basic cause of these differences? Are they compatible with constructing the planets by fragmentation of the nebula, followed by condensation to form the giant gaseous protoplanets? The major compositional variations, even among the giant planets, from that of our ideas of the primordial solar composition seem to argue against this. The planetesimal hypothesis in contrast, accounts naturally for these disparities. The crucial question, addressed in the next section, is the timing of the formation of Jupiter [68]. Some dispersal of the nebula must have occurred even before Jupiter accreted, for that planet has a higher rock/gas ratio than the solar values.

Removal of the gaseous nebula is most likely associated with early violent T Tauri and FU Orionis activity. These events occur within perhaps 10^6 years of the arrival of the Sun on the main sequence. Jupiter has to form before the gas is dispersed, so that early formation of Jupiter seems inevitable. This probable ancient age of the asteroid belt and the necessity to accrete gas to Jupiter while the nebula was still dominated by H and He restrict the formation of Jupiter to a period within about 10^6 years of the early active period of the Sun, since such activity is generally held to be responsible for the clearing of the nebula.

Saturn has to have formed comparatively rapidly, but this occurred at a slightly later stage, when more of the gaseous nebula had been lost. Uranus and Neptune must have accumulated at a time later than Jupiter and Saturn, after significant amounts of gas had been swept away. Ejection of planetesimals by Jupiter is also likely to result in a major, but presumably random, depletion of material from the solar system. This process might deplete the nebula if there were a higher initial mass in the nebula compared to the present mass of the planets.

It seems clear that the inner rocky planets accumulated in a gas-free environment, in contrast both to Jupiter and to the Hayashi-type models in which the terrestrial planets accumulate in the primordial gas-rich nebula [69].

1.9. FORMATION OF THE GIANT PLANETS

"An obvious... upper limit on the timescale for assembling... the giant planets is the age of the solar system... However,... some models of the formation of Uranus and Neptune have had difficulties in meeting even this modest constraint" [70].

None of the giant planets are of solar composition and thus do not reflect the composition of the original nebula. Instead, all the giant planets have rocky cores of about 10-20 Earth masses and gas/rock ratios that are not present in solar proportions. Apparently the giant planets accreted rock and ice from the nebula more efficiently than gas. This is the reverse of what would be expected from models that form the giant planets by fragmentation of the nebula. None of these compositional facts appears to fit the simple concept of giant gaseous planets, which might be expected to be rather similar in composition or to show some orderly evidence of zonation in the primitive nebula. Accordingly, a two-stage process to form the giant planets appears to be required rather than a single-stage gravitational collapse from the primordial nebula.

1.9.1. Formation Times

Early runaway accretion is needed to form the giant planets because of the very short lifetime (a few million years at most) of the gas-rich nebula. Even in the Hayashi gas-rich scenario, Jupiter requires 4×10^7 years and Saturn 6×10^8 years to form, with very long periods for Uranus and Neptune. Calculations based on Safronov [71] and Wetherill [72] models for planetesimal accretion produce growth times for the giant planets of 10^8-10^{11} years, clearly inconsistent with the age of the solar system (4.6×10^9 years) and the timescales for removal of H and He and the nebula of 10^6 years. All these times for planetary formation are too long and indicate that the models are wrong. The basic problem is not that growth does not occur quickly in the early stages, but that for low-mass nebula models, the nearby feeding zones are exhausted as the eccentricities of the planetesimals increase. The crux in models of runaway accretion is to supply a sufficient number of planetesimals through radial transport. Various explanations have been offered. These usually include the presence of a more massive nebula and low relative velocities of planetesimals, that enable faster accretion rates [73]. Other estimates for the times taken to form a 10-Earth-mass core are 700,000 years for Jupiter, 3.8 m.y. for Saturn, 8.4 m.y. for Uranus, and 23 m.y. for Neptune [74]. These also seem too long compared to the astrophysical requirements to dissipate the nebula on timescales of 10^6 years.

The argument for early growth of Jupiter depends on several observations. These include the absence of a planet in the asteroid belt, the small size of Mars, and the small number of planets. All appear to be consistent with the early formation of a dominating large planet containing most of the mass and angular momentum of the system.

1.9.2. Asteroid Belt

A first-order observation is the great depletion of matter in the asteroid belt, most reasonably attributed to the prior formation of Jupiter, since there seems no reason to postulate a hole in the nebula. The combined mass of the asteroids is 10^{-3} Earth masses, an order of magnitude less than the mass of the Moon. Since there is no reason to suppose that this region was depleted relative to the rest of the nebula, material must have been removed both from it and from the feeding zone for Mars. Subsequent mixing or stirring by high-speed planetesimals must have been minor, otherwise these differences would have been evened out and the zonal nature of the belt, a relic of a very early heating episode, would have been destroyed.

The clear implication is that Jupiter was present before protoplanets were able to complete their accretion. Since meteorites are derived at least from the inner portions of the belt, we have access to material that has been affected to some degree by Jupiter. The very old dates for meteorites make it likely that the asteroid belt has had its present population and structure since close to T_0.

1.9.3. Nebular Lifetimes

The major time constraint for the formation of Jupiter and Saturn is given by the observation that T Tauri stars clear their circumstellar disks on timescales of 10^6 years (section 1.11). The planetesimals in the inner solar nebula must have been so small (10^{-2} Earth masses) that they did not trap a primitive atmosphere from the nebula, but they must have been large enough to survive being swept away by strong stellar winds during the early violent T Tauri and FU Orionis stages of solar activity.

In order to produce runaway accretion to build a 10-Earth-mass core for Jupiter, a seed nucleus 2-3 times that of the next largest planetesimal appears to be required. Once such cores reach 10 Earth masses, rapid gravitational collapse of gas onto the nucleus will occur. However, the initial distribution of planetesimal sizes is unknown. In addition, protoplanet growth will slow down when it has accreted everything within reach.

1.9.4. Increased Nebular Density

Many dynamical problems remain and no completely self-consistent explanation for this early appearance of Jupiter on the scene in the emerging solar system has been offered until recently [75]. Much of the problem of rapidly growing a planetary embryo of 10 Earth masses, on which the H and He can collapse gravitationally, results from the assumption of a minimum mass nebula. One solution to this problem proposes that the density of material in the region of Jupiter was 5-10 times that needed to account for its core [76], which is about 15-30 Earth masses of rock and ice [77]. This core could have accreted in a few times 10^5 years. "A surface mass density of solids in the range 15-30 g/cm^2 appears to be sufficient to allow Jupiter's core to be formed by rapid runaway growth of planetesimals on a timescale of 5×10^5-10^6 years" [78].

The nominal density of rock and ice required to account for the present mass is about 4 gm/cm^2. This will lead to runaway growth of an Earth-sized core in 10^5 years, but at this stage eccentricities have increased and the planetesimals thin out so much that growth slows, and times of the order of 10^8 years are required to assemble a large enough core to cause the gas to collapse by gravitational attraction. However, by this time, the gas has gone. If the surface density can be increased by a factor of 5 or so, the problem goes away, and a massive core for Jupiter can form quickly, while the gas is still around [78].

In addition, density wave torques may operate, so that the planetesimals do not become stranded, and new material can be supplied once local sources are depleted. In this model, runaway growth can occur on timescales of 10^5 years [79].

1.9.5. "Snow Lines"

How might such an increased nebular density occur? If it was an original feature, then there would be large amounts of material in the inner solar system that have to be ejected. A simpler solution to the problem [80] is to condense water ice at the "snow line" around 5 AU effectively increasing the local density (Fig. 1.9.1). It is also necessary to add material to the region rather than by simply condensing water ice. Such a process can be linked plausibly to

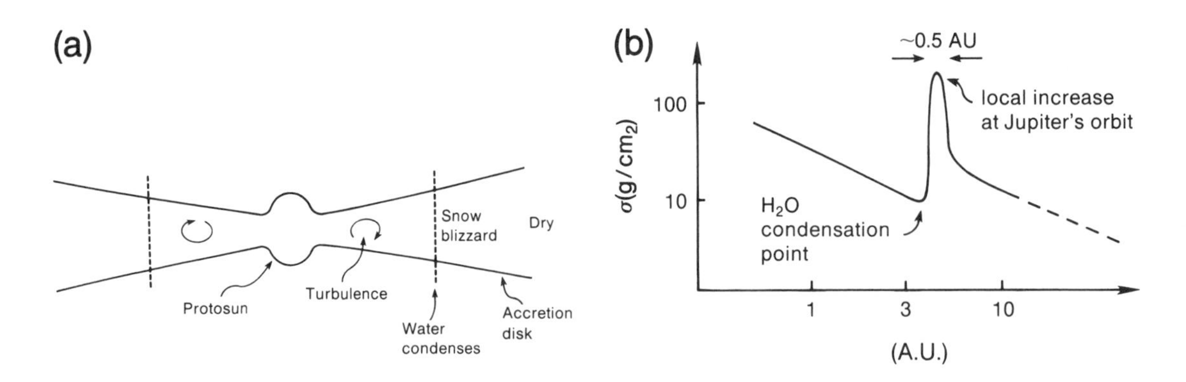

Fig. 1.9.1. *One possible scenario for increasing the surface density of solids at the location of Jupiter.* ***(a)*** *Physical model;* ***(b)*** *Surface density as a function of distance from the Sun. After Stevenson D.J. (1989) in The Formation and Evolution of Planetary Systems (H. A. Weaver and L. Danly, eds.), p. 85, Fig. 4, Cambridge Univ., New York.*

the sweeping out of volatiles, including water, from the inner nebula by early violent solar activity, and so occur at a usefully early stage. Processes associated with the early Sun can thus result in an increased density at 5 AU without needing to appeal to an initial heterogeneous nebula. In this context, it appears that the critical core mass to cause catastrophic collapse of gas from the nebula is about 10 Earth masses [81], although much smaller masses might suffice in a water-rich envelope [82].

If a more massive nebula is employed to ensure rapid growth of Jupiter, it is necessary to remove the excess solid material from the solar system. This problem has accounted for the popularity of low-mass nebulae, so that the excess mass problem is solved by definition, at the cost of unrealistically long timescales for producing the giant planets. However, once Jupiter has reached massive proportions, it becomes a major gravitational influence that perturbs the remaining planetesimals. Some of them would be accreted, but most would be tossed out of the solar system. In these models, Jupiter may have moved perhaps 1 AU closer to the Sun during this period. The planetesimals ejected from the nebula would carry some angular momentum.

Jupiter's effect on the asteroid belt in this scenario was to produce rather high relative velocities among the surviving planetesimals. Most of the mass in the Mars zone was also ejected, about 3% remaining compared with 0.02% of the original mass left in the asteroid belt. Jupiter will have scattered some planetesimals into Earth-crossing orbits, but only a few would reach Venus.

If the nebula was very massive and Jupiter had to eject about its own mass, then it may have formed at about 7 AU and moved Sunward to its present location at 5 AU. Such models remove Jupiter from the vicinity of the "snow line" and hence remove an attractive feature of the scenario. However, if the increased nebular density is tied to the sweepout of material from the inner nebula by early violent solar activity, it is unnecessary to form Jupiter further out in the nebula.

The gravitational forces of Earth and Venus are too weak to eject planetesimals from the solar system. Accordingly, the mass of the nebula in these regions was probably close to the minimum value required to grow Venus and Earth. If Venus accreted everything in a zone from 0.5 to 0.85 AU, then the surface density of the nebula was about 15 g/cm^2. Planetesimals scattered by Jupiter would have subjected Mars to a heavy bombardment, with lesser effects on the Earth and Moon, and would have done only minor damage to Venus and Mercury.

1.9.6. Uranus and Neptune

The growth periods for Uranus and Neptune are of the order 5×10^6-10^8 years, significantly longer than for Jupiter, and they have probably not moved far from their original locations. Uranus and Neptune managed to obtain only a few Earth masses of H and He. Uranus and Neptune either did not manage to accrete critical core masses for trapping gas, or by the time they had formed cores of 10 or so Earth masses, the gas was mostly gone. Although both effects may contribute, the latter explanation appears more probable since the core masses for all the giant planets are apparently very close to the point where they would cause gravitational collapse of H and He.

Thus, significant nebular erosion of gas occurred before the growth of Uranus and Neptune was complete, and most of this gas must have been gone before the cores of Uranus and Neptune reached critical mass to trap the fleeing gas. For this reason, these planets are ice giants compared with the gas giants of Jupiter and Saturn. Accretion in the region of the outer planets is also complicated by the low velocities needed to escape from the solar system.

1.9.7. Limits to Growth

Finally, it is necessary to address the reasons why Jupiter, in particular, but also the other giant planets, stopped growing, since they all have similar sized cores. What were the limits to the growth of the giant planets? Why are they not larger? For the ice giants, Uranus and Neptune, the answer is clear: the nebula ran out of gas. A similar scenario could be invoked for Saturn.

For Jupiter, it is probable that its large size eventually resulted in a tidal truncation of the nebula both Sunward and outward in the nebula. The runaway collapse of gas onto Jupiter ceased as that planet cleared a gap in the nebula and truncated its growth by tidal forces. Jupiter's tidal radius is about 0.36 AU. Thus, the formation of planets is self-limiting: they run out of material. This constitutes an argument for low- rather than solar-mass nebulae; Jupiter and Saturn might have grown much larger in a massive nebula. In the case of Uranus and Neptune, earlier dissipation of the nebula depleted the supply of H and He.

Most hypotheses for planetary accretion show that ejection of planetesimals from the solar system dominated over accretion. Both Uranus and Neptune would have ejected large amounts of mass in such models, which argues for a more massive nebula than that required to account for the present masses of the planets [83].

Much of the ejected material left at about escape speed and so probably was perturbed by stars and the remnants of the molecular cloud into the Oort comet cloud. Such a mechanism means that comets may not necessarily be primitive objects, but may have had a more complex history within the planetary regions before being ejected into the outer reaches. The composition of Comet Halley (sections 2.15 and 3.10) encourages this view, since its Fe/Si and Mg/Si ratios are far removed from our current ideas about primitive nebular compositions. The CH_4/CO ratio in Comet Halley is also high compared to estimates of the primordial ratio, perhaps consistent with some chemical processing of primitive material [84].

1.10. ACCRETION FROM DUST OR FROM A HIERARCHY OF OBJECTS

One might assume *a priori* that formation of planets from a dusty disk would occur by accretion of dust, coagulating around centers initially caused by gravitational instabilities, and perhaps related to the Titius-Bode spacing. This simple picture, which was a feature of some older models, does not accord, however, with the observational evidence that indicates that planets accrete from a hierarchy of objects rather than from dust. There are several lines of evidence in support of the concept that a whole variety of bodies of differing sizes grew in and populated the early solar system before being swept up into the present planets.

Planetary and satellite surfaces record the impact of many large objects (Chapter 4). Ringed basins up to 1000 km in diameter are common, caused by bodies of the order of 100 km in diameter. Collisions of even larger bodies are probably responsible for the Procellarum Basin, 3200 km in diameter, on the Moon, and perhaps for the dichotomy between the northern and southern hemispheres of Mars.

The obliquities of the planets are consistent with collisions of very large bodies. If the planets had accreted from dust, or indeed from small objects only, their rotational axes should be normal to the plane of the ecliptic, in the absence of any objects big enough to disturb this tidy arrangement. Even Jupiter has a 3° tilt, and the other planets show substantial tilts. This argues for the former presence of very massive objects. Collision with an Earth-sized object is required to knock Uranus on its side, while large collisions are required to account for the 23° obliquity of the Earth and the current 25° tilt of Mars. The slow backward rotation of Venus is an apparent anomaly in the solar system, but is consistent with collision with a Mars-sized object that was sufficiently energetic to reverse the direction of rotation [85]. The high density of Mercury is most rationally explained by the removal of part of its silicate mantle as a result of a large collision with an object about one-sixth of the original preimpact mass of Mercury. The collision of an object about 0.10-0.20 Earth masses with the Earth is the most satisfactory current hypothesis to explain the origin of the Moon, accounting for the several unique compositional, orbital, and dynamical features of our satellite.

Accretion from dust might be expected to produce rather uniform planetary compositions, particularly for the terrestrial planets. Alternatively, some regular zonation in planetary compositions might be observed with heliocentric distance. Neither of these signatures is apparent. The meteorite evidence is instructive. Individual meteorites are not uniform aggregates, but typically consist of mixtures of several components or are fragments of larger bodies. It is an interesting fact that the parent asteroids of the meteorites, although all quite different in composition, must have been individually very homogeneous. This is illustrated in Fig. 1.10.1, which indicates that selective accretion of silicate, metal, or sulfide particles, or of chondrules and matrix operated only at the earliest stages of aggregation, before the aggregates reached between millimeter and centimeter size, since gram-sized samples of chondrites are generally quite uniform in chemical composition.

The next question to address is whether the accreting objects were undifferentiated objects (of CI composition) or were differentiated. There is considerable evidence from the asteroid belt that most of the inner belt objects were differentiated [86]. The S-type asteroids, which dominate the inner belt, and which presumably most closely resemble those planetesimals that accreted to form the inner planets, are all differentiated objects. Curiously, they do not resemble the spectral signature of the ordinary chondrites.

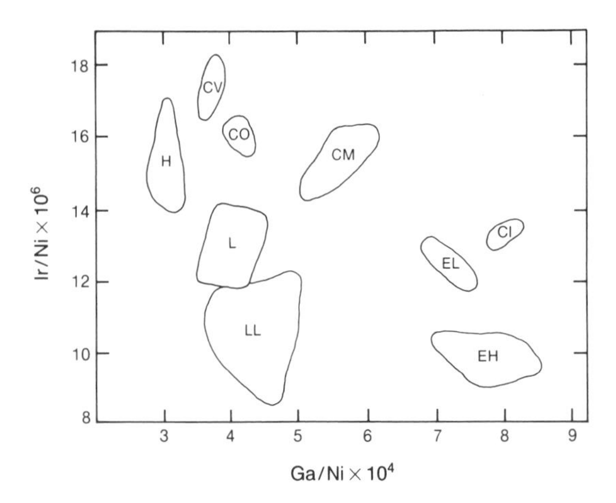

Fig. 1.10.1. *The nine chondrite groups do not define a simple compositional sequence on this plot of Ir/Ni vs. Ga/Ni, indicating that the nebula was neither homogeneous nor zoned in any simple manner. After Scott E. R. D. and Newsom H. E. (1989) Z. Naturforsch., 44a, 927, Fig. 2.*

In the present context, it seems likely that the planetesimals from which the terrestrial planets accreted, Sunward of the asteroid belt, were also differentiated. An important deduction from this is that the Earth and the inner planets accreted from objects in which metal-silicate fractionation had already occurred. This will occur at low pressures (typically less than 10-20 kbar) so that metal-sulfide-silicate equilibria will be established under low-pressure conditions. When this material accretes to a planet, melting during accretion, or melting induced by large impacts, will cause rapid segregation of these phases. High-pressure equilibration is unlikely to occur under conditions of catastrophic core formation; rather, there is likely to be a simple separation of metal and sulfide from silicate based on density differences.

In summary, the evidence from the present solar system is consistent with the existence of large precursor bodies during the planetary accretion stage. The obliquities of the planets and the slow retrograde motion of Venus are most reasonably explained by the late-stage impact of large planetesimals. Large bodies capable of producing these effects are thus apparently required to account for the observational evidence in the solar system. The heavily cratered surfaces present from Mercury to the satellites of the outer planets provide evidence for a saturation bombardment history of objects up to a few hundred kilometers in diameter that produced the ringed basins with diameters up to 2000-3000 km [87]. The asteroid belt presents us with evidence of objects as large as 1 Ceres (933 km in diameter), while the meteorites are mostly fragments of larger bodies; 9 at least for the chondrites and over 60 parent bodies for the irons. Accordingly, it appears reasonable to suppose the existence of large (up to Mars-sized, 0.1-0.2 Earth mass) bodies during the accretion of the terrestrial planets. The timescale required to build and differentiate these objects into mantles and cores may be at least 10^7 years, so that they accumulate into the terrestrial planets in a gas-free environment. In this scenario, the largest collisions, such as that of a Mars-sized object with the Earth, will be among the last accretionary events.

1.11. PLANETESIMALS

According to the current version of the planetesimal hypothesis, at a very early stage the dust in the rotating disk of the solar nebula began to clump together, so a hierarchy of bodies formed, beginning with grains and proceeding through meter-sized lumps to objects of kilometer size, finally reaching dimensions of hundreds to thousands of kilometers during the stage before the final assembly of the terrestrial planets. These early objects are termed planetesimals and current thinking regards them as the building blocks of the terrestrial planets and of the rocky cores of the giant planets.

The term planetesimal originated with T. C. Chamberlin and F. R. Moulton [88]. In their hypothesis, it referred to the small bodies that condensed from the filament pulled out of the Sun. The name has survived, in an altered context, as a useful term to describe, according to the planetesimal hypothesis, the hierarchy of small precursor bodies from which the inner planets were assembled.

In the current understanding, planetesimals form following the gravitational contraction of the original, more-or-less spherical mass of gas and dust to a rotating disk, the stage at which it is most commonly thought of as the solar nebula. This occurs rather rapidly on timescales of 10^4-10^6 years. In a turbulent nebula, kilometer-sized objects may grow in 10^4 years. The process must postdate chondrule formation, which must occur while the dust is still in a dispersed state.

Planetesimal accretion from millimeter-sized objects can begin when the dusty midplane region of the nebula reaches a density at which self-gravitation can occur, enabling growth to kilometer-sized bodies. Appropriate densities in the dusty layer are 4×10^{-7} gm/cm^3 at $pH_2 = 10^{-4}$ atm [89]. The initial kilometer-sized bodies have masses about 10^{17} gm and densities of 1 gm/cm^3. This process is proposed to take of the order of 10^4 years. Some growth of planetesimals must occur during the T Tauri or FU Orionis stages of stellar evolution [90] as gas, dust, volatile elements, and water are driven out of the inner solar system. Only material in the size range from meters to kilometers will survive during these early events, eventually accumulating into planetesimals that ultimately form the inner planets.

1.11.1. Settling Rates

A further factor to consider in the early stages of planetesimal growth concerns the settling time of dust toward the midplane of the nebula. Calculations indicate that millimeter-sized grains will settle in about 10^3 years, but a 1-μm-sized grain takes about 10^6 years. However, there are so many uncertainties, and the role of turbulence is so dimly understood that such timescales might be taken with a grain of salt [89].

The crucial parameter is the time at which the Sun begins its violent T Tauri and FU Orionis stage. The strong stellar winds will remove any material in the inner nebula that is not in condensed meter- to kilometer-sized bodies by that time. If there is a compositional difference between fine and coarse particles in the nebula, then the differences in settling rates will provide a mechanism for producing compositional differences. The collapse to the central plane of the nebula does help to solve part of the problem of the initial accretion or sticking together of particles by providing higher dust densities.

Do we have information on fine and coarse particle differences? The fine-grained matrix of CI is probably not representative, since it has been subjected to aqueous alteration, while the fine-grained matrix of chondrites does not represent primitive unaltered nebular or interstellar dust (see section 2.4).

Much of the fine material may never have reached the midplane before being dispersed by stellar winds, so providing a ready mechanism for fractionation. This raises questions about the uniformity of the composition of the nebula and its relation to both the canonical CI composition, and its relation to solar abundances, which have scarcely been addressed.

1.11.2. Compaction

Among the many other problems in this dimly understood field is the question of compaction of planetesimals in very weak gravitational fields. Most meteorites are relatively dense. If they are at all representative of planetesimals, then these must have been compacted. The most likely candidate (gravitational forces in the nebula are too small) is impact. Successive impacts during the next growth stage of planetesimals from kilometer to 100-1000-km size will provide sufficient energy to sinter and even induce some metamorphic recrystallization. It should be noted, however, that most chondrites have never resided in large planetesimals greater than about 100 km in diameter. Bodies of this size are not sufficiently massive to achieve a spherical shape by self-gravitation.

1.11.3. Mixing

Although much mixing of planetesimals subsequently occurs during planetary formation, it is an observational fact that the inner planets differ both among themselves and from the meteorites in volatile-element contents, as well as in oxygen isotopes. What seems to be fairly clear is that cross-contamination among the H, L, and LL planetesimals was minimal; xenoliths of other classes are not very common in meteorites. Mixing did not result in homogenization; perhaps the effect of the perturbations was the reverse and generated compositional differences.

1.11.4. Asteroids as Analogues of Planetesimals

The compositional and isotopic differences among the meteorites from the asteroid belt and the inner planets indicate that we do not have samples of their precursor planetesimals. What sort of hard evidence do we have for the existence of these now-vanished objects? There are several different converging lines.

The asteroids are analogues for planetesimals, even if they probably differ in composition from the population that went to make up the terrestrial planets. Phobos (Fig. 1.11.1), one of the martian satellites, appears to be a primitive object (section 6.6) and may be a captured asteroid. It may provide us with an example of a planetesimal, and the failure of the USSR Phobos mission to provide compositional information is a matter of deep scientific regret. The results obtained from the mission, however, indicate that Phobos is less hydrated than the martian surface [91], consistent with a primitive anhydrous mineralogy, and hence has a probable origin as a captured asteroid from the outer part of the asteroid belt.

Fig. 1.11.1. *A composite photo of Phobos, the larger satellite of Mars and an analogue for a planetesimal. From NASA Reference Publication 1109, (1984).*

These results are consistent with similar ground-based observations on Deimos, the other martian satellite [92].

1.11.5. Evidence from Cratering

The second direct piece of evidence for the former existence of large bodies comes from the observation that all the older preserved surfaces on planets and satellites are saturated with craters. The lunar surface is the classic example, but from Mercury, close to the Sun, out to the satellites of Uranus, the photographs reveal that a massive bombardment struck planets and satellites. Craters of all sizes are present, from micrometer-sized pits due to impact of tiny grains on lunar samples, up to giant ringed basins over 1000 km in diameter. Like the smile of the Chesire Cat in *Alice in Wonderland,* the craters record the previous existence of now vanished objects, in this case, the planetesimals.

The extent of this bombardment on the Moon after it had reached its present size ***and after the lunar crust had formed,*** is revealed by the evidence that at least 80 basins with diameters greater than 300 km and 10,000 craters in the size range from 30-300 km formed before the main bombardment ceased about 3800 m.y. ago. The last major ringed basin is Mare Orientale, formed on the Moon at about 3845 m.y. ago [93] (Fig. 1.11.2).

Since a similar but probably more intense barrage struck the Earth, this accounts for the absence of identifiable rocks older than that age. Current estimates indicate that 200 ringed basins with diameters greater than 1000 km formed on the Earth after it accreted due to the impact of bodies a few hundred kilometers in diameter. On the Earth, in contrast to the Moon, the evidence has been removed, since the smashed up breccias would have been easy meat for the terrestrial agents of erosion.

It is customary to marvel at the extent and fury of this bombardment, but it should be recalled that it occurred on already formed crustal surfaces (the lunar crust was probably in place by 4440 m.y. ago). Thus, the visible cratering record followed planetary accretion and differentiation. It may not represent the tail-end population of accreting objects, but results from more random stochastic events due to the late disruption of Moon-Mars sized objects. This interesting problem is discussed in Chapter 4. In summary, there is plenty of evidence for the existence of planetesimals up to 100 km or so in diameter in the early solar system, together with a few left-over survivors.

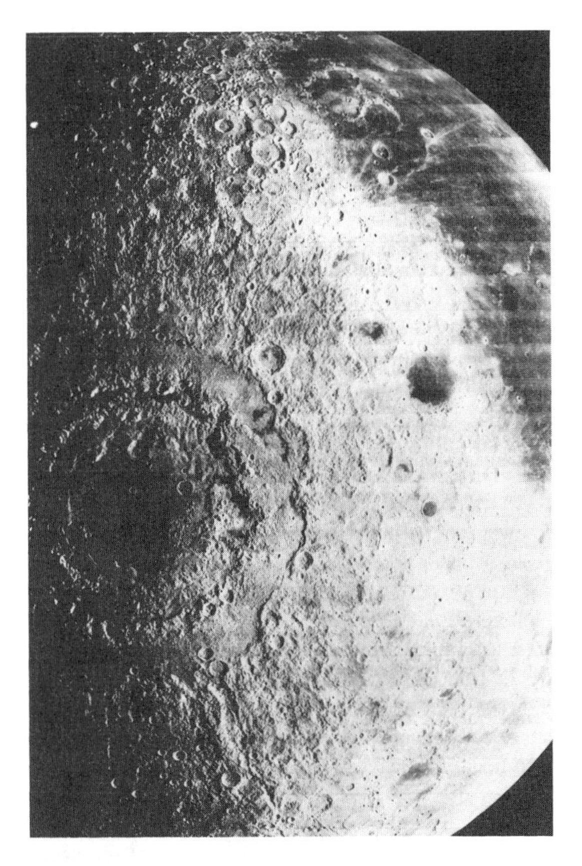

Fig. 1.11.2. *Mare Orientale, the classical example of a multiring basin. The diameter of the outer mountain ring, Montes Cordillera, is 900 km. The stuctures radial to the basin are well developed. The small area of mare basalt to the northeast is Grimaldi, while the western shore of Oceanus Procellarum fills the northeastern horizon. (NASA Orbiter IV 187 M.)*

Were the planets assembled from these 100-km diameter bodies, or were there Moon-Mars-sized bodies in the hierarchy of objects that accreted to form the terrestrial planets? The next major piece of evidence comes from the tilt or inclination of the planets to their axis of rotation. The largest impact is required to account for Uranus. A body the size of the Earth, crashing into the planet, would be needed to tip it through 90°. Smaller collisions are needed to account for the tilt of the other planets, but a few Mars-Earth-sized objects must have been responsible, since the impacts of many small bodies will average out. In summary, there is ample evidence for the existence of a variety of small and large precursor bodies, or planetesimals, in the early solar system [94].

1.12. FRACTIONATION IN PRECURSOR PLANETESIMALS

The planetesimal hypothesis envisages a hierarchical accretion process, with the final population of objects that were swept up into the terrestrial planets comprising 10-20% of the mass of the planet [95]. Recent calculations (Table 1.12.1) indicate the presence of many large precursor objects in the inner solar system. Some of the largest, of martian size, would have made respectable inner planets in their own right if their fate had taken a different course. Were they already differentiated into silicate mantles and metallic cores before they came to a violent end as they were swept up into Earth or Venus? It seems probable that the larger of these bodies may have already gone through a melting stage and core formation. In the large impact hypothesis for lunar origin, a crucial parameter is that the Mars-sized body has already formed a metallic core and a silicate mantle, from which the Moon is essentially derived. The iron core accretes to the Earth, adding perhaps 10% of core volume. If the Earth were put together from a large number of such objects, then metal-silicate fractionation occurred in relatively low-pressure environments before accretion.

Is there independent evidence that such melting and fractionation occurred? Perhaps the most dramatic information is provided by the achondritic and iron meteorites that indicate that differentiation occurred at about 4550 m.y. ago in small parent bodies. Even with our present limited sampling of meteorites, over 60 parent bodies are represented. The implication is that such melting and differentiation was widespread. The source of heat remains obscure, but such processes must have occurred effectively at T_0.

TABLE 1.12.1. Two possible inventories of planetesimals required to form the Earth.

(1)	(2)
1 Mars	5 Mars
10 Moons	3 Moons
10^2 Iapetus	12 Iapetus
•	•
•	•
•	•
10^9 1 km planetesimals	10^5 1 km planetesimals
TOTAL = 1 Earth Mass	TOTAL = 1 Earth Mass

1. Equal amount of mass in each mass decade.
2. Equal surface area in each mass decade.

Data from Stevenson D. J. et al. (1986) in *Satellites* (J. A. Burns and M. S. Matthews, eds.), p. 58, Table 2, Univ. of Arizona, Tucson.

The zonal nature of the asteroid belt, with the differentiated members occupying the inner belt and the more primitive types dominating the outer belt, indicates that melting and differentiation were controlled in some manner by distance from the Sun. It is thus reasonable to suppose that in the region of the terrestrial planets, Sunward of the inner belt, these processes were more extensive.

The ordinary chondrites contain reduced metal, sulfide, and silicate phases, and have also lost some of their inventory of volatile elements. These are the sort of geochemical fractionations that are observed in the terrestrial planets. A fundamental question then arises. Was the metal-silicate-sulfide fractionation carried out in the nebula before planetesimal, let alone planetary, accretion? If so, the consequences are significant: Core-mantle equilibration under terrestrial conditions becomes a meaningless concept if metal-silicate partitioning took place before chondrule melting.

To address these questions, it is necessary to consider the initial state of material in the nebula. A wide range of meteoritic evidence attests to a cool rather than a hot nebula. This includes preservation of isotopic anomalies, particularly for oxygen, rapidly cooled chondrules, presence of hydrated minerals in CI meteorites, and multiple evaporation and condensation episodes to account for CAIs. Metal, sulfide, and silicate phases existed already in the early nebula. As the dust settled to the midplane, physical and perhaps magnetic separation of the dust occurred. The silicate dust melted selectively during the chondrule-forming event, with some minor reduction and loss of iron, and depletion of sulfides and volatile elements. The planetesimals would consist of an intimate mixture of all these phases. Accordingly, if melting occurs, then the meteoritic evidence informs us that these phases will separate, even in the low-gravity fields of small planetesimals.

What were the sources of heat for melting small bodies in the solar system (see also section 3.12)? This question of heat supply for early solar system metamorphism and melting is essentially unresolved. In the current wisdom, there are two principal mechanisms. If ^{26}Al ($t_{1/2}$ = 730,000 years) was live in the early solar system, it could have constituted an important heat source. The second possibility is inductive heating during the early intense T Tauri and FU Orionis stages of solar activity. In any event, it is difficult to escape from a very early heating scenario, although the mechanism remains obscure.

A reasonably firm constraint on the heat source is that it varied with distance from the Sun. Heating appears to have fallen off rapidly beyond 2 AU and does not seem to have been effective beyond about 3 AU. This provides strong support for the Sun as the source of the energy; ^{26}Al would be expected to be more uniformly distributed in the nebula. Although the amount of ice in the outer reaches of the nebula might affect the amount of melting, the temperature gradient across the asteroid belt seems too steep for this explanation to be valid. There are other uncertain variables. Accretion proceeds more slowly with increasing radial distance, so that the outer asteroids might accrete a little later than the inner, acquiring less live ^{26}Al.

The observation that large excesses of ^{26}Mg are present in plagioclase crystals in an abraded clast in the unequilibrated Semarkona LL2 ordinary chondrite is important, in that it may constitute direct evidence for the presence of live ^{26}Al [96]. This clast is a fragment of an igneous rock and is not a chondrule. It has a REE pattern suggestive of igneous differentiation and appears to have been derived from a planetesimal that has accreted and differentiated on a very short timescale, of the order of the mean life of ^{26}Al. The planetesimal was then broken up, and the abraded fragment incorporated into the Semarkona chondrite within a few million years. The ^{26}Al heat source would certainly allow significant thermal metamorphism in such bodies, accounting for the higher meteoritic petrological types 4-6 [96]. Since there is no sign of excess ^{26}Mg in plagioclases from eucritic meteorites, probably derived from a differentiated asteroid (? Vesta), ^{26}Al does not seem to be the heat source in this case [97].

Problems with a solar source revolve about the timescale involved. Such electromagnetic inductive heating can occur only during the early T Tauri and FU Orionis stages, which last only about 10^6 years or perhaps less. In most versions of the planetesimal hypothesis, the planetesimals have not grown to an appreciable size at this stage, but the timescales for planetary accretion may be shorter than currently supposed, so that the terrestrial planets may have formed in 10^7 rather than 10^8 years.

This compression of the timescale does not alter the sequence of events, so that, for example, the gas was gone well before the accretion of the inner planets. Nevertheless, it does raise some interesting possibilities. If the heating is restricted to very short timescales, a requirement for both mechanisms, then the planetesimals were very small, less than 10 km in diameter. These differentiated objects could then be accreted into larger bodies, possibly with some accretional heating causing metamorphic effects. Two periods of heating, the first due to ^{26}Al or inductive heating, and the second, milder, heating due to accretion, may in fact explain the record. Cooling rates of the meteorite parent bodies may also be consistent with this scenario.

In summary, neither of the currently proposed heating mechanisms seems very realistic and "early planetesimal heating must be due to some effect we have not even been able to imagine," [98] except that the variation observed with heliocentric distance in the asteroid belt points to a cause connected with the early Sun. The principal effect for planetary growth is not only that the terrestrial planets mostly accreted from smaller differentiated bodies, but that these were probably hot. Melting of the terrestrial planets during accretion is thus likely.

1.13. ACCRETION OF PLANETESIMALS

The discussion throughout this book is heavily weighted toward the planetesimal hypothesis. As discussed in other sections, the planetesimal hypothesis explains the heavy cratering of surfaces, the obliquities of the planets, the anomalous cases of the Moon and Mercury, the compositions and history of the icy satellites of the giant planets, and the variable compositions of the terrestrial planets. In assessing theories for the origin of the Earth, the current view is that planets accrete from a hierarchy of planetesimals. Such bodies of course may be differentiated into cores and mantles, which may then be broken up by collisions and reaccreted in differing proportions of metal and silicate fractions, so that much diversity of composition among the accreting bodies can be expected.

1.13.1. Timescales for Accretion

Accretion times for the inner planets appear to be of the order of 10^7-10^8 years [99] and core formation is judged to occur simultaneously with accretion once the accreting planet is large enough for melting to occur.

1.13.2. How Many Objects?

How many bodies were involved and what were the relative sizes of the accreting planetesimals? The presence of many large precursor objects can be inferred from computer simulations for the region of the inner planets (Mercury, Venus, Earth, and Mars) [100]. These indicate the presence of over 100 objects of lunar mass (7.35×10^{25} g), 10 with masses exceeding that of Mercury (3.39×10^{26} g), and several exceeding the mass of Mars (6.42×10^{26} g), forming the final population of planetesimals existing just before the final sweep-up [94]. Venus and the Earth acquired most of them (Mars is only about 1/10 Earth mass while Mercury is only about 1/20 Earth mass).

Probably about 50-75% of present Earth mass was accreted from these massive planetesimals, with the remainder coming from a multitude of small bodies. The larger of these bodies probably had already gone through an intraplanetary melting stage, with core formation occurring before they were swept up by the inner planets.

Since there appears to be a plethora of objects of diverse chemistry, models that reduce the planetary precursors to two components appear unduly simplistic. Accordingly, there is some difficulty, for example, in identifying the two components in such models of planetary formation [101], particularly since they are proposed to come from separate parts of the nebula, one oxidized and the other highly reduced. The common identification of the oxidized component with CI chondrites (from the asteroid belt?) on the one hand, and the reduced component with the enstatite chondrites (postulated to form originally from within one AU?) on the other, does not accord with the oxygen isotope evidence. The real situation is undoubtedly much more complex. If the inner planets all accrete from only two such components, one might expect more uniformity, for example, in oxygen isotopes. These models tell us in fact that there is little evidence of lateral mixing or homogenization in the nebula. Thus, as noted in section 2.20, the zonation in the asteroid belt may be an analogue for conditions in the inner nebula.

1.13.3. Width of Planetary Feeding Zones

A crucial question for the terrestrial planets (it cannot be so readily addressed at present for the giant planets) is the width of the feeding zones from which they accumulated. Equilibrium condensation models predict a strict radial zoning outward from the Sun, with accretion of the planets from distinct compositional zones. The other extreme model suggests that there was major overlap between the planetary accretion zones [102].

An initial assumption in this model is that the planetesimals begin with semimajor axes between 0.7 and 1.1 AU and are subsequently dispersed to form the four inner planets that lie between 0.4 and 1.5 AU. Such mixing would be expected to erase any primordial differences and result in a rather uniform composition for the terrestrial planets. This is not observed. A more recent study [103] indicates that the inner planets retain some memory of local sources, and the meteoritic evidence is strongly suggestive of this scenario [104] (see section 5.5 on Mg/Si ratios in the Earth). The evidence seems mostly against simple condensation models related to heliocentric distance for the formation of the inner planets, and in favor of accretion in restricted feeding zones from differentiated planetesimals in a system in which stochastic processes played a leading role.

Wasson [105] lists the following sequence of meteorites in order of increasing distance from the Sun: E, H-L-LL, CV, CO, CM, and CI. However, they do not form a monotonic sequence and few of their properties correlate simply with heliocentric distance. New chondrite classes (e.g., Kakangari, ALHA 85085) do not fit simply into the sequence. However, it is also clear (see section 2.13) that we have only a very limited sampling. Chondrites are complex mixtures of very diverse components, so that chaotic conditions prevailed at a very early stage of solar nebula history. Several events were probably superimposed. In addition to early solar heating, local high-temperature events connected with falling of material into the nebular midplane occurred. The general perception that the enstatite, ordinary, and carbonaceous classes of chondrites are representative of a compositional sequence of increasing distance from their Sun is almost certainly in error. The evidence from the asteroid belt is that the individual feeding zones for individual asteroids may be <0.3 AU, but of course [106] mixing occurs during accretion. Nevertheless, the planets retain a memory, if somewhat blurred, of the initial location in the nebula, of the planetesimals from which they accreted, the differences in oxygen isotopes being the most significant evidence for this.

Several lines of evidence support the concept of little lateral mixing in the nebula, which has many implications for the accretion of the planets. Among these is the low abundance of foreign inclusions in meteoritic breccias, which implies little lateral mixing during meteorite formation. Typically, the volume of foreign clasts is less than 1% [107]. This indicates very little mixing between the regions from which the differing classes of chondrites accreted. It also indicates that collisional mixing and stirring of the belt must have been relatively minor and the separate chondrite classes accreted within quite narrow nebular zones, perhaps less than 0.1 AU wide. This is supported by the rather low collisional history in the belt, so that, for example, the basaltic surface of Vesta, which probably dates from 4.6 b.y., has apparently been preserved.

Mixing of planetesimals during accretion of the terrestrial planets seems to have been minimal and there seems to have been little mixing among the H, L, and LL groups of chondrites. These examples seem to be at variance with models that call for widespread scattering and mixing of material through the inner solar system.

1.13.4. Accretion of the Inner Planets

It is clear that the inner planets differ from one another in composition. In the planetesimal hypothesis, this is attributed essentially to random accumulation of planetesimals of differing compositions, perturbed from any initial zonation. This occurs because of gravitational perturbations once the planetesimals reach lunar size. Some of an initial compositional zoning is preserved, particularly for Mercury. Venus should represent the best average of the region between 0.6 and 1.2 AU (Fig. 1.13.1). Both Earth and Mars are biased toward accretion from the outer zones. From such models, any simple relationship with a strict radial zoning is not to be expected, and the compositions of the terrestrial planets are to some degree random [108].

During the accumulation of the planets, considerable mixing occurs. Nevertheless, in the computer

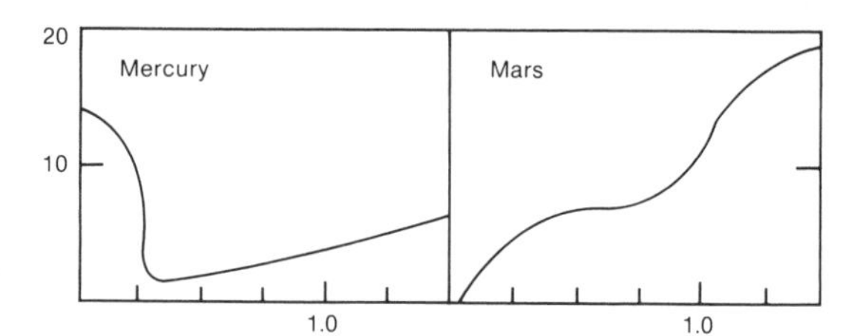

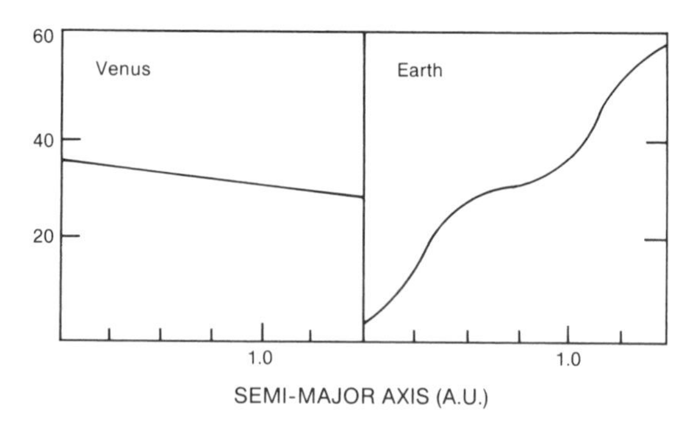

Fig. 1.13.1. *The variation in the semimajor axes of planetesimals, which subsequently accreted to form the terrestrial planets. After Wetherill G. W. (1988) in Mercury (F. Vilas et al., eds.), p. 683, Fig. 6, Univ. of Arizona, Tucson.*

simulations there generally remains a correlation between the provenance or source regions of the accreting planetesimals and the distance of the planet from the Sun. This is in accord with the various differences in composition, oxygen isotopes, and many other lines of evidence that preclude any thorough mixing or homogenization of the nebula [109]. Any initial or primordial zoning may of course be preserved merely due to the conservation of angular momentum and energy; these factors will also act against any homogenization of an initially compositionally zoned nebula.

Formation of the inner or terrestrial planets is assumed to occur through accretion of planetesimals in a largely gas-free environment. This raises another interesting question. What was the source of the volatile elements and water in the Earth and the other terrestrial planets? If the inner solar system was swept clear of this material, then the only sources were the volatiles trapped early enough ($<10^6$ years) in large enough planetesimals (>1 km?) to survive being removed by the violent solar activity, or alternatively, the volatiles came in late in comets or planetesimals from the far reaches of the solar system. It seems probable that the Earth accreted under relatively dry conditions, and was never swathed in a primitive steam atmosphere [110]. A further consequence of accretion of large planetesimals is that planetary melting is inevitable; the geochemical implications are addressed in Chapter 5.

Although the inner planets have a general "chondritic" composition, they do not match in chemical or isotopic parameters any specific classes of meteorites. Rare-gas data indicate that no known meteorite class has appropriate elemental and isotopic compositions close to those of the inner planets. Both Mars and the Earth are depleted in xenon, relative to CI and H chondrites. No presently known meteorite group has K/U ratios appropriate to constitute suitable building blocks for the Earth. Oxygen isotope data also show that, except for fractionated basaltic meteorites (ruled out on other grounds), no observed class matches the terrestrial data, nor do the more primitive meteorite groups match the oxygen isotope data for the SNC meteorites (= Mars).

The major-element compositions of the inner planets are not yet well enough understood to rule out whether there are variations in Fe, Mg, and Si relative to the more refractory elements (Al-Ca-Ti). Separation of Si from the more refractory elements by over a factor of 2 occurs among the chondrite groups. This is most probably due to the physical separation of mineral phases rather than to slight differences in volatility. For the Earth, a principal uncertainty lies in the composition of the lower mantle. If it is the same as the upper mantle, then the Earth is enriched in Mg, but not Al, relative to Si, or depleted in Si.

A curious attempt to resolve this problem asserts that the Mg/Si ratio in the upper mantle is in fact the same as the solar ratio, and that the CI value is enriched in Si relative to solar values [111]. Since the CI meteorite data match the solar data within 10% for nearly all elements, and no enrichments relative to solar abundances for other much more volatile elements are observed, there seems to be no logical basis for this suggestion [112]. This topic is discussed more fully in the section on the bulk composition of the Earth dealing with the composition of the upper and lower mantles and Mg/Si ratios (section 5.5).

If the meteorites provide us with an analogue for the terrestrial precursor planetesimals, then elemental and mineralogical fractionation was endemic in the early nebula. If silicate, sulfide, and metal phases, formed under low pressures, were already present in the accreting planetesimals, separation of these phases may occur concomitantly with accretion and thus there may be little high-pressure equilibration between core and mantle in the Earth. In such a scenario, sulfur becomes a viable candidate for the light element in the Earth's core.

1.13.5. Accretion of Giant Planet Cores

As discussed above in section 1.8, a basic problem with the planetesimal hypothesis is achieving a rapid growth for the cores (10-15 Earth masses) of the giant planets, particularly Jupiter. This requires runaway growth. Calculations show that for the sort of nebular densities postulated by Lissauer [113] a 10-Earth-mass core can accrete in 3×10^5 years. Such a core can then accrete H and He from the nebula before the gas is dispersed by early solar activity on timescales of 10^6 years [114]. The process of accreting the planets was very efficient. The region between Jupiter and Saturn is very clean; no objects such as Chiron, which wanders alone in the 10-AU void between Saturn and Uranus, have been detected. Recent calculations indicate that no objects in orbits of low inclination will survive in the region between Jupiter and Saturn for more than 10-30 m.y. [115]. Neither are orbits between Uranus and Neptune stable. However, some zones between Saturn and Uranus may be stable over periods comparable with the age of the solar system [116]. If any objects other than Chiron exist in this region, they are at least 3 orders of magnitude fainter [117].

1.14. AN ACCRETIONARY SEQUENCE

The chronology of planetary accretion is not well constrained. It is based almost entirely on the meteorite evidence discussed in section 3.7, although the lunar samples have provided valuable insights into the gap between about 4440 and 3800 m.y. [118].

In the present context, however, we are mostly concerned with the chronology of the accretion process. It seems to be becoming clearer that most of the events occurred on a very short timescale fairly close to T_0 and that the terrestrial planets were recognizably present by about 4450-4500 m.y. (i.e., 50-100 m.y. after T_0). On the basis of the interpretation of the evidence considered in this book, the following sequence of events is proposed.

The earliest discernible events seem to have been the production of the short-lived radioactive nuclides, ^{146}Sm, ^{244}Pu, ^{129}I, and ^{53}Mn in supernovae. ^{53}Mn, with the shortest mean life, was produced in such a catastrophic event perhaps 24 m.y. before the formation of the CAIs. A molecular cloud complex then formed over a period of perhaps 20 m.y. Quite early in the history of the molecular cloud complex, within a few million years, a slowly spinning fragment of the cloud separated. This began to collapse to a rotating disk of gas and dust, the primordial solar nebula. The first events that can be dated appear to have been the formation of the refractory inclusions or CAIs, during complicated and repeated episodes of condensation and evaporation. These events occurred at about 4560 m.y. ago, possibly in the molecular cloud or during the collapse of the detached fragment to the nebular disk. There are currently unresolved differences between the isotopic and astronomical timescales, the former indicating somewhat longer timescales. If the isotopic data (section 3.7) are correct, the formation of the CAIs must predate the full establishment of the rotating nebular disk.

Following the establishment of a separate rotating system [119], the dust began to settle to the midplane of the nebula, a process taking perhaps 10^4-10^6 years. During this formation of a dense dusty disk, which is only dimly understood, several important processes occurred.

Metal, sulfide, and silicate phases were already present. As these settled to the midplane of the nebula, segregation occurred; silicates clumped together selectively, so that during transient heating events, millimeter-sized, silicate droplets were melted, forming chondrules. In contrast to the many episodes responsible for the curious chemistry of the CAIs, most chondrules appear to have been the product of a single melting episode. The currently favored mechanism is flash melting by nebular flares. These events occurred after the formation of the CAIs. Independent evidence is available (section 3.5) to support the existence of CAIs prior to chondrule formation.

As the dense dusty rotating disk formed, instabilities developed, resulting in the clumping together of the metal, sulfide, and chondrules. These grew by mutual collisions, sticking together at low velocities, forming the chondritic meteorites and eventually reaching kilometer-sized objects or planetesimals.

During these events in the nebula, the Sun reached the hydrogen-burning stage. As it settled onto the main sequence, a period of intense and violent FU Orionis and T Tauri activity began, which lasted perhaps 10^6 years. During this time, reversal of the inward flow of material to the growing star occurred, and the strong outflowing solar winds cleared the inner nebula of H, He, and the other gases. Heating drove volatile elements out of the inner nebula. This major separation of volatile from refractory material is recorded by the low initial Sr ratios for the basaltic achondrites, and probably occurred at about 4555-4560 m.y. (section 3.7). Water ice was removed out to a "snow line" at about 4 AU. The material that was already coagulated into planetesimals survived in the inner nebula, ultimately to become the building blocks of the terrestrial planets.

Rapid accretion of planetesimals to form a 10-Earth-mass core occurred at about 5 AU, probably assisted by a higher local density in the nebula. This was most likely connected with the condensation of volatiles and water ice, driven from the inner nebula at a "snow line" beyond 4 AU. Since Jupiter is located just within the zone where water ice is stable, a rapid pile-up of this and other volatiles as the inner nebula was cleared may have triggered the initial formation of the jovian core onto which the gaseous components (mainly H and He) could collapse by gravitational attraction. Hydrogen, He, and the other gaseous and icy components collapsed gravitationally onto the proto-Jupiter core at 5 AU on a timescale of 10^6 years, this runaway accretion occurring before the gaseous constituents of the nebula were dispersed. Accordingly, Jupiter had to grow within 10^6 years before the gas was dissipated [120].

Such a rapid early growth of Jupiter was required to prevent large planets forming at the present location of Mars or in the asteroid belt. Rather slower growth of similar cores occurred at 10, 20, and 30 AU forming Saturn, Uranus, and Neptune. These planets, forming further out in lower-density regions of the nebula, accreted more slowly, Saturn taking perhaps twice as long as Jupiter.

Uranus accreted in about 10^7 years, Neptune again taking about twice that time [121], comparable to the timescales for the accretion of the inner planets. Both ice giants formed from icy planetesimals, accreting only token amounts of H and He. By this time, the remaining gases had been swept far into the outer reaches of the solar system.

In the inner nebula, the gas was completely gone from the nebula by about 10^6 years. The terrestrial planets accumulated from left-over planetesimals in a gas-free environment from relatively restricted feeding-zones on timescales of 10^7-10^8 years. During this period, a hierarchy of planetesimals formed, before being swept up into the four inner planets. Much of the material in the asteroid belt was either accreted to Jupiter or tossed out of the solar system, leaving a starved but informative remnant [122] about 5% of the mass of the Moon. Mars likewise grew in a poverty-stricken zone, while Mercury, in addition to losing significant mass by collision, also grew in a depleted portion of the nebula, close to the dominant Sun.

Many large bodies also populated the outer solar system. Massive collisions with them tilted the planets and produced planetary subnebulae from which the regular satellites of Jupiter, Uranus, and Neptune formed. Capture of left-over icy planetesimals completed the satellite inventory for the outer planets. The Moon formed by the impact of a large Mars-sized body on the Earth. It was mostly melted and the feldspathic highland crust crystallized at 4.44 b.y., providing a lower time limit for the event [118].

Similar massive impacts caused Venus to reverse its rotation and removed much of the silicate mantle of Mercury. Phobos and Deimos are probably captured asteroids. By 10^8 years after the initial separation of the nebula, the solar system was essentially complete.

NOTES AND REFERENCES

1. A frequently quoted example of unlikely events is the demise of the only elephant in the Leningrad Zoo. It was reputedly killed by the first shell, fired from a long-range railway gun, to hit the center of the city during the siege of Leningrad by the German army in the Second World War. The fate of this unfortunate beast may be apocryphal, for it is not mentioned in the three standard works on the investment of the city: Goure L. (1962) *The Siege of Leningrad,* Stanford Univ., 363 pp.; Pavlov D. V. (1965) *Leningrad 1941,* Univ. of Chicago, 186 pp.; Salisbury H. E. (1969) *The 900 Days: The Siege of Leningrad,* Harper and Row, New York, 635 pp.
2. For example, Cameron A. G. W. (1988) *Annu. Rev. Astron. Astrophys., 26,* 441; Wetherill G. W. (1989) in *The Formation and Evolution of Planetary Systems* (H. A. Weaver and L. Danly, eds.) p. 1, Cambridge Univ., New York.
3. See Jaki S. L. (1978) *Planets and Planetarians,* Halstead Press, Wiley, New York, 266 pp.
4. Morris D. E. and O'Neill T. G. (1988) *Astron. J., 96,* 1127.
5. For example, Ringwood A. E. (1966) *Geochim. Cosmochim. Acta, 30,* 41; Ringwood A. E. and Anderson D. L. (1977) *Icarus, 30,* 243.
6. For example, see discussion in Lewis J. S. and Prinn R. G. (1984) *Planets and Their Atmospheres: Origin and Evolution,* p. 193, Academic Press, New York.
7. Wasson J. T. (1985) *Meteorites; Their Record of Early Solar-System History,* Freeman, New York, 267 pp.
8. This classification can be found in the brief treatise by Williams I. P. (1975) *The Origin of the Planets,* Adam Hilger, London, 97 pp.
9. This stream flowed across the Gallipoli Peninsula into the Hellespont (Dardanelles). In 405 BC, the Spartan fleet under Lysander destroyed the Athenian fleet at the mouth of the stream. This defeat cut off the supply of corn to Athens and ended the Peloponnesian War.
10. See for example, Rawlins D. (1990) *Bull. Amer. Astron. Soc., 22,* 1040, and references therein. Ptolemy revised the star listings of Hipparchus, who worked between about 146 and 127 BC. It is not clear, however, that Ptolemy made original observations, and his scientific credibility is further weakened since he worked for the state religion, Serapism, which was heavily based on astrology.
11. Copernicus was reluctant to publish and is reputed to have received a copy of his masterpiece on May 23, 1543, the day he died. Few modern authors would care to wait so long. Copernicus was appointed to a lifelong sinecure at the age of 24 as a canon of Frauenberg, Poland, surely an argument for tenure.
12. The three laws are (1) Planets move around the Sun in elliptical orbits, with the Sun as one of the foci; (2) A radius vector between a planet and the Sun sweeps out equal areas in equal lengths of time; and (3) The squares of the periods of revolution of the planets are proportional to the cubes of their distance from the Sun. The first two laws were stated in 1609, the third in 1618.

13. Jaki S. L. (1978) *Planets and Planetarians,* Wiley, New York, p. 26.
14. Ibid, p. 27.
15. Buffon Georges-Louis Leclerc (1778) *Histoire naturelle, Vol. 1,* p. 161, de l'Imprimerie de F. Dufart, Paris.
16. Bickerton A. W. (1880) *Trans. Proc. New Zealand Institute, XIII,* 154.
17. Arrhenius S. (1908) *Worlds in the Making* (trans. H. Borns) Harper, New York, 229 pp.
18. See T. J. J. (1910) *Researches on the Evolution of Stellar Systems, Vol. 2, The Capture Theory of Cosmical Evolution,* Nichols and Sons, Lynn, Massachusetts, 734 pp.
19. Chamberlin T. C. (1905) *Carnegie Institution of Washington Yearbook, 3,* 195; Chamberlin T. C. and Salisbury R. D. (1906) *Geology, Vol. II,* pp. 1-81, Henry Holt, New York; Moulton F. R. (1905) *Astrophys. J., 22,* 165; Moulton F. R. (1906) *An Introduction to Astronomy,* Macmillan, New York, 557 pp.
20. Jeans J. H. (1917) *Mem. R. Astronom. Soc., 62,* 1; Jeans J. H. (1919) *Problems of Cosmogony and Stellar Dynamics,* Cambridge Univ., Cambridge, 298 pp.; Jeffreys H. (1917) *Sci. Progr., 12,* 52; Jeffreys H. (1924) *The Earth, Its Origin, History and Physical Constitution,* Cambridge Univ., Cambridge, 278 pp.
21. Russell H. N. (1935) *The Solar System and its Origin,* Macmillan, New York, 144 pp.; Spitzer L. (1939) *Astrophys. J., 90,* 675.
22. It is curious to record that there is no reference to the Jeans or Jefferies hypothesis in Chamberlin T. C. (1927) *The Two Solar Families: The Sun's Children,* Univ. of Chicago, 311 pp. French workers in turn appear to have ignored Chamberlin's views. Thus H. Poincare (1911) does not mention the Chamberlin-Moulton theory in *Leçons sur les Hypothèses Cosmogoniques,* A. Hermann et Fils, Paris, 294 pp.
23. Spitzer L. (1939) *Astrophys. J., 90,* 675; Russell H. N. (1935) *The Solar System and its Origin,* Macmillan, New York, 144 pp.; Lyttleton R. A. (1960) *Mon. Not. R. Astron. Soc., 121,* 551. Clerk Maxwell had also demonstrated last century that the tidal field of the Sun would prevent the formation of individual planets; see W. D. Niven, ed. (1890) *The Scientific Papers of James Clerk Maxwell,* Cambridge Univ., New York.
24. Lyttelton R. A. (1936) *Mon. Not. R. Astron. Soc., 96,* 559; Lyttelton R. A. (1938) *Mon. Not. R. Astron. Soc., 98,* 536.
25. Lyttelton R. A. (1940) *Mon. Not. R. Astron. Soc., 100,* 546.
26. Hoyle F. (1945) *Mon. Not. R. Astron. Soc., 105,* 175.
27. Woolfson M. M. (1960) *Nature, 187,* 47; Woolfson M. M. (1964) *Proc. R. Soc., A282,* 485; Dormand J. R. and Woolfson M. M. (1989) *The Origin of the Solar System—The Capture Theory,* Halstead, Wiley, New York, 230 pp.
28. Schmidt O. J. (1944) *C. R. Doklady Akad. Nauk. SSSR, 45,* 229.
29. Alfven H. (1942) *Stockholms Obs. Ann., 14 (2)*; Alfven H. (1943a) *Stockholms Obs. Ann., 14 (5)*; Alfven H. (1943b) *Nature, 152,* 721; Alfven H. (1954) *On the Origin of the Solar System,* Oxford Univ., 194 pp.
30. Pendred B. W. and Williams I. P. (1968) *Icarus, 8,* 129.
31. Rene Descartes: "... He was too wise a man to encomber himselfe with a Wife; but as he was a man, he had the desires and appetities of a man; he therefore kept a good conditioned hansome woman that he liked, and by whom he had some children ... He was so eminently learned that all learned men made visits to him, and many would desire him to shew them his Instruments (in those days mathematicall learning lay much in the knowledge of Instruments, and ... in doeing of tricks) he would drawe out a little Drawer under his Table, and shew them a paire of compasses with one of the legges broken; and then, for his ruler, he used a sheet of paper folded double ..." Dick O. L. (1958) *Aubrey's Brief Lives,* p. 94, Secker and Warburg, London. Aubrey's original spelling is retained. Tycho Brahe, in contrast to Descartes, was married, but was criticized for marrying a peasant's daughter!
32. For example, Kuiper G. P. (1951) *Proc. Nat. Acad. Sci., 37,* 1; Fig. 1.5.1b, after Kuiper G. P. (1951) in *Astrophysics A Topical Symposium* (J. A. Hynek, ed.), p. 374, McGraw-Hill, New York.
33. McCrea W. H. (1960) *Proc. R. Soc., A256,* 245.
34. For example, Williams I. P. (1969) *Mon. Not. R. Astron. Soc., 146,* 339; Williams I. P. (1972) *Astrophys. Space Sci., 18,* 223; Williams I. P. and Donnison J. R. (1973) *Mon. Not. R. Astron. Soc., 165,* 295; Williams I. P. and Handbury M. J. (1974) *Astrophys. Space Sci., 30,* 215; Williams I. P. and Handbury M. J. (1972) *Astrophys. Space Sci., 18,* 223.
35. Jaki S. L. (1978) *Planets and Planetarians,* p. 116, Wiley, New York.
36. Hastie W. (1900) *Kant's Cosmogony,* p. 26, J. Maclehose, Glasgow, quoted by Jaki S. L. (1978) *Planets and Planetarians,* Wiley, New York, p. 111.
37. Jaki S. L. (1978) *Planets and Planetarians,* Wiley, New York, p. 119.
38. See Brush S. G. (1987) in *History of Science, 25,* 245, for a detailed account of this celebrated exchange.
39. Egyed L. (1960) *Nature, 186,* 221.
40. Hoyle F. (1960) *Q. J. R. Astron. Soc., 1,* 28.
41. For example, Lewis J. S. (1973) *Annu. Rev. Phys. Chem., 24,* 339; Lewis J. S. (1974) *Science, 186,* 440; Goettel K. A. and Barshay S. S. (1978) in *The Origin of the Solar System* (S. F. Dermott, ed.), p. 611, Wiley, New York.

42. For example, Clayton R. N. et al. (1985) in *Protostars and Planets II* (D. C. Black and M. S. Matthews, eds.), p. 765, Univ. of Arizona, Tucson; Gehrels T., ed. (1978) *Protostars and Planets,* Univ. of Arizona, Tucson, 756 pp; Black D. C and Matthews M. S., eds. (1985) *Protostars and Planets II,* Univ. of Arizona, Tucson, 1293 pp; Wood J. A. and Morfill G. (1988) in *Meteorites and the Early Solar System* (J. F. Kerridge and M. S. Matthews, eds.), p. 329, Univ. of Arizona, Tucson.
43. Morfill G. E. (1988) *Icarus, 75,* 371.
44. For example, Grossman L. and Larimer J. W. (1974) *Rev. Geophys. Space Phys., 12,* 71.
45. For example, Turekian K. K. and Clark S. (1969) *Earth Planet. Sci. Lett., 6,* 346.
46. See especially comments in *Basaltic Volcanism on the Terrestrial Planets,* Lunar and Planetary Institute (1981), pp. 646-647.
47. Wood J. A. (1962) *Nature, 194,* 127.
48. Anders E. (1971) *Annu. Rev. Astron. Astrophys., 9,* 1.
49. For example, Wänke H. et al. (1984) in *Archean Geochemistry* (A. Kroner et al., eds.), p. 1, Springer-Verlag, Berlin.
50. For example, Ganapathy A. and Anders E. (1974) *Proc. Lunar Sci. Conf. 5th,* pp. 1181-1206.
51. Wänke H. et al. (1984) in *Archean Geochemistry* (A. Kroner et al., eds.), p. 1, Springer-Verlag, Berlin; Ringwood A. E. (1979) *Origin of the Earth and Moon,* Springer-Verlag, Berlin, 295 pp.
52. McCrea W. (1974) in *On the Origin of the Solar System* (H. Reeves, ed.), pp. 2-20. Nice Symposium, CNRS, Paris.
53. Torbett M. et al. (1982) *Icarus, 49,* 313.
54. Nieto M. M. (1972) *The Titius-Bode Rule of Planetary Distances: Its History and Theory,* Pergamon, Elmsford, 161 pp. See also Hills J. G. (1970) *Nature, 225,* 840; Dole S. H. (1970) *Icarus, 13,* 494; Prentice A. J. R. (1978) in *The Origin of the Solar System* (S. F. Dermott, ed.), p. 111, Wiley Interscience, New York.
55. Nieto M. M. (1972) *The Titius-Bode Rule of Planetary Distances: Its History and Theory,* Pergamon, Elmsford, p. 131.
56. For a contrary view, see Prentice A. J. R. (1978) in *The Origin of the Solar System* (S. F. Dermott, ed.), pp. 111, Wiley Interscience, New York.
57. For example, Kuiper G. P. (1954) *Proc. Nat. Acad. Sci., 40,* 1101.
58. Urey H. C. (1954) *Astrophys. J. Supp., 1,* 147.
59. Cameron A. G. W. (1978) *Moon Planets, 18,* 5.
60. Edgeworth K. E. (1949) *Mon. Not. R. Astron. Soc., 109,* 600; Safronov V. S. (1972) *NASA TTF-667*; Goldreich P. and Ward W. R. (1973) *Astrophys. J., 183,* 1051; see also Boss A. P. (1988) *LPI Tech. Rpt. 88-04,* p. 51.
61. Cameron A. G. W. (1978) in *The Origin of the Solar System* (S. F. Dermott, ed.), p. 49, Wiley, New York.
62. McCrea W. H. (1960) *Proc. R. Soc., A256,* 245; McCrea W. H. (1978) in *The Origin of the Solar System* (S. F. Dermott, ed.), p. 75, Wiley, New York.
63. McCrea W. H. (1974) "Origin of the solar system: Review of concepts and theories" in *On the Origin of the Solar System* (H. Reeves, ed.), Nice Symposium, CNRS, Paris, pp. 12-13.
64. Ozima M. and Podosek F. (1983) *Noble Gas Geochemistry,* p. 208, Cambridge Univ., New York.
65. Stevenson D. J. (1982) *Planet. Space Sci., 30,* 755.
66. The question of the origin of the sodium and potassium in the mercurian atmosphere is addressed in Chapter 5. The assessment is made there that a meteoritic or cometary source is less likely than an indigenous crustal, probably feldspathic, source.
67. Cameron A. G. W. (1988) *Annu. Rev. Astron. Astrophys., 26,* 441; Benz W. and Cameron A. G. W. (1990) in *The Origin of the Earth* (J. H. Jones and H. E. Newsom, eds.), p. 61, Oxford, New York.
68. Saturn is only 30% of the mass of Jupiter, so it can be ignored in a first-order approximation.
69. For example, Hayashi C. et al. (1985) in *Protostars and Planets II* (D. C. Black and M. S. Matthews, eds.), p. 1100, Univ. of Arizona, Tucson.
70. Pollack J. B. and Bodenheimer P. (1989) in *Origin and Evolution of Planetary and Satellite Atmospheres* (S. K. Atreya et al., eds.), p. 572, Univ. of Arizona, Tucson.
71. Safonov V. S. (1972) *NASA TT 677.*
72. Wetherill G. W. (1980) *Annu. Rev. Astron. Astrophys., 18,* 77.
73. Wetherill G. W. and Stewart G. R. (1987) *Lunar and Planetary Science XVIII,* p. 1077, Lunar and Planetary Institute, Houston.
74. Celotto E. et al. (1989) *Lunar and Planetary Science XX,* p. 149, Lunar and Planetary Institute, Houston.
75. Lissauer J. J. (1987) *Icarus, 69,* 249.
76. Ibid; Lissauer J. J. and Greenzweig Y. (1989) *Bull. Am. Astron. Soc., 21,* 915.
77. Pollack J. B. (1984) *Annu. Rev. Astron. Astrophys., 22,* 389.
78. Lissauer J. J. (1987) *Icarus, 69,* 257.
79. Ward W. R. (1989) *Astrophys. J., 345,* L99.

80. Stevenson D. J. and Lunine J. I. (1988) *Icarus, 74,* 146.
81. Mizuno H. (1980) *Prog. Theor. Physics, 65,* 544.
82. Stevenson D. J. (1984) *Lunar and Planetary Science XV,* p. 821, Lunar and Planetary Institute, Houston.
83. Fernandez J. A. and Ip W.-H. (1984) *Icarus, 58,* 109.
84. Prinn R. G. and Fegley B. Jr. (1989) in *Origin and Evolution of Planetary and Satellite Atmospheres* (S. K. Atreya et al., eds.), p. 78, Univ. of Arizona, Tucson.
85. It is generally assumed, following Laplace, that the prograde rotation of the planets is a normal consequence of accretion in a rotating nebula. If the inner planets accreted from a large number of small objects, a large impact might be needed to spin the planet up; see Giuli R. T. (1968) *Icarus, 8,* 301. This could constitute an argument for accretion from a hierarchy of bodies, with late impacts of large bodies spinning up the planets. Thus, in this scenario the slow backward rotation of Venus would be due not to a large impact, but to the absence of such an event; see discussion by Wood J. A. (1986) in *Origin of the Moon* (W. K. Hartmann et al., eds.), p. 35, Lunar and Planetary Institute, Houston.
86. Gaffey M. J. (1988) *Lunar and Planetary Science XIX,* p. 369, Lunar and Planetary Institute, Houston.
87. Wilhelms D. E. (1987) *The Geologic History of the Moon,* USGS Prof. Paper 1348, 302 pp.
88. Chamberlin T. C. and Salisbury R. D. (1906) *Geology, Vol. II,* Holt, New York, p. 1-81; Moulton F. R. (1906) *An Introduction to Astronomy,* Macmillan, New York, 557 pp.
89. Wasson J. T. (1985) *Meteorites: Their Record of Early Solar-System History,* Freeman, New York, 267 pp.
90. See DeCampli W. M. (1981) *Astrophys. J., 244,* 124.
91. Bibring J.-P. et al. (1989) *Nature, 341,* 591.
92. Bell J. F. et al. (1989) *Lunar and Planetary Science XX,* p. 58, Lunar and Planetary Institute, Houston.
93. Wilhelms D. E. (1987) *USGS Prof. Paper 1348,* 212.
94. See section 1.12.
95. Scott E. R. D. and Newsom H. E. (1989) *Z. Naturforsch., 44a,* 924.
96. Hutcheon I. D. and Hutchison R. (1989) *Nature, 337,* 238.
97. Schramm D. N. et al. (1970) *Earth Planet. Sci. Lett., 10,* 44.
98. J. A. Wood, personal communication, August 1989.
99. Wetherill G. W. (1988) in *Mercury* (F. Vilas et al., eds.), p. 670, Univ. of Arizona, Tucson.
100. Wetherill G. W. (1985) *Science, 228,* 877; Wetherill G. W. (1986) in *Origin of the Moon* (W. K. Hartman et al., eds.), p. 519, Lunar and Planetary Institute, Houston; Wetherill G. W. (1988) in *Mercury* (F. Vilas et al., eds.), p. 670, Univ. of Arizona, Tucson; Wetherill G. W. and Chapman C. R. (1988) in *Meteorites and the Early Solar System* (J. F. Kerridge and M. S. Matthews, eds.), p. 35, Univ. of Arizona, Tucson.
101. Wänke H. et al. (1984) In *Archean Geochemistry* (A. Kroner et al., eds.), p. 1, Springer-Verlag, Berlin.
102. Wetherill G. W. (1985) *Science, 228,* 877.
103. Wetherill G. W. (1988) in *Mercury* (F. Vilas et al., eds.), p. 670, Univ. of Arizona, Tucson.
104. Scott E. R. D. and Newsom H. E. (1990) *Z. Naturforsch., 44a,* 924; Taylor S. R. (1988) in *Meteorites and the Early Solar System* (J. F. Kerridge and M. S. Matthews, eds.), p. 512, Univ. of Arizona, Tucson.
105. Wasson J. T. (1985) *Meteorites: Their Record of Early Solar-System History,* Freeman, New York, 267 pp.
106. Scott E. R. D. and Newsom H. E. (1990) *Z. Naturforsch., 44a,* 924.
107. Scott E. R. D. (1988) *Meteoritics, 23,* 300.
108. See Wetherill G. W. (1988) in *Mercury* (F. Vilas et al., eds.), p. 670, Univ. of Arizona, Tucson.
109. See Wetherill G. W. (1978) in *Protostars and Planets* (T. Gehrels, ed.) p. 565, Univ. of Arizona, Tucson.
110. Matsui T. and Abe Y. (1986) *Nature, 319,* 303; Matsui T. and Abe Y. (1986) *Nature, 322,* 526; Kasting J. F. (1988) *Icarus, 74,* 472; Zahnle K. J. et al. (1988) *Icarus, 74,* 62.
111. Ringwood A. E. (1989) *Earth Planet. Sci. Lett., 95,* 1.
112. See comment by Palme H. (1990) *Nature, 343,* 23.
113. Lissauer J. (1987) *Icarus, 69,* 262; see also Ward W. R. (1989) *Astrophys. J., 345,* L99.
114. Wetherill G. W. (1988) *Lunar and Planetary Science XIX,* p. 1265, Lunar and Planetary Institute, Houston; Bodenheimer P. and Pollack J. B. (1986) *Icarus, 67,* 391.
115. Franklin F. et al. (1989) *Icarus, 79,* 223.
116. Duncan M. et al. (1989) *Icarus, 82,* 402.
117. Kowal C. T. (1989) *Icarus, 77,* 122.
118. For example, Carlson R. W. and Lugmair G. W. (1988) *Earth Planet. Sci. Lett., 90,* 119.
119. Morfill G. E. and Wood J. A. (1989) *Icarus, 82,* 225.

120. Stevenson D. J. (1984) ***Lunar and Planetary Science XV,*** p. 822, Lunar and Planetary Institute, Houston.
121. Celotto E. et al. (1989) ***Lunar and Planetary Science XX,*** p. 149, Lunar and Planetary Institute, Houston.
122. Among the many historical analogies that might be made, the present condition of the asteroid belt is reminiscent of the depleted state of La Grande Armée of Napoleon after the retreat from Moscow in 1812. The asteroids might have formed a planet, La Grande Armée might have had far greater influence on the course of world history; both were reduced to a remnant by external influences.

Chapter 2

The Solar Nebula

Illustration, preceding page—*The Rosette Nebula in Monoceros (NGC 2237). The cavity in the center has a diameter of a few light years and is probably due to strong stellar winds originating from a cluster of OB stars in the center of the nebula. This nebula is, of course, several orders of magnitude larger than the solar nebula from which the solar system evolved, but the illustration serves as a possible model for conditions in the solar nebula when strong winds from the T Tauri stage of the early Sun began to drive gas and volatile elements out of the inner nebula. (61-cm Schmidt photograph by Cerro Tololo Inter-American Observatory and Kitt Peak National Observatory. Reproduced with permission.)*

The Solar Nebula

2

2.1. THE INITIAL CONCEPT

The fundamental and perhaps the most obvious fact about the solar system is that the planets and satellites mostly lie close to the plane of the ecliptic (the Sun-Earth plane) and, with minor and informative exceptions, rotate in the same sense, both around the Sun and about their axes of rotation. In the eighteenth century, scientists were unaware of the retrograde revolution of Venus, of the existence of minor satellites with perturbed orbits, of the strange orbit of Pluto, and of other irregularities: what they saw appeared to be an orderly system.

This fortunate lack of confusing detail, for we can be overwhelmed by too much information if we lack a broad overview, enabled the French astronomer and mathematician Laplace to propose in 1796 that the Sun and planets arose from a rotating disk of dust and gas, the solar nebula. This elegant concept survived in its original form until late in the nineteenth century. The crucial and ultimately fatal flaw of the original laplacian theory was the failure to account for the concentration of angular momentum in the planets and of mass in the Sun [1].

The perceived problems with laplacian models led to the emergence of catastrophic hypotheses. These proposed that the planets arose as a result of tidal interactions with a passing star that pulled out a cigar-shaped filament from the Sun. Subsequently it was shown that such material would disperse rather than condense to form planets. These hypotheses also failed to account for the fact that the Sun, with 99.9% of the mass of the system, has only 2% of the angular momentum, the rest residing principally in Jupiter. The useful term planetesimal remains as a relic from these notions [2].

Laplace would doubtless be gratified that his concept of the solar nebula has ultimately survived: The view that the Sun and the planets formed from a rotating disk of gas and dust, the solar nebula, provides such an obvious explanation that it has become axiomatic. Nevertheless, it is wise to question established views. Perhaps the seeming planarity and symmetry of the system, rather than being an inherent property of the rotating disk, was imposed on an unstructured cloud by the early formation of Jupiter. Apparently, it is not so. Computer studies of planetary orbits show that the rotation of the planets in one plane results from the initial condition of the nebular disk rather than from being forced by Jupiter or by a passing star [3].

The term "solar nebula" is often used as though it were a fixed entity. However, several stages of development may be discerned, and some confusion arises when it is unclear exactly which stage is being discussed. This is more than a trivial problem in semantics. At an early stage, the nebula refers to a fragment that separated from a molecular cloud. The mass and angular momentum of this piece determines much of the subsequent evolution. As the fragment collapses to a disk, dust settles to the midplane and the Sun begins to condense. At a slightly later stage, which is the main topic of this book, the solar nebula refers to the disk of dust and gas remaining during and after the formation of the Sun, and from which the planets, satellites, and asteroids were formed. During these events, the nebula was continuously changing and the term thus has to cover a long sequence of events. An important concept, however, is that the nebula is a fragment that became detached from a larger molecular cloud and that it did not form by the accretion of fine dust and gas from the general interstellar medium. The isotopic information available (section 3.7) indicates a spread of 5-10 m.y. for the evolution of the nebula. Astronomical timescales are only about 1 m.y., but the problem may be resolved if the formation of the refractory inclusions (CAI), which provide the earliest dates, predated the disk period.

The solar nebula, as broadly defined, presents us with several interesting problems. What caused the initial collapse from a large molecular cloud? How did solid objects begin to accrete and finally accumulate into the planets? What was the interaction of the early Sun with the nebula? What processes were responsible for the major division of the solar system into two distinct regions comprising the terrestrial rocky planets and the giant gaseous planets? Other questions deal with nebular composition. Was it homogeneous, or were there zones with varying chemical or isotopic abundances?

Planetary formation is bound up with star formation. This is reasonably well understood; stars form from fragments of relatively dense molecular clouds in the galaxy, and evolve through rather violent stages (usually specified as T Tauri or FU Orionis activity) before settling down on the main sequence. The formation of stars can be studied during most of their evolutionary stages. Astronomers thus possess a considerable advantage compared with planetary scientists, and even with students of history: They can look backwards in time. Since the universe is large enough, they can find samples of stars and galaxies at more primitive stages in their evolution. They can also look forward in time. For example, the nearby star Beta Hydri, which is very similar to the Sun, is about 9 b.y. old, twice the age of the Sun, an observation that provides us with the comforting assurance of a relatively long life expectancy for the solar system. In our investigation of planetary systems, however, we reach a dilemma. The planetary scientist, like the historian, has only one example, the present scenario, together with those few relics that have survived from previous epochs (not, we hope, as shrouds of Turin) to tell the tale of former events.

Single-star systems are not very common, perhaps even rare, compared to binary-star systems, which constitute the majority of observable stars. Even triple-stellar associations are frequently observed. In contrast, we do not know whether any other planetary systems exist. They are exceedingly difficult to detect by present techniques. The best evidence comes from observations of relative radial velocities of stars [4]. Several stars showed variations consistent with the presence of companion objects in the range of 1-9 Jupiter masses. None was in the range of 10-80 Jupiter masses expected for brown dwarfs. These objects seem to be very elusive, and perhaps even nonexistent, for only a few possible candidates have been found [5].

This apparent evidence for low-mass companions approximately the size of Jupiter and for the apparently common occurrence of disks around young, pre-main-sequence stars strengthens the case for the existence of other planetary systems. If so, would they resemble our own? Would we see something like the Galilean satellite system of a few equal-sized planets, systems with one giant planet, or a single brown dwarf companion comprising a failed binary-star system? Dust disks around mature single stars appear to be common, but may be due to cometary or asteroidal dust, and they do not necessarily imply the existence of a planetary system. Although it is commonly supposed that the solar system represents a failed binary-star system, this concept neglects the fundamental difference between the processes responsible for planetary as opposed to stellar formation. Jupiter is not a failed star formed by fragmentation of a molecular cloud, but a true planet, built up bit by bit in the nebula until the core mass was sufficient to cause the gravitational collapse of nebular gas.

Returning to our quest for other planetary systems, would they consist of a dusty disk, a sequence of asteroid belts spaced out like the rings of Laplace's original concept, something bearing a close resemblance to our own familiar system, or something totally bizarre, reminiscent of the scenarios of science fiction? In short, should we seek to establish some universal principles of planetary formation that once in operation in a nebula of appropriate mass would produce something recognizable as the solar system, or should we consider our system unique? This is a recurring theme in the history of attempts to explain the solar system.

Hypotheses are often driven by a wish to find intelligent life elsewhere in the universe and thus to suppose that Earth-like planets are common [6,7]. This book adopts the different thesis that the evolution of our system was dominated by chance events. Unlike stars, which condense from fragments of molecular clouds from the top down and so exhibit considerable uniformity, planets are built up bit by bit from solid particles in circumstellar disks. If the planetary embryos reach sufficient size before the gaseous components in the nebula are dissipated by early violent stellar activity, the gas may collapse by gravitational attraction, forming Jupiter-type gas giants. If the gas departs before this can occur, a system composed of rocky, gas-depleted planets might result. The details of these processes are likely to be variable, so that other planetary systems, if they exist, will differ considerably from our own. Probably all physically realistic models that have been proposed have formed somewhere around one of the 10^{11} suns in each of the 10^{11} galaxies in the presently observable universe.

In this work, the position is adopted that the gaseous giant Jupiter formed very early, within about a million years of the initiation of the system, and that the terrestrial rocky planets accreted from planetesimals much later (over periods of 10^7-10^8 yr) in a gas-free environment. Other planetary systems may show all possible variations around this theme. Accordingly, I do not seek some grand unified theory, but examine our solar system as it is, attempting to understand how it achieved its present state, why it is so diverse in detail, and why, for example, the four satellites of Jupiter, which might have been expected to be very similar, all turned out so different.

2.2. THE ORIGIN OF THE UNIVERSE

The current understanding is elegantly described by Steven Weinberg and it would be superfluous to repeat or to paraphrase the evidence so assembled for the Standard Hot Big Bang hypothesis [8]. A very readable account of the intellectual problem of the origin of the universe is given by Stephen Hawking [9], particularly with respect to the question whether the Big Bang represents a singular point or beginning. Only a few comments will be made concerning this interesting philosophical question. One is that the Standard Hot Big Bang hypothesis does not account in any obvious manner for the present state of the universe, in which most of the observable matter is irregularly distributed in galaxies. The notion that the universe began at some definable point has always been philosophically unsatisfactory, and is coming increasingly under question [10]. As Hawking has remarked, "in the end our work became generally accepted and nowadays nearly everybody assumes that the universe started with a Big Bang singularity. It is perhaps ironic that, having now changed my mind, I am trying to convince other physicists that there was in fact no singularity at the beginning of the universe ... it can disappear once quantum effects are taken into account" [11].

The age of the universe, that is, the time since the Big Bang, given by the reciprocal of the Hubble "constant," is most probably 15-20 billion (10^9) years. Estimates as low as 10^{10} years continue to appear, although they seem to be in conflict with the observational evidence for the apparently older ages of globular clusters in the spherical shell of our own galaxy. Our knowledge of the scale of the universe is still limited, but it is worth recalling that the most distant, and hence oldest, galaxies do not show the expected high abundance of blue stars indicative of early stages of both stellar and galactic evolution.

Although the Big Bang model is conventionally accepted within the scientific community, it should be realized that other possibilities may be available. Thus, the Big Bang may represent a local event in a larger system, and the Standard Hot Big Bang may not be the only candidate. As another author has commented, "In all respects save that of convenience, this view of the origin of the universe is thoroughly unsatisfactory" [12].

Clearly, we should keep open the possibility that we are still observing a small corner of the universe. Our cosmic horizons have continually expanded with each advance in technology, and further significant advances in cosmology may be expected. Some of the alternatives are given by Ellis [13] in a review that may be read with interest by those with a sufficient background in theoretical physics and cosmology [13]. More Earth- or planetary-bound readers should be content with Weinberg [10] or Hawking [9]. In the meantime we might recall the comment that "the steady state theory has a sweep and beauty that for some unaccountable reason the architect of the universe appears to have overlooked. The universe is in fact a botched job, but I suppose we still have to make the best of it" [14].

2.3. BIG BANG SCENARIOS AND ELEMENT SYNTHESIS

Two pieces of evidence support the presently accepted Standard Hot Big Bang cosmological model. One is the 3-K microwave background and the other is the success of the hypothesis in accounting for the large cosmic abundance of He [15]. Another test is to compare the pregalactic abundances of the other light elements with the predictions of their primordial abundances from the standard model.

Which elements are synthesized in the Big Bang? In addition to He, candidates include D and Li. Galactic halo stars, among the earliest to form in our galaxy, should be the best candidates to provide evidence of the primitive compositions resulting from the Big Bang. This is not a simple problem, since the abundances now observed are changed by nuclear reactions within stars, but the abundances of the light elements D, He, and Li are generally consistent with the predictions of the Standard Hot Big Bang theory [16].

2.3.1. Deuterium

Deuterium abundances in stars are difficult to determine, since the nuclide is destroyed in stellar interiors at temperatures above 6×10^5 K, being converted to ^{3}He. Probably the element is consumed during the highly convective contraction stage in the early stages of stellar evolution as the star moves toward the main sequence. The estimate for the primordial D/H ratio is about 10^{-5}, but there is considerable scatter, up to a factor of 2.

2.3.2. Helium

One of the major successes of the Big Bang theory is that it can account for the high abundance of ^{4}He in the universe. Other nucleosynthetic theories fail this test by an order of magnitude. Its abundance in stars generally appears to be within about 20% of the solar value, which points to a uniform origin, but much better precision (2%) is needed before this can be considered a real constraint. The value of the primitive He abundance (Y_p) measured in extragalactic H II regions is 0.245 ± 0.003 [17], but this value may be too high because of the contribution of ^{4}He synthesized more recently in stars [18]. The value for Y_p seems to be becoming established at 0.235 ± 0.004 [19].

It might appear at first sight that Jupiter and Saturn would preserve the primitive solar system abundance, but their observed atmospheric values are quite different. For Jupiter, the atmospheric fraction of He varies between 0.17 and 0.24. For Saturn, the analogous value is 0.14, very different from the solar value of 0.274 [20]. The explanation is that He has probably been separated from H by fractionation at high pressures deep within the gaseous planets, where hydrogen becomes metallic. The variations from the primordial values observed in the jovian and saturnian atmospheres must represent the results of intraplanetary processes since it is difficult to envisage processes that would separate these gaseous elements at nebular pressures and temperatures.

Values for ^{3}He are difficult to estimate because it is both produced and destroyed in stars. There is considerable scatter in the interstellar values of ^{3}He from $4\text{-}40 \times 10^{-5}$ relative to H [21]. The primordial ^{3}He/H ratio is estimated at 1.2×10^{-5} [22].

2.3.3. Lithium

There is fairly good agreement between the observed abundance data and the predictions by the standard Big Bang model. However, there is still uncertainty about the amount of Li produced in the Big Bang and some revision of the standard model may be required. The rather uniform distribution of Li over the galactic halo argues for synthesis of this element during the Big Bang. Lithium is readily destroyed in stellar interiors at temperatures greater than 2×10^6 K, so that deep convection will deplete Li. A fairly constant Li/H ratio of 1×10^{-10} is observed in old Population II stars and this has been interpreted as the primordial abundance [23]. However, the Li abundance in Population I stars shows a depletion depending on mass for stars of the same age, and the maximum value of 10^{-9} may represent the primordial value. The matter is still unresolved [24]. The latest value for ^{7}Li/H is $1.2 \pm 0.3 \times 10^{-10}$ [25].

2.3.4. Synthesis of the Heavy Elements

One of the most significant results of scientific inquiry in this century has been the explanation of the origin of the chemical elements, a discovery resulting from an integration of nuclear physics, astronomy, and astrophysics. There is not space here to refer to the extensive literature on element synthesis in stellar interiors [26,27], but some comments on supernovae are relevant in view of the recent event observed in the Large Magellanic Cloud.

Supernovae are generally classified into two types: Type I, which are due to the explosion of white dwarf stars, in which elements up to the iron peak are synthesized, and Type II supernovae, which are the result of the explosion of red giants and supergiants. Blue giants may also explode, as occurred in the 1987a supernova in the Large Magellanic Cloud.

2.3.5. Supernova 1987a

The concept of the synthesis of the heavy elements in stars has received striking proof in the observation of heavy elements (Ni, Co, Cl, and S) produced in the Type II supernova (1987a) observed in the Large Magellanic Cloud [28]. Supernova 1987a was caused by the explosion of a comparatively small blue giant star, Sanduleak [Sk] -69°202, which was about 50 times solar radius. It was only about 10 m.y. old and had apparently passed through a He-burning red giant evolutionary stage during the last 10% of its rather brief life span. When its He was exhausted, it shrank by a factor of 10 to a blue giant, commencing C burning and the rapid run down to the explosion [29].

The light decay curve observed after the explosion matched the 77.1-day half-life of ^{56}Co. Initially, ^{56}Ni was formed. This decayed with a half-life of 6.1 days to ^{56}Co, which then decayed with a half-life of 77.1 days to stable ^{56}Fe. Supernova 1987a thus provided direct evidence of element synthesis in supernovae, the abundances of Co, Ni, S, and Cl currently observed being far in excess of normal stellar abundances (about 0.07 solar masses of ^{56}Ni were produced). The Fe, Ni, and Cl abundance ratios are close to solar, although this may be a coincidence. Sulfur abundances are below solar levels [30].

2.4. THE SCALE AND STRUCTURE OF THE UNIVERSE

"On the galactic scale of things, the solar system is a rather small place" [31].

In order to obtain some perspective on the solar system, it is useful to contemplate the scale of the universe as we perceive it at present. The distance between the Sun and the Earth is 149.6×10^6 km or one astronomical unit (AU). The diameter of the Sun is 1.392×10^6 km, close to 0.01 AU. The planetary system extends out to the orbit of Neptune at about 30 AU. There is probably a Kuiper Cloud of icy comets not far beyond Neptune [32], an inner Oort Cloud extending from 10^3-10^4 AU, and the classical Oort cloud of comets, marking the outer bound of the solar system at about 10^4-10^5 AU [32]. Dusty disks around other stars extend out to a few hundred AU. All these distances are very small on a galactic scale.

The nearest star (Proxima Centauri, an 11th magnitude M5 red dwarf, the faintest of a triple-stellar system of which Alpha Centauri is the brightest) is about 4.3 light years (270,000 AU or 40×10^{12} km) or about 1.3 parsec distant from Earth (1 parsec = 3.262 light years).

The nearest star-forming clouds are in the galactic arm in Taurus, about 150 parsec distant. The dark cores of the clouds, which are currently regarded as the site of star formation, are about 0.1 parsec in diameter, while the giant molecular clouds reach up to 100 parsec in diameter. Our galaxy is about 85,000 light years (about 26,000 parsec) in diameter, the Sun being situated about 28,000 light years (about 8500 parsec) from the center. The nearest major galaxy, M.31 or Andromeda, is 2×10^6 light years (about 6×10^5 parsec) distant, and forms one of the 17 members of the local group of galaxies, outside which the expansion of the universe occurs. One of the most distant galaxies is 0902+34. It has a redshift (z) = 3.395, placing it at a distance of 15 billion light years [33]. It contains stars more than a billion years old and thus has an old stellar population. The total mass of this galaxy is 10^{12} solar masses and its radius is 10-25 kparsec, similar in size to nearby galaxies. This very distant galaxy, although it might have been expected to be different, thus seems to be unremarkable [34].

The most distant objects are quasars with redshifts exceeding 4. The currently most distant object (quasar PC 1158 + 4635, in Ursa Major) has a redshift of 4.73 [35]. Quasars at $z > 4$ seem to have a relatively normal heavy-element abundance. This suggests that there was nucleosynthesis at this remote epoch, with consequent enrichment of the interstellar medium in heavy elements. Although over 5000 quasars are known, there is a sharp decline in their number beyond a redshift of 2.5 (Fig. 2.4.1). It is not yet clear whether the cutoff is real or is some kind of observational artifact. Until this problem is resolved and we have more observations of galaxies at remote

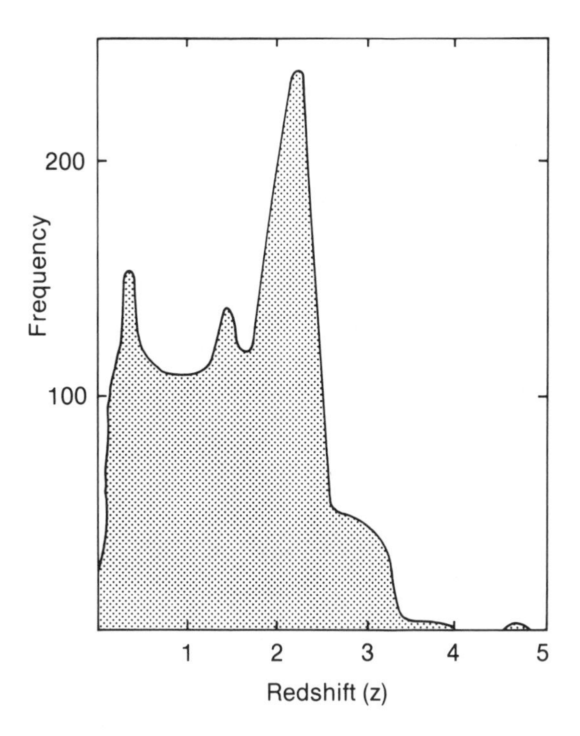

Fig. 2.4.1. *The fall-off in objects with redshift greater than $z = 2.5$. The extreme redshift known at present is 4.73.*

distances ($z > 5$), it will not be possible to resolve questions about the time of galactic formation. However, the discovery of what appear to be very early galaxies, the so-called "blue fuzzies," also shows a redshift cutoff at about 3 and suggests that we are indeed seeing a first epoch of galaxy formation (see section 2.5) [36]. Although the formation of galaxies is commonly assigned to an early period, they may be still forming, and have shorter lifetimes than generally supposed.

2.4.1. The Structure of the Universe

A major shift in our understanding of the structure of the universe has occurred within the past few years. Before about 1980 it was generally accepted that the universe was uniform, with galaxies distributed evenly out to the limits of vision. This view has changed dramatically and the structure is now known to be far from random. Galaxies are distributed along chains, sheets, filaments, and in knots. The largest sheet-like structure currently observed is the "Great Wall," which contains thousands of galaxies and is over 500×10^6 light years long [37].

In many cases, galaxies appear to occupy the surfaces of spherical shells surrounding dark regions apparently devoid of galaxies [38] (Fig. 2.4.2). Such structures have been compared to soap bubbles. Two-dimensional pictures of galactic distribution look like filaments apparently because the galaxies congregate around the edges of the very large empty bubbles. These great bubbles are about 150 million light years across, and apparently contain little luminous matter. Whether these enormous voids are empty of dark matter as well as being deficient in visible galaxies is unclear, although some irregular galaxies may be present. Thus, we do not know whether dark matter is uniformly distributed or follows the sporadic patterns of the luminous galaxies. The problem of the dark matter, which may contain most of the mass of the universe, cannot be addressed here except to note that much of the "missing mass" may, in the apparent absence or scarcity of very faint stars, brown dwarfs, and such creatures, consist of the remnants of massive stars formed at an early stage [39].

Curiously, the 3.1-K microwave background, generally thought to be a remnant from the Big Bang, is very uniform and does not appear to show temperature fluctuations. Since this dates from the period in the Big Bang cosmology when the universe was about 10^5 years old, the density fluctuations now apparent in the visible matter arose subsequently.

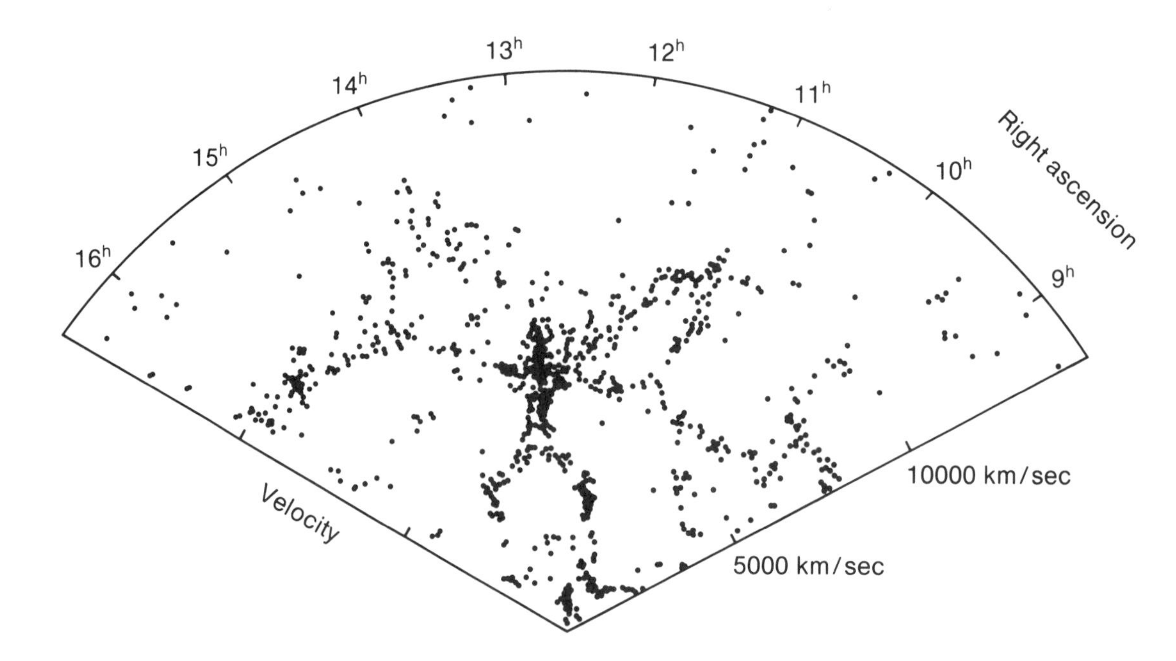

Fig. 2.4.2. *The distribution of 1060 galaxies with magnitudes less than 15.5 and velocities less than 15,000 km/s for a segment near the north galactic pole. After de Lapparent V. et al. (1986) Astrophys. J., 302, L2, Fig. 1.*

Our own galaxy, the local group, and all nearby galaxies and groups of galaxies, are being pulled through space at 600-700 km/sec to a distant (about 40 Mparsec) region of mass concentration, sometimes known as the "Great Attractor," which lies beyond the Hydra-Centaurus Supercluster, a part of the Virgo cluster. Recent studies suggest that there is a yet more distant, larger concentration of galaxies at about 140 Mparsec, lying in the same direction, which may be responsible for the local motion [40].

2.4.2. The Darkness of the Night Sky

Related to the scale of the universe is one of the more interesting cosmological problems: why is the sky dark at night? The question is simple but has a long history and many attempts have been made to explain the problem. Shakespeare noted in 1599 in *As You Like It* that "a great cause of the night is the lack of sun." However, if the universe is infinite, then every line of sight must eventually intercept a star. Accordingly, the night sky should be ablaze with stars. The problem became famous as "Olbers's paradox" after Heinrich Olbers, the German astronomer who pointed out the problem in 1823. He considered that starlight was absorbed in space, so that the universe resembled a foggy forest: only the nearest trees are visible. He was anticipated in this notion by the Swiss astronomer, Jean-Phillipe Loys de Cheseaux, in 1744. J. Herschel, however, in 1848 pointed out that the radiation would heat up the gas, and eventually emit as much as it received, so that the night sky should be bright. Kepler, already aware of the problem in 1610, considered that the universe was finite, like a grove of trees in the midst of a plain, an analogy that is still frequently invoked.

There is not space here to list all the ingenious solutions proposed for this problem, but a final resolution appears to have been reached [41]. The night sky is dark because the universe does not contain enough energy to create a bright sky. Because of the finite lifetimes of stars, there are not enough visible stars to cover the sky. "The stars needed to cover the sky cannot be seen because their light has yet to reach us" [41, p. 203]. Early insights to the correct solution were proposed both by selenographer Johann Heinrich Madler, by Edgar Allan Poe in his essay "Eureka" (1848), and by Kelvin. "He took the rationalist attitude that paradoxes are the result of misunderstandings; they lie in ourselves and not in the external world. It seems ironic that he was the first to solve with rigor and utmost lucidity a riddle that later, when his work lay forgotten, became known as Olbers's paradox" [41, p. 165].

Bright sky universes can exist, but have to be many orders of magnitude more dense than our own. If all the matter in the universe is converted to energy, the average temperature is only 20 K. A bright-sky universe at 6000 K would contain radiation with a mass 10^{10} times greater than that of the stars, which convert only about 0.1% of their mass into radiation.

2.5. GALAXIES

The structure of our galaxy is shown in Figs. 2.5.1 and 2.5.2. The first figure shows the spiral structure, while Fig. 2.5.2 shows the central bulge as revealed

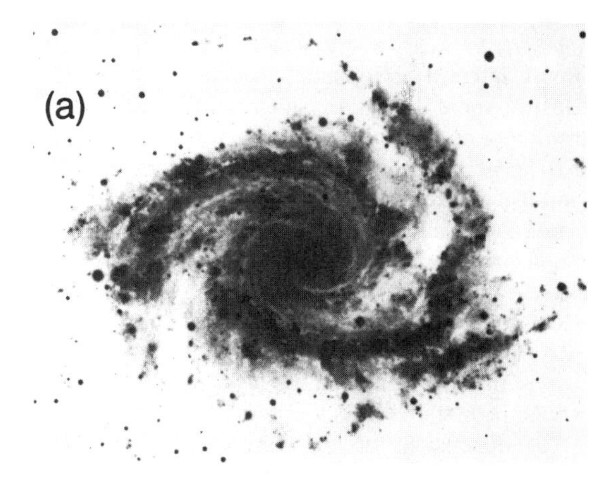

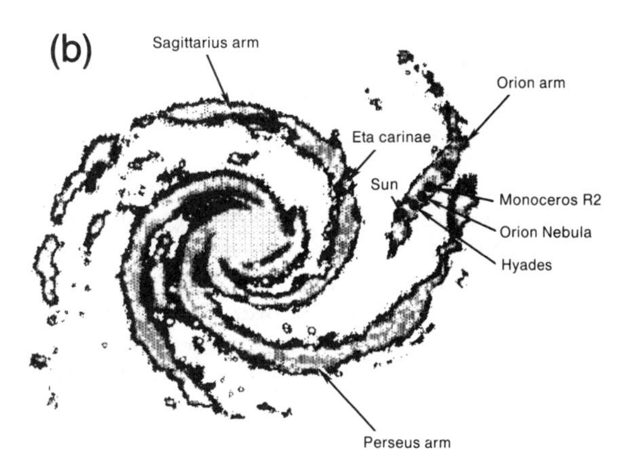

Fig. 2.5.1. *(**a**) A typical spiral galaxy, NGC 2667 (courtesy Anglo-Australian Telescope). (**b**) A sketch of the probable structure of the Milky Way galaxy showing three nearby arms, the position of identified regions of new star formation, and the Sun.*

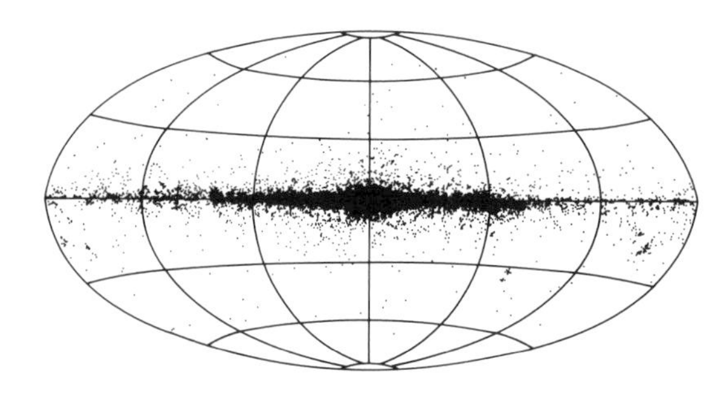

Fig. 2.5.2. *The disk and central bulge of the Milky Way galaxy are well shown in the distribution of stars detected in the infrared by the IRAS. After Beichman C. A. (1987) Annu. Rev. Astron. Astrophys., 25, 542, Fig. 9.*

by the Infrared Astronomical Satellite (IRAS). Perhaps the most dramatic result from the IRAS investigation of our own galaxy has been this delineation of the central bulge of the galaxy, from stars detected emitting infrared radiation between 12 and 25 μm (Fig. 2.5.2). This observation again raises some interesting questions with respect to stellar formation and galactic evolution. The galactic bulge has generally been supposed to be about as old as the galaxy (15-20 b.y.). However, the observed infrared sources suggest that star formation has occurred in the bulge within the past 10^9 years.

The dimensions of our own galaxy have been scaled down by about 15% from earlier estimates. The distance from the Sun to the galactic center is now estimated at 8.5 kparsec, rather than 10 kparsec. The rotation speed of the solar system about the galactic center is 220 km/s, about 10% less than the previously used value of 250 km/s [42].

2.5.1. Origin

The classification of galaxies into elliptical, spiral, and barred galaxies is shown in Fig. 2.5.3. This, however, is not an evolutionary sequence and it is still not clear how galaxies form in an expanding universe; their existence is not predicted in a simple way from the Big Bang cosmology. Since they are the dominant visible structures in the universe, the failure of contemporary cosmology to account for them must be considered less than desirable.

The standard model for spiral galaxy formation, including our own, begins with a spherical mass of gas that collapses to a rotating disk within a few hundred million years, leaving a halo of globular clusters to outline its original extent. This would imply that the globular clusters should all be about the same age. However, this does not seem to be the case, and wide variations in ages have been reported, casting doubt on the classical model [43].

As stars evolve, they synthesize the heavy elements (or "metals" [44]), which are eventually dispersed into the interstellar medium. As the galactic disk forms, it becomes enriched in heavy elements. Accordingly, the enrichment of heavy elements is lower in the halo stars than in the disk, where the process of star formation is still continuing.

Clearly, it appears that galaxies may have a much more complex history than previously imagined. Like terrestrial continents, they may be composed of many separate terranes swept together. Thus, galaxies probably do not evolve in isolation, but may have undergone many interactions (possibly a better term than collisions). For example, the halo of our galaxy may have been acquired from the accretion of small companion galaxies. Alternatively, the collapse from a sphere to a disk occurred over timescales of billions, rather than a few hundred million, years. This would greatly complicate the problems of galactic evolution from a sphere to a disk. In the conventional explanation, the galaxy collapses rather quickly from a spherical form to a rotating spiral disk. Evidence usually cited in support of this scenario is the less-

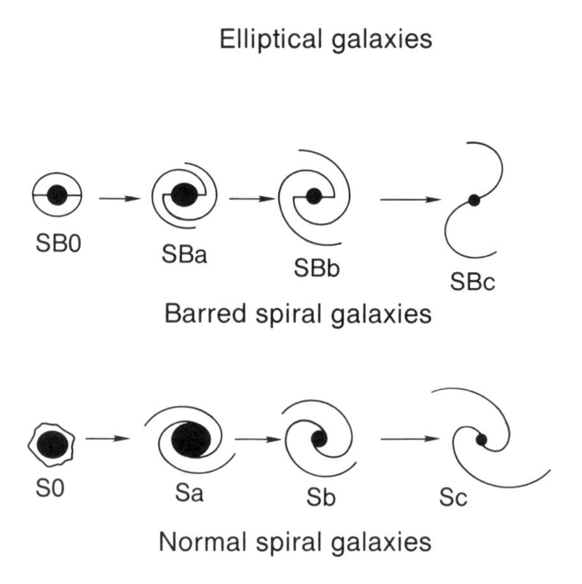

Fig. 2.5.3. *The standard Hubble classification of galaxies.*

evolved, heavy-element-poor stellar Population II in the halo areas, compared with the younger, blue, heavy-element-rich Population I stars characteristic of the spiral arms. This is consistent with the expected age sequence. Curiously, however, very few halo stars have very low heavy-element abundances. Even the most primitive examples contain heavy elements such as Ba, perhaps indicating that element synthesis in stars has continued for a longer period than currently imagined. Most galaxies contain stars with ages of 10^{10} years, but the globular cluster ages, as noted above, show a wide spread. In the standard model of galactic collapse, they should exhibit a close range in ages (~200 m.y.) [45].

One view of galaxy formation is that the distinction between elliptical and spiral galaxies seems to be related to the rate of star formation. If this process occurs rapidly, then an elliptical mass of stars results; if it is slow, the gas in the galaxy collapses to a rotating disk (in a large-scale analogue of the solar nebula). Star formation then occurs in the spiral arms of the rotating galaxy. Perhaps the elliptical galaxies possessed greater gas density, while those that became spirals possessed an initially higher angular momentum [46]. A second view regards the disk galaxies as primordial, forming from the collapse of protogalactic gas clouds. Elliptical galaxies in this model are the secondary result of the merging of disk galaxies, rather than being of more fundamental significance in cosmology [47].

The major significance of this view, which is supported by a considerable amount of observational evidence [48], is that galaxy formation did not occur during an early fixed epoch, but is continuing, and the splendid visual galaxies are not permament, but fleeting, like everything else in nature.

2.5.2. Age

There has been some confusion over the age of galaxies. It was previously thought that most galaxies formed very early, around the same time, at a redshift of about 3.0, with the collapse of large gas clouds. If galaxies are all old objects and are not forming in the present epoch, one might expect their formation to be a consequence of some early stage of the Standard Hot Big Bang model. Most galaxies now observed appear to be mature well-evolved systems, so there has been a general consensus that they formed in a much earlier epoch. Youthful galaxies should therefore be very distant in time and space, remote from our own mature systems. Recent evidence makes this view much less certain. One such galaxy, ESO 400-G43, a blue compact galaxy that is quite heavy-element poor (about 1/8 of the solar value), contains no old stars and appears to be forming at present [49].

What appear to be very faint blue galaxies, referred to as "blue fuzzies," have been detected using solid-state charge-coupled detectors, searching regions that appear blank on the most sensitive photoplates [50]. The "blank" areas are covered with very faint, very blue objects that appear to be early galaxies. The blue color is due to UV radiation from new stars redshifted into the blue region of the spectrum. There is a sharp cutoff in redshifts greater than about 3, although the technique could detect redshifts out to 7. These are the oldest visible objects found to date and they apparently reflect the earliest recorded period of galaxy formation.

2.5.3. IRAS Data

The IRAS opened new windows on our view of the universe. The principal finding outside the Milky Way galaxy was the very large number of galaxies that were emitting up to 95% of their luminosity in the infrared (10-100 μm). About 25,000 such sources were discovered, mainly in the 60-μm band. Many galaxies appear to have episodes of very rapid star formation. These are referred to as "starbursts" and are very prominent in the infrared. Such activity cannot persist for more than 10^8 or 10^9 years without using up all the available interstellar material. The data raise many obvious questions, and major advances in our understanding of star formation in galaxies will probably emerge [51].

2.6. MOLECULAR CLOUDS AND INTERSTELLAR DUST

The material now in the solar system came from the interstellar medium, which typically has a density of about 1 atom/cm^3, but much material is present in denser aggregates that appear to be precursor stages for star formation, and from which the solar nebula separated as a fragment. Various types of molecular clouds have been identified.

1. The *diffuse clouds* are partly transparent to UV (>1200 Å) and visible light, since they contain low gas densities (10^2 atom/cm^3). Gas temperatures in the diffuse clouds may reach 200 K. Only rather simple molecules exist since UV radiation is strong enough to dissociate large, weakly bonded molecules [52].

2. *Dwarf molecular clouds* (DMC) are widely scattered through the galaxy and probably have life-

times of 10^8 years. Molecular cloud densities range from 10^2 to $>10^6$ H_2/cm^3.

3. *Giant molecular clouds* are the most massive objects in the galaxy. They are composed of many smaller clumps, which are 2-5 parsec in diameter, and contain 1-100 solar masses. The classic example of a giant molecular cloud is the Orion nebula, visible due to the high flux of ionizing radiation. Typically, the giant molecular clouds are about 100 parsec in diameter, with masses 10^5-10^6 solar masses, temperatures about 10-20 K, and density about 100-1000 H_2/cm^3. These molecular assemblages are not in thermodynamic equilibrium. In these giant clouds, about 70 different molecular species have been identified. A full list of species currently known is given in Table 2.6.1.

The production of the molecules in interstellar clouds is now generally ascribed to gas-phase ionic reactions. In the diffuse clouds H_2 is ionized by cosmic rays, while C is ionized by UV photoionization. The resulting ions combine to form various molecules [53]. In dense molecular clouds, H_2 and He are ionized by cosmic rays, but in contrast to the diffuse clouds, C is ionized mainly by the reaction of He^+ with CO [53].

Discussions about the comparison between the solar nebula and the interstellar medium are complicated by several factors. A major problem is the timescale difference. Was the interstellar medium 4.6 b.y. ago the same as we now observe in the solar neighborhood? The isotopic and chemical compositions will have evolved since that time due to nucleosynthesis.

The most distinct interstellar fingerprints are the D/H ratios, which may be up to 3 orders of magnitude higher than the solar system ratio in molecules in cold clouds, due to fractionation at low temperatures [54].

The reaction

$$D^+ + H_2 = H^+ + HD$$

TABLE 2.6.1. Interstellar molecules.

Simple Hydrides, Oxides, Sulfides, Halides, and Related Molecules				
H_2	CO	NH_3	CS	NaCl
HCl	SiO	SiH_4	SiS	AlCl
H_2O	SO_2	CC	H_2S	KCl
	OCS	CH_4	PN	AlF
Nitriles, Acetylene Derivatives, and Related Molecules				
HCN	HC≡C—CN	H_3C—C≡C—CN	H_3C—CH_2—CN	H_2C=CH_2
H_3CCN	H(C≡C)$_2$—CN	H_3C—C≡CH	H_2C=CH—CN	HC≡CH
CCCO	H(C≡C)$_3$—CN	H_3C—(C≡C)$_2$—H	HNC	
CCCS	H(C≡C)$_4$—CN		HN=C=O	
HC≡CCHO	H(C≡C)$_5$—CN		HN=C=S	
H_3CNC				
Aldehydes, Alcohols, Ethers, Ketones, Amides, and Related Molecules				
H_2C=O	H_3COH	HO—CH=O	H_2CNH	
H_2C=S	H_3C—CH_2—OH	H_3C—O—CH=O	H_3CNH_2	
H_3C—CH=O	H_3CSH	H_3C—O—CH_3	H_2NCN	
NH_2—CH=O	$(CH_3)_2CO$?	H_2C=C=O		
Cyclic Molecules				
C_3H_2	SiC_2	C_3H		
Ions				
CH+	HCO+	HCNH+		
HN_2+	HOCO+	SO+		
	HCS+			
Radicals				
OH	C_3H	CN	HCO	C_2S
CH	C_4H	C_3N	NO	NS
C_2H	C_5H	H_2CCN	SO	
	C_6H			

Data from Irvine W. M. and Knacke R. F. (1989) in *Origin and Evolution of Planetary and Satellite Atmospheres* (S. Atreya et al., eds.), p. 5, Fig. 1, Univ. of Arizona, Tucson.

is exothermic to the right, so that the reverse reaction is hindered at low temperatures. Accordingly, most of the D in interstellar clouds will reside in HD molecules. Some of this material may be preserved in carbonaceous chondrites, explaining the high D/H ratios observed [55].

However, the situation for C isotopic ratios is less clear cut. Thus, the claim that the difference between the $^{12}C/^{13}C$ ratios in carbonaceous chondrites and the present terrestrial value reflects the preservation of interstellar material in the meteorites [56] must be viewed with caution on account of the wide variations observed in the astronomical data. Although an accepted value for the interstellar $^{12}C/^{13}C$ ratio is 43 [57], there is much variation between values ranging from 40 to over 100. The terrestrial value is 90. Although Comet Halley has a ratio of 65 ± 8, this need not imply an interstellar source for Halley, in view of the uncertainty in the value noted above [57]. It does indicate, however, that Comet Halley is not a pristine sample of the solar nebula.

The reaction

$$^{13}C^{+} + {}^{12}CO = {}^{12}C^{+} + {}^{13}CO$$

proceeds to the right at low temperatures, which could explain an enrichment of ^{13}C in CO observed in the dense clouds.

Most of the C in interstellar clouds is in the form of CO, with CH_4/CO ratios less than 10^{-2} in the gas phase, but most of the C in planets is present as CH_4. At some stage during or following nebular collapse the gas chemistry was changed. Carbon monoxide is apparently scarce on Triton and Pluto. These bodies probably formed in the primordial nebula. Methane is the dominant form of C in the satellites of the outer planets, which most likely formed in the higher-temperature and higher-pressure environment of planetary subnebulae.

2.6.1. Cores

Many high-density cores have been observed in molecular cloud clumps. These frequently contain infrared objects, some of which are T Tauri stars [58]. The cores have densities of 10^4 H atoms/cm^3. Their masses are of the order of solar masses, and they have radii of perhaps 0.1 parsec. Temperatures are about 10 K. They are thus cold and will collapse due to gravity, forming stars on timescales of 10^5 years. A good example exists in Taurus. How these cores are fragmented from the molecular clouds is not well understood, since the gas has overcome the forces associated with the interstellar magnetic fields [59].

2.6.2. Interstellar Dust

It is useful to separate the discussion on interstellar dust from that other variety of extraterrestrial or "cosmic" dust derived principally from comets and the asteroid belt (see section 3.11). The question of the previous history of the dust component is both important and obscure. Interstellar grains, typically 0.1 μm in diameter, mostly originated in stellar atmospheres or condensed during supernova events. These are the ultimate building blocks of the planets, but clearly the planets accreted from a hierarchy of objects, not from dust. As turbulence develops in the nebula, clumps of grains are likely to form by sticking together in low-velocity collisions, growing to centimeter size in the outer nebula [60].

What stuck the grains together? Clumps of silicate dust must have been in existence before chondrule formation (see section 3.5) and must have already been separated from metal and sulfide phases. Accordingly, these three principal phases must have been present in the interstellar dust. Table 2.6.2 lists possible refractory components of interstellar dust, and Table 2.6.3 lists possible volatile components. The distribution of abundant elements in the dust and gas components of interstellar clouds is given in Fig. 2.6.1.

TABLE 2.6.2. Probable refractory components in interstellar dust.

Component	Spectral Signature	Abundance
Silicate	9.8, 18μm	100% Si, 10-20% O
Carbonaceous ultraviolet absorber	2175 Å	25% C
hydrogenated amorphous carbon (HAC)	7000 Å	5-15% C
quenched carbonaceous composite (QCC)	3.28, 3.4, 6.2, 7.7, 8.6, 11.3 μm	
polycyclic aromatic hydrocarbon (PAH)		1-2% C
organic refractory	3.4 μm	25% C
Silicon carbide	11.4 μm	

Data from Irvine W. M. and Knacke R. F. (1989) in *Origin and Evolution of Planetary and Satellite Atmospheres* (S. Atreya et al., eds.), p. 30, Table 5, Univ. of Arizona, Tucson.

TABLE 2.6.3. Volatile components of interstellar dust.

Component	Spectral Signature	Abundance
H_2O	3.07, 6.0, 12 μm	5-15% O
CO	4.67 μm	few % of C
Carbonaceous		
ice band "wing"	3.4 μm	10-25% C
hydrocarbon or alcohol	6.8 μm	
"XCN, XNC"	4.62 μm	
C_3, CN	4.9 μm	
Nitrogen	2.97 μm	30% N
NH_3		
Sulfur		
H_2S	3.9 μm	few % of S
"XS"	4.9 μm	

From Irvine W. M. and Knacke R. F. (1989) in *Origin and Evolution of Planetary and Satellite Atmospheres* (S. Atreya et al., eds.), p. 31, Table 6, Univ. of Arizona.

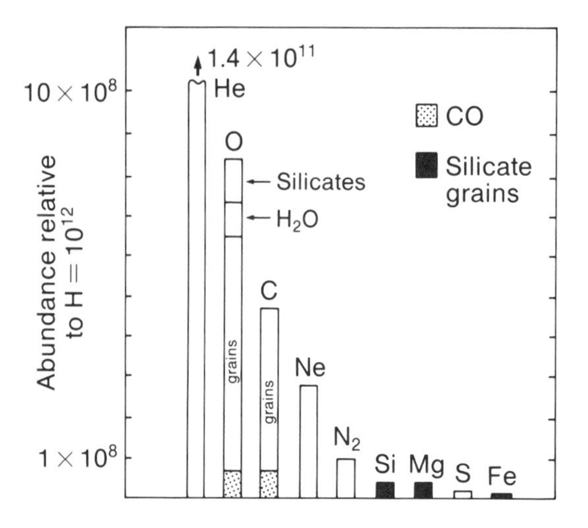

Fig. 2.6.1. *The distribution of common elements in gas and dust in molecular clouds. After Irvine W. M. and Knacke R. F. (1989) in Evolution of Planetary and Satellite Atmospheres (S. Atreya et al., eds.), p. 34, Fig. 10, Univ. of Arizona, Tucson.*

There is a great variety of components in interstellar dust. A possible scenario for their evolution is shown in Fig. 2.6.2. The inorganic components (carbon and silicates) of the dust are formed in stars in the later stages of stellar evolution, when massive outflows occur, particularly during the red-giant and subsequent stages. The organic and icy components, on the other hand, seem to be formed in the lower-temperature environments of the interstellar medium by accretion and reactions induced by ultraviolet radiation. Dust is destroyed or altered by shock waves from supernovae.

It is commonly supposed that C in the interstellar medium is present as graphite. This is based on the identification of an absorption line at 2175 Å as due to graphite [61]. Graphite is, however, probably not a very important component of interstellar dust [62] although some secondary graphite, about 1 μm in diameter, may be present. Based on the mineralogy of primitive meteorites and interplanetary dust particles, the main forms of carbonaceous material that accreted to the solar nebula were hydrocarbons or poorly crystallized and amorphous carbon rather than graphite [63]. Microdiamonds and SiC grains, with large rare gas, Si, and C isotopic anomalies, are present in C2 carbonaceous chondrites. These grains represent interstellar material [64].

2.6.3. Diamonds

The diamonds, averaging 26 Å in diameter, are of interstellar origin. They contain r- and s-process Xe and are the main form of elemental C in primitive meteorites. The diamonds may have formed in high-velocity grain-grain collisions in supernova shock waves, where pressures and temperatures are high enough to convert graphite into polycrystalline diamond. Evidence for this is the similarity between meteoritic diamond and samples produced in detonation soot by high explosives detonated in argon-filled chambers. The size, shape, and degree of crystallinity of the diamond clumps produced are almost identical to the meteoritic diamond, consistent with the production of the latter by shock synthesis [65].

Formation of 1-10-nm diamond crystallites is envisaged [65] as resulting from collisions between charged interstellar carbon grains (about 100 nm in size) moving at high velocity (10-100 km/s) in the magnetic field behind supernova shock waves. The preferred site for the production of diamond appears to be in the atmospheres of late-type carbon-rich stars. The Xe component, with its signature of r- and s-processes, must have another source. One interesting solution to this problem is that the diamonds are formed in the atmosphere of a carbon star that has a white dwarf companion. This eventually becomes a Type 1 supernova, producing Xe with s- and r-process signatures, which are implanted in the diamonds [66]. However, it is also possible to form diamond films from reactions in the gas phases under

Evolution of interstellar dust

STARDUST: Silicates, graphite, amorphous carbon, PAHs, SiC

FORMATION PROCESSES IN CIRCUMSTELLAR SHELLS: Nucleation, condensation, coagulation

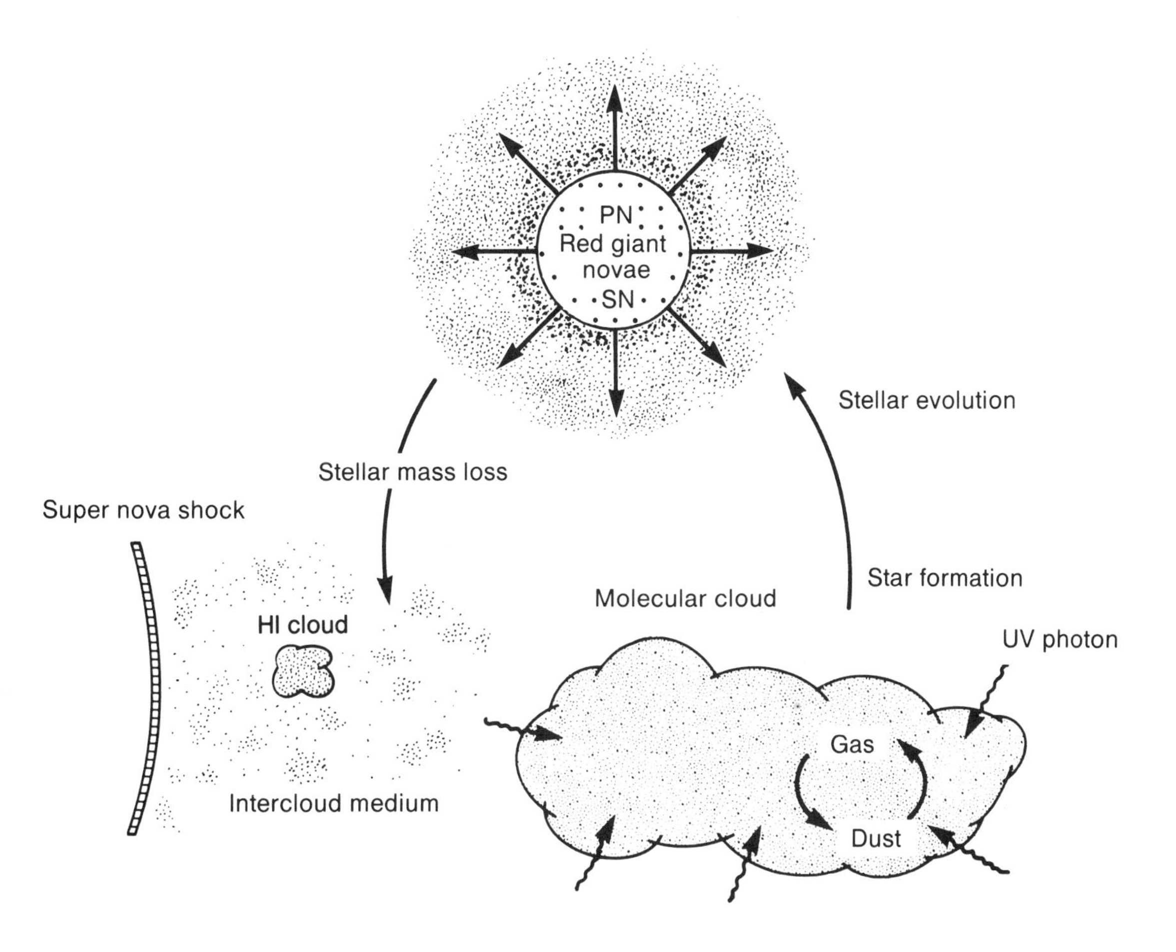

DUST IN THE INTERSTELLAR MEDIUM: Stardust, icy grain mantles, organic refractory mantles

FORMATION PROCESSES IN MOLECULAR CLOUDS: Accretion, reaction, UV photolysis, transient heating

DUST DESTRUCTION PROCESSES: Sputtering and vaporization in shocks

Fig. 2.6.2. *The processes that affect the evolution of interstellar dust are shown in diagrammatic form. After Tielens A. G. G. M. (1988) LPI Tech. Rpt. 88-04, p. 16, Fig. 5.*

low pressures [67] so that there may have been ample opportunity, even within the solar nebula, for diamond films to have been formed by reactions with CH_4.

2.6.4. Silicon Carbide

Another type of presolar carbon is SiC. Several varieties, ranging in size from 0.03 to 4 μm, occur in association with the diamond. Amorphous carbon occurs along with diamond and SiC. Their identification as interstellar rests upon the presence of isotopically anomalous C, N, Si, and noble gases (e.g., Ne-E). The isotopic anomalies for C, N, and Si are extreme, and have been suggested to require at least six components [68].

Anomalous isotopic Si in coarser single grains from the Orgueil (CI) and Murchison (CM2) carbonaceous chondrites presents a simpler picture of derivation by mixing of two sets of Si atoms derived from distinct isotopic sources and produced by C-, Ne-, and O-burning stages in supernovae [69]. They contain no evidence of nuclear particle tracks or radiation defects [70].

Carbon, nitrogen, and silicon isotope data on individual grains show wide differences, apparently indicating separate sources. Silicon carbide may form in the atmospheres of C-rich stars; diamond most likely formed in supernova shock waves. The isotopic anomalies in diamond and SiC are typically orders of magnitude greater than those observed in oxides, a fact attributable to the resistance of these phases to aqueous alteration or heat [71]. The survival of these anomalies, particularly for the noble gases, indicates that the particles were not heated during or since their arrival in the solar system. The mean age of the particles is about 40 m.y., much less than the estimated age of about 1000 m.y. for refractory interstellar grains [72]. The significance of this discrepancy, if real, is not clear. It could mean formation of the solar system from anomalously young material, with consequently young estimates of the age of the elements based on solar system material. This important paradox remains to be understood.

Curiously, SiC grains from the E4 enstatite chondrite, Indarch, have normal isotopic Si. This suggests that isotopically anomalous Si is not found in the highly reducing nebular environment [C/O > 1] in which the enstatite chondrites formed [73].

Both SiC and diamond are rare in meteorites, SiC constituting only 4 ppm of the total carbon content. This is surprising, since SiC is thought to be relatively common in the interstellar medium. These examples of interstellar material are very resistant, surviving both natural and laboratory processing. Possibly they do not represent an average sample of interstellar dust, and the solar nebula did not receive much material derived from carbon stars [74].

2.7. SEPARATION OF NEBULAE

Giant molecular clouds seem to be the environment in which massive stars of the OB association (20 times solar mass) form, although some solar-mass stars are formed as well. Although there is a continuum of masses in molecular clouds, OB star formation thus seems to be confined to the massive clouds, and these stars seem to have an origin distinct from that of smaller-solar-mass-type stars.

Accordingly, the large molecular clouds are of less immediate interest in the present context than the smaller dark clouds, about 2-5 parsec in diameter, with 10^3-10^4 atoms/cm^3 and ages of 10^7-10^8 years, which seem to be the principal site of star formation of about solar size. Hence they are of most interest in our investigation of the formation of planetary systems around such stars. These clouds contain both T Tauri stars and infrared objects detected by the IRAS. The clouds are supported against gravitational collapse by a combination of thermal pressure, turbulence, and possibly magnetic fields. However, magnetic fields will not support a molecular cloud indefinitely because only a small proportion (10^{-5}-10^{-8}) of the atoms and molecules are ionized.

What causes the clouds to fragment? What determines the size, rotation, and angular momentum of the fragments? Why do some finish up as binary stars, and what distinguished our own nebula? These questions remain a fertile field for investigation. Some ideas previously thought to be important constraints are no longer central to the models. Thus, there seems to be no evidence for a hierarchical fragmentation of large (1000-solar-mass) clouds [75]. The collapse of clouds, resulting in star formation, appears to occur from the inside-out due to gravitational instability. External mechanisms for cloud collapse such as compression by supernova shock waves appear a less likely mechanism. Once star formation begins in molecular clouds, the clouds have very short lifetimes [76] (Table 2.7.1). There appears to be a lower limit to cloud breakup so that "the minimum protostellar mass for Population I stars produced through cloud collapse and fragmentation is about 0.01 (solar masses)" [77].

TABLE 2.7.1. Lifetimes for molecular clouds.

Mass Relative to Solar	Cloud Ages Before Star Formation (m.y.)	After Star Formation (m.y.)
10^6-10^7	50-200	20-100
10^4-10^6	20-40	10-20
10^1-10^4	1-10	1-10 (T Tauri)

Data from P. Solomon, personal communication, 1989.

2.8. DUST DISKS AROUND STARS

2.8.1. Disks Around Pre-Main Sequence Stars

The chief interest in the context of planetary formation is in the existence of disks around pre-main sequence stars. Although the evidence is mainly circumstantial, it seems possible that at least half of young stars (with ages less than 3 m.y.) are surrounded by dusty disks similar in size to the solar system. These disks typically seem to have radii of about 10-100 AU and masses about 0.01 to 0.1 solar mass. The best evidence is the disk around HL Tauri. The disks disappear in less than 10^7 years, placing a stringent astrophysical constraint on the formation of planetary systems [78]. Figure 2.8.1 shows the frequency distribution for the excess infrared for stars with ages less than and greater than 3 m.y., indicating that disks dissipate on timescales of a few million years.

Dust and gas disks around T Tauri stars appear to be of the order of 0.1-1.0 solar mass. Many have intrinsic luminosities, for which accretion of gas and dust onto the protostar is the likely source. Their large infrared excesses are most probably due to thermal emission from a circumstellar dust disk.

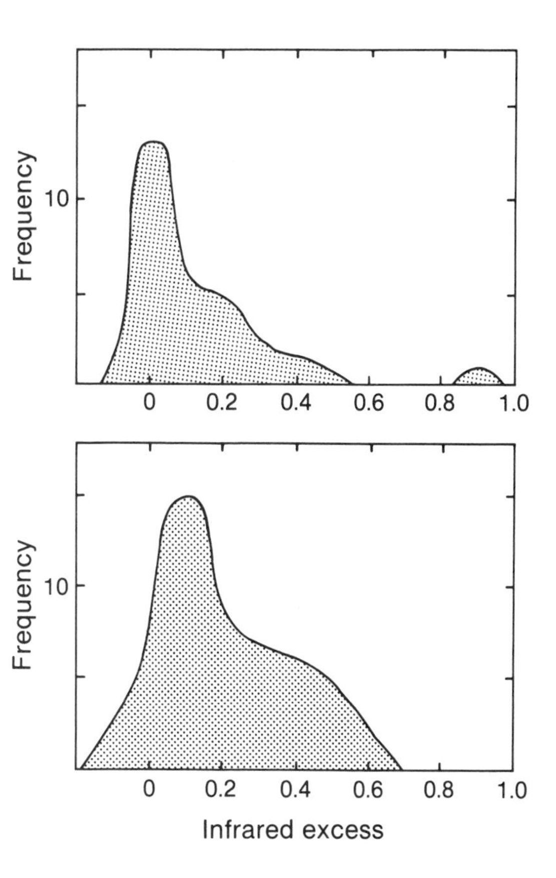

Fig. 2.8.1. *The distribution of the infrared excess for pre-main-sequence stars older than 3 m.y. (upper) and younger than 3 m.y. (lower), showing that the the infrared excess decreases for stars older than 3 m.y. This is the best astrophysical constraint for the lifetime of dusty circumstellar disks, and hence the building of planetary systems. After Strom S. E. et al. (1989) in The Formation and Evolution of Planetary Systems (H. A. Weaver and L. Danly, eds.), p. 102, Fig. 6, Cambridge Univ., New York.*

2.8.2. HL Tauri Disk

The disk that surrounds HL Tauri does appear to be related to planet and star formation. Its extent has been mapped from molecular line emissions from molecules such as CO and NH_3. Two critical observations are that the gas component forms an elongated structure, and that the gas is rotating around the star in keplerian motion as predicted [79].

The disk around HL Tauri has a deep 3.1-μm absorption feature attributable to water ice, and one at 10 μm consistent with the presence of silicates. The disk radius is of the order of 2000 AU. The mass of the disk is perhaps about 0.1 solar masses, and "it seems reasonable to claim that in HL Tauri we may be witnessing the beginnings of a solar system like our own" [80].

2.8.3. Disks Around Mature Stars

Possible protoplanetary disks were discovered around Alpha Lyrae (Vega), Beta Pictoris, Epsilon Eridani, and Alpha Pisces (Formalhaut). The typical size of the infrared emitting disk is 200 AU. Another 20 stars are possible candidates. Thus, it appears that dusty disks may be very common features around main-sequence stars [81]. The emission is mainly from 10-30-μm dust grains. Table 2.8.1 [82] gives basic information on the circumstellar dust shells around the first four stars that were detected. The dust shell around Beta Pictoris has been photographed (Fig. 2.8.2). The visible disk, apparently seen edge-on, extends out from an inner clear zone of about 20-30-AU radius to about 400 AU, and is about 50 AU thick at 300 AU. Thus, it is likely that the dust shell does not extend all the way into the star. Possibly the inner 20-30-AU gap, corresponding to the orbits of Uranus and Neptune in our own system, is kept clear by large planetary bodies [83].

The Beta Pictoris disk has a mass of approximately 200 Earth masses, and could be an analogue for the inner Oort comet cloud. It is apparently composed of silicate grains, perhaps coated with water ice, for the temperatures (Table 2.9.1) are low enough for this phase to condense. Figure 2.8.2 provides some direct evidence for the existence of circumstellar disks around stars. It also substantiates the Cameron model of an extended accretion disk [84,85]. Since the dust particle sizes are mostly in the micrometer range, the Poynting-Robertson effect can be expected to deplete them. Thus, there is probably a population of larger-sized bodies to resupply the fine grains [86].

TABLE 2.8.1. Circumstellar dust shells detected by IRAS.

Star	α Lyr	α PsA	β Pic	ϵ Eri
Stellar type	A0	A3	A5	K2
Star mass ($M_{\odot}$)	2.0	1.75	1.5	0.8
Luminosity ($L_{\odot}$)	58	13	6.5	0.37
Distance (pc)	8.1	7.0	16.6	3.3
Shell temperature (K)	85	55	100	45
Shell radius (AU)	85	94	20	23
Orbital period at shell (yr)	554	689	73	123

Data from Weissman P. R. (1986) in *The Galaxy and the Solar System* (R. Smoluchowski et al., eds.), p. 234, Table 6, Univ. of Arizona, Tucson.

Despite these interesting discoveries, it is not clear how relevant the dusty disks around Beta Pictoris and Alpha Lyrae (Vega) are to conditions in the early nebula, since both stars are relatively old. Vega is 3×10^8 years and Pictoris 10^9 years old. What is being observed at Vega is a remote dust cloud, without gas, that extends out to 85 AU. This does not seem to be evidence for planetary formation as is sometimes claimed.

The search for other planetary systems is of course of considerable antiquity, but has been stimulated by these discoveries. One candidate, where a possible giant planet may be in close orbit, about equivalent to that of Mercury, is the star HD114762 in the constellation of Coma Berenices [87]. The star has one-tenth of solar abundance of heavy metals and is accordingly 2-3 times older than the Sun. The minimal mass of the potential companion is at least 10 times that of Jupiter, but is probably much larger. It seems too massive to form as a planet so close to the primary star, so that it is a real candidate for the long-sought, but rarely confirmed, class of brown

Fig. 2.8.2. *The accretion disk around Beta Pictoris, which extends out to about 400 AU. The disk is about 50 AU thick at 300 AU from the star. Courtesy B. A. Smith, Univ. of Arizona, Tucson.*

dwarf stars, formed by fragmentation of a molecular cloud, but with a mass too low to initiate H burning. Other possible candidates in the range of 1-9 Jupiter masses would be consistent with the observed variations in relative radial velocities of several solar-type stars [88]. These observations raise the possibility that many single stars have previously undetected companions.

2.9. NEBULAR COLLAPSE, NEBULAR LIFETIME, AND ANGULAR MOMENTUM TRANSFER

The nebula is not a fixed entity; rather, three stages of nebular evolution may be distinguished:

1. Infall of gas begins. The density and the temperature of the nebula increase. The principal effect is to increase the mass of the nebula, rather than the mass of the primitive Sun in the center.

2. The nebula is in its maximum dissipative phase. Mass flows into the forming Sun. The nebula reaches its highest temperature toward the end of this stage and the beginning of the third stage. Stars are not observed until after this stage

3. The nebula becomes isolated, and infall of material to the Sun becomes less as the gas in the nebula becomes depleted. By this time, the Sun has formed and is beginning to affect the nebula causing (among other effects) the reversal of flow in the nebula. Stars are commonly observed toward the end of this stage as the nebula clears due to T Tauri activity. FU Orionis still appears to be accreting material from its disk and so is in the early phase of the third stage. Planetary formation also begins at this stage. Depletion of H and He, at least in the inner nebula, also occurs at this stage. Table 2.9.1 gives a list of the properties of the solar nebula [89]. Figure 2.9.1 shows a diagrammatic picture of the collapse of a cloud into a rotating nebula [90].

The free-fall collapse time of an interstellar molecular cloud core is 10^5-10^6 years [91]. Direct evidence of cloud collapse has recently been obtained for M49A, a cloud about 100 light years in diameter, about 5×10^4 light years distant from the Sun. Gas from both sides of the cloud is moving toward the center. About a dozen bright massive O-type stars have formed in a ring in this cloud, indicating that at least for this class of stars, multiple star formation occurs [92].

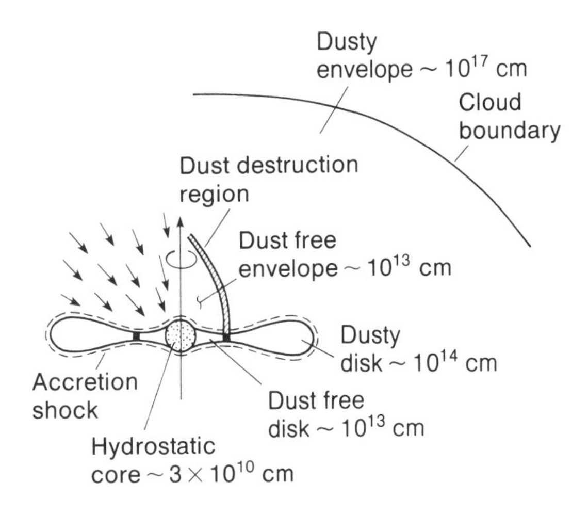

Fig. 2.9.1. *A diagrammatic representation of the collapse of a rotating cloud into a disk, with a single central condensation. After Tielens A. G. G. M. (1988) LPI Tech. Rpt. 88-04, p. 9, Fig. 2.*

TABLE 2.9.1. Properties of a low-mass solar-type nebula in the Jupiter-Saturn region.

Temperature	200 K
Orbit radius	$\sim 10^{14}$ cm
Orbit frequency	$\sim 10^{-8}$ s^{-1}
Surface density, solids	3 g cm^{-2}
Surface density, gas	200 g cm^{-2}
Scale height of gas disk	10^{13} cm
Gas density	2×10^{-11} g cm^{-3}

Data from Stevenson D. J. et al. (1986) in *Satellites* (J. A. Burns and M. S. Matthews, eds.), p. 65, Table 3, Univ. of Arizona, Tucson.

2.9.1. Nebular Lifetimes

What was the lifetime of the nebula as a rotating disk of dust and gas, from its separation from the molecular cloud until it broke up into proto-Sun and planetesimals? There is some convergence of opinion that it was about 3×10^5 years. The timescale from initial cloud collapse to proto-Sun ignition and nebular removal is about 10^5-10^6 years, on the same order as the free-fall time and similar to the lifetime of T Tauri stars [93]. The actual formation time of the nebular disk is much less than the time needed for core collapse, and probably occupies less than 10^5 years.

2.9.2. The Angular Momentum Problem

The ultimate origin of the angular momentum of the solar system remains obscure [94]. There is fairly convincing astronomical evidence that the angular momentum of binary stars is not derived from the general galactic rotation [95], but probably results from turbulence or disordered motions in clouds within the galaxy [91].

Binary systems of low-mass stars have high angular momenta, about 10^3 times that of the solar system and 10^5 times that of the Sun. Stars of about 1 solar mass have less angular momentum than more massive stars and the values for the Sun seem to be typical of other low-mass single stars, so it may be an intrinsic property of such stars. Pre-main-sequence stars of solar mass rotate slowly; young T Tauri stars rotate at about one-tenth of their breakup velocity. As the star approaches the main sequence, the radiative core shrinks, and the surface spins up [96].

If the angular momentum of the planetary system is put back into the Sun, then that for the Sun is raised to the level of more massive stars. Although this has been cited as evidence for the common occurrence of planetary systems around solar-mass stars with low angular momenta, there are objections to this "cheerful idea" [97].

The problem of the outward transfer of angular momentum has been a vexing dilemma for models attempting to explain the origin of the solar system. The Sun has 99.9% of the mass of the solar system, but contains only about 2% of the angular momentum, most of the rest residing in Jupiter. It is generally assumed that the solar nebula originated from the collapse of a dense interstellar cloud, rotating rather slowly. The rate of rotation needs to be such that the cloud collapses to a disk in order to explain the present solar system. The cloud must thus possess sufficient specific angular momentum ($J/M = 10^{19}$-10^{20} cm^2/s) to prevent collapse, forming a central Sun rather than a disk. At some stage during the formation of the Sun and planets, angular momentum is transferred outward. This is the rock on which most theories for the formation of the solar system have foundered. This problem caused the eclipse of the laplacian models and led to the rise of tidal theories, which in turn succumbed. It was long ago pointed out that the angular momentum of the passing star in such tidal hypotheses was insufficient to account for that of the solar system [98].

Just as the failure to account for the high angular momentum of the Earth-Moon system led to the collapse of all theories for lunar origin except the giant impact hypothesis, which was specifically tailored to cope with the problem, so theories for the origin of the solar system have, in general, failed to deal with this fundamental question. It is not clear whether we have a complete answer yet. Mechanisms are needed to transport material inward and angular momentum outward. Protostellar collapse begins when magnetic fields cease to be dynamically significant. Hence, they cannot be expected to transport angular momentum during the later collapse phase (see below). Outward transport must occur within the nebula. The viscous accretion disk concept was introduced to overcome the angular momentum problem, but has run into difficulties itself.

Formation of a disk presents its own special problems; mass must be transported inward to form the proto-Sun, while angular momentum must be transferred outward. Many workers have addressed this problem and a wide variety of physical processes have been suggested. Part of the problem appears to be an artifact of the models, since most workers envisage an initially symmetrical disk (as in Fig. 2.9.1).

Most naturally formed disks, of which the spiral nebulae are the most beautiful examples, are, however, nonaxisymmetric. In such disks, differential rotation can simultaneously transfer angular momentum outward and mass inward through gravitational torques. Even rather minor differences in radial symmetry can give rise to gravitational torques that can clear the nebula on timescales of 10^5-10^6 years, in accord with astronomical requirements [99]. Such a model obviates the requirements for turbulent viscosity, frequently appealed to as a physical mechanism for outward transfer of angular momentum [100].

Magnetic fields may become important if the proto-Sun or nebula generated a field through dynamo action [101]. A principal piece of evidence for the existence of such fields is that carbonaceous chondrites have been magnetized in strong magnetic fields with intensities in the range 0.1-10 oersted. Since the fields may have declined before the separation of protostellar disks occurs, they are unlikely to be a significant mechanism in transferring angular momentum outward [102,103]. An advantage of the gravitational torque mechanism is that large-scale mixing is avoided. Accordingly, any presolar heterogeneities can be preserved.

2.10. STAR FORMATION AND EVOLUTION

At this stage in our discussion of nebular evolution, it is appropriate to consider the problems of the formation of stars. The formation and composition of the Sun is addressed in the following sections. After this discussion, we return to consider the next stages in the evolution of the nebula, in which the role of the central Sun now plays a dominant role.

The formation of stars is one of the basic problems in astrophysics. Among unsolved problems dealing with the birth of stars is the question of size. The upper mass limit is very uncertain, since some supermassive stars appear to be, in fact, unresolved stellar clusters rather than single stars [104]. What factors limit the growth of stars apart from the mass of the initial fragment that broke away from the molecular cloud? Do nebulae collapse to a sphere or a disk?

High-mass stars (OB) originate in giant molecular cloud complexes in the spiral arms of galaxies. These clouds are warm (>20 K), whereas the cold clouds (<10 K) in which solar mass stars may form occur throughout the galactic disk. It is curious that star formation leads to so many objects of the appropriate mass to burn H and begin to form elements heavier than He. Most stars are between 0.1 and 2 solar masses. Such simple conundrums (why are most stars about the same size?), like other similar questions (why is the sky dark at night, or why are chondrules all about the same size?), often conceal fundamental truths. Are stars limited in size by chance or is there some process, such as outflowing stellar winds induced by the onset of thermonuclear reactions, that limits the mass of the stars [105]?

It is not yet clear how the molecular cloud fragments into clumps of the appropriate mass to form stars. However, once the cloud core forms, it collapses under gravity within 10^5 years or so to form a star. The size of the star is controlled by the initial size of the clump that became detached from the molecular cloud. The smallest object likely to form from the collapse of a fragment of a molecular cloud is about 0.01 solar masses [106].

There is a clear association of young stars with the spiral arms of galaxies, but controversy continues over the significance of this observation and whether the spectacular spiral arms are the result of star formation or induce the birth of stars. The spiral arms are generally considered to result from the passage of density waves, which are traveling waves orbiting the galaxy. It is possible that the passage of the density waves concentrates molecular clouds into the dust lanes in the spiral arms and sets the stage for star formation. In this model, the passage of the density wave triggers star formation in the molecular cloud [107]. The alternative view is that supernova shock waves compress molecular clouds (OB star formation may require external triggers), precipitating star formation.

The next stage occurs when slowly rotating cloud cores form, typically with dimensions of about 0.1 parsec and about 1 solar mass. Then the cloud core collapses and forms a central, rapidly rotating protostar surrounded by a disk, perhaps 30 AU in radius. This is deeply embedded within an infalling envelope of gas and dust (Fig. 2.10.1) [108].

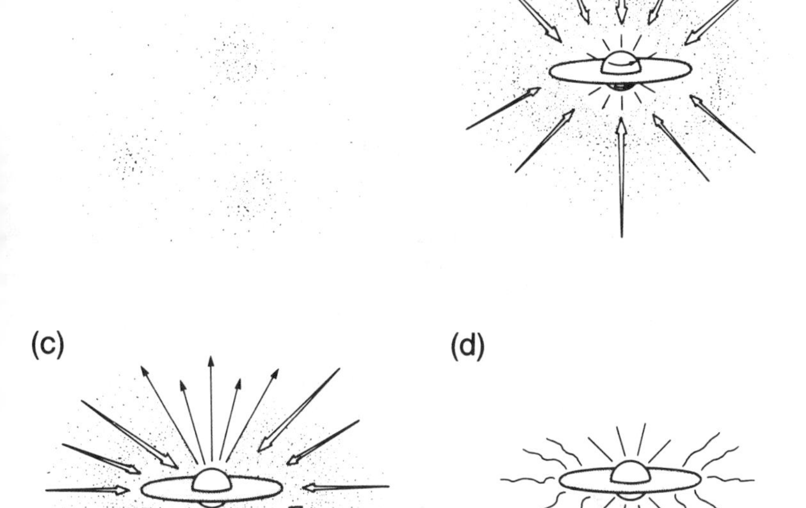

Fig. 2.10.1. *Formation stages for single stars: **(a)** core formation within molecular clouds; **(b)** a disk forms with a central condensation; **(c)** bipolar outflows begin along the rotational axis of the system; **(d)** at the end of the infall stage, the central star becomes visible, surrounded by a circumstellar disk. After Tielens A. G. G. M. (1988) LPI Tech. Rpt. 88-04, p. 6, Fig. 1.*

A shock front separates the accreting material from the disk and central core. Radiation from the central star is absorbed by the dusty envelope during the infall stage, and is reradiated in the infrared (30-100 μm). The IRAS survey found evidence of many such objects in dark cloud cores [109]. The time-scales involved are of the order of 10^5 years. The overall efficiency of star formation appears to be about 2% [110]. The total production rate within the galaxy is 3-5 solar masses per year. The rate of return of material to the interstellar medium is about 1-2 solar masses per year. This rate would deplete the galaxy of gas in about 10^9 years, which is a very short timescale. Since star formation is still proceeding in our galaxy, which is at least an order of magnitude older, perhaps it is accreting new material or undergoing bursts of star formation.

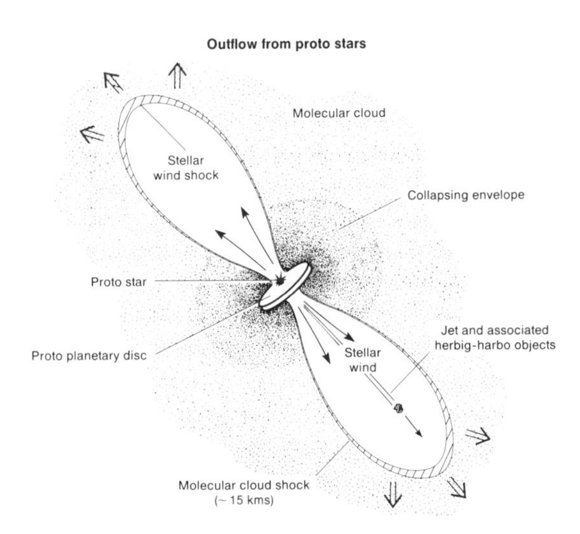

Fig. 2.10.2. *The outflow from protostars mostly occurs along the rotational axis. After Tielens A. G. G. M. (1988) LPI Tech. Rpt. 88-04, p. 11, Fig. 4.*

2.10.1. Pre-Main Sequence Evolution

As the mass of the protostar increases due to infalling material, ignition of D occurs as the stellar mass reaches about 0.3 solar masses. By the time the protostar has grown to 0.5 solar masses, it becomes convective, and thus well mixed. Protostars may lose large amounts of mass (0.2 solar mass in extreme cases) in very short timescales (10^5 years) through the mechanism of strong stellar winds. Mass loss rates of 3×10^{-6} solar masses per year are typical. The central star becomes visible, surrounded by a nebular disk. The surface temperature of the protostar is about 4000 K [111].

The general absence of effects on elemental abundances attributable to ionization and radio observations of lack of ionization indicate that these outflowing stellar winds are mainly neutral. Evidence for the existence of neutral winds as the main driving source for bipolar outflows from low-mass protostars has recently been obtained. The mechanisms involved seem to be connected with centrifugal forces "from the surface of a magnetically active and rapidly rotating protostar" [112].

In the next stage, strong bipolar outflows of gas and jets of material (Herbig-Haro objects) break out, mainly along the rotational axis of the star (Fig. 2.10.2). These bipolar-mass outflows of cold gas and Herbig-Haro objects are characteristic of the evolution of young stars. Infalling gas does not fall directly onto the rotating protostellar objects, but forms a disk around it, perpendicular to the axis of rotation. Theoretical studies of disk formation indicate that both central condensations and a trailing spiral structure are produced [113].

The importance of such spiral (nonaxisymmetric) structures is that gravitational torques can transfer angular momentum outward and mass inward to the growing star [114]. The outflowing material may also transfer angular momentum via magnetic activity, accounting for the slow rotation rates of T Tauri stars. Thus, material can be transferred into the star and angular momentum transferred outward to the disk by these processes. Eventually, a central star is left, surrounded by a disk.

More and more of the infalling material accretes to the disk rather than to the star. Radiation from the disk adds excess infrared to the energy spectrum. The infrared spectrum at this stage should show a broad emission hump (60-100 μm) with absorption at 10 μm from silicate grains and at 3.1 μm from water ice. Finally, the disk disappears. A planetary system may form or the disk may be swept away by massive stellar winds.

What terminates the infall stage? Probably a variety of mechanisms operate, but the onset of strong stellar winds is a major factor. As the parent cloud becomes depleted, stellar winds reverse the decreasing flow of material into the star. The onset of stellar winds is probably determined by D burning. This drives convection, which, together with rotation, generates dynamo action and so magnetic fields can arise. These may power the stellar winds, which reverse infall of matter from the disk. The mass of stars is thus limited and they cannot grow indefinitely.

The dimensions of the growing Sun are always much less than those of the nebula, so that most of the mass of the growing Sun comes in through the nebula. How long does this period of gas infall to the Sun last? The astrophysical evidence shows that stars of the mass of the Sun first appear with about 4 solar radii, and are still contracting and accreting gas from their accretion disk. FU Orionis is a good example, and it is estimated that the accretion times are of the order of 10^4 years [115]. T Tauri and FU Orionis stars represent the last stages of pre-main sequence stellar evolution. There is general agreement that such stellar outbursts are a feature of early stellar evolution [116].

2.10.2. Binary and Multiple Stars

Single stars are rare and most stars are binary or form members of triplets. "Most normal main-sequence stars are members of both short-period and long-period systems" [117]. Possibly over 80% of all stars belong to binary or multiple systems [118]. The major interest here in multiple-star systems is to examine whether there is any connection with planetary formation. Several modes of formation have been proposed. In a manner somewhat analogous to theories of lunar formation, these classical explanations may be divided into capture, separate nuclei, and fission. All suffer from various defects [119].

There are many dynamical objections to capture, and other properties such as the similar masses of many binaries remain unexplained by this model. The formation of stars from separate nuclei, close enough to be bound gravitationally, scarcely constitutes an explanation. Fission, which requires a rapidly rotating system to break in two, suffers from severe dynamical problems [120]. Although it was widely argued that binaries with short periods formed by fission, while those with longer periods were captured, the classical fission hypotheses have not survived. Binary-star systems thus do not appear to form from disks, but from direct collapse of rotating molecular cloud fragments.

The orbits of binary stars become more ellipitical with increasing distance between the stellar pair. If the smaller of the pair formed in an accretion disk, a circular orbit should result. This constitutes evidence for formation of both objects by direct collapse from the molecular cloud fragment rather than from an accretion disk. The formation of an accretion disk instead of a binary-star system must turn on the amount of angular momentum inherited by the the cloud fragment. This must be essentially random, leading to the sober conclusion that our presence in the solar system was an accidental consequence of the amount of angular momentum possessed by the initial cloud fragment that became the solar nebula.

Binary- and multiple-star systems form during the fragmentation of collapsing molecular clouds [121]. What makes a binary system form rather than a single star and a planetary system? The answer seems to be connected with both the angular momentum of the fragment of the molecular cloud and, perhaps more importantly, the degree of central condensation of the cloud core. Those systems that are highly centrally condensed are less likely to form binary systems [122]. It is on such matters that the ultimate existence of the solar system, and of ourselves, turns.

2.10.3. Brown Dwarfs

Brown dwarf stars seem to be scarce, possibly because they fall below the limit of star formation by fragmentation. They are expected to be about the size of Jupiter, but at least 80 times more massive, with surface temperatures of a few thousand Kelvin. A lower limit of about 0.08 solar masses (80 Jupiter masses) is set by the threshold for H burning, just sufficient for hydrogen slowly burning for about 10^9 years to reach surface temperatures of perhaps 3000 K. Since low-mass stars are common, one might expect many such small brown dwarfs to occur by simple extrapolation. However, this does not seem to be the case and they remain very elusive, none being detected in the IRAS survey.

One possible candidate is a companion to the white dwarf Giclas 29-38 [123]. The object has a temperature of 1200 K, a mass between 0.04 and 0.08 solar masses, and its luminosity is 5×10^{-5} times solar luminosity, but the identification is uncertain. Another very cool (2100 K) candidate is associated with the white dwarf GD165 [123]. Perhaps another is associated with the star HD 114762, where a possible giant planet or brown dwarf may be in close orbit [124].

This scarcity or near absence of brown dwarfs seems to indicate that there is a real gap between planets and stars, indicative of fundamental differences in mode of formation. Stars form by fragmentation and condensation of cloud masses; planets form by accretion of material in dissipating circumstellar disks. The apparent observational gap between about 10 Jupiter masses, the possible upper limit for planets, and 0.08 solar masses (80 Jupiter masses), the lower limit for hydrogen burning, may thus represent a real discontinuity [125].

Since binary- and multiple-star systems are so common, and since the giant planets mostly have regular satellite systems, most stars may have other bodies in orbit. Whether they would resemble our present solar system is unlikely in view of the many random events involved in the formation of our present system. Thus, both brown dwarfs and super-Jupiters may be equally scarce.

2.11. EARLY VIOLENT STELLAR ACTIVITY: T TAURI AND FU ORIONIS STARS

Young stars in the same mass range as the Sun lose large amounts of gas as they evolve toward the main sequence [126]. These include major bipolar outflows of cold gas, as well as the emission of Herbig-Haro objects and visible jets. Enormous energies are involved, in the range 10^{43}-10^{47} ergs.

2.11.1. T Tauri Stars

T Tauri stars have masses similar to that of the Sun, and possess spectral evidence of a steady mass loss and stellar winds. The photospheric temperatures of these very active objects are cool (about 4000 K), the mass ranges are 0.2-3 solar masses, and the "contraction ages" are from 2×10^5 to 2×10^7 years. T Tauri stars possess a high abundance of Li and so must be young. Many are surrounded by dusty disks that apparently contribute most of the infrared excess observed in their spectra. The infrared excesses in passive disks are caused by absorption of photons from the star onto dust grains, which then reradiate at longer wavelengths. The mass loss from T Tauri stars is of the order of 10^{-8} solar masses per year. FU Orionis outbursts are much more violent, with mass losses 100-1000 times those of T Tauri, so that the loss rates may go as high as 10^{-5} solar masses per year [127]. Herbig-Haro objects (see below) are associated with T Tauri stars (Fig. 2.10.2). Bipolar gas outflows are also observed from central, but optically invisible, embedded infrared objects. All these phenonoma are probably linked.

While classical T Tauri stars have circumstellar disks of dust and gas, "naked T Tauri stars" have much less circumstellar disk material and exhibit very high levels of surface activity, including intense magnetic activity. They apparently represent stellar evolution at ages of 10^5-10^7 years. If the Sun went through this stage, it covers a crucial period in the development of the solar system.

2.11.2. FU Orionis Stars

FU Orionis stars are T Tauri stars that appear to go through a violent eruptive phase, possibly ejecting a dust ring. V1057 Cygnis is an example of a T Tauri star that has undergone an FU Orionis outburst, increasing in luminosity by about 6 magnitudes in 1936-37. The radiation field of V1057 Cygnis would raise the black-sphere temperature to above 1000 K within a radius of about 4 AU [128]. FU Orionis stars are not rare. At least five examples have been observed in the past 50 years, following the initial observation of FU Orionis. Probably such behavior, with multiple FU Orionis-type flares, may be common during the early stages of stellar evolution. The violent activity appears to be due to rapid accretion of material from a protostellar disk onto the growing star. The times associated with such activity are of the order of 10^4 years [129]. It has been suggested that infall of Jupiter-sized objects might be responsible for the violent outbursts typical of FU Orionis activity. This might be adduced as circumstantial evidence for the early formation of such bodies [130].

Associated with such activity, as also in the case of the T Tauri stars, are the Herbig-Haro objects, which have very high proper motions reaching 100-400 km/s. They appear to be gaseous objects shocked by very high velocity stellar winds, and their presence implies very high velocity energetic outflows early in stellar evolution. How they came to be accelerated to such high velocities is unclear. If they are accelerated by stellar winds, then the wind must have velocities in the range 100-400 km/s.

2.11.3. Bipolar Outflows

The other extraordinary occurrence is the high-velocity bipolar flows of cold molecular gas. They are massive (1-100 solar masses), cold (10-90 K), and of rather low density (1000 H_2/cm^3). These bipolar outflows are very energetic (10^{43}-10^{47} erg) and are associated with Herbig-Haro objects. The gas outflows are moderately collimated. Inflow and outflow appear to be taking place simultaneously. Bipolar ejection is probably associated with processes internal to the growing star. Although spectacular, it is likely to have only a minor effect on nebular evolution, and so is only of marginal interest to this inquiry [131]. Such energetic flows are common and appear to be a characteristic of early stellar evolution. Over 50 sources are known within 1 kparsec of the solar system. "The molecular outflow phase may be the earliest observationally identifiable stage of stellar evolution . . . associated with the first 10^5 years of stellar life" [132].

The outflows may provide an important source of energy to giant molecular clouds. It is generally inferred that the early Sun went through similar violent episodes, although definitive evidence has still not been obtained from meteorite studies. If, in fact, the outflows occur in the first 10^5 years of the Sun's life, such activity might well occur prior to the formation of the meteorites. Primitive meteorites might be expected to show, for example, high contents of ^{21}Ne produced by spallation reactions induced by energetic flare particles. Some grains with solar flare tracks do indeed contain excess spallogenic ^{21}Ne, notably in the carbonaceous chondrite, Murchison. These levels would require exposure to normal galactic cosmic rays in excess of 100 m.y. Since most other evidence suggests very short precompaction ages, this indicates that the grains underwent an early exposure to energetic particles [133]. Most likely these effects were due to an early active Sun, and record events on timescales of 10^5-10^6 years.

2.12. THE FORMATION AND COMPOSITION OF THE SUN

The formation of the Sun must have resembled that of the low-mass stars discussed above. It must have predated that of the terrestrial planets, since these formed in an environment from which the H and He gas, the main solar constituents, had been swept away. The most likely mechanism for this is strong stellar winds, blowing during the T Tauri and FU Orionis stages of stellar evolution. The giant planets contain higher than solar abundances of the heavy elements and thus are not primordial fragments of the nebula; their formation must postdate that of the Sun and reflect a later stage of nebular evolution. However, Jupiter at least predates the formation of the inner planets by a time sufficient to affect the growth of the latter, and to inhibit the formation of a planet in the asteroid belt.

2.12.1. Composition

It is only comparatively recently that it has been realized that the Sun and other stars are composed dominantly of H and He. Indeed, the first worker who suggested this in 1925 suspected some unrecognized error in the spectral line analysis [134]. The high abundance of H and He in the Sun and in other stars was confirmed shortly thereafter, but it is interesting that such a fundamental fact about the universe is such a recent discovery [135].

The composition of the solar photosphere is given in Table 2.12.1 [136]. The solar data remain the source for the solar nebular abundances of H, C, N, and O, which are depleted in meteorites. Helium is difficult to determine, and the best estimate for the He mass fraction [Y] is 0.275 with a range 0.26-0.29.

The iron data are of particular interest on account of the long history of apparent discrepancies between the solar and meteoritic data. This problem was first noted in the 1960s. Although ingenious solutions to the problem of fractionating Fe but not the other siderophile elements [137] were proposed, the Fe values in the Sun were eventually shown to be in error. Nevertheless, some recent determinations [138] showed Fe values in the Sun to be consistently higher by about 40% than the CI values (7.67 relative to $\log N_H = 12.00$, compared to the well-established meteoritic value of 7.51 on the same scale). The companion siderophile elements Ni and Co, however, are similar in both sets [138]. This posed a paradox, requiring some way of separating Fe from Ni and Co. In addition to the solar photospheric data, values are obtainable from the corona and the solar wind or solar energetic particles (Table 2.12.2).

The estimation of solar abundances of Fe is problematic, and depends heavily both on models of the solar photosphere and on models of the formation of spectral lines. A recent reassessment of the solar spectral abundances by Holweger and coworkers [140] in which the Fe II lines (which are in local thermal equilibrium) are used, confirms the equivalence of the solar and meteoritic Fe abundances.

The position is adopted in this book that the best estimates of the solar Fe values agree with CI Fe data and that there is no case to justify moving the overall solar system Fe abundance from the meteoritic value, in agreement with the opinion of Grevesse and Anders [141]. This problem is further discussed in section 2.15.

2.12.2. Solar Corona

The abundance data for the solar corona reveal that many elements are fractionated with respect to the photospheric data. These data give us the composition of the outer layers of the Sun, and show some interesting digressions from the photospheric values, taken as representative of the bulk solar (and nebular) values. This comparison is shown in Fig. 2.12.1.

TABLE 2.12.1. Element abundances in the solar photosphere (log $N_H = 12.00$).

Element	Photosphere*	Meteorites†	Phot.-Met.*	Element	Photosphere*	Meteorites†	Phot.-Met.*
1 H	12.00	[12.00]	—	44 Ru	1.84	1.82	+0.02
2 He	[10.99]	[10.99]	—	45 Rh	1.12	1.09	+0.03
3 Li	1.16	3.31	-2.15	46 Pd	1.69	1.70	-0.01
4 Be	1.15	1.42	-0.27	47 Ag	(0.94)	1.24	(-0.30)
5 B	(2.6)	2.88	(-0.28)	48 Cd	1.86	1.76	+0.10
6 C	8.56	[8.56]	—	49 In	(1.66)	0.82	(+0.84)
7 N	8.05	[8.05]	—	50 Sn	2.0	2.14	-0.14
8 O	8.93	[8.93]	—	51 Sb	1.0	1.04	-0.04
9 F	4.56	4.48	+0.08	52 Te	—	2.24	—
10 Ne	[8.09]	[8.09]	—	53 I	—	1.51	—
11 Na	6.33	6.31	+0.02	54 Xe	—	2.23	—
12 Mg	7.58	7.58	0.00	55 Cs	—	1.12	—
13 Al	6.47	6.48	-0.01	56 Ba	2.13	2.21	-0.08
14 Si	7.55	7.55	0.00	57 La	1.22	1.20	+0.02
15 P	5.45	5.57	-0.12	58 Ce	1.55	1.61	-0.06
16 S	7.21	7.27	-0.06	59 Pr	0.71	0.78	-0.07
17 Cl	5.5	5.27	+0.23	60 Nd	1.50	1.47	+0.03
18 Ar	[6.56]	[6.56]	—	62 Sm	1.00	0.97	-0.03
19 K	5.12	5.13	-0.01	63 Eu	0.51	0.54	-0.03
20 Ca	6.36	6.34	+0.02	64 Gd	1.12	1.07	+0.05
21 Sc	3.10	3.09	+0.01	65 Tb	(-0.1)	0.33	(-0.43)
22 Ti	4.99	4.93	+0.06	66 Dy	1.1	1.15	-0.05
23 V	4.00	4.02	-0.02	67 Ho	(0.26)	0.50	(-0.24)
24 Cr	5.67	5.68	-0.01	68 Er	0.93	0.95	-0.02
25 Mn	5.39	5.53	-0.14	69 Tm	(0.00)	0.13	(-0.13)
26 Fe	7.51	7.51	0.00	70 Yb	1.08	0.95	+0.13
27 Co	4.92	4.91	+0.01	71 Lu	(0.76)	0.12	(+0.64)
28 Ni	6.25	6.25	0.00	72 Hf	0.88	0.73	+0.15
29 Cu	4.21	4.27	-0.06	73 Ta	—	0.13	—
30 Zn	4.60	4.65	-0.05	74 W	(1.11)	0.68	(+0.43)
31 Ga	2.88	3.13	-0.25	75 Re	—	0.27	—
32 Ge	3.41	3.63	-0.22	76 Os	1.45	1.38	+0.07
33 As	—	2.37	—	77 Ir	1.35	1.37	-0.02
34 Se	—	3.35	—	78 Pt	1.8	1.68	+0.12
35 Br	—	2.63	—	79 Au	(1.01)	0.83	(+0.18)
36 Kr	—	3.23	—	80 Hg	—	1.09	—
37 Rb	2.60	2.40	+0.20	81 Tl	(0.9)	0.82	(+0.08)
38 Sr	2.90	2.93	-0.03	82 Pb	1.85	2.05	-0.20
39 Y	2.24	2.22	+0.02	83 Bi	—	0.71	—
40 Zr	2.60	2.61	-0.01	90 Th	0.12	0.08	+0.04
41 Nb	1.42	1.40	+0.02	92 U	(<-0.47)	-0.49	—
42 Mo	1.92	1.96	-0.04				

*Values in parentheses are uncertain.
† Values in brackets are based on solar or other astronomical data.

Data from Anders E. and Grevesse N. (1989) *Geochim. Cosmochim. Acta, 53,* 199, Table 2, except for Fe values, which are from Holweger H. (1990) *Astron. Astrophys., 232,* 510.

TABLE 2.12.2. Element abundances in the solar corona.

Element	Corona	Photosphere	Corona-Phot.
1 H	—	=12.00	—
2 He	10.14	10.99	-0.85
6 C	7.90	8.56	-0.66
7 N	7.40	8.05	-0.65
8 O	8.30	8.93	-0.63
9 F	(4.00)	4.56	(-0.56)
10 Ne	7.46	8.07	-0.61
11 Na	6.38	6.33	+0.05
12 Mg	7.59	7.58	+0.01
13 Al	6.47	6.47	0.00
14 Si	=7.55	7.55	0.00
15 P	5.24	5.45	-0.21
16 S	6.93	7.21	-0.28
17 Cl	4.93	5.5	-0.57
18 Ar	5.89	6.58	-0.69
19 K	5.14	5.12	+0.02
20 Ca	6.46	6.36	+0.10
21 Sc	(4.04)	3.10	(+0.96)
22 Ti	5.24	4.99	+0.25
23 V	(4.23)	4.00	(+0.23)
24 Cr	5.81	5.67	+0.14
25 Mn	5.38	5.39	-0.01
26 Fe	7.65	7.67	-0.02
28 Ni	6.22	6.25	-0.03
29 Cu	(4.31)	4.21	(+0.10)
30 Zn	4.76	4.60	+0.16

Data from Anders E. and Grevesse N. (1989) *Geochim. Cosmochim. Acta, 53,* 203, Table 4.

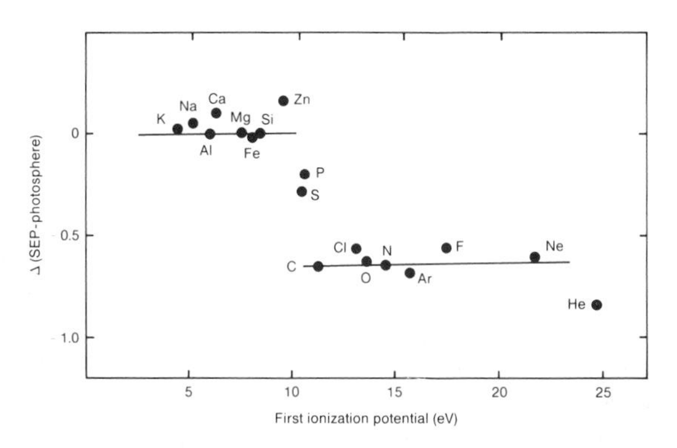

Fig. 2.12.1. *The abundances of solar energetic particles (SEP) compared with those in the solar photosphere, showing the strong depletion of elements with first ionization potentials above 10 eV. After Anders E. and Grevesse N. (1989) Geochim. Cosmochim. Acta, 53, 205, Fig. 3.*

The principal difference between the photospheric and coronal values is that elements with high first ionization potentials are depleted in the corona. The more easily ionized elements are present at the abundance levels seen in the solar photosphere. This fractionation of elements in the corona appears to take place in the lower chromosphere where the temperature range is 5800-6800 K [142].

This is one of the few examples in the solar system where element distribution depends on ionization potentials and where element abundances are controlled by high-temperature (several thousand Kelvin) plasma conditions. It is the lack of such correlations in solar system, meteoritic, or planetary abundances that has led cosmochemists to query the Alfven-Arrhenius model in which plasma conditions dominate planetary formation processes [143].

2.13. MASSIVE VS. SMALL NEBULAE

The further evolution of the nebula, in which the central Sun is shining, is now considered. Many models of planetary evolution begin by considering the nebula at this stage, when the Sun is a principal actor on the scene. Historically, there have been two competing models for the original size of the solar nebula following the formation of the Sun.

The large nebula models contain about one solar mass. In various scenarios, this nebula fragments into a number of giant gaseous protoplanets about the size of Jupiter. A principal reason for postulating a nebula of this size is that it removes the timescale problems of forming the giant gaseous planets during the brief lifetime of the nebula. In smaller nebulae, the accretion of Uranus and Neptune in Safronov-type models can easily exceed the age of the solar system. In the large mass models, the planets can be formed very rapidly by breaking up the nebula into fragments that condense into giant gaseous protoplanets [144].

However, the requirement for massive nebulae that will fragment due to gravitational instabilities evaporates if planets are not formed by such a mechanism. The small-mass nebulae contain only 0.01-0.04 solar masses, the lower limit being set by the present planetary masses enhanced to solar abundances by the addition of the lost H and He [145].

Such small-mass nebulae do not fragment due to self-gravity, but the dust grains separate from the gas and settle toward the midplane of the nebula (Fig. 2.13.1). Planetesimals accrete from this dust in the midplane, forming, in the case of the giant planets, 10-20 Earth-mass cores, onto which the gaseous components collapse under gravitational attraction. The planetesimal model, adopted here, effectively builds the planets brick by brick, rather than by fragmentation of the nebula into giant gaseous protoplanets. This concept effectively removes the rationale for the large nebulae.

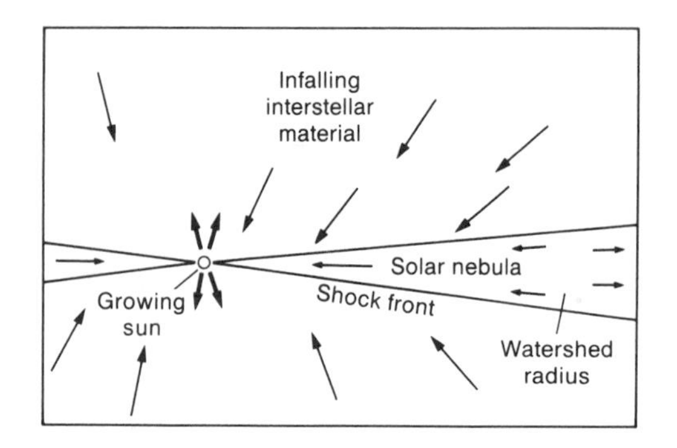

Fig. 2.13.1. *The formation of an accretionary disk by the infall of interstellar material. After Wood J. A. and Morfill G. E. (1988) in Meteorites and the Early Solar System (J. F. Kerridge and M. S. Matthews, eds.), p. 334, Fig. 6.3.1, Univ. of Arizona, Tucson.*

Somewhat larger nebulae of about 0.1 solar masses have been advocated to produce sufficient density in the regions of Jupiter and Saturn so that runaway accretion of their planetary cores is promoted before the gaseous components of the nebula are dissipated. The excess mass is removed by ejecting it from the solar system, using Jupiter and Saturn to pump up the velocities and eccentricities of the surplus planetesimals. In the course of these events, these planets must move from their original formation distances at 8 and 16 AU respectively, closer (to 5 and 10 AU) to the Sun [146]. This poses new problems, however, since the original accretion zone for Jupiter is now 3 AU past the "snow line."

The dilemma can be resolved by returning to a smaller mass nebula, and building up the nebula density at 5 AU by sweeping out material from the inner nebula during the early violent T Tauri and FU Orionis stages. In summary, excluding the Sun, a small nebula mass of about 0.04 solar masses seems adequate with which to proceed to construct the solar system.

2.14. NEBULAR STRUCTURE AND TEMPERATURE

Most models of the nebula show a symmetrical form, with little activity outside that of the early Sun in the center. Such concepts are probably far from reality. If the early nebula were not axisymmetric, but had a spiral structure, and was turbulent, a number of otherwise puzzling features of early solar system history become clearer. Thus, gravitational torques in a nonsymmetrical nebula can transport angular momentum outward, as mass flows inward to the growing Sun. These merely require that the nebula is significantly nonsymmetric. In such systems, angular momentum transport can occur on timescales of 10^4 years or less [147], while dissipation of energy in a turbulent nebula can provide for high energy heating events to produce refractory inclusions and chondrules.

This question of nebular temperatures has caused a considerable amount of controversy, much of it misdirected. The basic problem is that the nebula is not static, but evolves with time. The nebula was not a cold, inert disk, but a dynamic, turbulent, evolving system with much energy to dissipate. It is therefore necessary when discussing the problem of nebular temperatures to discriminate between the various stages of nebular development and also to allow for variations both in the horizontal and vertical planes. When all these factors are taken into account, much of the controversy over whether the nebula was hot or cold disappears.

In the initial stages of nebular development, material streams in to accrete to the Sun. High gas temperatures due to compressional heating result during this hot short inflow to form the Sun. Timescales are of the order of 10^5 years. Temperatures reach 1500 K at distances of 2.5 AU (i.e., in the region subsequently occupied by the asteroid belt) while nebular infall is occurring. This compressional heating falls off sharply with height so that the temperature at 0.1 AU above the midplane is only about 500 K. Temperatures become significantly lower after 10^5 years as compressional heating decreases, so that the nebula is cool during planetesimal growth [148].

At this point, no solid condensed material is likely to be present within 2-3 AU of the growing Sun. Material that accretes early and is heated to high temperatures will mostly finish up in the Sun. It is possible that some of the volatile species do not condense in the inner nebula because the temperatures are too high and are accreted to the Sun in a gas phase. When the flow reverses and the winds from the T Tauri and FU Orionis stages sweep out the inner nebula, the temperature falls. It is interesting to speculate on the cause of the depletion of volatile elements, which is such a distinctive feature of the terrestrial planets and some of the meteorites. Either the volatile elements were present in the inner nebula and were swept away with the gas, or else temperatures during the solar infall stage were too high for them to condense, so that they were accreted to the Sun along with the infalling gas. In either event, those elements with condensation temperatures below 1100 K are depleted in the inner nebula.

The rapid cooling times experienced by the chondrules (section 3.5) demand a thin, cold, and uninsulated place in the nebula. Probably they formed well above the midplane of the nebula and so shed little light on temperatures at the midplane. At the site of formation of the chondritic meteorites, the nebula was cool, not hot, about a few hundred Kelvin at the present location of the asteroid belt. The chondrites are mixtures that have not been heated significantly since their accretion, although they bear evidence of previous high-temperature events [149].

The carbonaceous chondrites, spectrally similar to the C-type asteroids, contain hydrated silicates, which would decompose to anhydrous silicate plus H_2O at >700 K. Water ice, which is stable beyond 5 AU, condenses at 160 K at nebular pressures (10^{-3} bar). The oxygen isotope evidence and the general differences among the planets also suggest low nebular temperatures. This apparently conflicts with the petrographic evidence from meteorites for high temperatures in the asteroid belt. Is there a resolution to this paradox? The high-temperature mineralogy is not ubiquitous; the primitive meteorites are complex mixtures of low- and high-temperature phases. In contrast, the differentiated meteorites form in planetesimals or asteroids in local high-temperature environments. Transient high-temperature processes seem best fitted to resolve the dilemma. In this context, crucial evidence comes from the chondrules, which record heating, element fractionation, and cooling on very short timescales (minutes to hours). Nebulawide heating is not consistent, for example, with the quenching observed in chondrules. Thus, transient, localized, high-temperature events seem necessary to explain the meteoritic evidence. An origin for chondrules in a turbulent nebula by flash-melting of silicate clumps of dust by nebular flares well above the nebular midplane seems consistent with the evidence.

Nebular models are still in a developmental stage and no doubt will change to accommodate new observations or previously overlooked or little understood factors. For example, recent models indicate that a rather broad zone extending beyond 1 AU might reach a temperature of 1500 K [150]. Such temperatures would evaporate small grains, recondensing the material on larger grains, leading to a preferential growth of larger grains.

In summary, our ideas about temperatures in the nebula either in time or space are in a rudimentary state. It seems reasonably well estabished that water ice condensed at temperatures of 160 K at 5 AU, which was a major factor in precipitating the growth of Jupiter. The depletion of volatile elements (with condensation temperatures less than 1100 K) in the inner nebula also occurred early. The compositions of the outer planets seem more related to their relative timescales of formation, so that Uranus and Neptune grew in a nebula that had already lost much of its H and He. The overall planetary compositions do not display much evidence of a steady decrease in temperature outward from the Sun; the zonation in the asteroid belt is due to a later stage in the evolution of the nebula.

2.15. NEBULAR COMPOSITION: CI CHONDRITES AND COMETS AS SAMPLES

"There is no a priori reason that a rock falling from the sky should have the average composition of the solar system" [151].

The term "solar system abundances" refers to the composition of the solar system, including the Sun. Such abundance estimates were formerly often referred to as "cosmic," "primitive," or "primordial," but refer strictly to the composition of the solar system. They are not universal since the galaxy is being continuously enriched in heavy elements due to nucleosynthesis in stars. The abundances in the Sun are referred to as "solar," and presumably represent a typical sample of the local interstellar medium 4.6 b.y. ago. The present interstellar medium will be enriched in heavy elements over that which segregated in the solar nebula.

In the present context, the term "gaseous" is used to refer to H, C, N, O, and the noble gases. The other elements are classified as "very volatile" (e.g., Bi, Tl); "volatile" (e.g., Rb, Cs); "moderately volatile" (e.g., K, Mn); "moderately refractory" (e.g., V, Eu); "refractory" (e.g., Ca, Al, U, La), and "super-refractory" (e.g., Zr, Sc).

The question of the composition of the nebula depends on which stage of nebular evolution is meant. It is clear, particularly from the evidence of zonation in the asteroid belt and from the differing compositions of the planets, that much differentiation and segregation of elements has occurred since fragmentation from the molecular cloud. In the present context, it is customary to attempt to obtain the composition of the nebula from the comparison between the solar and meteoritic abundances. This is a very useful approach, but its limitations should be recognized.

There has been a general consensus that the composition of the original nebula is given by that of the CI class of carbonaceous chondrites. There are two rationales for this. One is that the odd-mass nuclei fall on smooth curves of abundance vs. mass number, and so retain a memory of nucleosynthetic events [152] (Fig. 2.15.1). The other is that there is quite a good match for the nongaseous elements between the composition of the CI meteorites and that of the solar photosphere. The CI abundances fall mostly within the error bars of the photospheric abundances [153] (Fig. 2.15.2). What is surprising is that there is so little evidence of element correlations among the small, but analytically significant, variations in the most precise sets of data [154]. This is perhaps due to the very complex mineralogy of the CI meteorites, coupled with the degree of alteration they have experienced.

However, the lack of correlation may represent a more primitive feature. There are two contrasting opinions about the ultimate source of the CI material. The first supposes that it results from condensation down to 300 K or so from a gas of solar composition. The second recognizes that some unaltered interstellar material is present, since isotopically anomalous noble gases are present [155].

Although CI carbonaceous chondrites are frequently employed to provide the composition of the primordial solar nebula, they are not primary, but have been subjected to aqueous alteration with the production of hydrated mineral phases. There is general agreement (section 3.6) that such hydrated minerals are not primary nebular phases. "It is paradoxical that the CI chondrites, whose compositions appear to be the most primitive, have experienced the most extensive alteration" [156].

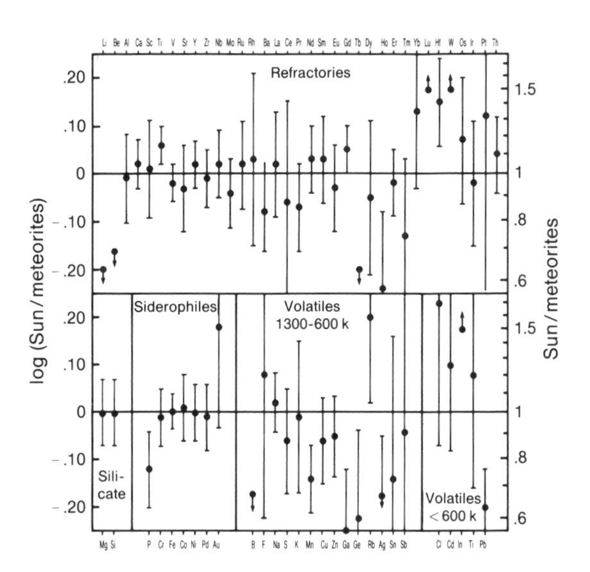

Fig. 2.15.2. *The abundances of the elements in CI carbonaceous chondrites relative to those in the solar photosphere, showing the close correspondence for most elements. After Anders E. and Grevesse N. (1989) Geochim. Cosmochim. Acta, 53, 206, Fig. 4. The Fe abundance is from Hollweger H. et al. (1990) Astron. Astrophys., 232, 510.*

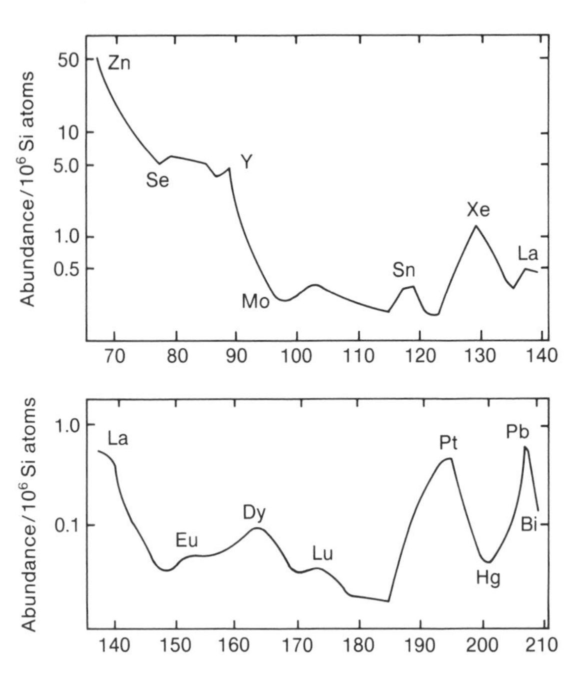

Fig. 2.15.1. *The abundances of odd-mass nuclei, relative to Si = 10^6 atoms, fall on relatively smooth curves. After Anders E. and Grevesse N. (1989) Geochim. Cosmochim. Acta, 53, 197, Figs. 5 and 6.*

The latest compilation provides both an evaluation of the CI abundance data, and also a comparison with the solar photospheric values [157]. There is some fractionation in the solar coronal values, and the photospheric values provide a truer estimate of the solar composition (Fig. 2.12.1). The CI data are given in Table 2.15.1. Although the match for most elements between the CI and solar photospheric abundances is very good (Fig. 2.15.2), there are apparent discrepancies for Mn, Ge, Pb, and W. This raises the interesting question of whether the resemblance between the abundances in the CI chondrites and the solar photosphere is fortuitous [158].

The iron abundance problem has been discussed above (section 2.12). Since iron is a key element in most geochemical models of the solar system this raises a basic and important question about its nebular abundance. There is a long history of difficulty in measuring the iron abundance in the Sun. For several years prior to 1967 [159] the calculated solar

TABLE 2.15.1 Solar system abundances of the elements (atoms/10^6 Si).

Element	Solar System	Mean CI Chondrite	Orgueil	Element	Solar System	Mean CI Chondrite	Orgueil
1 H	2.79×10^{10}	—	2.02 %	44 Ru	1.86	712 ppb	714 ppb
2 He	2.72×10^9	—	56 nL/g	45 Rh	0.344	134 ppb	134 ppb
3 Li	57.1	1.50 ppm	1.49 ppm	46 Pd	1.39	560 ppb	556 ppb
4 Be	0.73	24.9 ppb	24.9 ppb	47 Ag	0.486	199 ppb	197 ppb
5 B	21.2	870 ppb	870 ppb	48 Cd	1.61	686 ppb	680 ppb
6 C	1.01×10^7	—	3.45 %	49 In	0.184	80 ppb	77.8 ppb
7 N	3.13×10^6	—	3180 ppm	50 Sn	3.82	1720 ppb	1680 ppb
8 O	2.38×10^7	—	46.4 %	51 Sb	0.309	142 ppb	133 ppb
9 F	843	60.7 ppm	58.2 ppm	52 Te	4.81	2320 ppb	2270 ppb
10 Ne	3.44×10^6	—	203 pL/g	53 I	0.90	433 ppb	433 ppb
11 Na	5.74×10^4	5000 ppm	4900 ppm	54 Xe	4.7	—	8.6 pL/g
12 Mg	1.074×10^6	9.89 %	9.53 %	55 Cs	0.372	187 ppb	186 ppb
13 Al	8.49×10^4	8680 ppm	8690 ppm	56 Ba	4.49	2340 ppb	2340 ppb
14 Si	1.00×10^6	10.64 %	10.67 %	57 La	0.4460	234.7 ppb	236 ppb
15 P	1.04×10^4	1220 ppm	1180 ppm	58 Ce	1.136	603.2 ppb	619 ppb
16 S	5.15×10^5	6.25 %	5.25 %	59 Pr	0.1669	89.1 ppb	90 ppb
17 Cl	5240	704 ppm	698 ppm	60 Nd	0.8279	452.4 ppb	463 ppb
18 Ar	1.01×10^5	—	751 pL/g	62 Sm	0.2582	147.1 ppb	144 ppb
19 K	3770	558 ppm	566 ppm	63 Eu	0.0973	56.0 ppb	54.7 ppb
20 Ca	6.11×10^4	9280 ppm	9020 ppm	64 Gd	0.3300	196.6 ppb	199 ppb
21 Sc	34.2	5.82 ppm	5.83 ppm	65 Tb	0.0603	36.3 ppb	35.3 ppb
22 Ti	2400	436 ppm	436 ppm	66 Dy	0.3942	242.7 ppb	246 ppb
23 V	293	56.5 ppm	56.2 ppm	67 Ho	0.0889	55.6 ppb	55.2 ppb
24 Cr	1.35×10^4	2660 ppm	2660 ppm	68 Er	0.2508	158.9 ppb	162 ppb
25 Mn	9550	1990 ppm	1980 ppm	69 Tm	0.0378	24.2 ppb	22 ppb
26 Fe	9.00×10^5	19.04 %	18.51 %	70 Yb	0.2479	162.5 ppb	166 ppb
27 Co	2250	502 ppm	507 ppm	71 Lu	0.0367	24.3 ppb	24.5 ppb
28 Ni	4.93×10^4	1.10 %	1.10 %	72 Hf	0.154	104 ppb	108 ppb
29 Cu	522	126 ppm	119 ppm	73 Ta	0.0207	14.2 ppb	14.0 ppb
30 Zn	1260	312 ppm	311 ppm	74 W	0.133	92.6 ppb	92.3 ppb
31 Ga	37.8	10.0 ppm	10.1 ppm	75 Re	0.0517	36.5 ppb	37.1 ppb
32 Ge	119	32.7 ppm	32.6 ppm	76 Os	0.675	486 ppb	483 ppb
33 As	6.56	1.86 ppm	1.85 ppm	77 Ir	0.661	481 ppb	474 ppb
34 Se	62.1	18.6 ppm	18.2 ppm	78 Pt	1.34	990 ppb	973 ppb
35 Br	11.8	3.57 ppm	3.56 ppm	79 Au	0.187	140 ppb	145 ppb
36 Kr	45	—	8.7 pL/g	80 Hg	0.34	258 ppb	258 ppb
37 Rb	7.09	2.30 pm	2.30 ppm	81 Tl	0.184	142 ppb	143 ppb
38 Sr	23.5	7.80 ppm	7.80 ppm	82 Pb	3.15	2470 ppb	2430 ppb
39 Y	4.64	1.56 ppm	1.53 ppm	83 Bi	0.144	114 ppb	111 ppb
40 Zr	11.4	3.94 ppm	3.95 ppm	90 Th	0.0335	29.4 ppb	28.6 ppb
41 Nb	0.698	246 ppb	246 ppb	92 U	0.0090	8.1 ppb	8.1 ppb
42 Mo	2.55	928 ppb	928 ppb				

Data from Anders E. and Grevesse N. (1989) *Geochim. Cosmochim. Acta, 53,* 198, Table 1.

value was only 20% of the meteoritic value, until the oscillator strength of the spectral lines was revised. Could Fe in fact be fractionated from the other siderophiles? The CI meteorites have been pervasively altered by water. Magnetite has formed as a secondary product (section 3.6) and contains only very low abundances of Ni and Co, so that selective removal of magnetite could deplete the CI meteorites in iron without affecting the abundances of the other siderophile elements. However, such scenarios are unnecessary. New determinations of the Fe abundance in the Sun match the CI data [160, 161] and this confirms the CI value [162] adopted in this book, despite the siren-like attractions of the higher Fe abundance for planetary compositions [163].

Some anomalies remain. All carbonaceous groups (CI, CM, CO, and CV) are enriched in ^{50}Ti relative to both the Earth and the other meteorites [164] (see Table 2.15.2). Does this mean that the CI meteorites sampled a special part of the nebula? Although

TABLE 2.15.2. Average Ti isotopic patterns for carbonaceous chondrites.

Class	ϵ (47/46)	ϵ (48/46)	ϵ (50/46)
CI	-0.5 ± 0.3	+0.4 ± 0.7	+1.7 ± 0.5
CM	-0.1 ± 0.4	+0.2 ± 0.6	+2.4 ± 0.5
CO	-0.1 ± 0.5	+0.5 ± 0.7	+3.0 ± 0.5
CV	-0.4 ± 0.3	+0.5 ± 0.5	+3.2 ± 0.4

Data from Niemeyer S. (1988) *Geochim. Cosmochim. Acta, 52,* 2942, Table 2.

four distinct reservoirs are required to account for the ^{50}Ti isotopic anomalies [165] the anomalies are small and are probably due to the presence of small refractory Ti-containing grains rather than representing some truly basic inhomogeneity on the nebula.

Broader questions are raised in this "holy grail"-type search for the true primitive solar nebula composition. As we shall see, other hoped-for primitive compositions in both interplanetary dust or comets, for which apparently impeccable cases can be presented, fail to provide the hoped-for solution. The inference is clear that we have reached the limits of this kind of inquiry, and other sets of questions need to be addressed concerning the chemical evolution of the nebula.

Although we can use the CI data as a general guide to solar nebula composition, elemental fractionation has affected the inner portions of the solar nebula, because the terrestrial planets are not of CI composition. It should be emphasized at this stage that it is not possible to construct the inner planets from CI meteorites, despite the clear attractions of such a move [166]. Nor is it possible to construct a smoothly varying nebular composition with heliocentric distance, the result of decreasing temperatures, once again despite the obvious advantages [167].

The inner planets formed from material that was already depleted in elements volatile at about 1100 K. The evidence is shown clearly by the relationships between volatile elements such as K and refractory elements such as U or La, which will be discussed at length later. In addition, it has proven difficult to account for the noble gas abundances, particularly for Kr and Xe, in the terrestrial planets by using CI as one of the components [168].

Could comets, which are principally discussed in section 3.10, represent pristine solar nebular or "holy

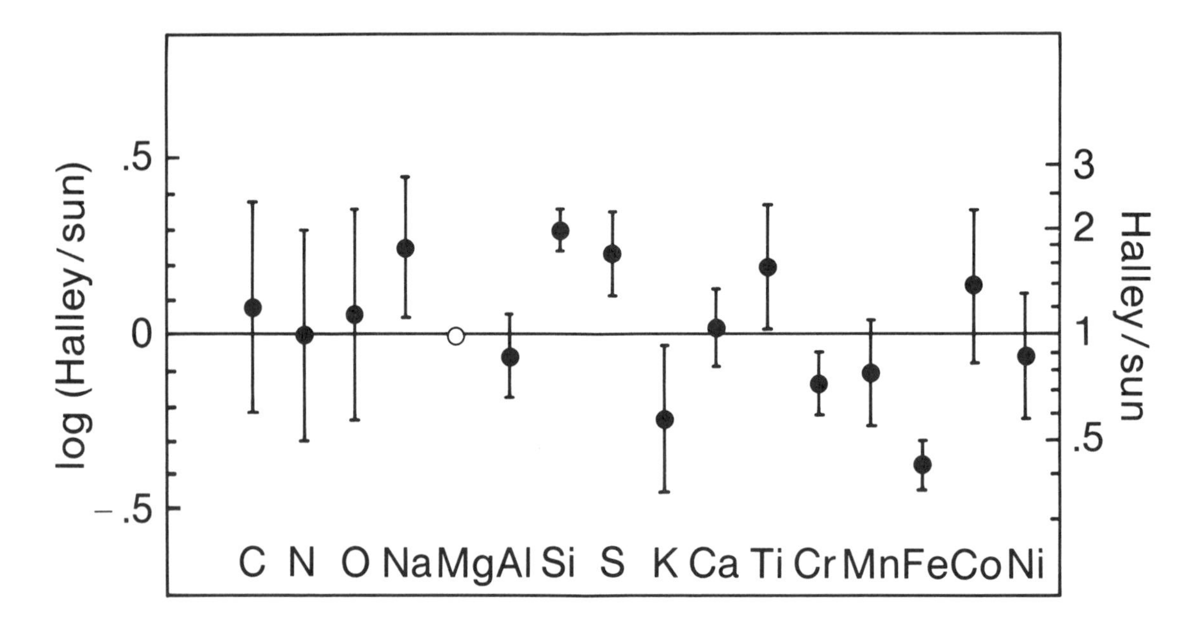

Fig. 2.15.3. *The abundances of the elements in Comet Halley relative to those in the solar photosphere, normalized to Mg, and showing the wide variations from solar abundances for Si and Fe. After Anders E. and Grevesse N. (1989) Geochim. Cosmochim. Acta, 53, 197, Fig. 9.*

grail" samples? The experience with Comet Halley is less than reassuring. Figure 2.15.3 shows that while there is a general match with the solar photospheric abundances, two critical elements for terrestrial planetary compositions, Si and Fe, show wide differences. The Fe/Si and Mg/Si ratios for Halley are the lowest in the solar system, except for that unique object, the Moon. Comet Halley does not appear to be composed of "pristine" solar nebula matter, at least for its dust component. Finally, the composition of interplanetary dust offers little comfort to seekers after the "primitive nebular composition" (see sections 2.6 and 3.11) [169].

One might suppose that Jupiter, at least, and possibly Saturn, might be pristine samples of the nebula, as predicted if they originated as giant gaseous protoplanets from fragments of the nebula. However, they contain more than the solar complement of heavy elements, this component increasing outward from Jupiter to Neptune. Their atmospheric compositions, particularly Saturn's, do not have solar ratios of H to He. Although this has been suggested as due to "an inhomogeneity in the radial distribution of He" [170], the difficulty of separating He from H under nebular conditions makes it certain that we are looking at fractionation within the planets as the cause, rather than a separation of H from He in the nebula (see section 5.12 for a discussion of possible mechanisms).

The position is adopted here that an approximate first-order composition for the solar nebula is available from the CI abundances, but that, because of the inherently heterogeneous nature of the early nebula, we are unlikely to make much further progress along that particular track. This is of most concern in attempts to obtain overall compositions for the terrestrial planets.

2.16. THE NOBLE GASES

Two basic patterns are commonly identified. The so-called "solar" pattern is that of the solar wind, and is customarily—although mistakenly—taken as representative of both the Sun and of the primordial solar nebula. The overall variation is shown in Fig. 2.16.1 (Table 2.16.1). Little of this material seems to have been incorporated in the Earth. Thus, "no obvious evidence for a major component of primordial or captured solar-composition gas can be found on (the terrestrial) planets" [171]. Likewise, "if the earth's noble gases were ever, in any fundamental sense, more primitive than they are now, the traces of these primitive gases have vanished or remain hidden" [172]. This evidence is generally interpreted to indicate that the primordial solar nebula gases had dissipated by the time the Earth, and the other terrestrial planets accreted from planetesimals, the gas having been swept away by earlier violent solar activity. Some "solar" gas may have been trapped in planetesimals that subsequently accreted to the Earth.

The so-called "planetary" pattern is that found in the chondrites, and is distinguished by showing a depletion of the light gases (e.g., Ne) relative to

TABLE 2.16.1. Noble gas ratios.

	He^4/Ne^{20}	Ne^{20}/Ar^{36}	Ar^{36}/Kr^{84}	Kr^{84}/Xe^{132}
Solar system	850	37	3320	20.6
Solar wind	570 ± 70	45 ± 10		
Solar corona, spectr.	—	17 + 19,-9		
SEP, observed	480 ± 90	43 ± 9		

Data from Anders E. and Grevesse N. (1989) *Geochim. Cosmochim. Acta, 53,* 204, Table 5.

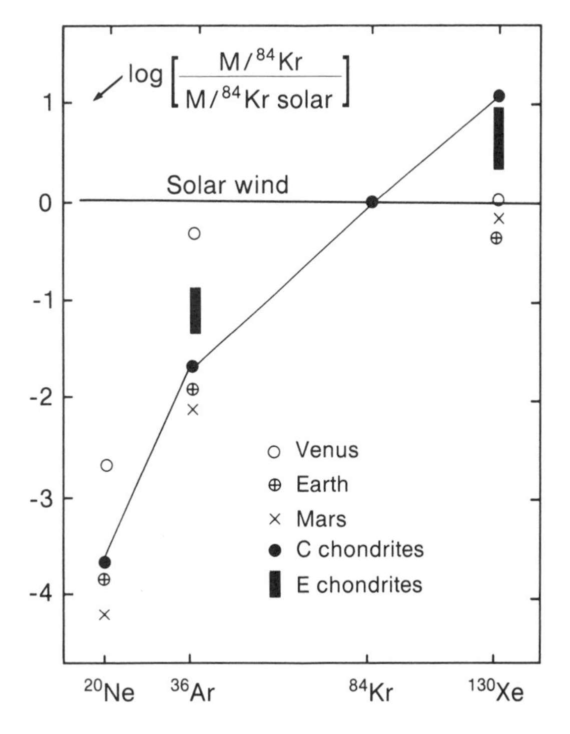

Fig. 2.16.1. *The abundances of the noble gases in terrestrial planetary atmospheres and in meteorites relative to solar wind abundances. After Hunten D. M et al. (1988) in Meteorites and the Early Solar System (J. F. Kerridge and M. S. Matthews, eds.), p. 570, Fig. 7.10.1, Univ. of Arizona, Tucson.*

the solar wind. However, there is so much variation among the various classes of meteorites and the terrestrial planets as shown in Fig. 2.16.1 that the term "planetary gases" should be abandoned. Whether this desirable objective can be reached is problematical.

The terrestrial noble gas budget is depleted in the light gases (e.g., Ne) relative to the solar wind. Even if traces of a "solar" noble gas abundance pattern are found (M. Honda, personal communication, 1990), this would not necessarily imply that accretion of the Earth occurred in the presence of a gaseous primitive nebula. It could mean that some of the precursor bodies that accreted to form the Earth trapped solar gas at a very early stage, and retained it until accretion to the Earth. Such a noble gas signature might be a useful index of the amount of volatile material received by the Earth from local planetesimals.

If the Earth had been immersed in the gas-rich solar nebula during accretion, it is sufficiently massive to have captured a primary atmosphere. Such an atmosphere would have a surface pressure of 10^3 atm, with a surface temperature of 4000 K [173]. Absorption of gas from such an atmosphere would result, for example, in a Ne budget 10-100 times greater than the present atmospheric content [174]. "Present noble gas observations do not support such a scenario: present abundances are too low and the isotopic compositions are wrong as well" [175].

The general question of planetary atmospheres is addressed later, but a few comments may be made at this stage. There is a general resemblance between the relative abundances of the noble gases in chondrites and in the atmospheres of both the Earth and Mars for Ne, Ar, and Kr. However, there is a significant difference in the abundances of the heavier noble gases. Xenon is relatively depleted by a factor of about 20. A number of explanations have been advanced to account for this discrepancy. It has been proposed, for instance, that the "missing" Xe is trapped in Antarctic ice, in shales, or has not been outgassed from the mantle to the same extent as Ne, Ar, or Kr. None of these suggestions appears adequate. The most probable explanation is that the noble gases in the Earth and Mars were not derived from typical chondrites. Of the present classes, the only chondrites with high Ar/Xe and Kr/Xe ratios are the CO3 and E4 types. These are, however, ruled out as major contributors to the compositions of the inner planets on other grounds.

A plot of $^{36}Ar/^{132}Xe$ vs. $^{84}Kr/^{132}Xe$ shows that most meteorite data form a distinct population, which may pass through the solar value (Fig. 2.16.2). The terrestrial, martian, and SNC (= Mars?) data form a separate linear array consistent with the existence of distinct reservoirs and thus with an inhomogeneous nebula.

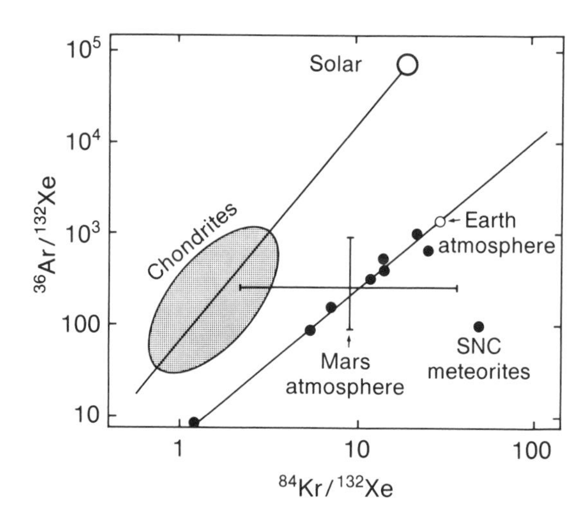

Fig. 2.16.2. *A comparison between meteoritic and planetary noble gas data, showing the separate arrays for Earth atmosphere, Mars atmosphere, and SNC meteorites on the one hand, compared with the solar wind and chondritic meteorites on the other. After Ott U. and Begemann F. (1985) Nature, 317, 509.*

The noble gas abundances in Venus and Mars present another unresolved problem. The concentration ratios are 75/1/0.1 for Venus, Earth, and Mars, which are not readily accommodated by current models. These interesting problems of the atmospheric compositions of the terrestrial planets are addressed further in sections 4.6 and 5.10.

Overriding all these differences is the problem that the planetary atmospheres are mostly secondary or have been altered by fractionation following accretion. Removal of all or part of primitive atmospheres by massive planetesimal and asteroidal collisions, and additions from or removals by comets (section 3.10 and 4.6) have to be taken into account. Thus, any primitive terrestrial atmosphere would have been removed by the giant lunar-forming impact. Although the noble gas abundances, both isotopic and elemental, have a rich tale to tell about later atmospheric evolution, stochastic effects obscure much of the crucial evidence relevant to the present inquiry into the earliest state of the solar system.

2.17. VOLATILE ELEMENTS (C, H, O, N) AND NEBULAR CHEMISTRY

The primitive solar nebula consisted mainly of H and He, which are the principal constituents of the Sun, Jupiter, and Saturn. The abundances of these elements in the solar nebula are given in Table 2.15.1. The terrestrial planets and meteorites are depleted in these elements, and also in a wide range of elements that are volatile below about 1000-1100 K. The carbonaceous chondrites (CI) have retained only 5-10% of the solar abundances of C, N, and H_2O. Nevertheless, these elements are much more abundant than the rare gases, and must have been accreted to the Earth in condensed rather than gaseous (N_2 and CO) phases.

The volatile elements are widespread in the outer solar system. Water ice has been identified spectroscopically on most satellites and the densities of these satellites are low enough to suggest that a substantial fraction of their mass is composed of water ice (see Chapter 6).

2.17.1. Oxygen Isotope Variations

The variation in the distribution of the three oxygen isotopes provides significant information about the early solar nebula (Fig. 2.17.1). Whatever the origin of these differences, Fig. 2.17.1 clearly indicates the existence of separate reservoirs. These cannot be related by simple fractionation. In that case, the data would lie along a line with a slope of 0.5. The values for anhydrous minerals from Allende CAI lie along a line with a slope of 1.0, pointing toward a separate ^{16}O-enriched source, which is distinct from the terrestrial fractionation line (Fig. 2.17.2). Such lines of unit slope cannot be produced by fractionation. The existence of distinct ^{16}O reservoirs is often attributed to addition of material formed by nucleosynthesis in differing regions. However, non-mass-dependent fractions may also be produced by chemical effects [176]. The principal interest in the present context is less with the origins of the isotopic variations, for which a vast, confused, and acrimonious literature exists, than with the demonstration that the nebula was heterogeneous rather than homogeneous with respect to oxygen and probably other components as well.

The oxygen isotope plot shown in Fig. 2.17.1 clearly separates various oxygen isotope reservoirs. Among the many observations that may be made, it is clear that the present population of meteorites mostly do not have oxygen isotope signatures similar to those of the terrestrial planets. Only the EH and EL groups overlap with the terrestrial values. H, L, and LL chondrites (all long-standing favorite candi-

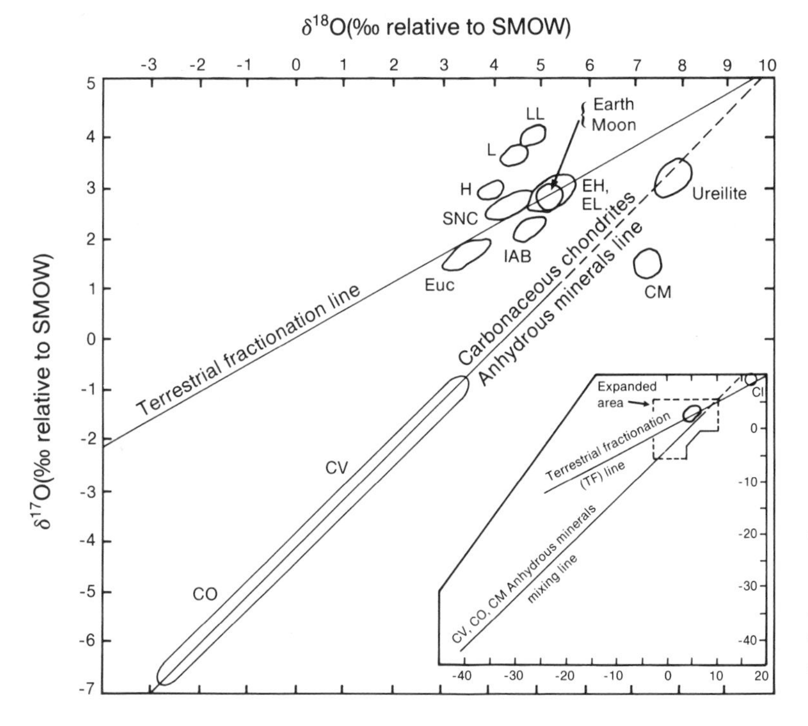

Fig. 2.17.1. *The relationship between $\delta^{18}O$ and $\delta^{17}O$ for the Earth, Moon, and various meteorite classes. After Taylor S. R. (1988) in Meteorites and the Early Solar System (J. F. Kerridge and M. S. Matthews, eds.), p. 527, Fig. 7.8.9, Univ. of Arizona, Tucson.*

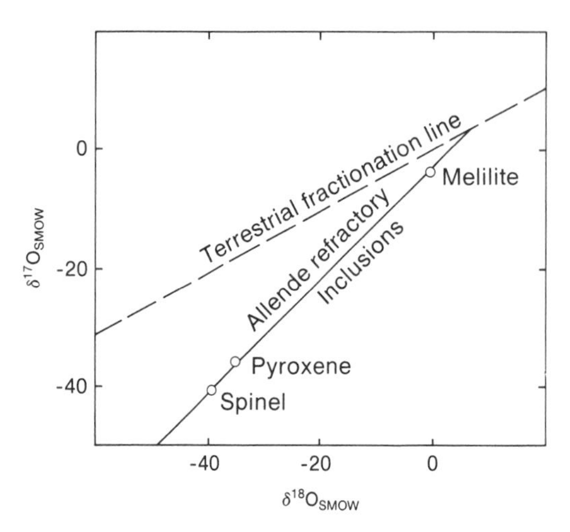

Fig. 2.17.2. *The oxygen isotopic composition of minerals from Allende refractory inclusions (CAI), which define a line with a slope of 1, relative to the terrestrial fractionation line with a slope of 0.5. After Thiemens M. H. (1988) in Meteorites and the Early Solar System (J. F. Kerridge and M. S. Matthews, eds.), p. 900, Fig. 12.1.1, Univ. of Arizona, Tucson.*

dates as planetary builders) and 1AB iron groups show no overlap with terrestrial values.

The carbonaceous CO, CM, and CV groups are clearly very distinct from the terrestrial value. This makes them an unlikely source for the popular late-accreting veneers used to account for everything from excess siderophiles in the upper mantle to the presence of oceans. It has been suggested from the oxygen isotope data that the CO, CM, and CV groups of carbonaceous chondrites formed in the outer solar system, but the general lack of mixing between the outer and inner regions makes this unlikely [177]. Assuming that the SNC meteorites come from Mars, they form a separate group, which again does not overlap with any of the other meteorite groups or with that of the Earth-Moon system. This is additional evidence for lack of homogeneity within the nebula at distances of 1-2 AU.

2.17.2. Chemical Reactions in the Nebula

Chemical reactions within the nebula are driven by the thermal energy of the nebula; other possible sources (e.g., radioactive decay from ^{26}Al, lightning, shock waves, and solar photons) are of secondary importance [178].

The opacity of the nebula caused by dust means that heating by intense solar ultraviolet photon fluxes will be restricted to the inner parts of the nebula, probably within 4 AU in the solar nebular disk. It is unlikely that such material processed in the inner nebula will diffuse outward. A flux between 10^2 and 10^3 times the present flux would be needed before this process becomes important in nebular chemistry. This means that most chemical processing of nebular gases and grains will occur within the inner nebula, or within planetary subnebulae where temperatures and pressures may rise sufficiently to initiate reactions.

The state of carbon and nitrogen in the nebula is critical for models for the formation of the outer planets and their satellites (see section 6.2). If the primitive nebula reflects its parental source in the giant molecular clouds, carbon is present mainly as CO, not as methane (CH_4), with CH_4/CO ratios less than 10^{-2}, and nitrogen is principally present as N_2 not as NH_3. This assumes that no processing has occurred during nebular separation and collapse, and this picture is probably overly simplistic. The presence of abundant CH_4 and NH_3 in the outer solar system satellites and giant planetary atmospheres points to extensive chemical processing in the planetary subnebulae (see section 6.2). The existence of these compounds in Comet Halley also raises some awkward questions [178]. Only about 10% of carbon is present as a condensed phase. Studies of the composition of Pluto and Neptune suggest that carbon is not present as CO or CH_4 ices, or in clathrate-hydrates, but principally resides in complex organic compounds [179], a conclusion consistent with the composition of the carbonaceous chondrites (section 3.6).

2.18. HOMOGENEITY OR HETEROGENEITY?

This topic has had an interesting history that reflects a common approach to many scientific problems. What we cannot see or measure, we imagine to be simple and homogeneous. The distribution of galaxies or the composition of the Earth's mantle are recent examples, but increasing knowledge has revealed unsuspected complexities. The approach is related to the human tendency to underrate what we do not understand. This intellectual safety valve allows us to live with the unknown, and is no doubt responsible for the popularity of religions, astrology, and mysticism in general.

One might confidently have expected the primordial solar nebula, a rotating disk of gas and dust fragmented from a molecular cloud, to be both symmetrical and uniform in composition and indeed this was the common perception. The first glimmer of a new complexity was the discovery of neon E, pure ^{22}Ne, derived from ^{22}Na (with a half-life of 2.6 years) as a component of meteoritic neon [180].

The next significant observation, that oxygen, an abundant species, showed variations in ^{16}O relative to ^{17}O and ^{18}O, revealed decisively that the nebula could not have been well mixed [181]. This notion was soon supported by the host of isotopic anomalies reported in CAIs and other minor constituents in meteorites [182].

Two significant advances in our understanding resulted. First, some interstellar material had survived the conditions in the early solar nebula because the isotopic anomalies probably were not produced by processes within the nebula. Second, this preservation places limits on the amount of heating of the nebula. If all the material in the primordial nebula had been vaporized, then it would have been homogenized. Accordingly, homogenization did not occur on a large scale, and some presolar material survived conditions in the early nebula. This whole question is complicated by the relative timing of the various events. Thus, the nebula may have reached high temperatures very early, and may have been cooler during chondrule formation and accretion of meteorites. Presolar or interstellar grains may have continued to fall into the nebula throughout the planetesimal and accumulation stages. Accordingly, it could be difficult to identify the time of arrival of a particular grain displaying isotopic anomalies. Such late-arriving material falling into a cool well-mixed nebula might account for what we observe.

Since many of the isotopic anomalies in meteorites could be ascribed to survival of dust grains, it might be assumed that at least the gas in the nebula was well mixed. However, the presence of "exotic" rare gas components in meteorites does not encourage this speculation. They indicate that the nebula was not well mixed, even with respect to the gaseous components. The processes of inflow of mass to the proto-Sun and outward transfer of angular momentum are unlikely to promote mixing and are aided by gravitational torques in a nonaxisymmetric nebula. Accordingly, we seem to require both an unsymmetrical nebula and an imperfectly mixed one [183].

This conclusion is dependent on the scale at which one views the problem. Possibly the nebula was homogeneous on a broad scale for most elements, and differs only on the scale of meteoritic inclusions. However this view does not seem to be correct. There are significant differences in composition and density among the terrestrial planets and meteorites. The question whether these represent initial differences in the nebula is addressed in the next sections.

2.19. HELIOCENTRIC AND VERTICAL ZONING

The components in the present solar system display significant, although not systematic, variations in composition with distance from the Sun. Some of this results from early solar activity and some from late events in planetary formation. The fundamental division of the system into the inner small rocky and outer giant gaseous and icy planets is here considered to be both a consequence of the clearing of gas and volatile elements from the inner solar system by early T Tauri and FU Orionis activity, and of the more rapid growth of the giant planets. Water ice condenses at nebular pressures at 160 K at about 5 AU. This "snow line" effect is likely to assist in the rapid growth of Jupiter as water and other volatiles are swept out of the inner solar system and pile up at the "snow line." As a corollary, the giant planets must reach their present large sizes before the early violent solar activity erodes the solar nebula in the region of their accretion.

In contrast, the high density of Mercury probably does not result from high temperatures due to its proximity to the Sun, a tempting false trail responsible for much wasted effort on grand solar system models. It can be rationally ascribed as due to a late large collision, stripping away much of the silicate mantle of the planet (sections 4.8 and 5.3), so that Mercury is not, as frequently supposed, an end member of a compositional sequence. Earth and Mars show significant compositional differences, as do Jupiter, Saturn, Uranus, Neptune, and their satellites. For example, the saturnian satellites are in general less dense than either those of Uranus or Jupiter, a fact difficult to accommodate without some *ad hoc* adjustments to any theory predicting smoothly varying compositional gradients with heliocentric distance.

One problem is to disentangle the effects of planetary accretion and fractionation from that due to primordial heliocentric zoning. The most significant result seems to be that the observed differences are based principally on volatility. Thus, Mars has a higher K and Rb content than the Earth. The zoning in the asteroid belt (see next section), as well as the great diversity of meteorites, informs us of the variety of compositions of planetesimals available from which to construct the terrestrial planets. Lateral mixing of planetesimals during accretion is likely to smear out any original differences based on heliocentric distance. We can get some measure of this from two sources: first, from the frequency of inclusions of different classes of meteorites within one another, and, second, from the distribution of different types within the asteroid belt. Both indicate rather restricted lateral mixing.

2.20. ASTEROID BELT ZONATION

The composition and structure of the asteroid belt are vital pieces of evidence in our attempts to understand the solar system, and are discussed at several points in this book, specifically in sections 3.13 and 5.11 [184].

The compositions of the terrestrial planets are mostly distinct. Thus, there does not appear to be much evidence for homogenization during planetary accretion, and it seems likely that the planets were assembled either from rather restricted zones or from a random sampling of planetesimals of differing compositions if we are to account for the variations in oxygen isotopes, chemistry, and density. In this context, it is interesting that the asteroid belt is zoned. Since this zoning apparently predates the accretion of the planets, it indicates that lateral mixing of planetesimals, which formed very early, and from which the planets most likely were accreted, was limited.

Is the present zonal structure of the asteroid belt primary or secondary? That is, does it reflect an initial inhomogeneity within the nebula, or did the structure arise through secondary interactions due to heating, resonance, or tidal effects following condensation of the solar nebula? If the zoned structure of the asteroid belt is primary and not due to secondary tidal or resonance interactions, it may serve as an analogue for the primitive nebula following gas loss and the accretion of Jupiter. This requires knowledge of the initial condition of the nebula at the present site of the asteroid belt. It is exceedingly unlikely to suppose that the original solar nebula was deficient in mass at its present condition (Fig. 2.20.1).

The conventional explanation is that the growth of the belt was starved by early growth of Jupiter. Pumping up of the eccentricities of the asteroids by interaction with Jupiter will limit their growth to about 10^{24} g bodies [185]. Such growth must occur before the H and He are dispersed from the nebula; present ideas about timescales indicate periods of 10^6 years. If one applies the standard planetesimal model in the primordial solar nebula, then runaway accretion should produce 1/3 Earth-mass bodies in about 3×10^5 years, and ultimately a terrestrial-sized planet [186]. Since this has not occurred, loss of the material in the belt must have occurred very early, implying that Jupiter was present soon after T_0 (4560 m.y.) [187].

Thus, the belt in its present state apparently dates back nearly to the origin of the solar system, and there is much evidence suggesting that the present low mass in the region extends back almost to T_0. This view is supported by the survival of the basaltic surface of 4 Vesta, since basaltic achondrite ages are almost this old, and its surface, coated with basaltic lavas, is not expected to survive massive collisions.

The oxygen isotopic evidence suggests that the nebula was not well mixed, and that the planets accreted from relatively small zones, or at least preserved initial heterogeneities if lateral mixing occurred. The material in each zone varied, but how did this variation in composition arise? In the case of the asteroid belt, the answer seems fairly clear (see sections 3.13 and 5.11 for a detailed discussion of the evidence).

Heating of the inner belt induced melting and differentiation of the asteroids, resulting in the present zonation. Since the sizes of the parent bodies differed considerably, many distinct compositional varieties can be expected to arise. The accretion of

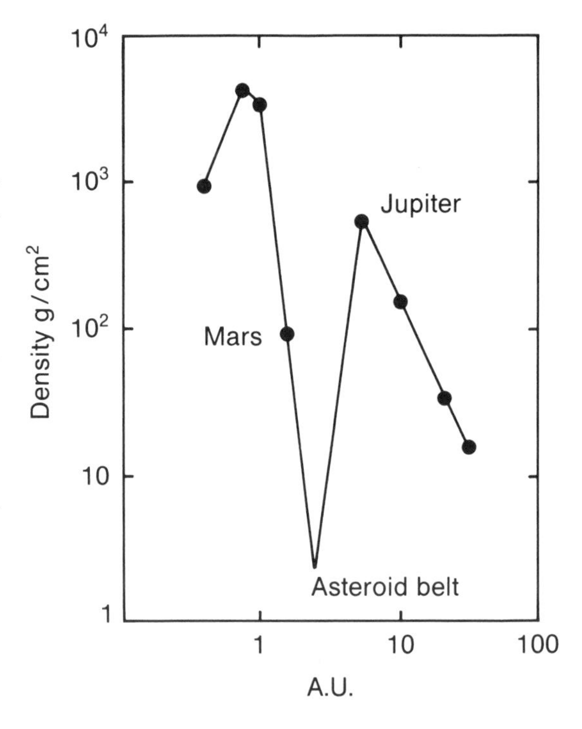

Fig. 2.20.1. *The density at a late stage (post-Jupiter formation) in the solar nebula inferred from the present distribution of mass, after restoring the gaseous component to the inner solar system. After Weidenschilling S. J. (1977) Astrophys. Space Sci., 51, 153.*

planets from an essentially random set of fractionated planetesimals seems adequate to account for the resulting diversity of the inner planets.

The problem of early heat sources for the asteroids must be considered unsolved. A first-order observation is that the zonal structure of the belt implies that heating was related to distance from the Sun. A second deduction from this observation, since the heating must have occurred at the earliest times by any mechanism, is that the zonal arrangement of the belt retains this primordial structure and has not been seriously perturbed since that time, implying a scarcity of belt-crossing objects.

The source of the heating remains to be identified. Neither heating by long-lived radionuclides with half-lives $>10^9$ years nor tidal heating is adequate. Gravitational potential energy for small objects is likewise inadequate as a heat source. Melting by impacts requires high velocities (>1 km/s), but the growth of small planetesimals requires lower velocities, otherwise they will be disrupted rather than growing by accretion.

The problem of the distribution of ^{26}Al remains a puzzle. If it is the source for the heat, why should it be zoned with respect to distance from the Sun? Aluminum-26 can heat objects of the order of 10 km in diameter. Magnesium-26 excesses correlated with Al are mainly restricted to refractory inclusions in meteorites, and would need to be radially distributed through the belt to explain the zonal arrangement. However, such excesses have also been found in an unequilibrated LL Type-3 chondrite, so ^{26}Al remains as a possible early heat source for planetesimals [188].

Perhaps the strongest argument in favor of an ultimate energy source from the Sun is this zonal arrangement of compositions in the belt correlated with heliocentric distance, albeit somewhat smeared out by collisions. When this evidence is correlated with that of gas loss in the inner solar system, it seems clear that the early Sun is the prime suspect. However, direct heating by an early superluminous Sun is likely to cause only minor surface melting [189].

Electromagnetic inductive heating is another possibility. This requires both strong solar winds and magnetic fields, which could induce electric currents in asteroidal bodies, causing heating. Such effects are associated with T Tauri or FU Orionis conditions as the Sun settles onto the main sequence, and thus occurs effectively at T_0 since these active stages last only a few million years [190]. Inductive heating is not effective in small bodies, but is likely to melt only larger asteroids. Whatever the source of heat, it has to be early, implying the presence of asteroidal bodies large enough to be melted also at times close to T_0. Once melted, cooling and metamorphic effects can occur on a protracted timescale.

The overall question of the source of early heating remains unanswered (see section 3.12), since neither inductive heating nor ^{26}Al appears to be adequate. The basic problem with ^{26}Al as a heat source is that the basaltic achondrite ages are of the order of 20 m.y. younger than T_0 so ^{26}Al will have decayed long before. The absence of ^{26}Mg anomalies in eucrites is additional evidence against this source [191].

2.21. GAS LOSS FROM THE INNER NEBULA

Many meteorite constituents formed in a high O/H environment so that O was enriched relative to H by at least 2 orders of magnitude at the time of their formation [192]. Since the meteorite constituents form very early, the H (and He) that form the bulk of the nebula must have been lost from the zone in which the meteorite minerals formed very close to T_0, probably within 10^6 years of that time. After that time, more oxidizing conditions prevailed in the region of meteorite formation.

The solar system has been swept clean of gas. The present solar wind has only about 10 atoms/cm^3 at the location of the Earth's orbit at present, compared with a concentration in the original nebula of about 10^{15}/atoms cm^3. When did this drastic decline happen? There is strong observational evidence from young stars of the removal of gaseous envelopes within 10^6-10^7 years following the formation of the protostar. Probably the smaller estimate of a few million years is more likely. These observational values are judged to be rather more reliable than estimates based on theoretical models [193,194].

Energetic mass loss thus starts at about 10^5 years with complete loss of dust and gas by 10^6-10^7 years, as appears to be the case for the young stars HL Tauri and R Monocerotis [196]. This implies that the giant planets must have formed within 10^6-10^7 years of the proto-Sun while the gas was still around. The noble gases are strongly depleted in the Earth, implying final accretion of this planet later in a gas-free environment. Accordingly, early formation of the inner planets, which is the only time during which accretion in the primordial gas-rich solar nebula could occur, seems to be ruled out. By the time the Earth, Venus, Mars, and Mercury accreted, a process taking perhaps 10-100 m.y., the gas was gone.

Although removal of gas is usually ascribed to strong solar winds during the T Tauri phase, it may in fact occur somewhat earlier, associated with the

FU Orionis stage. How much mass can be removed? This depends on the coupling between the solar winds and the nebula, which may not be very efficient, particularly if the winds are preferentially concentrated along the rotation poles rather than in the midplane of the nebula. As well as mass loss outward to the nebula, much of the dust in the inner nebula will be carried into the Sun along with the instreaming gas. Since the Sun has the canonical "solar system" composition and resembles that of the outer asteroid belt, as presumably sampled by the CI meteorites, there must be little elemental fractionation involved [196].

It seems likely, therefore, that the volatile depletion in the inner nebula is mainly due to a sweepout of the volatile components by early violent activity. This was perhaps coupled with an initial depletion in the later stages of the accretion of the Sun as planetesimals became large enough to retain the more refractory components, while uncondensed volatiles were swept along with the gas into the Sun. Clearly, it will be very difficult to distinguish between these two stages.

What were the means for the dispersal of the nebula and what was the timing of the event? The cause of this early volatile loss in the inner portions of the solar nebula is not clearly established, but must be connected with intense solar activity as the Sun settled onto the main sequence of stellar evolution. Mechanisms for the removal of the H and He from the region of the inner planets are apparently restricted to intense solar activity (T Tauri- and FU Orionis-type outbursts) as the early Sun arrives on the main sequence and the inflow of material to the growing Sun is reversed. Current hypotheses for stellar evolution and observations in regions of young star formation indicate that strong stellar winds, flare activity, etc. operate only for timescales of the order of 10^6 years or less.

Accordingly, dispersal of H and He and other gases from the inner regions of the nebula by these mechanisms occurs within a few million years of T_0. The noble gas evidence from the "nonsolar" abundance signatures in meteorites indicates that such events preceded the formation of many groups of meteorites, dated at about 4.55 b.y.

Meter-sized objects in the inner nebula will resist removal by T Tauri winds [197], but finer material (dust and gas) will be swept away. Kilometer-sized objects would be resistant even to very strong stellar winds. Growth to centimeter-sized objects, during sedimentation to the midplane of the nebula, can occur within about 10^3 years [198]. However, quiet sedimentation to the nebular midplane is unlikely to occur in the presence of the turbulence expected in the early nebula [199]. Clearly, rapid accretion of kilometer-sized objects must have occurred very early, within a few million years of T_0. An important corollary is that accretion formed such objects within 10^6 years of proto-Sun formation. Otherwise all material, gas and fine dust alike, would have been swept out of the inner solar system and we would not be discussing the problem.

In summary, (1) energetic mass loss starts at 10^5 years and there is complete loss of gas and dust by 10^6-10^7 years and (2) the noble gas evidence indicates accretion of the Earth in a gas-free environment. This rules out early planetary formation (within 10^6 years of T_0), which appears to be the only time during which accretion in a primordial solar nebula could occur. This is perhaps not a surprising conclusion when the overall volatile-depleted composition of the inner planets is contrasted with the gas-rich composition of Jupiter and Saturn.

2.22. VOLATILE ELEMENT DEPLETION IN THE EARLY NEBULA

Widespread depletion of moderately volatile elements (400-1100 K) occurs in the terrestrial planets and to a lesser extent in meteorites. Does this occur at the same time and by the same process as the loss of the gaseous elements, or are the processes unrelated? This is a fundamental question, the answer to which depends on the interpretation of the Rb- and Pb-isotopic data from the meteorites.

The best evidence for the widespread depletion of volatile elements in the inner solar system comes from the abundances of the moderately volatile element K relative to the refractory element U measured in the terrestrial planets (Fig. 2.22.1). Since these elements are both gamma ray emitters, this is one of the few geochemical measurements available for Earth, Venus, and Mars, as well as for the meteorites, and it provides some crucial information. Potassium and U, although distinctly different in chemical properties, ionic radius, and valency, nevertheless share a common characteristic that makes them useful as geochemical indices of planetary compositions. They are both excluded from the common rock-forming minerals in basalts (e.g., are incompatible elements), and so are concentrated together in residual melts during crystallization of basaltic silicate melts; accordingly, they tend to preserve their bulk planetary ratios during planetary differentiation. Their measurement, therefore, can provide information about the depletion of volatile relative to refractory elements in their parent body.

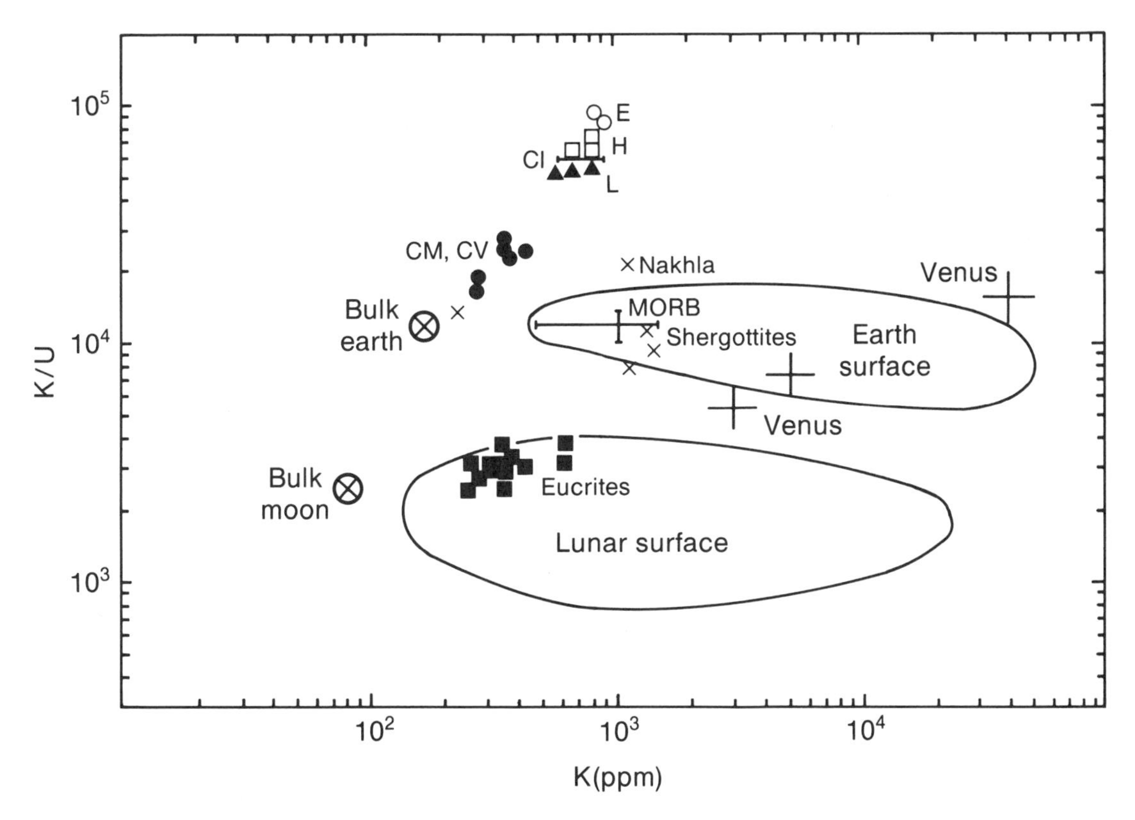

Fig. 2.22.1. *The variation in the abundances of K (a moderately volatile element) and U (a refractory element) in meteorites and the inner planets, illustrating the depletion of volatile compared to refractory elements in the inner nebula relative to their abundances in CI carbonaceous chondrites, which are taken as representative of the primordial abundances.*

The primordial ratios, as given by the CI meteorites, are appproximately 60,000, while terrestrial ratios are about 10,000. Potassium was not depleted in ordinary (H, L) chondrites, as shown by their high K/U ratios. This indicates, along with the noble gas data, that they did not form an important part of the planetesimal population from which the terrestrial planets were assembled. Nevertheless, the ordinary chondrites have lost large fractions of more volatile elements [200]. If the ordinary chondrites are derived from the asteroid belt, then this region experienced a less severe heating than the planetesimals from which Earth, Venus, and Mars were assembled, indicating that temperatures were higher nearer the Sun.

An initial question is whether the depletion of K observed in terrestrial samples is a bulk planetary effect or is local and surficial. This question is addressed in Chapter 5. The generally low values of the K/U ratios in Venus, Earth, and Mars (assuming the SNC connection [201]) indicates that the precursor planetesimals for all three planets formed from a volatile-depleted suite. It is shown in subsequent sections (e.g., section 5.8) that alternative hypotheses (that K is buried in planetary cores, or is boiled off during a high-temperature stage of planetary accretion) are unlikely.

In addition to the chemical evidence, the Rb/Sr isotopic systematics also indicate that the Earth is depleted in volatile Rb relative to refractory Sr. Since Rb has closely similar properties to K, it is unlikely that either K or Rb is present in the mantle in their primordial solar nebular concentrations and that Rb and the other volatile elements were depleted in the precursor material from which the Earth accreted. The most reasonable explanation is that these volatile elements were depleted in the precursor planetesimals from which the inner planets accumulated.

The implication from these observations is that not only were the noble gases cleared from the inner solar nebula, but a large fraction of volatile elements was also lost. A small-scale analogue of conditions near the early Sun is provided by the Galilean satellites of Jupiter. The decrease in density outward from Io to Callisto is widely interpreted as due to an increase in the water-ice/rock ratio, caused by mild heating of the proto-Jupiter nebula. The widespread nature of the depletion both in the terrestrial planets and meteorites must indicate that such conditions operated throughout the inner nebula out to a "snow line" at about 4-5 AU.

It is argued in Chapter 3 that the age and initial Sr isotopic data from meteorites indicate that they record a massive loss of volatile Rb relative to refractory Sr close to T_0, so that this was a nebulawide event rather than being connected with isotopic evolution in individual parent bodies.

This depletion must have occurred very early, since very low initial $^{87}Sr/^{86}Sr$ ratios are observed in the achondritic meteorites, which indicate a major separation of Rb from Sr at this early stage of solar system formation. The terrestrial Rb/Sr ratio of 0.03 is nearly a factor of 10 lower than the CI ratio, implying a massive preterrestrial loss of volatile elements, with the Earth accreting from already depleted planetesimals.

Some further loss of volatiles may occur during heating of material in high-energy collisions. Hence, the lunar ratios for both K/U and Rb/Sr are lower than terrestrial, but not by such a large factor. If the terrestrial ratios of $K/U = 10^4$ and $Rb/Sr = 0.03$ are typical of the inner planetary regions and of Mars-sized impactors (as discussed later), then it is possible that the major separation of Rb from Sr occurred at T_0, with a secondary lesser depletion during lunar formation.

The most primitive initial $^{87}Sr/^{86}Sr$ ratio so far measured is that given by some refractory inclusions (CAIs) in the Allende meteorite. These were heated much more severely than the flash melting episodes experienced by chondrules. The Allende data appear to represent early Rb loss in isolated heating events and to record the earliest events we can trace, which probably occurred in a turbulent predisk stage. The next datum is provided by the initial $^{87}Sr/^{86}Sr$ ratios observed in the basaltic achondrites (chondrite data is too imprecise to provide a useful marker). It is interpreted here as indicating a massive nebulawide loss of Rb and K relative to Sr and U, occurring during the violent solar activity.

In summary, the depletion of volatile K relative to refractory U informs us of the nebulawide extent of the volatile depletion, and the information from the Rb/Sr system tells us the date of the volatile loss (see main discussion under section 3.7). The meteorites provide us with the time of volatile depletion in the inner nebula because the Pb-Pb and Rb-Sr ages give the time of separation and depletion of volatile Pb and Rb relative to refractory U and Sr from the primordial solar nebula values. Accordingly, volatile depletion in meteorites, and presumably also in the inner nebula closer to the Sun, in the region of the inner planets, occurred close to T_0.

The inner nebula was thus cleared of the gaseous elements and depleted in volatile elements very early, probably within about 1 m.y. of the arrival of the Sun on the main sequence. The cause of this clearing of the inner regions of the nebula is most probably connected with early intense solar activity, with strong stellar winds, and flare outbursts of the type observed with the very young T Tauri and FU Orionis stars. The CAI depletions are interpreted as occurring in an early turbulent conditions either during the separation of the nebula from the molecular cloud, or perhaps a little earlier if the apparent 10-m.y. gap between ALL and BABI (section 3.7) is confirmed.

NOTES AND REFERENCES

1. An extended discussion of the history of this problem is given in Chapter 6 of Jaki S. L. (1978) *Planets and Planetarians*, Wiley, New York, 266 pp.
2. Chamberlin T. C. (1905) *Carnegie Institution of Washington Yearbook, 3,* 195.
3. Morris D. E. and O'Neill T. G. (1988) *Astron. J., 96,* 1127.
4. Campbell B. et al. (1988) *Astrophys. J., 331,* 902.
5. Marcy G. W. et al. (1986) in *Astrophysics of Brown Dwarfs* (M. C. Kafatos et al., eds.), Cambridge Univ., New York, p. 50; Campbell B. et al. (1988) *Astrophys. J., 331,* 902; Zuckerman B. and Becklin E. E. (1987) *Nature, 330,* 138; Latham D. W. et al. (1988) *Science, 241,* 790; Latham D. W. et al. (1989) *Nature, 339,* 38.
6. See Epilogue.
7. Sagan C. (1988) *Planetary Report, VIII,* 16.
8. Weinberg S. (1977) *The First Three Minutes: A Modern View of the Origin of the Universe*, Basic Books, New York, 188 pp.
9. Hawking S. A. (1988) *A Brief History of Time*, Bantam, New York, 198 pp. The most aesthetic statement is *The Creation* by J. Haydn (Vienna, 1798).
10. Lynden-Bell D. et al. (1989) *Mon. Not. R. Astron. Soc., 239,* 201.

11. Hawking S. A. (1988) op. cit., p. 50.
12. Maddox J. (1989) *Nature, 340,* 425.
13. Ellis G. F. R. (1984) *Annu. Rev. Astron. Astrophys., 22,* 157.
14. Sciama D. (1967) quoted by Harrison E. R. (1981) *Cosmology, The Science of the Universe*, Cambridge Univ., New York, p. 320.
15. The 3-K cosmic background radiation has a spectrum that is a perfect match for that predicted for black-body radiation; see Lindley D. (1990) *Nature, 343,* 207, and Hogan C. J. (1990) *Nature, 344,* 107. How galaxies can arise in such a well-behaved system remains a mystery. None of the five current theories accounts for all the evidence; see review by Peebles P. J. E. and Silk J. (1990) *Nature, 346,* 233. See also Weinberg S. (1977) *The First Three Minutes*, Basic Books, New York, 188 pp.; Truran J. W. (1984) *Annu. Rev. Nucl. Part. Sci., 34,* 53; Boesgaard A. M. and Steigman G. (1985) *Annu. Rev. Astron. Astrophys., 23,* 319.
16. Kunth D. (1986) *Advances in Nuclear Astrophysics*, p. 25, Editions Frontières, Gif-sur-Yvette, France.
17. Boesgaard A. M. and Steigman G. (1985) *Annu. Rev. Astron. Astrophys., 23,* 319.
18. The correction for this additional He is not simple. Another estimate for Y_p is 0.230 ± 0.015; Steigman G. et al. (1989) *Comments Astrophys., 14,* 97. A value of 0.24 ± 0.01 is given by Kunth D. (1986) *Advances in Nuclear Astrophysics*, p. 25, Editions Frontières, Gif-sur-Yvette, France.
19. Pagel B. E. J. (1989) in *Evolutionary Phenonoma in Galaxies* (J. Beckman and B. E. J. Pagel, eds.), p. 368, Cambridge Univ., New York.
20. Anders E. and Grevesse N. (1989) *Geochim. Cosmochim. Acta, 53,* 197.
21. Rood R. T. et al. (1984) *Astrophys. J., 280,* 629.
22. Steigman G. (1989) *Am. Inst. Phys. AIP Conf. Proc., 183,* 314.
23. Spite M. and Spite F. (1982) *Astron. Astrophys, 115,* 357; Spite M. and Spite F. (1985) *Annu. Rev. Astron. Astrophys., 23,* 225.
24. Boesgaard A. M. and Steigman G. (1985) *Annu. Rev. Astron. Astrophys., 23,* 319.
25. The solar system abundances (mass ratios) for H, He, and the "heavy" ($Z > 2$) elements are (X) H = 70.683; (Y) He = 27.431, and (Z) Li-U = 1.886. X, Y, and Z are the conventional symbols used in the astronomical literature for the three groups of elements. Steigman G. (1989) *Am. Inst. Phys. AIP Conf. Proc., 183,* 317.
26. The most celebrated paper is Burbidge E. M., Burbidge G., Fowler W. A., and Hoyle F. (1957) *Rev. Mod. Phys., 29,* 547, usually referred to as B^2FH.
27. A standard reference is Clayton D. D. (1968) *Principles of Stellar Evolution and Nucleosynthesis*, McGraw-Hill, New York, 612 pp. There is a steady enrichment of heavy elements with time in the interstellar medium, so that younger stars (and galaxies) are enriched in heavy elements. Individual supernova ejecta are halted in the interstellar medium after about 10^5 years.
28. Rank D. M. et al. (1988) *Nature, 331,* 505.
29. Woosley S. E. and Phillips M. M. (1988) *Science, 240,* 750.
30. See review by Arnett D. W. et al. (1988) *Annu. Rev. Astron. Astrophys., 27,* 629.
31. Weissman P. R. (1986) in *The Galaxy and the Solar System* (R. Smoluchowski et al., eds.), p. 204, Univ. of Arizona, Tucson.
32. Weissman P. R., op. cit., p. 219; Weissman P. R. (1990) *Nature, 344,* 825; Weissman P. R. (1991) in *Comets in the Post-Halley Era* (R. L. Newburn et al., eds.), Kluwer, Amsterdam, in press.
33. Redshift is customarily expressed as z, which is the difference between the measured wavelength of an observed spectral line and the terrestrial reference value, divided by the terrestrial reference wavelength in angstroms.
34. Lilly S. J. (1988) *Astrophys. J., 333,* 161.
35. Schneider D. et al. (1989) *Astron. J., 98,* 1951.
36. Tyson J. (1988) *Astron. J., 96,* 1.
37. Geller M. J. and Huchra J. P. (1989) *Science, 246,* 897.
38. For example, Bhavsar S. and Ling N. (1988) *Astrophys. J., 331,* L63; de Lapparent V. et al. (1986) *Astrophys. J., 302,* L1; see also Shandarin S. F. and Zeldovich Ya. B. (1989) *Rev. Mod. Phys., 61,* 185.
39. Larson R. B. (1985) *Mon. Not. R. Astron. Soc., 218,* 409; Larson R. B. (1987) *Am. Sci., 75,* 376.
40. Scaramella R. et al. (1989) *Nature, 338,* 562.
41 The problem has been discussed in a very entertaining book that also finally resolves the paradox: Harrison E. R. (1987) *Darkness at Night: A Riddle of the Universe*, Harvard Univ., Cambridge, 293 pp.
42. Trimble V. (1986) *Comments Astrophys., 11,* 257.
43. Bolte M. (1989) *Astron. J., 97,* 1688.
44. The astronomical term for all elements heavier than He, otherwise referred to as "Z," not to be confused with "z," the redshift measure.
45. Beichman C. A. (1987) *Annu. Rev. Astron. Astrophys., 25,* 542, Fig. 7.
46. Lake G. and Carlberg R. G. (1988) *Astron. J., 96,* 1581 and 1587.
47. Barnes J. E. (1989) *Nature, 338,* 123.

48. For example, Hickson P. et al. (1991) *Astrophys. J. Supp.*, in press.
49. Bergvall N. and Jorsater S. (1988) *Nature, 331*, 589.
50. Tyson J. (1988) *Astron. J., 96*, 1.
51. Soifer B. T. et al. (1987) *Annu. Rev. Astron. Astrophys., 25, 187.*
52. A review is given by Smith D. (1987) *Philos. Trans. R. Soc., A323*, 269.
53. The paths by which the many molecular species are produced are given in Fig. 1, p. 275, for diffuse clouds and Fig. 2, p. 277, for dense clouds in Smith D. (1987) *Philos. Trans. R. Soc., A323.*
54. Clayton D. D. (1985) *Astrophys. J., 290*, 428.
55. Kerridge J. F. and Chang S. (1985) in *Protostars and Planets II* (D. C. Black and M. S. Matthews, eds.), p. 738, Univ. of Arizona, Tucson; Prinn R. G. and Fegley B. Jr. (1989) in *Origin and Evolution of Planetary and Satellite Atmospheres* (S. Atreya S. et al., eds.), p. 78, Univ. of Arizona, Tucson.
56. Kerridge J. F. and Chang S. (1985) in *Protostars and Planets II* (D. C. Black and M. S. Matthews, eds.), p. 738, Univ. of Arizona, Tucson.
57. See discussion in Lunine J. I. (1989) in *The Formation and Evolution of Planetary Systems* (H. A. Weaver and L. Danly, eds.), p. 238, Cambridge Univ., New York. See also p. 233.
58. Beichman C. A. et al. (1986) *Astrophys. J., 307*, 337.
59. Cameron A. G. W. (1988) *Annu. Rev. Astron. Astrophys., 26*, 441.
60. Cameron A. G. W. (1973) *Icarus, 18*, 407; Field G. B. and Cameron A. G. W., eds. (1975) *The Dusty Universe*, Smithsonian Institution, 323 pp.
61. Mathis J. S. et al. (1977) *Astrophys. J., 217*, 425; see also Draine B. T. (1984) *Astrophys. J., 277*, L71.
62. Rietmeijer F. J. M. (1988) *Icarus, 74*, 446.
63. See also Nuth J. A. (1985) *Nature, 318*, 166; Nuth J. A. (1988) in *Meteorites and the Early Solar System* (J. F. Kerridge and M. S. Matthews, eds.), p. 984, Univ. of Arizona, Tucson.
64. Tang M. and Anders E. (1988) *Astrophys. J., 335*, L31; Tang M. et al. (1989) *Nature, 339*, 351; Lewis R. S. et al. (1987) *Nature, 326*, 160; Tang M. and Anders E. (1988) *Geochim. Cosmochim. Acta, 52*, 1235. A useful scale was given by Ed Anders (personal communication, 1988): these diamonds would be the appropriate size if bacteria wore engagement rings.
65. Blake D. et al. (1988) *Lunar and Planetary Science XIX*, p. 94, Lunar and Planetary Institute, Houston; Badziag P. et al. (1990) *Nature, 343*, 244.
66. Jorgensen U. G. (1988) *Nature, 332*, 702.
67. There is an extensive literature in this field. For example, Tsuda M. et al. (1986) *J. Amer. Chem. Soc., 108*, 5780; Namba Y. et al. (1989) *J. Vac. Sci. Tech., A7*, 36.
68. Tang M. et al. (1989) *Nature, 339*, 351; Zinner E. et al. (1989) *Geochim. Cosmochim. Acta, 53*, 3273.
69. J. Stone, personal communication, November 1990.
70. Bernatowicz T. et al. (1988) *Meteoritics, 23*, 257.
71. Zinner E. et al. (1987) *Nature, 330*, 730.
72. Tang M. and Anders E. (1988) *Astrophys. J. Lett., 335*, L31.
73. J. Stone, personal communication, February 1991.
74. Tang M. et al. (1989) *Nature, 339*, 354.
75. "Bok globules" are dense cloud cores that have become separated from the surrounding molecular clouds and represent the earliest stage of star formation. Yuh J. L. and Clemens D. P. (1990) *Astrophys. J., 365*, L73.
76. Boss A. P. (1987) in *Interstellar Processes* (D. Hollenbach and H. Thronson, eds.), pp. 331-348, Reidel, Dordrecht.
77. Boss A. P. (1988) *Astrophys. J., 331*, 375.
78. Sargent A. I. (1989) in *The Formation and Evolution of Planetary Systems* (H. A. Weaver and L. Danly, eds.), p. 111, Cambridge Univ., New York; Strom S. E. et al. (1989) in *The Formation and Evolution of Planetary Systems* (H. A. Weaver and L. Danly, eds.), p. 91, Cambridge Univ., New York.
79. Sargent A. I. and Beckwith S. (1987) *Astrophys. J., 323*, 294.
80. Sargent A. I. (1989) in *The Formation and Evolution of Planetary Systems* (H. A. Weaver and L. Danly, eds.), p. 119, Cambridge Univ., New York.
81. Norman C. A. and Paresce F. (1989) in *The Formation and Evolution of Planetary Systems* (H. A. Weaver and L. Danly, eds.), p. 151, Cambridge Univ., New York.
82. From Weissman P. R. (1986) *Galaxy and the Solar System* (R. Smoluchowski et al., eds.), p. 234, Table VI, Univ. of Arizona, Tucson.
83. Smith B. A. and Terrile R. J. (1984) *Science, 226*, 1421; Norman C. A. and Paresce F. (1989) in *The Formation and Evolution of Planetary Systems* (H. A. Weaver and L. Danly, eds.), p. 151, Cambridge Univ., New York. Viewed from Vega, our solar system would probably be invisible in the infrared.
84. Cameron A. G. W. (1978) in *The Origin of the Solar System* (S. F. Dermott, ed.), p. 49, Wiley, New York.
85. An extensive review of dust disks around low-mass stars is given by Wolstencroft R. D. and Walker H. J. (1988) *Philos. Trans. R. Soc., A325*, 423; see also Sargent A. I. (1989) in *The Formation and Evolution of Planetary Systems* (H. A. Weaver and L. Danly, eds.), p. 111, Cambridge Univ., New York.

86. Lissauer J. J. and Griffith C. A. (1989) *Astrophys. J., 340*, 468.
87. Latham D. W. et al. (1988) *Science, 241*, 790.
88. Campbell B. et al. (1988) *Astrophys. J., 331*, 902.
89. A particularly perceptive review is given by Cassen P. and Boss A. P. (1988) in *Meteorites and the Early Solar System* (J. F. Kerridge and M. S. Matthews, eds.), p. 304, Univ. of Arizona, Tucson. See also a later review by Boss A. P. et al. (1989) in *Origin and Evolution of Planetary and Satellite Atmospheres* (S. K. Atreya et al., eds.), p. 35, Univ. of Arizona, Tucson; Cameron A. G. W. (1988) *Annu. Rev. Astron. Astrophys., 26*, 441.
90. Tielens A. G. G. M. (1988) *LPI Tech. Rpt. 88-04*, Fig. 3.
91. A. P. Boss, personal communication, November 1989.
92. Jackson J. M. (1988) *Astrophys. J., 333*, L73; Cameron A. G. W. (1985) in *Protostars and Planets II* (D. C. Black and M. S. Matthews, eds.), p. 1073, Univ. of Arizona, Tucson; Morfill G. E. et al. (1985) ibid., p. 493; see also Fegley B. (1988) *LPI Tech. Rpt. 88-04*, p. 61.
93. Cameron A. G. W. (1985) in *Protostars and Planets II* (D. C. Black and M. S. Matthews, eds.), p. 1073, Univ. of Arizona, Tucson; Cameron A. G. W. (1989) in *The Formation and Evolution of Planetary Systems* (H. A. Weaver and L. Danly, eds.), p. 277, Cambridge Univ., New York.
94. Boss A. P. (1988) *Lunar and Planetary Science XIX*, p. 122, Lunar and Planetary Institute, Houston; Boss A. P. (1989) *Astrophys. J., 345*, 554.
95. Herbig G. H. (1978) in *The Origin of the Solar System* (S. F. Dermott, ed.), p. 219, Wiley, New York.
96. Stauffer J. R. and Soderblom D. R. (1989) *Abstracts Presented at the Conference on The Sun in Time*, p. 47, Tucson, Arizona.
97. Herbig G. H. (1978) in *The Origin of the Solar System* (S. F. Dermott, ed.), p. 226, Wiley, New York.
98. Russell H. N. (1935) *The Solar System and Its Origin*, Macmillan, New York, 144 pp.
99. Boss A. P. (1988) *Lunar and Planetary Science XIX*, p. 122, Lunar and Planetary Institute, Houston.
100. Cameron A. G. W. (1978) *Moon Planets, 18*, 5.
101. Stepinski T. F. and Levy E. H. (1988) *Astrophys. J., 331*, 416.
102. Boss A. P. (1985) *Icarus, 61*, 3.
103. Heyer M. H. (1988) *Astrophys. J., 324*, 311.
104. Heydari-Malayeri M. et al. (1989) *Astron. Astrophys., 222*, 41.
105. Shu F. H. et al. (1987) *Annu. Rev. Astron. Astrophys., 25*, 21 and 72 (Fig. 7).
106. Boss A. P. (1987) *Astrophys. J., 319*, 149; Boss A. P. (1988) *Astrophys. J., 331*, 370.
107. Vogel S. N. et al. (1988) *Nature, 334*, 402.
108. Shu F. H. et al., op. cit.; Adams F. C. et al. (1987) *Astrophys. J., 312*, 788.
109. Beichman C. A. et al. (1986) *Astrophys. J., 307*, 337.
110. Myers P. C. et al. (1986) *Astrophys. J., 301*, 398.
111. From Bodenheimer P. (1989) in *The Formation and Evolution of Planetary Systems* (H. A. Weaver and L. Danly, eds.), p. 246, Cambridge Univ., New York.
112. Lizano S. et al. (1988) *Astrophys. J., 328*, 774.
113. For example, Boss A. P. (1986) *Astrophys. J. Supp., 62*, 519.
114. Larson R. B. (1984) *Mon. Not. R. Astron. Soc., 206*, 197.
115. Hartmann L. and Kenyon S. J. (1985) *Astrophys. J., 299*, 462.
116. An excellent review of star formation in molecular clouds is given by Shu F. H. et al. (1987) *Annu. Rev. Astron. Astrophys., 25*, 23.
117. Abt H. A. (1983) *Annu. Rev. Astron. Astrophys., 21*, 343.
118. Halbwachs J. L. (1987) *Astron. Astrophys., 183*, 234.
119. See review by Boss A. P. (1988) *Comments Astrophys., 12*, 169.
120. Boss A. P. (1988) *Comments Astrophys., 12*, 176; Durisen R. H. et al. (1986) *Astrophys. J., 305*, 281.
121. Hunter C. (1962) *Astrophys. J., 136*, 594, following an earlier suggestion by Hoyle F. (1953) *Astrophys. J., 118*, 513; see review by Boss A. P. (1988) *Comments Astrophys., 12*, 178.
122. Boss A. P. (1987) *Astrophys. J., 319*, 149.
123. Zuckerman B. and Becklin E. E. (1987) *Nature, 330*, 138; Stringfellow G. S. et al. (1990) *Astrophys. J., 349*, L59; Becklin E. E. and Zuckerman B. (1988) *Nature, 336*, 656.
124. Latham D. W et al. (1988) *Science, 241*, 790; Latham D. W. et al. (1989) *Nature, 339*, 38.
125. Zuckerman B. and Becklin E. E. (1987) *Nature, 330*, 138.
126. Lada C. (1985) *Annu. Rev. Astron. Astrophys., 23*, 267.
127. See review by Basri G. (1988) in *Cool Stars, Stellar Systems and the Sun* (J. L. Linsky and R. E. Stencel, eds.), p. 411, Springer-Verlag, New York.
128. Herbig G. H. (1978) in *The Origin of the Solar System* (S. F. Dermott, ed.), p. 219, Wiley, New York.
129. Hartmann L. and Kenyon S. J. (1985) *Astrophys. J., 299*, 462.
130. Herbig G. H. (1977) *Astrophys. J., 217*, 693.
131. Cameron A. G. W. (1985) *Astrophys. J., 299*, L83.

132. Lada C. (1985) *Annu. Rev. Astron. Astrophys., 23*, 301.
133. Caffee M. et al. (1983) *J. Geophys. Res., 88*, B267; Caffee M. et al. (1987) *Astrophys. J., 313*, L31; Wieler R. et al. (1989) *Geochim. Cosmochim. Acta, 53*, 1441, 1449; Olinger C. T. et al. (1989) *Abstracts Presented at the Conference on The Sun in Time*, p. 36, Tucson, Arizona.
134. Payne C. H. (1925) *Proc. Nat. Acad. Sci., 11*, 192; Payne C. H. (1925) *Stellar Atmospheres*, pp. 56, 185, 188, Harvard Univ., Cambridge.
135. Russell H. N. (1929) *Astrophys. J., 70*, 11; Stromgren B. (1932) *Zeit. Astrophys., 4*, 118; Stromgren B. (1938) *Astrophys. J., 87*, 520.
136. From Anders E. and Grevesse N. (1989) *Geochim. Cosmochim. Acta, 53*, 197, Table 2.
137. Notably Ni and Co; Taylor S. R. (1965) *Nature, 208*, 886.
138. Anders E. and Grevesse N. (1989) *Geochim. Cosmochim. Acta, 52*, 203.
139. Breneman H. H. and Stone E. C. (1985) *Astrophys. J., 299*, L57.
140. Holweger H. (1988) in *The Impact of Very High S/N Spectroscopy on Stellar Physics* (G. Cayrel de Strobel and M. Spite, eds.), p. 411, International Astronomical Union; Holweger H. et al. (1990) *Astron. Astrophys., 232*, 510.
141. E. Anders, personal communication, May 1989, January 1990.
142. Maltby P. et al. (1986) *Astrophys. J., 306*, 284.
143. Alfven H. and Arrhenius G. (1976) *NASA SP-345*; see the evaluation of Alfven's work by Brush S. G. (1990) *Rev. Mod. Phys., 62*, 87.
144. The argument, despite its polarization into the two camps of large vs. small nebulae, may be somewhat of a red herring since the mass differences may not be readily distinguishable for current models; Cameron A. G. W. (1988) *Annu. Rev. Astron. Astrophys., 26*, 441; Hayashi C. et al. (1985) in *Protostars and Planets II* (D. C. Black and M. S. Matthews, eds.), p. 1100, Univ. of Arizona, Tucson.
145. Safronov V. S. (1972) *Evolution of the Protoplanetary Cloud and Formation of the Earth and Planets*, NASA TT-667; Wetherill G. W. (1980) *Annu. Rev. Earth Planet. Sci., 18*, 77; Hayashi C. et al. (1985) in *Protostars and Planets II*, p. 1100, Univ. of Arizona, Tucson.
146. Boss A. P. et al. (1989) in *Origin and Evolution of Planetary and Satellite Atmospheres* (S. Atreya et al., eds.), p. 64, Univ. of Arizona, Tucson.
147. Boss A. P. et al., op. cit., p. 35.
148. Boss A. P. (1988) *Science, 241*, 505.
149. For example, Wasson J. T. and Kallemeyn G. W. (1988) *Philos. Trans. R. Soc., A325*, 535.
150. Boss A. P. (1988) *Science, 241*, 505; Morfill G. E. (1988) *Icarus, 73*, 371.
151. Burnett D. S. et al. (1989) *Geochim. Cosmochim. Acta, 53*, 471.
152. Suess H. and Urey H. C. (1956) *Rev. Mod. Phys., 58*, 53; Burnett D. S. et al. (1989) *Geochim. Cosmochim. Acta, 53*, 471; Anders E. and Grevesse N. (1989) *Geochim. Cosmochim. Acta, 53*, 197.
153. Anders E. and Grevesse N. (1989) *Geochim. Cosmochim. Acta, 53*, 197.
154. Burnett D. S. et al. (1989) *Geochim. Cosmochim. Acta, 53*, 479.
155. Jungck M. H. A. and Eberhardt P. (1985) *Meteoritics, 20*, 677. The difficult question of the nature of the composition of the fine-grained matrix material is addressed in section 3.4; see also Tomeoka K. and Buseck P. R. (1988) *Geochim. Cosmochim. Acta*, 52, 1627.
156. McSween H. Y. (1979) *Rev. Geophys Space Phys., 17*, 1074.
157. Anders E. and Grevesse N. (1989) *Geochim. Cosmochim. Acta, 53*, 197.
158. Rietmeijer F. J. M. (1988) *Chem. Geol., 70*, 33.
159. Urey H. C. (1967) *Q. J. R. Astron. Soc., 8*, 23.
160. Holweger H. (1988) in *The Impact of Very High S/N Spectroscopy on Stellar Physics* (G. Cayrel de Strobel and M. Spite, eds.), p. 411, International Astronomical Union.
161. Holweger H. et al. (1990) *Astron. Astrophys., 232*, 510.
162. Anders E. and Grevesse N. (1989) *Geochim. Cosmochim. Acta, 53*, 198, Table 1.
163. Anderson D. L. (1989) *Science, 243*, 367.
164. Niemeyer S. (1988) *Geochim. Cosmochim. Acta, 52*, 2941.
165. Niemeyer S. and Lugmair G. W. (1984) *Geochim. Cosmochim. Acta, 48*, 1401.
166. For example, Ringwood A. E. (1966) *Geochim. Cosmochim. Acta, 30*, 41.
167. For example, Lewis J. S. (1973) *Annu. Rev. Phys. Chem., 24*, 339; Lewis J. S. (1974) *Science, 186*, 440.
168. Dreibus G. et al. (1988) *Lunar and Planetary Science XIX*, p. 283, Lunar and Planetary Institute, Houston.
169. Rietmeijer F. J. M. (1988) *Chem. Geol., 70*, 33.
170. Prinn R. G. and Fegley B. (1989) in *Origin and Evolution of Planetary and Satellite Atmospheres* (S. Atreya et al., eds.), p. 78, Univ. of Arizona, Tucson.
171. Lewis J. S. and Prinn R. G. (1984) *Planets and Their Atmospheres*, p. 383, Academic, New York.
172. Ozima M. and Podosek F. A. (1983) *Noble Gas Geochemistry*, p. 208, Cambridge Univ., New York.
173. Hayashi C. et al. (1979) *Earth Planet. Sci. Lett., 43*, 22.

174. Mizuno H. et al. (1980) *Earth Planet. Sci. Lett., 50*, 202.
175. Ozima M. and Podosek F. A. (1983) *Noble Gas Geochemistry*, pp. 290-291, Cambridge Univ., New York.
176. An excellent and well-balanced review is given by Thiemens M. H. (1988) in *Meteorites and the Early Solar System* (J. F. Kerridge and M. S. Matthews, eds.), p. 899, Univ. of Arizona, Tucson.
177. Wasson J. T. (1985) *Meteorites: Their Record of Early Solar-System History*, Freeman, New York, 288 pp.
178. Prinn R. G. and Fegley B. (1989) in *Origin and Evolution of Planetary and Satellite Atmospheres* (S. Atreya et al., eds.), p. 78, Univ. of Arizona, Tucson.
179. Simonelli D. et al. (1989) *Icarus, 82*, 1.
180. Black D. C. and Pepin R. O. (1969) *Earth Planet. Sci. Lett., 6*, 395.
181. Clayton R. N. et al. (1973) *Science, 182*, 485.
182. For example, Clayton R. N. et al. (1988) *Philos. Trans. R. Soc., A325*, 483.
183. Stevenson D. J. (1988) *Lunar and Planetary Science XIX*, p. 1123, Lunar and Planetary Institute, Houston.
184. Basic references are Gehrels T., ed. (1979) *Asteroids*, 1181 pp. and Binzel R. et al., eds. (1989) *Asteroids II*, Univ. of Arizona, Tucson, 1258 pp.
185. Wetherill G. W. (1988) *Meteoritics, 23*, 310.
186. Wetherill G. W. and Stewart G. R. (1989) *Icarus, 77*, 330.
187. Lissauer J. J. (1987) *Icarus, 69*, 249; Wetherill G. W. (1989) in *Asteroids II* (R. Binzel et al., eds.), p. 661, Univ. of Arizona, Tucson.
188. Semarkona meteorite: Hutcheon I. D. and Hutchison R. (1989) *Nature, 337*, 238.
189. Sonett C. P. and Reynolds R. T. (1979) in *Asteroids* (T. Gehrels, ed.), p. 822, Univ. of Arizona, Tucson.
190. Mercer-Smith J. A. et al. (1984) *Astrophys. J., 279*, 363.
191. Wood J. A. and Pellas P. (1991) in *The Sun in Time*, Univ. of Arizona, Tucson, in press.
192. Rubin A. E. et al. (1988) in *Meteorites and the Early Solar System* (J. F. Kerridge and M. S. Matthews, eds.), p. 488, Univ. of Arizona, Tucson.
193. Shu F. (1987) *Annu. Rev. Astron. Astrophys., 25*, 23.
194. Guseinov K. M. (1988) *Earth, Moon, Planets, 43*, 1.
195. Sargent A. I. and Beckwith S. (1987) *Astrophys. J., 323*, 294.
196. The amount of material now in the inner solar system is so small in comparison with the mass of the Sun that it has no effect on the bulk composition of the system; Gehrels T., ed. (1978) *Protostars and Planets*, Univ. of Arizona, Tucson; D. C. Black and M. S. Matthews, eds. (1985) *Protostars and Planets II*, Univ. of Arizona, Tucson.
197. A. G. W. Cameron, personal communication, November 1989.
198. Nakagawa Y. et al. (1986) *Icarus, 67*, 375.
199. Weidenschilling S. J. (1988) in *Meteorites and the Early Solar System* (J. F. Kerridge and M. S. Matthews, eds.), p. 348, Univ. of Arizona, Tucson.
200. For example, H chondrites are depleted in Sb, F, Cu, Ga, Ge, Sn, S, Se, Te, and Ag by a factor of 0.23 compared to CI abundances; Krahenbuhl U. et al. (1973) *Geochim. Cosmochim. Acta, 37*, 1353.
201. For example, McSween H. Y. (1985) *Rev. Geophys., 23*, 391.

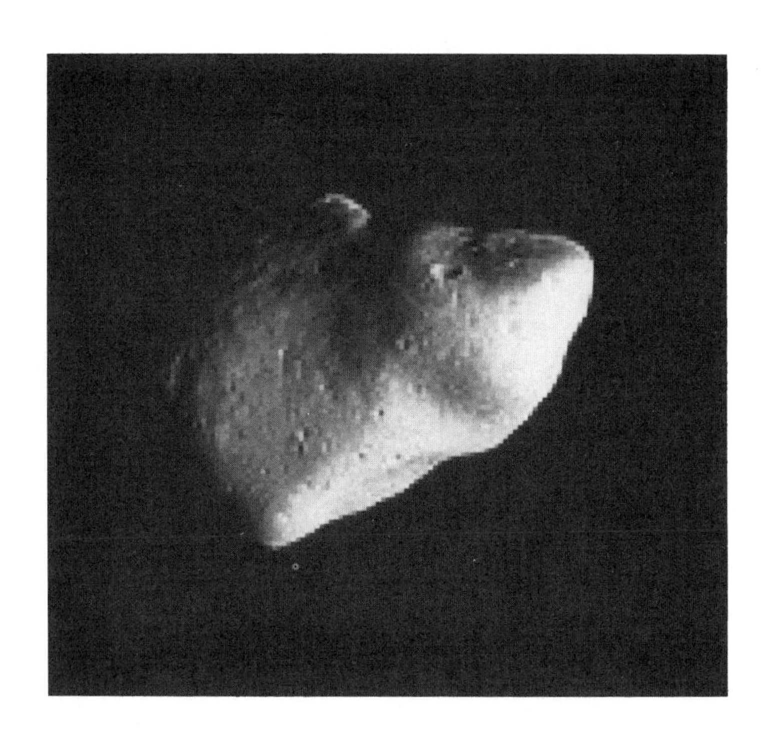

Chapter 3

The Meteorite Evidence

Illustration, preceding page—*Asteroid 951 Gaspra. Meteorites are mostly derived from the asteroid belt, and the first close-up view of an asteroid was of 951 Gaspra, photographed by the Galileo spacecraft on 29 October, 1991, from 16,200 km. 951 Gaspra is a main-belt S-class asteroid with a semimajor axis of 2.21 A.U. and dimensions of 18 × 11 × 10 km. The visible area of the asteroid is 16 × 12 km and the asteroid has a rotation period of 7 hours. This asteroid has suffered a number of collisions, as evidenced by the presence of many craters. (NASA JPL Galileo P-39432 GLL/GA2.)*

The Meteorite Evidence

3

3.1. THE MOST ANCIENT SAMPLES

"Many know much about this stone, everyone knows something, but no one knows quite enough" [1].

Meteorites present us with tantalizing evidence concerning the origin of the solar system and they provide us with the best evidence for the time when the solar system formed. Even though the CI carbonaceous chondrites were somewhat altered in an aqueous environment, the elemental abundances of these meteorites resemble those of solar photospheric spectra for the nongaseous elements, providing a basis for estimating the composition of the solar nebula [2].

Some of the grains in primitive chondrites retain a memory of presolar conditions. The refractory inclusions in chondrites tell us of the earliest recorded events in the solar system. Much chemical processing occurred very early, which resulted in the formation of chondrules, separation of metal, sulfide, and silicate phases, and depletion of volatile elements in the inner parts of the solar nebula. Differentiated meteorites such as the eucrites and the iron meteorites tell us of the existence of both planetesimals and of heat sources that induced melting. This differentiation also occurred very early, with the production of true igneous rocks close to T_0.

Since the cessation of the turbulent conditions accompanying planetary formation, meteorites that reach Earth have had an interesting dynamical history. This includes varying exposure to cosmic rays, which enable us to date both the breakup of parent bodies and their time of arrival on Earth. The exposure history of meteorites is, however, not of immediate concern here. What we seek in this account is the information that meteorites can provide on the composition and history of the solar nebula. They are the most ancient objects we can study in terrestrial laboratories and they preserve a unique record. Without their evidence we would have great difficulty in deciphering the history of the early solar system.

Nevertheless, meteorites need to be kept in perspective. They are rocks broken off asteroids and so are derived from a small, unique part of the system, from which we do not have a representative sample collection. Compositional zoning is evident in the asteroid belt, with more primitive carbonaceous material apparently increasingly abundant with distance from the Sun. This zonation is probably a secondary effect due to early heating of the inner belt, rather than representing a primary heliocentric zoning in the nebula. The elemental and isotopic composition of meteorites differ in many respects from the terrestrial planets, and the present population of meteorites is not left-over building blocks, although it does provide important analogues.

A decisive advantage is that we can sample, analyze, and, most importantly, determine the ages of meteorites in terrestrial laboratories. The difficulties in deciphering information from remote sensing of planetary surfaces, or in interpreting data from automated instruments (e.g., the Viking Landers on Mars) will ensure a continuing study of these accessible specimens of the asteroid belt. The length of this chapter is in itself a tribute to the unique importance of meteorites to studies of the early solar system.

3.2. PRESOLAR MATERIAL

In one respect, all the material in the solar nebula was presolar. Hydrogen and He with some D and Li were formed in the Big Bang. The heavier elements

TABLE 3.2.1. Isotopic ratios in meteoritic silicon carbide (SiC).

Ratio	Solar System	Silicon Carbide	
		Absolute	$\odot \equiv 1$
C^{12}/C^{13}	89.9	10-160	0.12-1.78
N^{14}/N^{15}	272.2	65-6300	0.24-23
Si^{29}/Si^{28}	0.0506	0.043-0.056	0.85-1.11
Si^{30}/Si^{28}	0.0336	0.026-0.037	0.78-1.10

Data from E. Anders, personal communication, 1990.

were synthesized from an immense number of stellar sources over the lifetime of the galaxy (>10^{10} years) and dispersed mainly by supernovae. The degree of homogenization of this material in the interstellar medium is unknown.

Most models suggest that the nebula fragmented from a small molecular cloud, and did not grow by accretion of interstellar dust. Accordingly, minimal heating by gravitational infall preceded the beginning of the rotational phase of the solar nebula. Little is known of the conditions to which the gases and grains were subjected on this long journey, and the search for "pristine" material has proven difficult.

In general, interstellar material is identified by the presence of isotopic anomalies that do not appear to result from solar nebula or solar system processes as we understand them. By this means, several forms of presolar carbon, notably SiC and diamond, have been identified in meteorites (Table 3.2.1). These have been discussed at length in section 2.6. The detection of such isotopic anomalies immediately raises the question of the preservation of presolar material in the nebula, and whether the strange isotopic anomalies are perhaps due to little-understood physical or chemical processes within the nebula.

TABLE 3.2.2. Enrichments observed in deuterium.

	D/H ($\times 10^5$)	Enrichment Factor
Interstellar	0.5-2	
Protosolar nebula	2 ± 1	1
Earth (SMOW)	15.7	8
Venus	$1{,}600 \pm 200$	800
Jupiter (methane)	3.6 ± 1.2	1.8
Saturn (methane)	2.0 ± 1.5	1
Uranus (methane)	$9^{+9}_{-4.5}$	4.5
Titan (methane)	$16.5^{+16.5}_{-8.8}$	8
Interstellar molecules		
HCN	80-680	340
HCO^+	40-1,000	500
	1,000-10,000	5,000
HNC	≥50,000	≥25,000
C_3H_2	2,900	1,500
Meteorites		
carbonaceous chondrites	8-60	30
ordinary chondrites	8-105	50
Interplanetary dust	12-160	80

Data from Zinner E. (1988) in *Meteorites and the Early Solar System* (J. F. Kerridge and M. S. Matthews, eds.), p. 970, Fig. 13.2.2, Univ. of Arizona, Tucson.

3.2.1. Deuterium/Hydrogen Ratios

Much organic matter in meteorites has high D/H ratios. Since high D/H ratios are commonly observed in interstellar material, this raises the possibility that such meteoritic material is preserving an interstellar signature [3] (Table 3.2.2). Thus, many organic compounds in interstellar clouds show marked enrichment in D. Cold molecular clouds show D/H enrichments up to several thousand per mil for molecules such as DCN/HCN [4].

With only two stable isotopes, it is not possible to disentangle the effects of mass-dependent fractionation for H and D from effects due to nucleosynthesis. However, the enrichment in D is connected with the chemistry of formation of the organic compounds and is not due to nuclear reactions. Probably it occurred in the interstellar medium where a combination of low temperatures and an ionizing medium permitted ion-molecule reactions to proceed in molecular clouds [5]. A possible pathway is shown in Fig. 3.2.1.

Extremely high values for D/H ratios, ranging up to 2500 per mil, have been found in interplanetary dust particles [6]. These are not induced by laboratory handling and are not instrumental artifacts. They are not produced during atmospheric entry, even though the upper atmosphere has a D/H value of +310 per mil compared to Standard Mean Ocean Water (SMOW). The anomalies are associated with a carbonaceous phase. Other particles have very exotic oxygen isotope compositions, unlike any known meteorites.

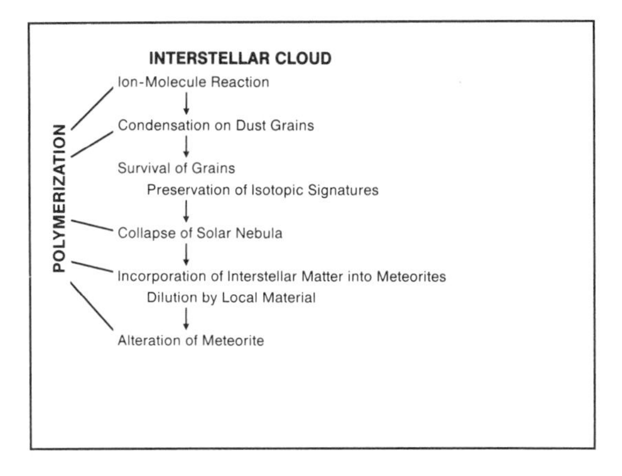

Fig. 3.2.1. *Diagram illustrating the evolution of deuterium-rich organic material from its synthesis in molecular clouds to its incorporation in meteorites. After Zinner E. (1988) in Meteorites and the Early Solar System (J. F. Kerridge and M. S. Matthews, eds.), p. 977, Fig. 13.2.6., Univ. of Arizona, Tucson.*

As noted earlier, large D excesses have been found in primitive meteorites. The highest value (D/H = 5470 per mil) is observed in the unequilibrated chondrite, Semarkona [7]. These anomalous D/H values in primitive meteorites appear to be associated with carbonaceous material, in contrast to the anomalies in the oxygen isotopes that seem to occur in the refractory grains such as hibonite. The D enrichments seen in organic material in meteorites probably originate in precursors in molecular clouds. The survival of such material in some meteorites indicates that temperatures were not sufficient to vaporize and homogenize all interstellar material in the nebula; consequently, much interstellar material may survive, particularly in the outer parts of the nebula.

3.2.2. Isotopic Anomalies

Three types of isotopic anomalies can be distinguished:

1. Those that can be attributed to the former existence of a short-lived parent. Examples include ^{26}Mg from ^{26}Al, ^{107}Ag from ^{107}Pd, ^{129}Xe from ^{129}I, and ^{53}Mn from ^{53}Cr. A requirement is that there should be a correlation between the magnitude of the anomaly and the inferred abundance of the parent. Many such systems have been disturbed, so, for example, some inclusions high in ^{27}Al show ^{26}Mg anomalies, but others do not. In such cases, it is unclear whether the ^{26}Al parent was present or whether the anomalous isotopic abundance is due to some other process.

2. Anomalies in stable isotopes that are generally associated with mass-dependent fractionation effects (e.g., enrichments in neutron-rich isotopes: ^{26}Mg, ^{48}Ca, ^{50}Ti, ^{54}Cr). Of course, the magnitude of these enrichments is due to the choice of the isotope used for normalization, and might conceivably reflect instead depletions in neutron-poor isotopes. These are the so-called FUN (fractionation and unknown nuclear) anomalies.

3. Very large isotopic anomalies in refractory grains such as hibonite and SiC. Considerable effort has been made to identify presolar material that carries the signature of anomalous isotopic compositions, which may have survived in meteorites. Attention has been focused on the calcium-aluminum-rich inclusions (CAIs) that represent the oldest solid material found in the solar nebula [8].

There is voluminous literature on the little-understood question of the origin of isotopic anomalies. Two effects are superimposed: mass-dependent fractionation associated with evaporation and condensation, and nucleosynthetic processes. Fractionation effects can be ascribed to multiple episodes of evaporation within the solar nebula, but residual non-mass-dependent anomalies may require a source outside the nebula. In the present context we are concerned not so much with the origin of the anomalies as with the evidence they provide for the survival of presolar material.

Prominent isotopic anomalies are found in neutron-rich isotopes of the iron-peak elements for ^{48}Ca, ^{49}Ti, ^{50}Ti, ^{54}Cr, and ^{64}Ni. These anomalies are usually attributed to neutron-rich nucleosynthesis occurring in supernovae. There is some correlation between the ^{48}Ca and ^{50}Ti anomalies that would indicate that these isotopes were produced together, as is required by the supernova hypothesis (Fig. 3.2.2). However, it is possible that these non-mass-dependent anomalies arise from nonnuclear processes such as sputtering [9,10]. A basic problem has been to isolate the general effects of galaxywide nucleosynthesis from those due to injection into the solar nebula of a freshly synthesized spike of isotopically anomalous material shortly before the condensation of the solar system.

Clayton [11] has pointed out that the bulk isotopic composition of the interstellar medium changes with time in a linear fashion due to stellar nucleosynthesis. He distinguishes "primary" from "secondary" products of nucleosynthesis. "Primary" isotopes are produced even in stars consisting only of H and He. Such

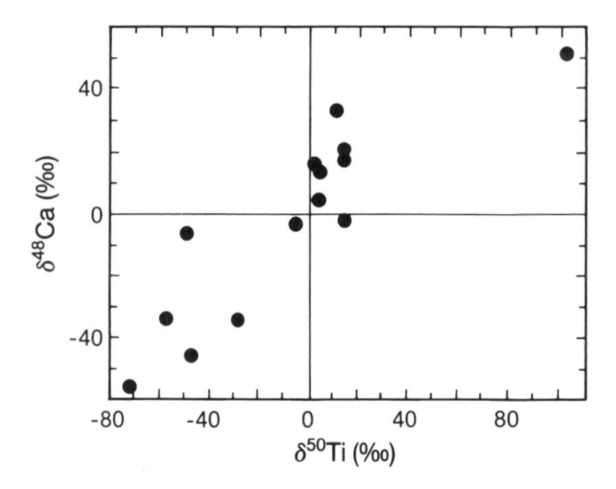

Fig. 3.2.2. *The correlation of the nucleosynthetic anomalies in ^{48}Ca and ^{50}Ti in Allende refractory inclusions. After Clayton R. N. et al. (1988) Philos. Trans. R. Soc., A325, 493.*

species include ^{12}C, ^{16}O, ^{20}Ne, ^{24}Mg, ^{28}Si, ^{32}S, ^{36}Ar, ^{40}Ca, ^{44}Ca, ^{48}Ti, ^{52}Cr, and ^{56}Fe. "Secondary" products are produced in proportion to the heavier elements (C,O) present initially in the star. These include ^{13}C, ^{17}O, ^{18}O, ^{21}Ne, ^{22}Ne, ^{25}Mg, ^{26}Mg, ^{29}Si, and ^{30}Si. Since the initial C and O content of stars has grown with time, so too will these "secondary" products. Thus, the present interstellar medium will be isotopically different from that from which the solar system condensed. Any mechanism that separates older from younger supernova ejecta will produce regions of isotopically anomalous material.

One of the most important variations in isotopic abundances in meteorites occurs for oxygen. This evidence is discussed mainly in section 2.17, but some important additional points may be noted here. The variations in oxygen isotope composition are due mainly to large differences in the ^{16}O abundance, and appear to be due to exchange between ^{16}O-poor gases and ^{16}O-rich solids. It is not clear whether these reservoirs are inherited from nucleosynthetic processes in the galaxy, or are due to chemical processes, producing non-mass-dependent fractionation in the nebula [12].

Variations in oxygen isotopes are observed among both planets and individual chondrules and indicate a lack of thorough mixing or homogenization in the nebula. The variations in individual chondrules and refractory grains (CAIs; see section 3.3) mostly fall on mixing rather than fractionation lines, indicating that the variations arose from exchange with a gaseous oxygen-bearing phase (H_2O or CO). It is curious that there is, in general, no correlation between the isotopic anomalies for oxygen and for other elements. Oxygen is present in the nebula both as a gas and as a condensed phase in minerals. In contrast, the other elements that display isotopic anomalies are present in solid phases. This difference may account for the observed lack of correlation (Fig. 3.2.3).

The persistence of isotopic anomalies in meteorites testifies to incomplete mixing in the solar nebula, but gives no information about the time when these variations arose. However, several chronometers are available. These include fission products from ^{244}Pu, ^{129}I that produces ^{129}Xe, ^{107}Pd that decays to ^{107}Ag, ^{53}Mn that decays to ^{53}Cr, and, important in the present context, ^{26}Al that decays to ^{26}Mg and that may be significant as a heat source for melting planetesimals, although considerable doubt exists about this [13]. The ^{26}Al/^{27}Al ratio was about 5×10^{-5}, providing an adequate heat supply if ^{26}Al was live in the early solar system. Live ^{26}Al has been detected, but whether it is present in the interstellar medium, continually supplied from novae or coming from the galactic center, is uncertain [14]. The ^{26}Al/^{27}Al ratio is of the order of 10^{-5}, perhaps adequate to account for the meteoritic evidence [15].

It is a curious fact, even in a topic so opaque as the FUN inclusions, that inclusions with ^{50}Ti anomalies lack ^{26}Mg evidence of the former presence of ^{26}Al, and vice versa. Although the conventional explanation is that the Ti anomalies originate in a source distinct from the ^{26}Al, it is puzzling that the refractory inclusions are so selective in acquiring one or the other of these anomalies.

The refractory inclusions contain opaque nuggets of refractory siderophile elements (Ru, Os, Re, Pt, Ir, W, and Mo). The micrometer-sized metallic nug-

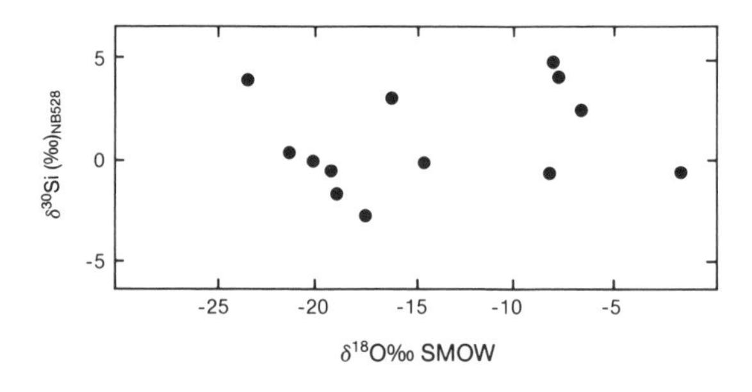

Fig. 3.2.3. *The lack of correlation of oxygen and silicon isotopic abundances in Allende refractory inclusions (CAI) indicates the operation of distinct cosmochemical processes such as condensation and evaporation for Si and exchange with a gaseous reservoir for O. After Clayton R. N. et al. (1988) Philos. Trans. R. Soc., A325, 484, Fig. 1.*

gets are surrounded by Ni-Fe metal, Fe-Ni sulfides, V-rich magnetite, and minor molybdenite, tungstate, and phosphate. These opaque assemblages, up to 1000 μm in diameter, have been referred to as *Fremdlinge* (German for "little strangers"), and were proposed to result from condensation at very high temperatures (1500-2000 K) from a nebular gas, or in a supernova envelope followed by reactions during cooling, and eventual incorporation in molten CAIs during a second (1700 K) brief heating event. On the basis of this scenario, *Fremdlinge* record much information about temperature and pressure conditions in the primordial nebula and have preserved a record of conditions in the early nebula, before formation of the CAIs [16]. They could thus be the oldest preserved solid materials to which we have access.

However, the alternative view that they formed as molten alloy droplets within the CAIs has received strong support from experimental studies that indicate that these opaque assemblages, a term probably preferable to *Fremdlinge,* originate within the CAI's liquids as homogeneous metal droplets [17]. It appears that after the homogeneous metallic droplets cooled and solidified, they exsolved into the various immiscible phases, which were subsequently partially oxidized and converted to sulfides. Accordingly, it appears that the *Fremdlinge* are not exotic, but have a domestic origin. This removes the embarassing geochemical problem of concentrating such rare elements as Ru, Os, Re, or Pt into nuggets in low-density nebulae. While the opaque assemblages record important information about the cooling history of the CAIs, they do not provide any insights into presolar conditions [18].

3.3. REFRACTORY OR CALCIUM-ALUMINUM INCLUSIONS (CAIs)

"The topic is hard to come to grips with. The literature of refractory inclusions... is mostly descriptive, consisting in large part of long, densely packed petrographic treatises. The properties of [refractory inclusions] are so diverse that it is difficult to generalize about them. No conclusion can be drawn about one subset... that is not inconsistent with observations made in other inclusions. It is hard to isolate the scientific issues addressed by [refractory inclusion] research, and it is hard even to define the class of objects referred to" [19].

In this text, refractory inclusions are referred to as such or as CAIs [20]. Although this latter term will not satisfy purists, it is in common usage, and is so used rather than RI (for refractory inclusion, a better term if it had come into use earlier) to avoid cluttering up the literature with more acronyms in this very confusing topic. Most CAIs are about the same size as chondrules, although they range from vanishingly small to 5-10 cm in size. They have formed in space, not in a planetary environment, and it is tempting to seek some analogies or relationships with chondrules; we will see later that there is a possible connection, although it seems clear that the conditions that gave rise to CAIs were both more extreme with respect to temperature and predated chondrule formation.

The CAIs have been classified into six groups on the basis of variations in their rare earth element (REE) patterns [21] (Tables 3.3.1, 3.3.2). The abundance patterns of the refractory siderophile elements in CAIs are related to their volatility, with more refractory elements (e.g., Re) being enriched relative to the more volatile elements (e.g., Pt, Ru). Molybdenum and W do not fit this simple pattern, but their distribution is consistent with high-temperature oxidizing conditions, with oxygen fugacities up to factors of 10^3 to 10^4 times that of the primitive solar nebula [22]. This enrichment is independent of pressure [23].

The refractory inclusion mineral suite includes perovskite, melilite, spinel, and hibonite, and matches the "condensation sequence" (see section 3.9). This led initially to a general perception that the CAIs had condensed from a hot nebula under equilibrium conditions. However, various factors led to a revision of a simple monotonic cooling scenario. It is clear that although the nebula could have reached temperatures of 1500 K at 2.5 AU during an early stage of infall of gas into the Sun, it is not clear that such high temperatures were maintained until the later formation of either the CAIs or the chondrules [24].

Various empirical observations argue against a hot nebula at this time. The mineral sequence does not follow the theoretical condensation sequence in detail, and differs from inclusion to inclusion, arguing for local rather than regional heating. Astonishingly complex REE patterns have been discovered that record multiple episodes of condensation and evaporation in localized environments. It is curious and paradoxical to record that the isotopic anomalies are best preserved in the most refractory material, which has undergone many episodes of high-temperature evaporation and condensation. One might have expected the isotopes to have been homogenized. Mass-dependent isotopic fractionation is to be expected under these conditions, but the survival of non-mass-dependent anomalies raises questions that have yet to be answered.

TABLE 3.3.1. Major-element compositions of refractory inclusions from CV3 and CM meteorites.

	CV3				CM		
	Type B-1	Type B-2	Type A	Spinel-rich Fine-grained	Corundum-Hibonite	Spinel Hibonite	Melilite-rich
SiO_2	29.1	32.9	25.1	33.7	—	—	~6.7
Al_2O_3	29.6	26.7	37.6	26.6	91.5	75.1	55.7
TiO_2	1.3	1.5	1.0	1.3	<1.9	2.5	1.8
FeO	0.6	0.9	1.7	2.3	0.6	0.4	0.3
MgO	10.2	11.5	4.3	13.1	~0.5	17.1	12.3
CaO	28.8	25.9	29.4	21.6	6.2	4.8	23.0
Na_2O	0.18	0.50	0.8	1.1	0.03	<0.03	0.02
K_2O	0.01	0.02	—	0.05	—	—	—
Cr_2O_3	0.04	0.05	0.02	0.1	<0.01	0.08	0.05
MnO	—	—	0.01	—	<0.01	<0.01	<0.01
V_2O_3	—	—	0.09	—	<0.03	0.29	0.12
NiO	0.06	0.04	0.03	0.08	<0.01	<0.01	0.02
Sum	99.89	100.01	100.05	99.93	100.73	100.27	100.01

Data from MacPherson C. J. et al. (1988) in *Meteorites and the Early Solar System* (J. F. Kerridge and M. S. Matthews, eds.), p. 765, Table 10.3.2, Univ. of Arizona, Tucson.

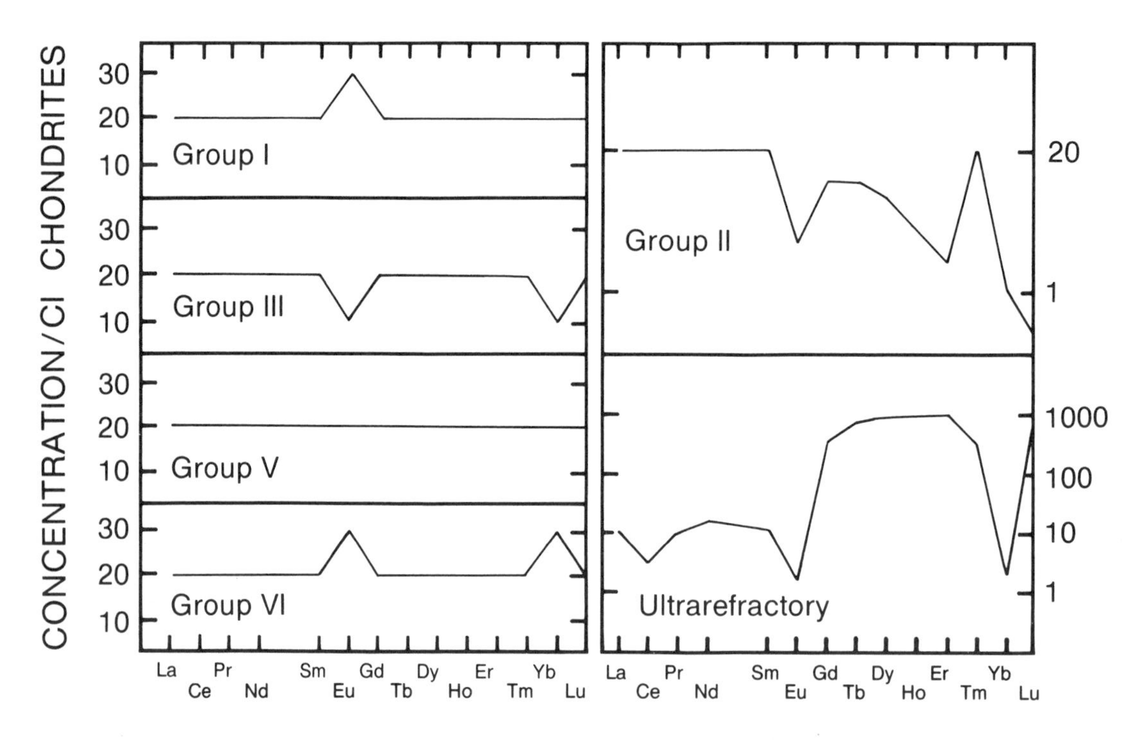

Fig. 3.3.1. *Typical rare earth element (REE) patterns seen in refractory inclusions (CAI). Group III and VI patterns show variations due to loss or gain of the most volatile REE, Eu, and Yb. Group II patterns have lost both the most volatile (Eu and Yb) and the most refractory REE, while the "ultrarefractory" patterns are enriched in the latter REE and depleted in the volatile and moderately refractory REE. After MacPherson G. J. et al. (1988) in Meteorites and the Early Solar System (J. F. Kerridge and M. S. Matthews, eds.), p. 756, Fig. 10.3.4, Univ. of Arizona, Tucson.*

The fine-grained CAIs appear to be mainly condensates, while the coarse-grained CAIs are evaporation residues, but multistage events of evaporation and condensation are required to account for the observed complexity; single-stage cooling from a hot nebula is inadequate to account for the observed mineralogy and geochemistry. Although most CAIs are enriched in refractory elements (Ca, Al, REE, Ti, Zr, Hf, etc.) some objects (amoeboid olivine aggregates) that are not particularly refractory are also included in the classification.

Evaporation presents an alternative to condensation; in this scenario the CAIs represent unvaporized residues. The high contents of refractory elements (20 × CI abundance levels) become rather readily understood. There is a basic problem in simple condensation sequences that is seldom addressed: The low density of matter means that timescales of the order of years are required for millimeter-sized objects to accrete even the common and abundant elements in the low-density nebula [25]. The growth of nuggets of rare noble metals or *Fremdlinge* seems impossible in such an environment.

Thus, CAIs contain evidence both of nebular condensates and refractory residues resulting from evaporation. Both processes are required to account for the Group II REE patterns (Fig. 3.3.1) that are depleted in both the volatile REE, Eu and Yb, and the most refractory REE, Er and Lu. The latter must have been removed from a vaporized system, which then condensed the less refractory elements, but failed to condense the most volatile elements. In order to account for the many diverse features of CAIs, it is clear that multiple cycles of evaporation and condensation are required [22] (Fig. 3.3.1).

Hibonites [$Ca(Al,Mg,Ti)_{12}O_{19}$], which are common constituents of CAIs, exhibit highly anomalous Ca and Ti isotope abundances and refractory element patterns. They appear to have formed by several episodes of melting of refractory dust aggregates [26]. It is also clear that the environment was rich in oxygen, perhaps up to 100 times the primordial O/H ratio, both from mineralogical and chemical evidence. This significant piece of evidence is consistent with very early depletion of H in the region of meteorite formation [27]. The environment was also enriched in ^{16}O, the CAIs being very enriched with $\delta^{18}O$ of -40 per mil and $\delta^{17}O$ of -42 per mil. Only hibonites have higher δO enrichments.

The ubiquitous thin rims, about 50 μm thick on CAIs, have caused much debate. The rims are sintered and are very different from chondrule rims, which are volatile-rich coatings. The CAIs rims are almost certainly refractory residues formed by flash heating, followed by some metasomatism. The principal evidence for this is that the REE patterns of the rims match those of the interior, but are enriched 3-5 times in absolute concentration levels. Two elements are exceptions: the volatile REE, Eu and Yb, which display relative depletions in the rims. Apart from Eu and Yb, the rim REE patterns mimic those of the interior, displaying all the complexities of the diverse patterns exhibited by the CAIs. Subsequent to the flash melting, some metasomatism by Si and Mg has

TABLE 3.3.2. Trace element compositions of Allende refractory inclusions (CAI).

Sample No.	3655 B	3529 Z	3529 0	3655 A	3529 Y
Group	II	I	III	V	VI
SiO_2 %	32.8	30.9	26.5	29.5	30.3
TiO_2 %	0.98	1.21	2.32	1.78	1.50
Al_2O_3 %	36.5	28.9	36.3	29.7	28.7
FeO %	2.63	0.30	3.52	1.00	0.87
MgO %	9.83	9.25	8.86	11.4	10.8
CaO %	17.2	29.4	23.4	26.5	27.6
Na_2O %	0.57	0.18	0.21	0.19	0.23
Ba ppm	52	75	42	37	37
Pb ppm	1.52	7.2	—	1.24	2.6
Th ppm	0.57	0.58	1.10	0.62	0.48
U ppm	0.025	0.09	0.10	0.092	0.20
Zr ppm	7.3	97	95	112	90
Hf ppm	0.18	2.35	3.20	2.74	1.97
Nb ppm	1.82	3.75	3.64	5.41	7.80
Y ppm	3.43	34.8	41	42	38
La ppm	9.21	5.36	7.51	4.89	4.43
Ce ppm	22.7	12.8	19.4	13.3	13.6
Pr ppm	3.44	2.03	2.68	1.99	1.65
Nd ppm	18.7	9.74	14.1	10.4	8.53
Sm ppm	5.38	2.77	4.12	3.00	2.65
Eu ppm	0.52	1.37	1.09	1.01	1.14
Gd ppm	3.57	3.62	6.15	4.19	3.24
Tb ppm	0.58	0.67	1.13	0.78	0.62
Dy ppm	3.12	4.62	7.80	5.36	4.51
Ho ppm	0.26	1.17	1.95	1.30	1.06
Er ppm	0.28	3.55	5.94	4.12	3.45
Tm ppm	0.74	0.45	0.85	0.62	0.46
Yb ppm	1.36	3.38	4.52	3.85	4.44
Σ REE ppm	69.9	52	77	55	50
Pt ppm	0.62	7.8	10.7	11.8	15.1
Ir ppm	0.35	7.4	8.0	10.7	9.1
Os ppm	0.27	7.0	7.5	9.9	9.6
Re ppm	—	0.57	0.62	0.80	0.97
Pd ppm	0.27	1.0	1.55	1.17	1.84
Rh ppm	0.15	1.09	0.94	1.14	1.34
Ru ppm	0.73	13.4	8.9	13.3	10.7

Data from Mason B. and Taylor S. R. (1982) *Smithson. Contrib. Earth Sci.*, *25*, 6, Table 1.

occurred. Perovskite is a major constituent of the rims, as expected, since major components of that mineral, such as Ti, are refractory. Melting times are of the order of one second, a time derived from the fact that melilite inside the CAIs remains unmelted 200 μm from the rim. Based on the refractory element enrichments, a thickness of about 100 μm of material was volatilized. Energy equivalent to 300 W/cm^2 is required. Whatever energy source was responsible, it was brief and localized. Many of the CAIs are distillation residues and must have started with volumes at least an order of magnitude larger. Curiously, they display ^{26}Mg anomalies.

A link between the CAIs and chondrule formation is preserved in some chondrules from the Felix CO3 carbonaceous chondrite. These chondrules appear to be remelted CAIs with highly fractionated REE patterns resembling Group II patterns. This indicates that CAI production preceded chondrule formation and could thus constitute evidence for formation of CAIs in a presolar environment. Alternatively, a very complex scenario of thermal processes in a very turbulent early nebula, probably before the disk formed, must be constructed. Their ages and initial Sr isotope ratios do not suggest a remote origin.

Perhaps the most significant observation is the very primitive initial ^{87}Sr/^{86}Sr ratio (ALL; see section 3.7) found in CAIs. This ratio is measurably lower than that found in achondrites (BABI). The primitive ratio observed in the Allende CAIs is interpreted as being caused by very local thermal episodes that resulted in the production of the CAIs and caused the local separation of volatile Rb from refractory Sr. Its wider significance is that it marks the oldest measurable event in the solar system, and so is taken as a convenient marker (or T_0) from which to measure solar system time. It may well mark a turbulent stage somewhere between the molecular cloud and the formation of the nebular disk.

3.4. MATRIX

The fine-grained matrix material in chondrites tells an important story, for it reveals the presence of heterogeneities on a micrometer scale in the nebula. In this study, we are interested in the least altered material; this effectively means that only the unequilibrated Type 3 meteorites are useful (see section 3.6). Types 1 and 2 have undergone aqueous alteration, and little evidence of primitive material remains, while Types 4-6 have undergone some thermal processing that has redistributed at least some elements.

The matrix is typically dark, FeO-rich, opaque, and truly fine-grained, mainly composed of micrometer-sized grains [28] (Table 3.4.1). This material is also found as rims on chondrules and as undigested lumps within chondrules (Fig. 3.4.1). The rims on chon-

TABLE 3.4.1. Mineralogy and abundance of matrix in chondrites.

Chondrite	Matrix (vol%)	Minerals
CI1	>95	phyllosilicates (serpentine), magnetite, dolomite, pyrrhotite, sulfates
CM2	55-85	phyllosilicates (serpentine), tochilinite, calcite, aragonite, magnetite, epsomite, pentlandite, pyrrhotite
CO3	30-40	olivine (Fa 30-60), phyllosilicates
CV3	35-50	olivine (Fa 40-60), high-Ca pyroxene (Fs 10-50, Wo 45-50), nepheline, sodalite, pentlandite, troilite, magnetite, phyllosilicates
C4-5	50-80	olivine (Fa 30-40), plagioclase (An 20-90) high-Ca and low-Ca pyroxene, magnetite, pentlandite, pyrrhotite
H3, L3, LL3	5-15	olivine (Fa 20-70), low-Ca pyroxene (Fs 1-20), glass, troilite, Fe, Ni, magnetite
Kakangari		Enstatite (Fs 1-10), troilite, Fe, Ni

Data from Scott E. R. D. (1988) in *Meteorites and the Early Solar System* (J. F. Kerridge and M. S. Matthews, eds.), p. 721, Table 10.2.1, Univ. of Arizona, Tucson.

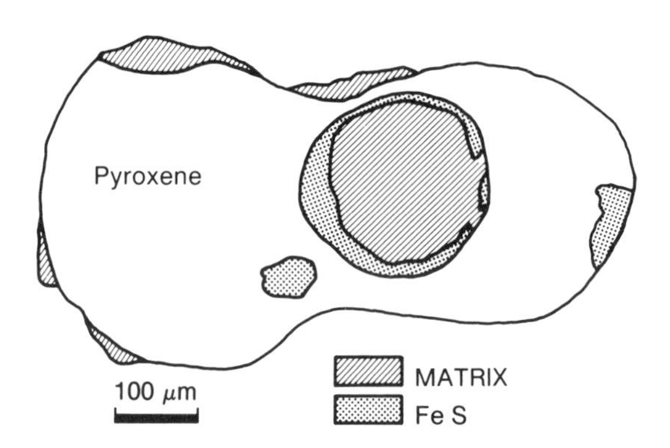

Fig. 3.4.1. *Diagram of a pyroxene chondrule from the Inman (LL3) chondrite that contains a 200-μm lump of opaque matrix, similar to that present on the rim, but distinct from the composition of the chondrule. After Scott E. R. D. et al. (1984) Geochim. Cosmochim. Acta, 48, 1745.*

drules from Type 3 chondrites are 10-30 μm wide, but their composition is distinct from that of the interior of the chondrule, and they must represent a later coating. They occur on chondrules from all types, CM, CO, and CV, as well as on chondrules in the ordinary (H, L, and LL) chondrites. Matrix also occurs as lumps, apparently entering the chondrule before crystallization (Fig. 3.4.1). The significant differences in composition are shown in Fig. 3.4.2, which shows that the matrix is much richer in FeO than the chondrules (Table 3.4.2).

Even though the grain size is very small (micrometer or submicrometer), the matrix shows wide variations in composition. Compositional variations occur over distances of 15 μm; fivefold variations occur between matrix lumps in individual chondrites. Matrix compositional variations up to an order of magnitude occur in Mg/Si, Al/Si, and Na/Si between different chondrites. These are primary differences caused by dust fractionation in the nebula, and are not secondary metamorphic effects.

The variations in composition of the matrix are due to differing amounts of olivine (Fe-rich) and the other phases. The major matrix minerals seem to be Fe-poor, low-Ca pyroxene (Fs_{2-10}) and Fe-rich olivine (Fa_{20-50}), accompanied by amorphous material containing Si, Al, Ca, Na, and K. The existence of heterogeneous fine-grained material as precursor phases makes it unnecessary to postulate planetary processes to establish such diversity [29].

TABLE 3.4.2. Comparison of mean composition (wt%) of chondrules in Type 3 ordinary chondrites with that of matrix material from which 75% of the FeO has been subtracted.

	Matrix Material		Chondrules
	Mean	Recalc.*	Mean
SiO_2	40.3	52.6	50.5
TiO_2	0.11	0.14	0.14
Al_2O_3	3.2	4.1	3.3
Cr_2O_3	0.42	0.55	0.54
FeO	31.3	10.2	10.2
MnO	0.36	0.47	0.33
MgO	19.6	25.6	29.9
CaO	1.50	1.96	2.2
Na_2O	1.4	1.8	1.4
K_2O	0.43	0.56	0.17
P_2O_5	0.28	0.37	0.06
NiO[5]	0.76	0.99	0.09
S	0.48	0.63	0.18
Total	100.1	100.0	99.0

* Composition recalculated after sufficient FeO removed to give an FeO concentration equal to that of the chondrule mean, with total normalized to 100%.

Data from Scott E. R. D. et al. (1984) *Geochim. Cosmochim. Acta, 48,* 1745, Table 7.

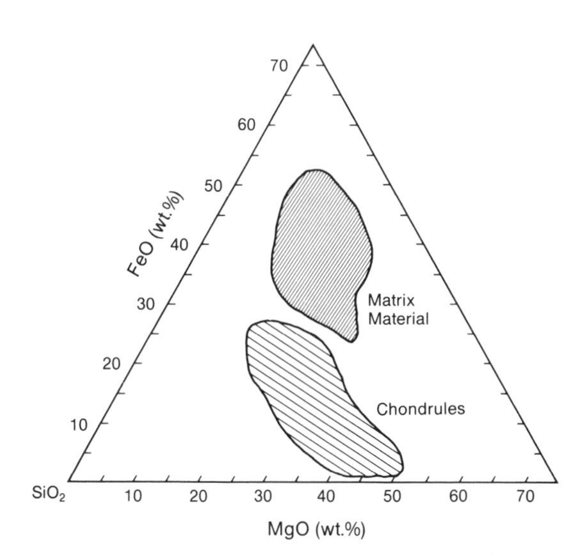

Fig. 3.4.2. *The compositions of matrix lumps, either within chondrules or on chondrule rims for six L-group chondrites show a complementary relationship to those of the bulk chondrules for the major elements Si, Fe, and Mg. After Scott E. R. D. et al. (1984) Geochim. Cosmochim. Acta, 48, 1745.*

The matrix lumps within chondrules differ from the matrix in the host chondrule, but are similar to that of the rim. They are thus foreign inclusions and not samples of the precursor material of the host chondrule. The matrix rims are different in composition from the chondrules and formed on the chondrules before they were incorporated into the host chondrite. The rims were acquired soon after solidification began.

Chondrule and matrix compositions show some complementary relationships. Figure 3.4.3 illustrates the complementary relationships between SiO_2, MgO, and FeO for matrix, chondrules, and bulk chondrites. The differences in composition between matrix and chondrules are due to variations in the amounts of Fe-rich olivine, Mg-olivine, Ca-pyroxene, and feldspathic material. The matrixes that constitute 10-20% of H, L, and LL chondrites show twofold differences in Mg/Si, Al/Si, and Na/Si, but the bulk chondrite compositions vary by less than 5%; the chondrule compositions must balance out the differences.

Early models sought to form chondrites from two components: chondrules, depleted in volatiles, and primitive dust [30]. These and similar models en-

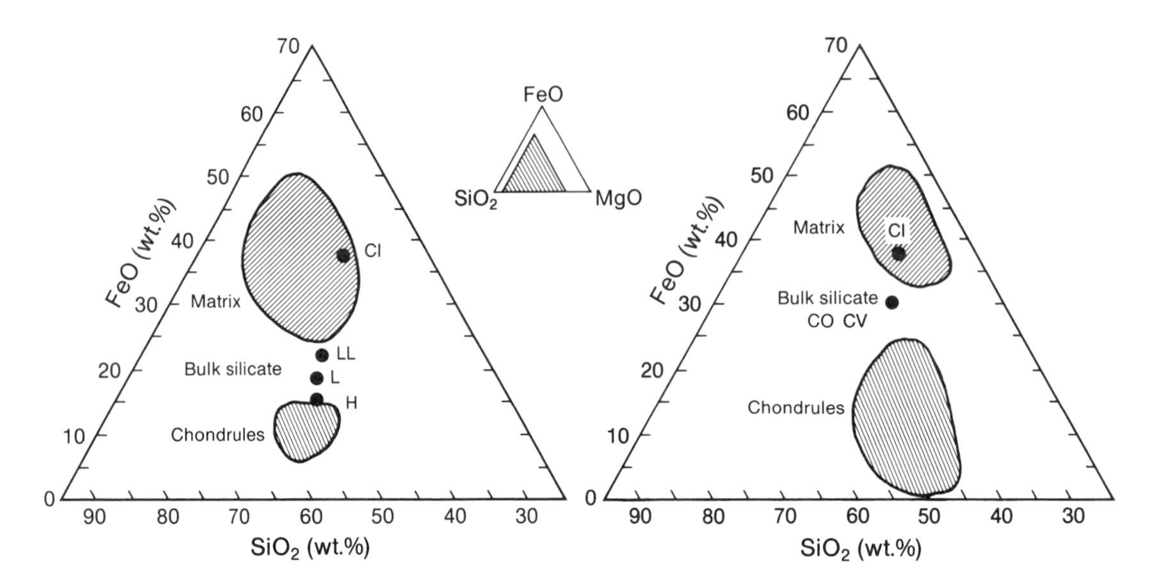

Fig. 3.4.3. ***The complementary relationship between matrix and chondrules holds for the LL, L, and H chondrites (left), as well as the CO and CV groups. The bulk silicate compositions for the bulk chondrites and for the CI chondrule-free carbonaceous chondrites are shown. After Scott E. R. D. et al. (1982) in Lunar and Planetary Science XIII, p. 705, Lunar and Planetary Institute, Houston.***

visaged a crucial role for the matrix as pristine material or "holy smoke." The matrix in most chondrites, however, is altered. This is particularly true of the matrix in CI carbonaceous chondrites, which has been subject to aqueous alteration.

As noted earlier, Fe-rich olivine is a prominent component of the least-altered matrix, but it is not clear whether it is primary or secondary in origin. Some evidence in favor of a pristine source for this mineral lies in the stacks of olivine plates, commonly observed in anhydrous Type 3 matrixes. These olivine plates are typically 1-10 μm thick and the whole assemblage is porous. These FeO-rich olivines (Fo_{50-65}) appear to have grown from a vapor phase [31] and a wide variety of models have been proposed for them.

Much of the confusion over the matrix of chondrites stems from their fine-grained nature; only recently have appropriate techniques become available to study them adequately. A detailed study of a large (700×800 μm) matrix lump in an ordinary chondrite (ALHA 77299,41; Type H3.7) clarifies many of the problems associated with matrix and chondrules [32]. This matrix lump essentially has CI composition, except for some depletion in K and Eu, attributable to a deficiency in the feldspathic component. The compositional differences between the chondrules and the matrix (Fig. 3.4.2) suggest that the chondrules are derived from the matrix by melting and loss of siderophiles and volatiles. The material in the matrix lump appears to have formed "by annealing of amorphous presolar or nebular condensates" [33].

The matrix in ALHA 77299 is heterogeneous in composition, mineralogy, and texture, but is dominated by Fe-rich olivine. The ALHA lump "consists of clastic angular olivine grains (1-2 μm) embedded within a very fine-grained (<0.5 μm) densely packed, nonclastic groundmass of Fe-rich olivine grains" [32]. From the texture of the olivines, the matrix could not have been annealed at temperatures higher than 400°C for any extended period, so the matrix cannot have formed by high-temperature crystallization. The ALHA lump has an oxygen isotopic composition that is distinct from both the bulk chondrite and the chondrules (Fig. 3.4.4).

The Semarkona meteorite matrix, which has undergone some aqueous alteration, has a very different oxygen isotope signature. The rim material is not as primitive as the other matrix, but has been modified during chondrule formation; "chondrule rims are moderately enriched in siderophile and volatile elements and the chondrule interiors exhibit a corresponding depletion" [32,34].

Many models have been proposed to account for the origin of the fine-grained matrix, driven partly by the search for the truly primitive nebular material ("holy grail" or "holy smoke"). Models that derive

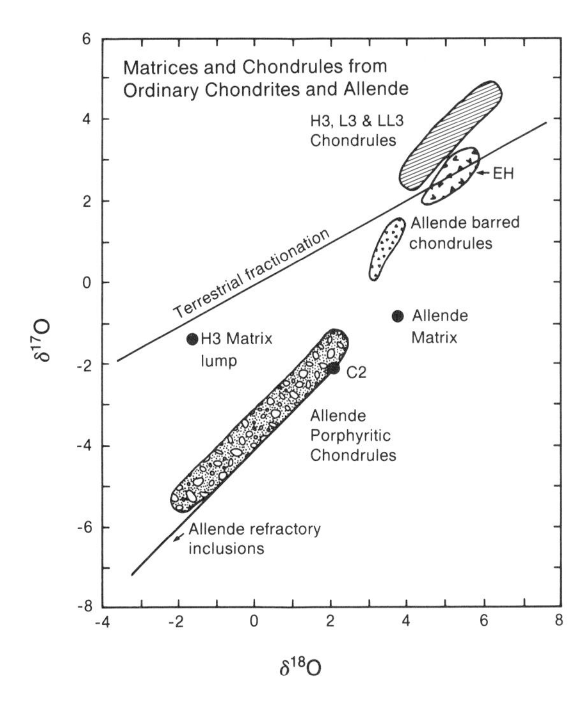

Fig. 3.4.4. *The oxygen isotope composition of chondrules and matrix in H3, L3, LL3, and EH chondrites and Allende. The H3 matrix lump is distinct in composition from the chondrules, as is also the case for Allende. After Scott E. R. D. et al. (1988) in Meteorites and the Early Solar System (J. F. Kerridge and M. S. Matthews, eds.), p. 735, Fig. 10.2.10, Univ. of Arizona, Tucson, and Rubin A. E. et al. (1990) Earth Planet. Sci. Lett., 96, 248, Fig. 1.*

matrix by grinding up chondrules do not account for the relatively unfractionated composition of the matrix, compared to the chondrules, especially for Fe and the other siderophile elements, which are notably depleted in chondrules [35]. Moreover, the matrix is complementary in composition to the chondrules and so cannot either have been derived from the fractionated chondrules, or represent truly primitive material. Oxygen isotopic compositions differ as well.

Another hypothesis proposes that the matrix forms by direct equilibrium condensation during cooling from the nebula at elevated temperatures [36]. However, it is not clear that the temperatures (1200°C) or the times of heating (6-8 hr) are relevant to nebular conditions or to the phases that might have condensed initially.

Annealing of precursor material, such as presolar grains, nebular condensates, or material remaining following chondrule formation appears to reproduce many of the features of the matrix [37]. Temperatures can be as low as 600°C. Low-temperature annealing of aggregates of fine-grained crystalline or amorphous material thus appears to be the most likely mechanism for producing chondritic matrix [32]. Because of the complementary relationship to chondrule compositions, matrix does not appear to be pristine solar nebula material (the long-sought "holy smoke"), but has been heated to varying degrees before being incorporated into chondrites. Nevertheless, it retains some memory of the precursor material, and so is crucial to our understanding of the early physical and chemical conditions in the solar nebula. The complementary relationship in FeO content between chondrules and matrix might therefore represent some physical separation based on density within the nebula, particularly if chondrules are formed far from the midplane (see section 3.5).

The matrix of carbonaceous chondrites records the existence of paleomagnetic fields with field strengths of about 1 oersted, equivalent to the present strength of the Earth's magnetic field at the surface. This strong field appears to have been shortlived and to have decayed rapidly. The paleofield intensities recorded in the achondrites, formed only a few million years later than the carbonaceous chondrites, are only about 0.1 oersted.

3.5. CHONDRULES

Chondrules are small spherical objects, typically about 0.5-1.5 mm in diameter, that occur in chondritic meteorites and appear to have crystallized rapidly from molten or partly molten drops. It is worth making a significant observation about the distinction between CAIs and chondrules. CAIs record multiple events of evaporation and condensation. Chondrules, in contrast, were heated once and cooled quickly. Perhaps there is a gradual change from the conditions that gave rise to the formation of CAIs to those responsible for forming chondrules [38]. Sorby remarked over 100 years ago that chondrules had been "molten drops in a fiery rain" [39] and little progress had been made beyond that point until rather recently [40].

One basic question concerning chondrules is whether they are formed within the nebula, in which case they are of primary importance, or whether they form as a consequence of some secondary process such as impact on a planetary body. In the latter case their compositions are too far removed from that of the nebula to provide useful insights into early solar system elemental abundances, since they will have been subjected to planetary fractionation processes.

TABLE 3.5.1. Abundances of chondrules, matrix, and inclusions, and chondrule characteristics in EH, ordinary (H, L, LL), CV, CO, and CM chondrites.

	EH	OC	CV	CO	CM
Petrographic Constituents					
Vol% matrix	<5	10-15	40-50	30-40	~60
Vol% inclusions	<<1	<1	6-12	10-15	~5
Vol% chondrules	15-20	65-75	35-45	35-40	≤15
Chondrule Characteristics					
Diameters (mm)	0.2	0.3-0.9	1.0	0.2-0.3	0.3
% porphyritic (P)	81	81	94	96	≥90
% nonporphyritic (NP)	19	15	0.3	2	3-8
% barred (BO)	≤0.1	4	6	2	present
Chondrules w/relict grains:					
% with isolated dusty olivines	present	5-15	<1	1-2	0-1
% with poikilitic olivine	5-10	20-30	20-50	~30	10-30

Data from Grossman J. N. et al. (1988) in *Meteorites and the Early Solar System* (J. F. Kerridge and M. S. Matthews, eds.), p. 627, Table 9.1.1, Univ. of Arizona, Tucson.

Among other significant parameters, chondrules, like hail stones, fall into a very restricted size range; the range in ordinary chondrites is from 0.2 to 3.8 mm, and most are less than 1 mm. Macrochondrules larger than 5 mm are sufficiently rare to call for special comment. One golf-ball-sized chondrule (diameter about 5 cm) has been described in an L chondrite [41]. Chondrules constitute 80% of ordinary chondrites, 30-50% of CV and CO and 5% of CM, while CI have none (see Table 3.5.1, which gives relative amounts of matrix, inclusions, and chondrules in the various classes of chondrites). The number density of chondrules in chondrites is 10^8-$10^9/m^3$, between 100 and 10^8 times their abundance in the nebula. These nebular densities can be assessed from the number of cratered and compound chondrules. The collection mechanism for chondrules must have been reasonably efficient [42]. Probably the most interesting fact is that the relative abundances of different chondrule types is nearly constant in the different classes of chondrites; the variations are less between meteorite classes than occur within individual chondrites.

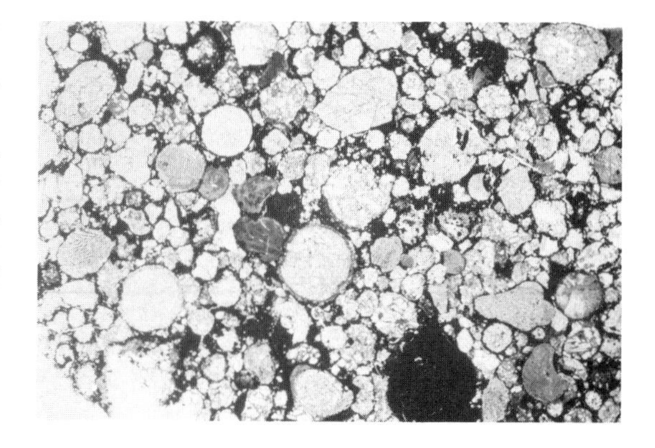

Fig. 3.5.1. *Chondrules in the unequilibrated Type 3 Semarkona chondrite. From Taylor G. J. et al. (1983) in Chondrules and Their Origins (E. A. King, ed.), p. 264, Fig. 1, Lunar and Planetary Institute, Houston.*

3.5.1. Textures

"When considered in detail, the texture of a particular chondrule may seem sufficiently unique to defy categorization" [43].

Chondrule textures are suggestive of rapid crystallization of free-floating objects (Fig. 3.5.1). The craters observed on chondrules may be dents caused by low-velocity collisions. Alternatively, they may be caused by loss of globules of immiscible metal-sulfide or by pressure-induced removal of material due to overpacking. No high-velocity impact craters are observed, such as are common on lunar spherules (Fig. 3.5.2). Compound chondrules are not uncommon, due apparently to collision and sticking together of chondrules while they are still plastic; hence, the density of these objects must have been high. As noted earlier, many chondrules are rimmed with fine-grained dark material that differs in composition from that of the chondrules. These rims are not alteration products. They are not restricted to any particular class of chondrites and occur on compound chondrules. The rims are a secondary coating postdating chondrule formation.

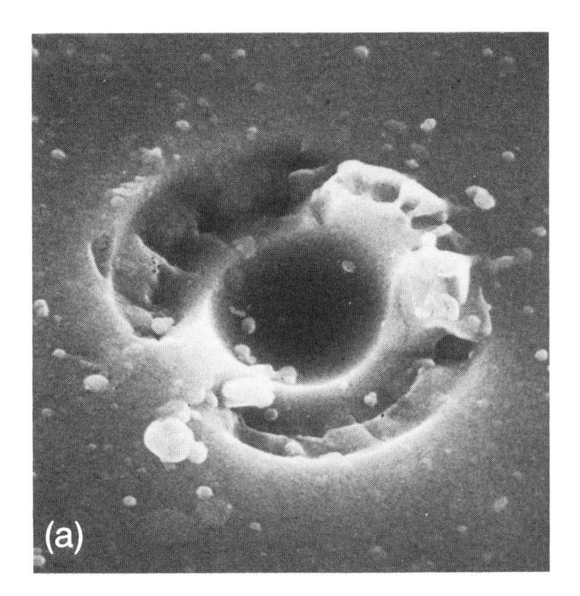

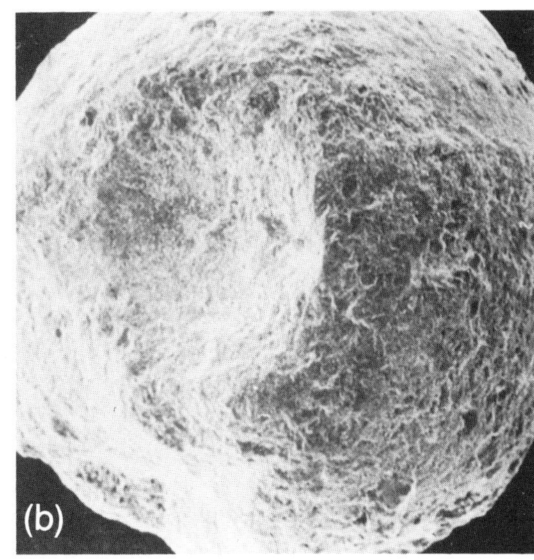

Fig. 3.5.2. *(**a**) Hypervelocity impact crater 5 μm in diameter on a lunar soil particle. (**b**) A 640-μm diameter crater on a chondrule from Ochansk. The lunar crater has a glass-lined pit surrounded by a spall zone, whereas the chondrule crater appears to be a simple dent caused by a low speed collision. After Taylor G. J. et al. (1982) in Chondrules and Their Origins (E. A. King, ed.), p. 266, Fig. 2, Lunar and Planetary Institute, Houston.*

The following textural classes have been established (where O = olivine and P = pyroxene): (1) porphyritic, which includes PO, PP, and POP, and which comprises 80% of all chondrules; (2) barred olivine or BO, 4%; (3) nonporphyritic, 15%, which includes (a) radial pyroxene or RP; (b) granular, including GP and GOP; (c) cryptocrystalline or C class; and (4) metallic or M, amounting to less than 1%.

The range of chondrule textures is a function of composition and heating time, and depends on the number of nuclei. Chondrules form within a narrow range of temperatures and cooling rates [44]. The textures have been reproduced under experimental conditions that included an initial temperature of 1556°C, heating times of 12 and 30 min, and cooling rates of 500°C/hr, with quenching below 1100°C [45]. The maximum temperature reached during chondrule formation appears to have been about 1700°C [46].

3.5.2. Composition

Chondrule compositions are shown in Tables 3.4.2, 3.5.2, and 3.5.3. The chemical and isotopic composition of chondrules is best described as chaotic. Individual chondrules show a factor of 2 variation in Si and Mg, a factor of 10 in refractory and volatile lithophiles, and of a factor of several hundred in siderophile and chalcophile elements. Within an individual meteorite that has not been subjected to postformation equilibration, there may be a wide spread in chondrule compositions. Thus, in the unequilibrated LL3 chondrite Semarkona, the range in Al/Si and Mg/Si in individual chondrules exceeds that found in the bulk chemical compositions of the major chondrite groups (Fig. 3.5.3).

TABLE 3.5.2. Comparison of chondrule populations in carbonaceous and ordinary chondrites.

Wt%	Carbonaceous Chondrites	H3 Chondrites	L3 Chondrites
SiO_2	45.2	50.3	46.0
TiO_2	0.19	0.14	0.13
Al_2O_3	4.64	3.23	3.35
Cr_2O_3	0.43	0.54	0.60
FeO	9.04	8.82	16.4
MnO	0.16	0.33	0.41
MgO	36.4	29.7	28.0
CaO	3.40	2.18	2.05
Na_2O	0.74	1.40	1.07
K_2O	0.05	0.18	0.22
No. of analyses	319	87	23

Data from McSween H. Y. (1982) in *Chondrules and Their Origins* (E. A. King, ed.), p. 206, Table 4, Lunar and Planetary Institute, Houston.

TABLE 3.5.3. Chondrule bulk compositions in unequilibrated Type 3 chondrites.

	Tieschitz (H3) $N = 17$		Hallingeberg (L3) $N = 24$	
	Mean	Geometric Mean	Mean	Geometric Mean
Al_2O_3 (%)	2.6 ± 0.6	2.5	3.2 ± 1.8	3.0
Cr_2O_3 (%)	0.60 ± 0.14	0.58	0.64 ± 0.24	0.60
Fe (total) (%)	12.9 ± 5.9	11.7	11.0 ± 4.9	10.0
MnO (%)	0.43 ± 0.16	0.39	0.38 ± 0.14	0.34
MgO (%)	29.5 ± 5.2	29.1	27.5 ± 3.4	27.3
CaO (%)	2.1 ± 0.7	2.0	2.6 ± 1.4	2.4
Na_2O (%)	1.08 ± 0.43	1.01	1.18 ± 0.35	1.11
Sc (ppm)	10.8 ± 3.8	10.3	13.1 ± 17.8	10.0
V (ppm)	77 ± 17	75	88 ± 17	86
Zn (ppm)	25 ± 16	20	30 ± 24	23
La (ppm)	0.38 ± 0.10	0.37	0.62 ± 0.77	0.49
Sm (ppm)	0.25 ± 0.06	0.25	0.42 ± 0.47	0.34
Eu (ppm)	0.11 ± 0.08	0.084	0.12 ± 0.11	0.094
Yb (ppm)	0.26 ± 0.08	0.25	0.34 ± 0.15	0.32
Lu (ppm)	0.038 ± 0.014	0.036	0.086 ± 0.164	0.057
Hf (ppm)	0.16 ± 0.06	0.15	0.29 ± 0.39	0.19
Co (ppm)	36 ± 162	80	142 ± 199	88
Ni (ppm)	3140 ± 3270	2030	4060 ± 8430	2020
Ir (ppm)	152 ± 208	74	432 ± 1390	71
Au (ppm)	49 ± 44	32	49 ± 57	32

Data from Gooding J. L. et al. (1980) *Earth Planet. Sci. Lett., 50,* 171, Table 1.

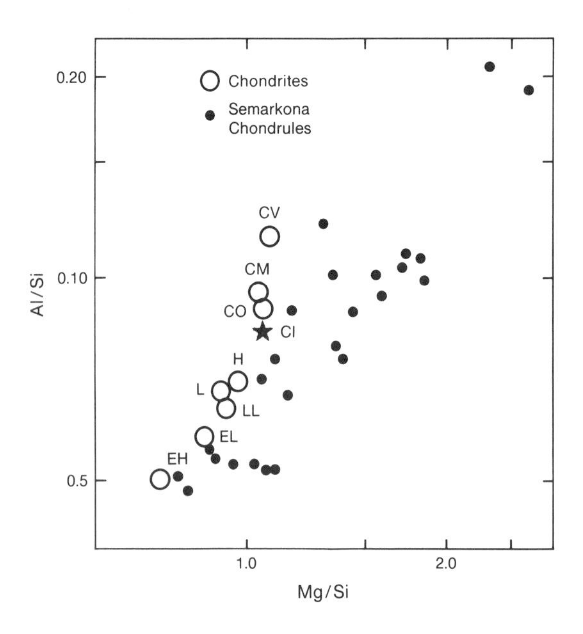

Fig. 3.5.3. *The variation in composition in 29 individual chondrules from the primitive LL3 chondrite Semarkona overlaps the mean compositions of the 9 major chondrite groups. After Scott E. R. D. and Newsom H. E. (1989) Z. Naturforsch., 44a, 929, Fig. 4.*

Chondrules are significantly depleted in siderophile and chalcophile elements and typically contain less than 20% of the bulk chondritic values. Since metallic (Fe-Ni) chondrules constitute less than 1% of chondrules (Fig. 3.5.4) and there are only rare primary sulfide chondrules, the chondrite matrix contains most of the siderophile and chalcophile elements. These elemental depletions in chondrules are not related to volatility. Most of the volatile alkali elements, Na, and K in unequilibrated ordinary chondrites are inside chondrules and are not depleted relative to bulk chondrite values. Precise data by isotope dilution analysis show that, in the unequilibrated CV3 Allende chondrules, there was no relative fractionation of K from Rb, indicating that there was little loss from vaporization during the chondrule-forming event.

Were the chondrules depleted in volatile elements during the chondrule melting process? Usually the presence of Na in chondrules is cited as evidence that volatiles are not depleted very much. Maybe the Na is due to secondary alteration, or equilibration, but since the chondrules in Type 3 unequilibrated chondrites contain Na in glassy mesostases within the chondrules, it appears to be original. The chondrules in the pristine unique chondrite ALHA 85085 are

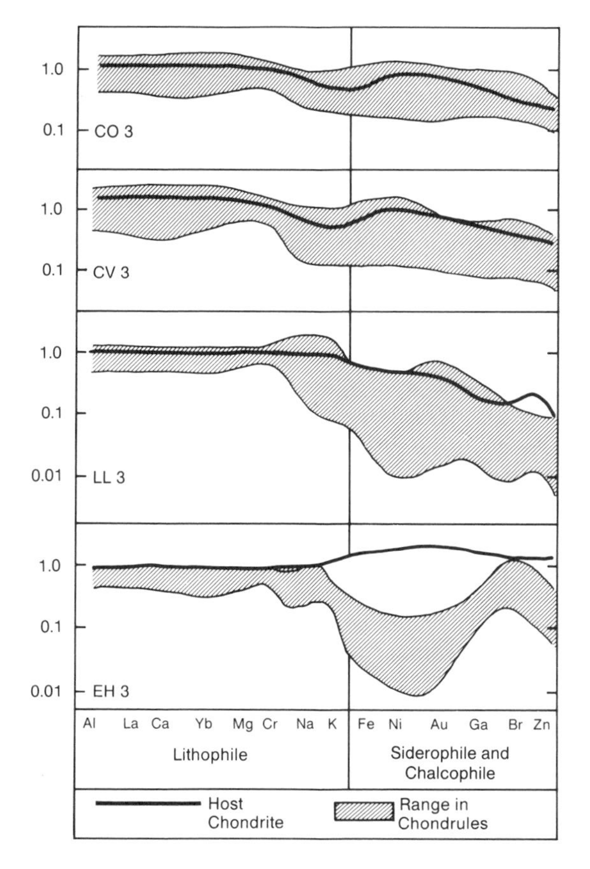

Fig. 3.5.4. *The distribution of lithophile, siderophile, and chalcophile elements in chondrules compared to those of their host Type 3 chondrites. The chondrules are in general depleted in siderophile and chalcophile elements relative to the bulk chondrites. After Grossman J. W. et al. (1988) in Meteorites and the Early Solar System (J. F. Kerridge and M. S. Matthews, eds.), p. 640, Fig. 9.1.5, Univ. of Arizona, Tucson.*

highly depleted in volatile elements [47]. Boron and Li, two volatile elements of unquestionable lithophile character, are "virtually absent from chondrules," as well as from metal and sulfide [48].

3.5.3. Rare Earth Elements

Generally the REE patterns in chondrules have been thought to be approximately chondritic with no significant Eu anomalies and mostly flat REE patterns (Fig. 3.5.5). However, recent precise determinations have revealed more complexity. REE data in chondrules from the unequilibrated Allende (CV3) chondrite [49] yield the important information that the REE in the precursor phases of chondrules were already fractionated. The precise Allende chondrule data [50] point unequivocally to REE fractionation in chondrule precursors. These involve variations in the abundances of Eu, Yb, and Ce due to gas-solid reactions based on relative volatility, rather than crystal-liquid fractionation based on ionic radius. Group II REE patterns found in some Felix chondrules indicate the presence of CAI precursor material. Equilibration also causes changes in the REE patterns in chondrules. As the chondrites become equilibrated, the REE patterns start to change; Eu depletions and enrichments reflect probable elemental redistribution during equilibration. Some HREE/LREE fractionation appears to be associated with equilibration, but may also reflect variations in precursor material. Since individual chondrules vary so much in composition, conclusions should not be based on single chondrule data.

The most fractionated REE patterns are found in chondrules from the equilibrated chondrites; the chondrules from UOC mostly follow the Hallingeberg and Semarkona flat REE patterns.

3.5.4. Uniformity of Composition?

Despite the great diversity of the compositions of individual chondrules, geometric means of chondrules from unequilibrated ordinary chondrites show

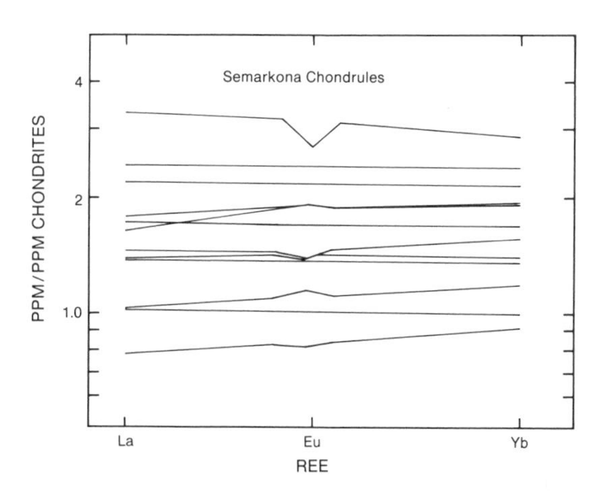

Fig. 3.5.5. *The rare earth element (REE) abundances in chondrules from the unequilibrated LL3 Semarkona chondrite, showing the generally unfractionated nature of the patterns. This indicates that the chondrules were not derived from precursor material that had experienced igneous fractionation processes. After Grossman J. N. et al. (1988) in Meteorites and the Early Solar System (J. F. Kerridge and M. S. Matthews, eds.), p. 644, Fig. 9.1.7, Univ. of Arizona, Tucson.*

no significant differences (distribution of most elements is log-normal). The Hallingeberg (L3) suite of chondrules is close to the average of unequilibrated chondrules, and is so used here (Table 3.5.3). Relative to CI nebular abundances, chondrules are enriched in lithophile elements and depleted in siderophile and chalcophile elements. This is consistent with melting of preexisting material that has undergone prior silicate-metal fractionation. Thus, the elemental variations in chondrules are due to the separation of metal-sulfide-silicate (Ni-Zn-Na) phases, not to volatility. This rules out models of chondrule formation by direct condensation from the nebula. We will return to this important point later.

Sizes, textural types, oxygen isotope composition, olivine and pyroxene compositions, and bulk compositions from the different groups of chondrites all show minor differences [51]. However, although there are many individual variations in composition among chondrules, there are close similarities among the mean compositions of chondrules from the H, L, and LL unequilibrated ordinary chondrite suites [52]. This uniform mean compositional nature of the chondrules does not reflect that of the various chondrite groups. Thus, total Fe, metallic Fe, and the Fe/Ni ratio all decrease in the sequence H, L, and LL; the chondrules show no corresponding differences related to the grouping of the parent chondrules based on iron content and oxidation state. Hence, the compositional variations among the chondrite groups must be due to the other components (matrix, metal) [53] (Table 3.5.2). Chondrules in equilibrated ordinary chondrites (Groups 4, 5, and 6) have experienced postformational changes that have obscured their primary compositional characteristics.

The important conclusion here is that these differences in chondrite composition demonstrate incomplete mixing in the nebula following chondrule formation.

3.5.5. Complementary Compositions

The complementary nature of chondrules and matrix is well illustrated in Fig. 3.4.3, which shows on an FeO-MgO-SiO_2 plot that matrix is much richer than chondrules in FeO. This diagram also makes the point that the bulk composition of chondrites is somewhat accidental, depending on the relative proportions of chondrules and matrix. These differences in bulk composition among the differing classes of chondrites also lend credence to the concept that the separate classes of chondrules were derived from separate regions and that mixing subsequent to chondrule formation was not thorough, otherwise all would have been homogenized. Accretion times for chondrules were probably very short relative to radial transport in the nebula, for otherwise these differences would have been smoothed out [54].

3.5.6. Depletion Mechanisms

A question of great interest concerns the mechanisms responsible for the depletion of chondrules in siderophile and chalcophile elements. Three processes have been suggested as responsible [55]:

1. Separate silicate, sulfide, and metal phases were present in the precursor material from which the chondrules were formed. In addition, the individual precursor lumps contained variable amounts of silicate, sulfide, and metal.
2. Loss of metal and sulfides as immiscible liquids during the melting process.
3. Reduction of oxidized precursor material during melting formed metal and sulfide phases.

Chondrules from the unequilibrated ordinary chondrites contain widely variable amounts of free metal and sulfide, generally showing depletion relative to the bulk chondrite values. Some metal-rich chondrules have Fe/Si and Ni/Fe ratios that exceed those of the bulk chondrites, in contrast to the general case.

Chondrules with high FeO, but low Fe metal contents, are depleted in siderophiles. This suggests the existence of precursor phases depleted in metal; reduction of FeO *in situ* would not deplete the trace siderophiles. Although it is clear that some reduction to produce metal occurs during chondrule melting, this does not appear to be the dominant process. Loss of metal-sulfide immiscible globules also appears to be responsible for the production of the dents and "craters" observed on chondrule surfaces [56]. It appears most likely that although minor reduction to metal and loss of volatile elements took place during chondrule formation, separate metal, sulfide, and silicate phases were present before that event. In this scenario, chondrules are mainly silicate because only the silicates had aggregated into millimeter-sized lumps before the melting event; the metal and sulfides remained finely dispersed. A corollory is that depletion of the volatile elements took place before chondrule formation, during the early T Tauri and FU Orionis stages that are probably responsible for gas and volatile element depletion in the inner nebula.

In summary, it is most likely that the precursor material of chondrules was heterogeneous and already depleted in volatiles, while the siderophile and chalcophile elements were already in separate phases. The chondrule-forming event selectively melted silicates.

3.5.7. Oxygen Isotopes

Data for individual chondrules do not plot near the values for their host chondrites (Fig. 3.5.6). The data, although widely scattered, appear to lie along trends dominated by two-component mixing. Chondrules from the equilibrated ordinary chondrites fall into two groups, but those from the unequilibrated ordinary chondrites, less affected by metamorphism, are more complex [57].

The Allende chondrules form a separate group, distinct from chondrules from both unequilibrated ordinary chondrites and CAIs, Allende dark inclusions, matrix, and bulk Allende, all of which lie on a separate (mixing? fractionation?) line (Fig. 3.4.4). In this context, it is interesting to record that Allende chondrules lie on a separate oxygen isotope line from Allende CAIs and require differing end members; they cannot have formed in any simple manner from the CAIs. Possibly the Allende chondrules formed from CAIs that had undergone another episode of mass fractionation.

The oxygen isotope heterogeneities are reminiscent of the compositional and mineralogical variations of the individual chondrules; clearly a very heterogeneous population of precursor material was sampled. The chondrules neither cluster about the host chondrite nor fall into distinct groups. Instead they form one grand population, with the same means for H, L, and LL groups.

The precursor material seems to have been neither well mixed nor homogenized by heating before the chondrule-forming event. It is also clear that the chondrules do not provide the whole oxygen isotope inventory, even for the ordinary chondrites that constitute 80% of chondrules; in this case the matrix must differ in oxygen isotope composition from the precursor phases, and must be ^{16}O-rich compared to the whole rock.

3.5.8. Ages of Chondrules

Individual chondrules seem to be contemporaneous with one another and with their host chondrite [59], to within the limits of precision of the U-Pb and Rb-Sr techniques (about 10 m.y.). Iodine-xenon relative ages for Allende and Bjurbole chondrules show some differences and do not appear to be isochronous, but the total spread is small. The total range of I-Xe ages appears to be about 5 m.y. but cannot be related to the other clocks. Argon-40/argon-39 ages show a spread of 200 m.y. almost certainly due to alteration. The matrix samples give I-Xe ages equal to or older than the chondrule ages, but give much younger Rb-Sr and Ar-Ar ages due to alteration or shock. Much better age resolution is needed before a meaningful interpretation can be made. Based on the concept that chondrules are formed in a turbulent stage of the nebula, it is probable that their formation ages span less than 10^6 years.

At the time of cooling, a magnetic field of a few (0.1-10) oersted was present, indicative of the presence of strong magnetic fields in the nebula at a very early stage.

3.5.9. Models for the Origin of Chondrules

"Chondrules likely formed by melting of preexisting materials that were both chemically and isotopically heterogeneous" [60].

The formation of chondrules was a significant event in solar system history. They constitute over half the mass of chondrites and are therefore not trivial. The chondrule production line was clearly efficient.

Two models are usually considered: the first forms chondrules in a "planetary" setting; the second forms them in the nebula, before the formation of large bodies. Our perception of the relative importance of chondrules for the history of the solar system depends on where we consider them to be formed. In the planetary setting, they are of secondary importance, a byproduct of some essentially stochastic process such as impact. In this case they convey no more information about the nebula than that derived from any other processed sample. If, on the other hand, they form in free space as an early nebular product, they have the potential to provide us with unique information about conditions and locations otherwise inaccessible [61].

Planetary origins for chondrules can be placed into four categories [61]:

1. Impacts on planetary surfaces. Although this scenario has been popular [62], a very large number of objections can be raised. A principal problem is that impact is very inefficient at producing glassy spheres. There are very few chondrulelike objects on the Moon, meteoritic achondritic breccias contain few chondrules, and suevite melts tend to be irregular masses of glass. Impacts on the lunar surface produce mostly agglutinates, which constitute about 50% of lunar soils. Chondrites, in contrast, contain virtually no agglutinates. Impacts are efficient at smashing up rock, producing more than 97% breccia and less than 3% melt. Meteoritic regolith (gas-rich) breccias likewise contain mostly broken-up fragments.

All the diverse glasses, rocks, and fragments in the lunar regolith are covered with microcraters caused by hypervelocity impacts. In contrast, the chondrules have only simple dents due to low-velocity collisions between plastic, partially molten specimens, to loss of immiscible metal or sulfide globules, or to deformation and element migration induced by overpacking (see section 3.6).

Fragments of solid precursor rock from the parent bodies, predictable in this hypothesis, appear to be very rare in the chondrites [63]. Many chondrules have rims of fine-grained material that differ in composition to the interior. Such rims are never observed on impact-produced spheres. Impact-produced melts at terrestrial craters tend to be uniform in composition; in contrast, individual chondrules are variable in composition [64].

All chondrules are old, with ages >4.4-4.5 b.y. Since there was a massive bombardment in the inner solar system probably extending down to 3.85 b.y., it might be expected that some chondrules would be younger than 4.4 b.y. Some chondrite ages might also be expected to be younger than the canonical 4.5 b.y.

2. Collisions during accretion of chondrite parent body. For accretion to occur, velocities must be low. In that case very little impact melting will occur [65]. If the velocities are high, no accretion will take place. These are fatal objections.

3. Collisions between molten planetesimals. This mechanism [66] has the apparent attraction of providing already molten material, which only has to be splashed about to form molten drops. However, there are several telling objections. Perhaps the first is the very restricted size range of chondrules, most lying between 0.5 and 1.5 mm. For a splash origin, one might expect a wide variety of sizes to result from the collision of molten bodies 10-20 km in diameter. Chondrules, like hailstones, are rather uniform in size, and there is no sign of any larger masses of melt. The model also does not account for the nearly ubiquitous rims.

The presence of lumps of dark matrix inside chondrules [67] is another problem not addressed by the model, as is the absence of fractionated compositions. Molten asteroids or planetesimals would be expected to differentiate, and igneous REE patterns could be expected, as well as wide variations from CI abundances in other elements. Chondrule compositions for the lithophile elements, however, are essentially unfractionated. Only separation of siderophile and chalcophile elements has occurred.

The most decisive evidence against the hypothesis comes from the oxygen isotope evidence. Individual chondrules show a wide range of isotopic compositions. If they were derived from the collision of two molten bodies, these should individually be homogeneous, and the oxygen isotopes of the resulting chondrules should mostly represent one or the other compositions. It is difficult to imagine efficient mixing on a millimeter scale during the collision, although some would lie along a mixing line. "These [collisional] models are not convincing, but are difficult to refute quantitatively" [68] although the Allende chondrule data (Fig. 3.4.4) seem very difficult to explain by such a mixing model. Although collisions between molten or partly molten planetesimals may not form chondrules, they may be responsible for producing the pallasites, the mesosiderites, and Shallowater, a unique enstatite achondrite [69].

4. Volcanism. This proposed mechanism [70] suffers from many obvious defects. Volcanism should produce fractionated compositions, particularly in incompatible element (e.g., K, Na, REE) abundances and Fe/Mg ratios, whereas chondrule compositions generally resemble solar nebular abundances. There are no chondritic compositions resembling the igneous achondrite compositions.

The production of spherules is not very efficient in volcanic processes except in fire fountaining; pyroclastic deposits are, however, dwarfed by lava flows in all observable volcanic processes, so that special conditions need to be invoked.

The oxygen isotope data are particularly difficult to explain by a volcanic hypothesis. As seen in Fig. 3.5.6, chondrules tend to fall on a line with a slope of 1, but igneous fractionation should produce trends along a line with a slope of 0.5 and, as in the terrestrial case, material from one body should lie along one line; in contrast, the chondrule data scatter widely.

All the "planetary" mechanisms proposed encounter insuperable difficulties, so a nebular origin becomes likely. A bonus is that chondrules become significant objects in the evolution of the nebula, rather than incidental products of some secondary process.

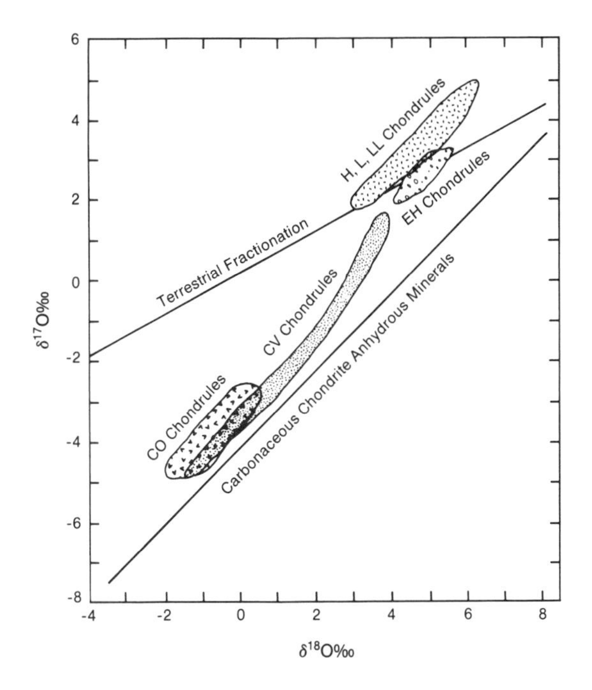

Fig. 3.5.6. *Oxygen isotopic composition of chondrules, showing the distinct fields occupied by chondrules from the different chondrite groups, indicating derivation from differing source regions. After Grossman J. W. et al. (1988) in Meteorites and the Early Solar System (J. F. Kerridge and M. S. Matthews, eds.), p. 649, Fig. 9.1.9, Univ. of Arizona, Tucson.*

3.5.10. Nebular Origin

The failure of "planetary" sources to account for chondrules focuses attention on origins within the nebula. Chondrules thus become of primary importance, reflecting the extended treatment given here.

An initial objection was concerned with the variation in chondrule compositions, since uniformity would be expected in the nebula. This led to the popularity of planetary processes that could be guaranteed to produce variable compositions [71]. This is a well-known philosophical trap; we ascribe uniformity to regions such as the solar nebula or the terrestrial mantle where there is a paucity of information.

It turns out that the matrix of chondrites is widely variable in composition [72]. Thus, the dust in the nebula was heterogeneous on a millimeter scale. Heterogeneities also must have existed on scales appropriate to the areal regions from which the individual chondrite groups were assembled. Planetary processes tend to homogenize isotopic anomalies. If these are carried by solids, their survival in the nebula is readily understood. Exchange with gaseous oxygen reservoirs in the nebula can produce the observed wide range of oxygen isotope variations in chondrules.

Evidence for an early origin for chondrules in the nebula is overwhelming. Localized processes (e.g., impacts on planetary, satellite, or asteroidal surfaces) do not account for them, since there is no sign of any large accompanying mass of partly fused and shocked material, the production of which would be expected to result from such processes.

Condensation from a cooling vapor is ruled out on several grounds. The chondrules would vaporize at temperatures of 1300-1900 K and gas pressures of 10^{-5}-10^{-6} atm expected at 3 AU in cooling nebula models [73]. The O/H ratio at the site of chondrule formation must be many times that of the nebula. How did this arise? Possibly dust was vaporized during the transient heating events close to the nebula midplane, releasing oxygen as a gas phase. Possibly hydrogen was already mainly gone at this stage of nebular evolution, leading to enhanced O/H ratios.

The alternative to condensation is that they formed by melting of preexisting solids. This model seems inescapable. The heating must have been short; otherwise these millimeter-sized objects would have vaporized and vanished. Mostly, they were not completely melted; relict grains of olivine are common [74]. Such grains occur in chondrules both from carbonaceous and ordinary chondrites, consistent with a single mode of formation for all chondrules.

The oxygen isotope data indicate that the chondrule compositions are quite different between the ordinary, carbonaceous, and enstatite chondrite classes, so that very little mixing of chondrules between different groups has occurred. This is an argument for lack of lateral mixing in the nebula, and an indication that the asteroid belt preserves some memory of individual compositional differences.

3.5.11. Chondrule Precursor Material

Does chondrule formation postdate metal-silicate and metal-sulfide fractionation? There appears to be distinct evidence that metal, sulfide, and silicate phases were present in the nebula before chondrule formation. Individual chondrules vary in major-element composition in unequilibrated chondrites, but the process of formation of chondrites achieves relative uniformity in composition. The process that remixes chondrules and matrix must act on material that was previously fractionated. Clearly the matrix and the dust that melted to form the chondrules must be of different composition. Why was the process that melted the chondrules so selective in choice of material [75]?

Calcium-aluminum inclusions and chondrules have traditionally been considered separately, being studied by separate groups of research workers. The melting that produced chondrules was so efficient that most evidence of the precursor material has been destroyed. Since the matrix of chondrites is complementary in composition to that of the chondrules, it does not provide direct information about the precursor material. However, some chondrules preserve evidence of CAI precursors [76].

Some chondrules in the Felix CO3 carbonaceous chondrite have strongly fractionated REE patterns resembling Group II CAIs, although they are more closely matched by those in hibonite grains from the Murchison CV3 carbonaceous chondrite (Fig. 3.5.7; Table 3.5.4). Such patterns are very uncommon in chondrules. These patterns are considered to represent those formed in precursor materials, since the temperatures reached during the chondrule-forming process were not sufficient to have caused fractionation among the REE. Evidence for this is also found in the retention of much more volatile elements such as K and Rb within the chondrules. Accordingly, the strongly fractionated REE patterns were established earlier, and chondrule formation reflects remelting of this material.

This important information indicates that the formation of CAIs preceded that of chondrules. The temperatures reached during CAI formation were higher, and reflected the complex sequences of condensation and evaporation required to produce the Group II REE patterns.

Other direct evidence of precursor phases has been found in the form of relict olivine grains [77,78]. This relict olivine "has numerous dusty inclusions and shows a dirty appearance" [79]. The dusty portions are metallic iron, providing evidence that some reduction was occurring *in situ* at the time of chondrule formation [80]. However, this seems to be a minor effect, and the expected interelement fractionation does not seem to be observed.

According to Scott [81] matrix material is the most likely precursor material for chondrules since they display appropriate variations in composition and highly variable oxygen isotope compositions. The formation of chondrules from matrix requires loss of both siderophiles and volatiles. Loss of Fe and siderophile elements, which are enriched in the rims,

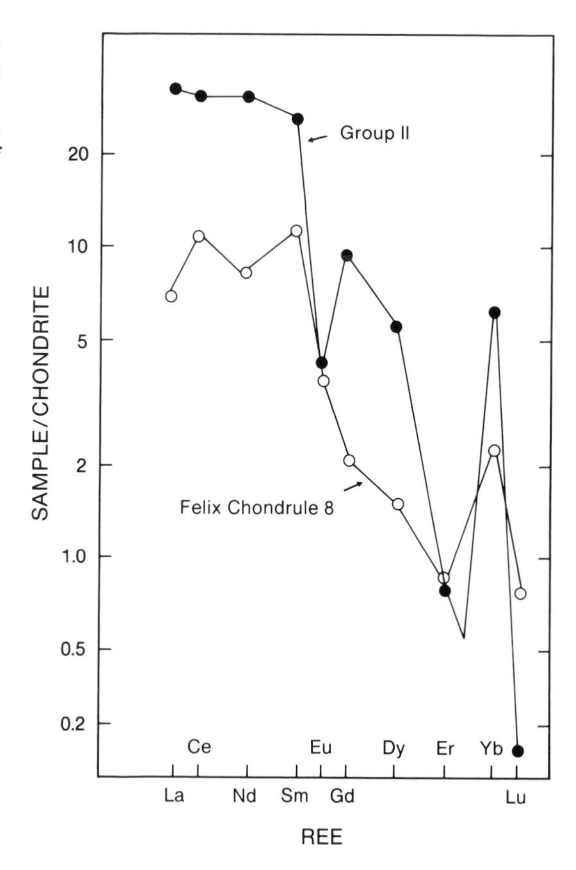

Fig. 3.5.7. *A rare example of a CAI Group II REE pattern preserved in a chondrule from the Felix chondrite, with a typical CAI Group II pattern shown for comparison. This is evidence that the formation of the refractory inclusions (CAI) present in Allende preceded that of the chondrule-forming event. After Misawa K. and Nakamura N. (1988) Nature, 334, 49, Fig. 2.*

TABLE 3.5.4. Composition of chondrules with Group II REE pattern from Felix CO3 chondrite.

Ca (%)	4.21
K (ppm)	1830
Rb (ppm)	7.02
Sr (ppm)	88.7
Ba (ppm)	13.5
La (ppm)	1.70
Ce (ppm)	6.76
Nd (ppm)	3.76
Sm (ppm)	1.74
Eu (ppm)	0.199
Gd (ppm)	0.414
Dy (ppm)	0.379
Er (ppm)	0.142
Yb (ppm)	0.374
Lu (ppm)	0.0191

Data from Misawa K. and Nakamura N. (1988) *Nature, 334,* 49, Table 2.

during chondrule formation has been demonstrated [82]. The lithophile-element composition of some chondrule rim material appears to be closely similar to that of the chondrules. Grossman and Wasson [82] argue that the rims form by melting during the chondrule-forming process, with metal and sulfide droplets migrating to the surface. The matrix-like rims appear to be accretionary. The metal-sulfide rims may well result from the chondrule-forming process since there appears to be sufficient variability in the matrix composition to encompass these separate explanations. In other words, some rims are formed during the chondrule production process; most, however, seem to be later coatings. The evidence from the similarity in element abundance patterns between the rim and interior of the chondrule from Semarkona [83] may indicate a genetic relationship between the parent material of the chondrules and matrix rather than derivation of the rim from the chondrule.

The question of chondrule precursor material is complicated by possible earlier loss of volatiles by solar-induced processes during T Tauri or FU Orionis activity. Thus, some matrix may be depleted in volatiles by these processes, while local volatile depletion occurs during chondrule formation.

Whatever scenario is adopted, it is clear that metal, sulfide, and silicate were present, and fractionation occurred either before or following the production of chondrules and before the accretion of the chondrites from these components. These events clearly preceded the planetesimal or planetary accretion stage. Accordingly, metal-sulfide-silicate fractionation was a nebular as well as a planetary process.

When did the volatile-refractory-element fractionation occur? Was it related or did it precede the chondrule production stage? Since chondrites are already depleted in volatile elements, it must precede their formation and take place effectively at T_0. If the volatiles were not lost during the chondrule-forming process, then they must have been lost before that event took place. The basis for this statement is that the chondrites are assembled from matrix and chondrules, and are depleted in volatiles. It does not seem realistic to appeal to later metamorphism. The chondrule-forming process appears to have taken place on preexisting dust that already possessed the chondritic class signatures [84].

How does metal-sulfide-silicate fractionation occur in the nebula? It might occur by gravity (as it does in planets) as the components settle onto the nebular midplane. What seems to be clear is that metal-sulfide-silicate equilibrium was mostly accomplished in the nebula, and that the planetesimals (and eventually the inner planets) were assembled from such materials. Since the precursor material for the planets was thus already separated into metal, sulfide, and silicate phases under low pressures, it makes little sense to talk of core-mantle equilibrium or high-pressure equilibration of such phases in the Earth; such processes were probably very limited [85].

Some interesting unanswered questions about chondrules remain. Most of the iron and sulfide resides in unmelted matrix, and does not appear to have been formed by reduction during chondrule formation. How important was the process? Was it responsible for depletion of volatiles? Chondrule production was common in the asteroid belt. Presumably it was more common in the inner nebula where temperatures were higher.

In summary, some general conclusions may be possible in this highly complex subject.

1. Chondrules melted from preexisting dust in the nebula in single events.
2. Chondrules formed during later single-stage melting events and at lower temperatures than CAIs, which record earlier multiple cycles of evaporation and condensation at higher temperatures.
3. Chondrule-forming events were rapid and local, and cooling occurred very rapidly.
4. The curious fact that chondrules are essentially composed of silicates and are depleted in chalcophile and siderophile elements is due to preferential aggregation of silicate dust into chondrule-sized lumps. Metal and sulfide remained dispersed, so they failed to form chondrules, except rarely during the heating episode. Possibly a mechanical effect is involved. Under nebular conditions, silicates may stick or clump together to make aggregates of chondrule size, but metal and sulfides do not aggregate. This would provide a simple explanation for much of the compositional evidence from chondrules. What is clear is that the metal, sulfide, and silicate phases were present in the nebula before chondrule formation.
5. Some minor reduction and loss of volatile elements occurred during the chondrule-forming process, but these were essentially secondary processes.
6. The major depletion of volatile elements (e.g., as shown by low K/U and Rb/Sr ratios) occurred in the inner nebula before the chondrule-forming event.

3.5.12. A Preferred Model for Chondrule Origin

The many suggestions that have been made for the origin of these enigmatic objects have been discussed in previous sections. All suffer from various defects as argued above. Here, the currently most plausible hypothesis, that of an origin by nebular flares, will be discussed.

It has been proposed that there were nebular flares, analogous to the presently observed solar flares. These highly energetic, short-lived (minutes to hours) phenonema result from the sudden release of perhaps 10^{32} ergs of energy stored in stressed magnetic fields [86]. The model assumes that similar brief, high-energy heating events occurred in the corona of the early nebula, high above the nebular midplane, and that chondrules resulted from the melting of silicate dust clumps during these events [87]. This mechanism (Figs. 3.5.8 and 3.5.9) explains a considerable number of features associated with chondrules. Rapid heating to high temperatures is achieved. Melting times were of the order of 0.1 s, accounting for the partly melted chondrules, and for their small size. The uniform size of chondrules although rarely commented upon, is a significant clue, and reinforces the hailstone analogy. Perhaps dynamic conditions in the nebula resembled the turbulence inside cumulonimbus clouds during production of hail in thunderstorms. The flares were only marginally able to melt the silicate dust lumps. Since the events took place high above the midplane of the nebula, in a thin and uninsulated region of low density, cooling is rapid. Metal, sulfide, and FeO-rich matrix were preferentially concentrated in the midplane of the nebula by gravitational settling. Strong magnetic fields of up to 10 oersted or more in the corona provide the magnetizing fields to account for the remnant magnetization measured in meteorites [88].

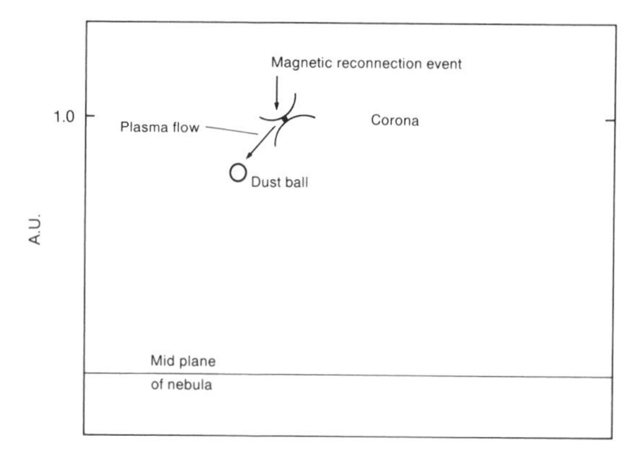

Fig. 3.5.8. *A schematic representation of the formation of nebular flares high above the mid-plane of the solar nebula. Chondrules form in this model by melting of fluffy dust balls about 1 AU above the nebular midplane from energetic particles produced in flares forming about 0.1 AU further out. After Levy E. H. and Araki S. (1989) Icarus, 81, 75, Fig. 1.*

The occurrence of the nebular flares high above the midplane allowed a gravitational separation of silicate dust clumps from sulfide and metallic iron that may have been concentrated in the midplane of the nebula, and so removed from the scene of melting. The nebula must have been turbulent to loft the silicate dust clumps up to 1 AU above the midplane. Since chondrules are so common, the nebula must have been a very dynamic place at this time, with chondrules forming like hailstones in a cumulus cloud in a thunderstorm ("molten drops in a fiery rain" as Sorby [39] observed), so that the nebula was not a cold passive disk, but had considerable energy to dissipate.

3.5.13. Summary

Metal, sulfides, and silicates existed prior to chondrule formation, and reduction of iron was only an incidental byproduct of the chondrule-forming process. Volatile-element depletion occurred in a separate event. Silicate dust was melted selectively during the chondrule-forming event, with some minor reduction and loss of iron, and minor depletion of sulfides and volatile elements. Although some reduction of iron and depletion of siderophiles, chalcophiles, and some volatiles may have occurred during chondrule formation, the major separation of these phases and elements apparently occurred within the dispersed material of the solar nebula, prior to the chondrule-melting event.

How was the silicate dust melted preferentially, without involving the metal and sulfide phases to more than a minor extent? Probably silicate dust was separated from metal and sulfide, either by differential gravitational settling or magnetically in the case of the metal, or perhaps the silicates stuck together more efficiently. As the dust settled to the midplane, physical and perhaps magnetic separation of the dust occurred, which could account for the complementary relationship between chondrules and the more iron-rich (denser?) matrix as well.

The most plausible scenario for the origin of chondrules is that they formed by melting of silicate dust by nebular flares, analogous to solar flares, high above the midplane of the nebula. In this scenario, the separation of metal from silicate might be accomplished by the preferential gravitational settling of metal to the midplane of the nebula, while the existence of strong magnetic fields associated with

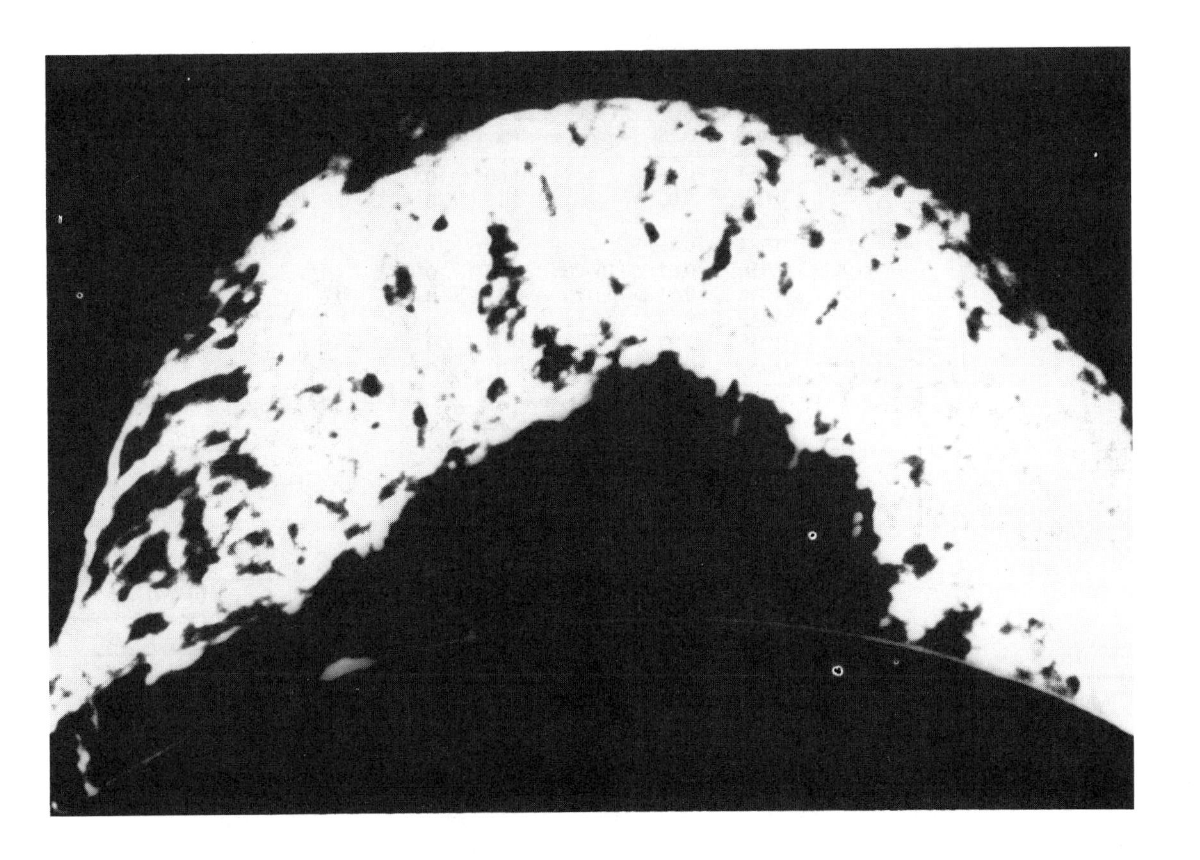

Fig. 3.5.9. *Photo of a solar flare, a possible analogue for the postulated nebular flares.*

the nebula flares could account for the magnetic signature in chondrites. Chondrules, which constitute over half the mass of ordinary chondrites, thus record the preferential rapid melting of silicate phases as one of the earliest events in the nebula after the formation of the CAIs.

3.6. CHONDRITES AND OTHER PRIMITIVE METEORITES

In this section we are mostly concerned with primitive meteorites and what they can tell us about the history of the nebula prior to planetary accretion [89].

3.6.1. Classification

Table 3.6.1 lists the basic properties of the nine presently known chondrite classes. Table 3.6.2 lists the criteria that are used to classify the chondrites into six petrographic types.

At least nine groups of primitive chondrites are recognized, but there are probably many more (e.g., Kakangari, Bencubbin, and ALHA 85085 [90]) representing additional types of primitive chondrites that do not fit into the present classes. Thus, Fig. 3.6.1 illustrates that the composition of the thumb-sized unique chondrite ALHA 85085 falls outside that of the presently known chondrites. Clearly, caution is called for in extrapolating the chondritic data to the wider field of planetary compositions.

3.6.2. Accretion

Most chondrites are microbreccias with uniform compositions, but composed of highly fractionated components, each of which has had its own individual genetic history. Preaccretion history includes many different processes such as low-temperature isotopic fractionation (e.g., in ^{12}C-^{13}C), formation of mineral aggregates, heating resulting in recrystallization, melting and evaporation followed by recon-

TABLE 3.6.1. Average properties of chondrite classes.

Group	Mg/Si	Ca/Si	Fe/Si	Fa (mol%)	Refractory Elements / Si	Fe_{met}/Fe_{tot}	$\delta^{18}O$ (‰)	$\delta^{17}O$ (‰)
CI	1.05	0.064	0.86	—	1.00	0	~16.4	~8.8
CM	1.05	0.068	0.80	—	1.13	0	~12.2	~4.0
CO	1.05	0.067	0.77	—	1.10	0-0.2	~-1.1	~-5.1
CV	1.07	0.084	0.76	—	1.35	0-0.3	~0	~-4.0
H	0.96	0.050	0.81	16-20	0.79	0.58	4.1	2.9
L	0.93	0.046	0.57	23-26	0.77	0.29	4.6	3.5
LL	0.94	0.049	0.52	27-32	0.76	0.11	4.9	3.9
EH	0.77	0.035	0.95	—	0.59	0.76	5.6	3.0
EL	0.83	0.038	0.62	—	0.60	0.83	5.3	2.7

Data from Sears D. W. G. and Dodd R. T. (1988) in *Meteorites and the Early Solar System* (J. F. Kerridge and M. S. Matthews, eds.), p. 15, Table 1.13, Univ. of Arizona, Tucson.

TABLE 3.6.2. Petrographic classes of chondrites.

	Petrographic Types					
	1	2	3	4	5	6
Homogeneity of olivine and pyroxene compositions	—	>5% mean deviations		<5% mean deviations to uniform	Uniform	
Structural state of low-Ca pyroxene	—	Predominantly monoclinic		Monoclinic >20%	Monoclinic <20%	Orthorhombic
Degree of development of secondary feldspar	—	Absent		<2-μm grains	<50-μm grains	>50-μm grains
Igneous glass	—	Clear and isotropic primary glass; variable abundance		Turbid if present	Absent	
Metallic minerals (maximum Ni content)	—	(<20%) Taenite absent or very minor	Kamacite and taenite present (>20%)			
Sulfide minerals (average Ni content)	—	>0.5%	<0.5%			
Overall texture	No chondrules	Very sharply defined chondrules		Well-defined chondrules	Chondrules readily delineated	Poorly defined chondrules
Texture of matrix	All fine-grained, opaque	Much opaque matrix	Opaque matrix	Transparent microcrystalline matrix	Recrystallized matrix	
Bulk carbon content	~3.5%	1.5-2.8%	0.1-1.1%	<0.2%		
Bulk water content	~6%	3-11%	<2%			

The strength of the vertical line is intended to reflect the sharpness of the type boundaries. Adapted from Van Schmus R. and Wood J. A. (1967) *Geochim. Cosmochim. Acta, 31,* 747, and Sears D. W. G. and Dodd R. T. (1988) in *Meteorites and the Solar System* (J. F. Kerridge and M. S. Matthews, eds.), pp. 23-24, Univ. of Arizona, Tucson.

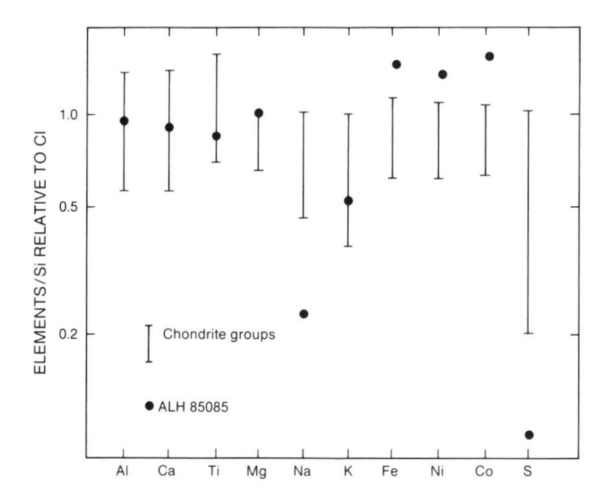

Fig. 3.6.1. *The element abundances in Antarctic meteorite ALHA 85085 fall outside the ranges in composition of the nine established chondrite groups. After Scott E. R. D. (1988) Earth Planet. Sci. Lett., 91, 13, Fig. 10.*

densation, and gas-solid reactions (metasomatism). Accretion of material to form chondrites took place at low temperatures that differed for each chondrite class. Textural changes such as sintering and grain growth occurred. Chaos on a micro scale seems to have been ubiquitous (Table 3.6.3).

TABLE 3.6.3. Xenoliths in meteorites.

Xenolith Type	Host Type
Carbonaceous	H
C2	C3
C2	H
C2	Howardite
C2	Howardite
C2, H	H
C3	C2
C3	C3
CAI	H, L, LL
CAI	E
Anom. chondrite	LL
Chondrules	Howardites
LL	H
LL	Mesosiderite
H	LL
H	L
Achondrite	L
Eucrite	Eucrite
Eucrite	Eucrite
Cristobalite	L
Cristobalite	L
Cristobalite	H

Data from Olsen E. J. et al. (1988) *Geochim. Cosmochim. Acta, 52,* 1615, Table 1.

3.6.3. Composition

Although the chondrites are presumably derived from a relatively small portion of the solar nebula, they exhibit a wide range of composition. They retain only a small component (5%) of solar abundances of H_2O, C, and N. While the H_2O content of CI meteorites may be as much as 10%, the ordinary and enstatite chondrites are highly depleted, with 100-1000 ppm H_2O. Among the ordinary chondrites, the H, L, and LL chondrites were all derived from separate parent bodies, as shown by a wide variety of parameters.

The elemental abundance patterns among the various chondrite classes, relative to CI abundances, are shown in Fig. 3.6.2 [91]. There are no significant compositional differences between Types 4, 5, and 6 among the ordinary chondrites except for the siderophile elements, including FeO [92]. However, there is very little sense to be made of the variations in composition within the chondrite groups (Fig. 3.6.3). Much energy was expended by geochemists in trying to link these changes to derivation from a uniform CI-type precursor, until the oxygen isotope evidence demonstrated that the groups were

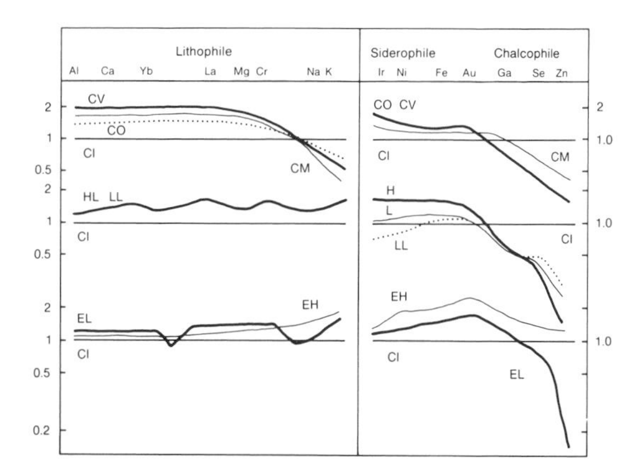

Fig. 3.6.2. *The elemental abundances in the major chondrite groups compared to those in CI carbonaceous chondrites, showing the general depletion of volatile, lithophile, and chalcophile elements. Each group has a distinct abundance pattern. After Sears D. W. G. and Dodd R. T. (1988) in Meteorites and the Early Solar System (J. F. Kerridge and M. S. Matthews, eds.), p. 21, Fig. 1.1.10, Univ. of Arizona, Tucson.*

unrelated. Thus, the volatile element depletions are not correlated with any other bulk chondrite property, such as degree of oxidation [93], and the view is adopted here that the process that depleted the inner nebula in volatile elements occurred at an earlier stage than the production of chondrules.

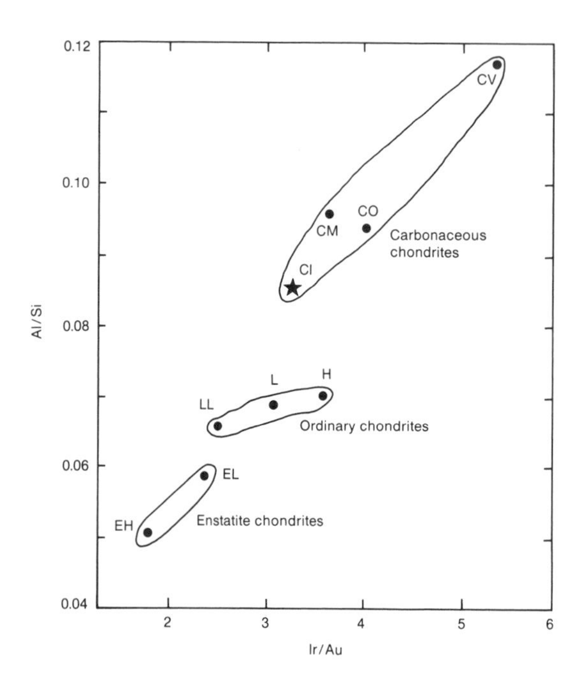

Fig. 3.6.3. *The variation in Al/Si and Ir/Au ratios for the nine chondrite groups results from physical separation of refractory phases before accretion of the chondrite parent bodies. After Scott E. R. D. and Newsom H. E. (1989) Z. Naturforsch., 44a, 927, Fig. 1.*

The most notable compositional feature of the ordinary chondrites is the evidence of depletions in volatile, siderophile, and chalcophile elements relative to CI. It is also noteworthy that the abundance pattern for each class is unique, suggesting accretion in a distinct region of the nebula (see section 5.11) with little overlap. However, there are no simple trends relating one group with another. This distinction also shows up very clearly on the oxygen isotope diagram (Fig. 3.6.4).

A significant observation concerning the composition of the chondrite groups is that the major elements (Fe, Mg, Si, and O) are fractionated among the various groups by factors of up to 2. Relative to CI, the ordinary and the enstatite chondrites are strongly depleted in Si and Mg. Since these elements are frequently used for normalization, such comparison procedures are misleading since metal/silicate fractionation was very extensive (see section 3.5). These differences are clearly shown in Fig. 3.4.3, which shows that the bulk chondrite compositions are mixtures of matrix and chondrules. Since there is considerable variation in the compositions of chondrules and matrix for SiO_2, FeO, and MgO, the differences among the chondrite groups (H, L, LL, and CO = CV) are to some degree accidental, and must reflect accretion from localized regions in the nebula. The nebula clearly did not become homogeneous following chondrule formation.

3.6.4. Antarctic Meteorites

The large sample return from Antarctica has raised the interesting question whether these meteorites, which record falls over the past half million years or so, represent the same population as those collected in historical times. Although some statistically significant differences have been claimed for H-group chondrites [94], most variations appear for mobile trace elements, so that weathering in the aqueous Antarctic environment cannot be excluded as a prime cause. Different populations might be expected to come from distinct parent bodies with differing cosmic-ray exposure ages. However, there is no discernible difference between Antarctic and non-Antarctic populations of H-group chondrites, suggesting that they come from the same sources (Fig. 3.6.5).

3.6.5. Carbon

Carbon occurs in various forms in carbonaceous chondrites. Organic carbon dominates, but is mostly due to reprocessing in the solar nebula. Most of the

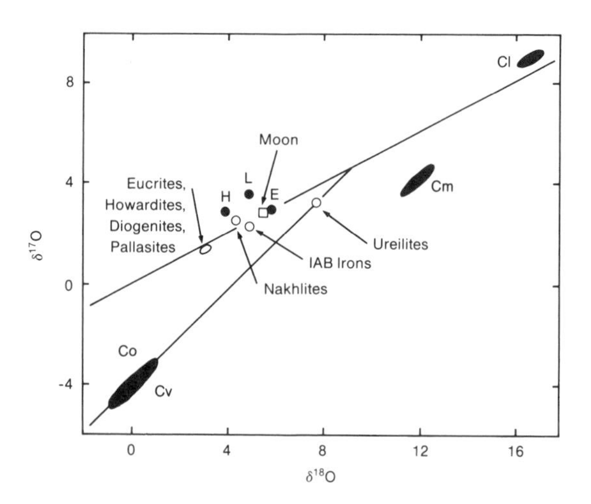

Fig. 3.6.4. *The oxygen isotope compositions of meteorites. The line passing through E (= Earth) is the terrestrial fractionation line with a slope of 0.5. The line with a slope of 1.0 (CO, CV) reflects mixing with an ^{16}O-enriched component. After Sears D. W. G. and Dodd R. T. (1988) in Meteorites and the Early Solar System (J. F. Kerridge and M. S. Matthews, eds.), p. 22, Fig. 1.1.11, Univ. of Arizona, Tucson.*

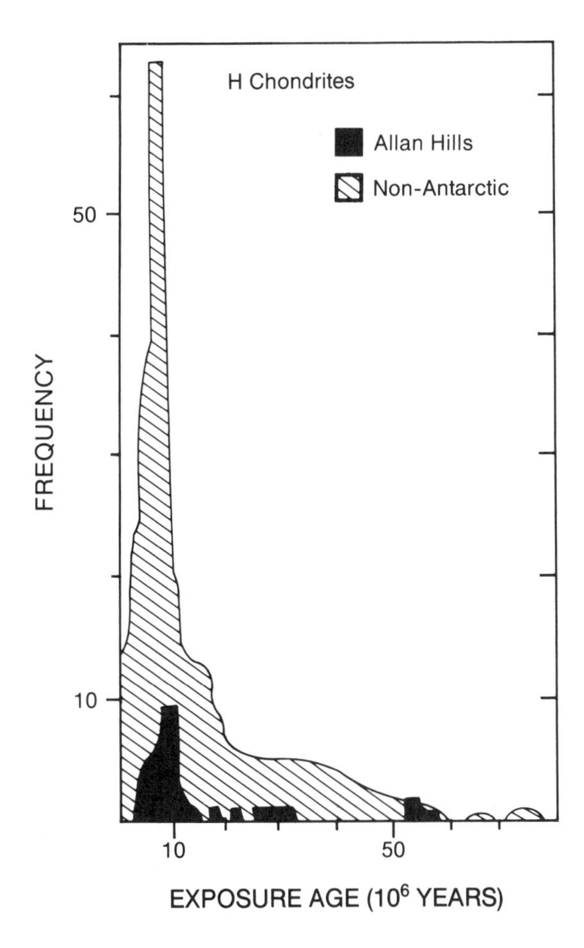

Fig. 3.6.5. *Similar cosmic ray exposure ages for Antarctic and non-Antarctic meteorite populations indicate that both are derived from a similar source. After Weber H. W. et al. (1988) Meteoritics, 23, 309.*

elemental carbon is present as diamond. Silicon carbide, which is of interstellar origin, as demonstrated by its highly anomalous isotopic composition [95], forms only a few parts per million of the total carbon. Graphitic carbon is much less abundant than diamond; apparently graphite is not abundant in interstellar matter. However, it is difficult to assess the interstellar abundances because of reprocessing in the solar nebula. Silicon carbide seems to be rarer in meteorites than might be expected by a factor of 10^4; perhaps the solar nebula was deficient in material from carbon-rich stars [96], but "the surviving interstellar component in meteorites may not be representative of average interstellar matter" [97].

However, carbon-rich stars are not expected to be abundant in the stellar associations most likely to have been associated with the formation of the solar system. Several processes, such as supernova shocks and heating in the solar system, are likely to destroy SiC, which should have been produced quite abundantly. Nevertheless, because of solar nebula processing, it is difficult to make an assessment of the relative abundances of the various forms of carbon in the interstellar medium.

3.6.6. Deuterium/Hydrogen Ratios

The most dramatic isotopic variations in chondrites are shown by hydrogen and deuterium (see section 3.2). Organic matter in meteorites has D/H ratios of 1×10^{-3} compared to 2×10^{-5} for cosmic hydrogen and 16×10^{-5} for terrestrial hydrogen. These high D/H ratios are of interstellar origin (see section 2.6) [98]. These are significant constraints on the amount, severity, and duration of thermal processing of chondritic material. High D/H ratios are indicative of unprocessed interstellar material. On the other hand, the hydrated silicates in CI and CM2 chondrites are secondary, produced on the parent body.

3.6.7. Siderophile Elements

The siderophile element abundances, including FeO, increase with metamorphic grade, so that the Type 3 chondrites are not simply related to Types 4-6 equilibrated chondrites by closed system metamorphism. However, the unequilibrated Type 3 chondrites are quite heterogeneous and consequently difficult to study. The trace siderophile elements correlate with Fe abundances, which is interpreted by Larimer and Wasson as evidence that the siderophiles were alloyed with metal "at the time of nebular fractionation" [99]. The siderophile elements differ widely in volatility [100].

Despite a widespread earlier assumption that the various chondrite classes were simply linked by variations in the oxidation state of iron, improved data now show that simple redox relations among the various chondrite groups can be ruled out (Fig. 3.6.6). Most chondrite groups are clearly resolved on this plot and do not form a simple linear relationship [101]. In this context, sulfur, generally regarded as a volatile element, is not depleted in chondrites of the higher "metamorphic grades."

3.6.8. Metamorphism

There is some dispute that full thermal equilibrium was attained in Types 4, 5, and 6 chondrites during metamorphism. For example, Allegan (H5) is chem-

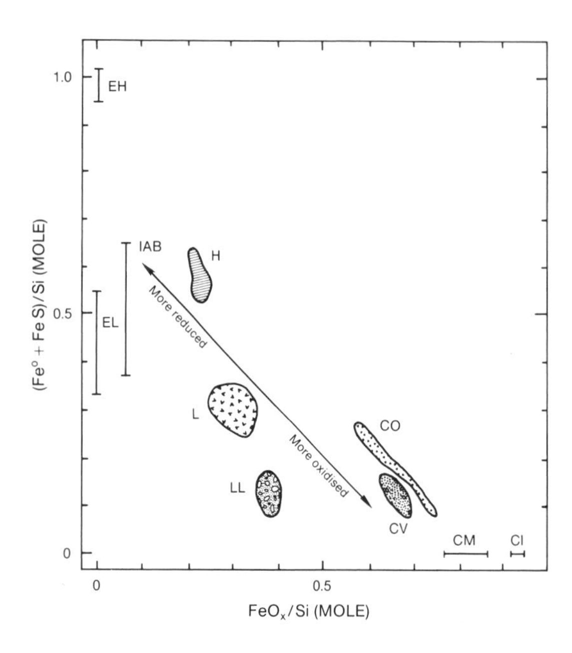

Fig. 3.6.6. *A plot of reduced Fe (Fe^0 and FeS) against oxidized Fe (FeO_x) using precise data indicates that most meteorite groups do not form linear arrays with slopes of -1. After Larimer J. W. and Wasson J. T. (1988) in Meteorites and the Early Solar System (J. F. Kerridge and M. S. Matthews, eds.), p. 419, Fig. 7.4.1, Univ. of Arizona, Tucson.*

ically unequilibrated [102]. However, the consensus among most workers on these complex objects is that the chondrite parent bodies were heated and the materials metamorphosed, but with little change in composition. Thus, in an extensive study of ordinary chondrites, "there is no evidence for systematic compositional differences among the Type 3-4-5-6 sequence . . . the data are consistent with isochemical thermal metamorphism of a common unequilibrated starting material" [103].

3.6.9. Hydration

CI meteorites are breccias composed of millimeter-sized clasts. The CI matrix in Orgueil, for example, consists of "an intimate mixture of Fe-bearing serpentine, saponite, and S- and Ni-bearing ferrihydrate" [104]. Carbonates and sulfates, formed by aqueous alteration, are also present, and aqueous alteration seems to have been pervasive [105].

The scenarios for alteration of meteorites are very complex, for the petrologically most primitive are C3, while CI, which have undergone aqueous alteration, have the most primitive compositions. CI chondrites are not, however, derived from CM or CV Types. The matrix mineralogy differs significantly among these classes and cannot have been caused by different amounts of aqueous alteration acting on a common precursor material [106]. Although the mineralogy of the CI meteorites is consistent with aqueous alteration of anhydrous material in CM chondrites [107], other criteria disallow this connection. However, leaching in CIs did not transport material over distances greater than 1 cm and the bulk meteorite analysis has not been affected by liquid transport. This has important consequences if we are to use CI compositions as analogues for that of the primordial solar nebula.

Hydrated minerals are also present in CM chondrites. The general consensus seems to be that the hydrated silicates are secondary, not primordial; their most probable cause is the melting of water ice. This scenario receives support from the absence of hydrated silicate spectra in the outer regions of the asteroid belt. Thus, the hydrated minerals are not nebular products, but are formed within planetesimals from reaction with water ice. The negative value of about -100 for δD for hydrated silicates in carbonaceous chondrites indicates a temperature of about 200 K for the condensation of the precursor ice [108,109]. Temperatures during alteration do not seem to have exceeded 25°C for the CM chondrites, and 150°C for the CI chondrites, based on the oxygen isotope data [110]. The presence of hydrated phases of course puts an upper limit of perhaps 350°C on the subsequent thermal histories. Whether the alteration occurs within the parent bodies or in a surficial regolith is still uncertain [111].

The hydrated silicates observed both in meteorites and on asteroidal surfaces are secondary products of low-temperature aqueous alteration [112] and are not primary phases. There is no evidence that the carbonate and sulfate phases in CI were primitive condensates from the nebula; they formed by aqueous alteration on the CI parent body. Such alteration must have occurred very early, since the lowest $^{87}Sr/^{86}Sr$ isotopic composition recorded in the carbonates is 0.69882 ± 20, coincident with that of the Allende refractory inclusions. Vein formation must have occurred within 10-100 m.y. of T_0, shortly after or perhaps during accretion [113].

Altered xenoliths of C3-type chondrites occur in Murchison (C2). The alteration has, however, preceded the incorporation of the xenoliths in the Murchison host, for the boundaries are sharp and the xenoliths display varying degrees of alteration, an argument against alteration in the nebula, which should produce uniformity. They appear to have experienced separate and individual alteration histories in some parent bodies before incorporation in the Murchison parent body [114].

3.6.10. Enstatite Chondrites

The enstatite chondrites represent an interesting class of meteorites. The two major classes, EH and EL, although they have identical oxygen isotope signatures (Fig. 3.6.4), are derived from separate parent bodies; there is a lack of EH clasts in EL and vice versa, although most are breccias. There are bulk compositional differences between them, but these were established by nebular, not planetary processes [115]. The occurrence of objects with identical oxygen isotope signatures, but with differing bulk chemistry and coming from distinct parent bodies must give pause to those who seek to use the oxygen data as grounds for a common origin (e.g., Earth and Moon). Such similarities merely indicate derivation from the same part of the nebula.

It is commonly stated that the enstatite chondrites contain a full solar complement of volatile elements and, although heavily reduced (all Fe is present as metal), have not been depleted in volatile elements to the same extent as the ordinary chondrites (H, L, and LL). Such comparisons are often made by normalizing the elemental abundances to silicon. However, compared to CI abundances, the E chondrites are depleted in Si, so they do not contain primordial solar nebular abundance patterns.

3.6.11. Chondrite Parent Bodies

The calculated orbits of three photographed chondrite falls (LL5-Innisfree, H5-Lost City, and H5-Pribram) are generally regarded as confirmatory evidence for an origin of the ordinary chondrites in the asteroid belt (see section 3.13). However, an outstanding anomaly is the apparent absence of parent bodies for the ordinary chondrites in the asteroid belt. The spectral signatures of the ordinary chondrites are distinct from those of the common S asteroids [116]. This apparent absence has been attributed to a failure to match laboratory spectra of meteorites with the observational data from the asteroid belt, due to the alteration of asteroidal surfaces by solar wind or other effects that produce a signature differing from the laboratory spectra, but this seems unlikely. The rather common occurrence of xenoliths of ordinary chondrites in meteoritic breccias has been used as evidence that they must be common objects in the main belt, possibly as S class [117]. However, a few objects in Earth-crossing orbit with an ordinary chondritic spectral signature have been detected (see section 3.13). At least three parent bodies are required; the possibility remains that ordinary chondrites are not very common, and may be derived from the rather rare Q asteroids. Future work will remove this problem, since there are strong dynamical arguments for deriving the ordinary chondrites from the asteroid belt. The oxygen isotope data suggest a minimum of 20 parent bodies for the observed population of stony meteorites. Cometary origins for gas-rich meteorites are not viable, since implantation of solar wind gases in these regolith breccias took place within a few AU of the Sun [118]. Samples of P and D asteroids (so-called ultracarbonaceous) are lacking in our present meteorite collections (see section 5.11).

Most data indicate parent bodies of the order of 100 km diameter [119]. Onion-shell models have been popular for the parent bodies of the chondritic meteorites. In such models, however, the hottest material would be the most deeply buried [120]. Hence, there should be a correlation between cooling rate and the petrographic type (3-6), but the cooling rates for metal grains in chondrites do not show such a correlation. Probably the parent bodies of the H and L chondrites were repeatedly broken up by collisions, and the onion-shell model is incorrect [121].

Chondrites do not seem to have originally been incorporated in much larger parent bodies, although such claims continue to be made. Thus, in a discussion on the oxidation state of chondrites, the evidence is used to conclude that the chondrite parent bodies were 30-70 km in diameter [122]. This conclusion is suspect, since the authors themselves comment on the large variations in the data and note that "individual equilibrated chondrites within a single chondrite group can have different oxidation states . . . Type 2 and 3 chondrites are highly disequilibrated assemblages that can preserve materials formed under different fO_2 conditions . . . The ordering of chondrite groups by oxidation state is based principally on the nature of the interchondrular material, some of which, e.g. magnetite in CI and CM chondrites, probably formed by secondary alteration. If we restrict our attention to chondrules (that are nebular products), a more muddled sequence emerges" [123].

3.6.12. Accretion Temperatures for Chondrites

There is a current debate over whether chondrites accreted hot (>800°C) or cold (<400°-500°C). The evidence usually adduced for hot accretion is that chondrules appear to have been plastic and suffered deformation during the accretion process [124]. There is a considerable body of textural evidence that appears to support this view [124], but it is not supported by the chemical and mineralogical evidence, which indicates low temperatures. Thus, magnetite is absent and troilite is present. Closer examination of the textural evidence indicates that the interiors of the chondrules are not deformed, and the many indented and close-packed chondrules appear to have been produced by "a form of diffusive mass transport... termed pressure induced diffusion... by removal of material at contacts between contact grains (chondrules and fragments)" [125]. Similar textures are observed in overclose packing of ooids in compacted limestone. If chondrites accreted at lower temperatures, this removes the constraint that they need to form while the chondrules are still hot, i.e., very quickly after the chondrule-forming process [126]. It should also be recalled here that "despite a common perception of carbonaceous chondrites as primitive nebular condensate material, no individual mineral phase in CI chondrites shows convincing evidence for such an origin" [126].

In summary, as was always emphasized by Harold Urey, chondrites are conglomerates. They are composed of essentially random mixtures of refractory inclusions, matrix, chondrules, sulfides, and Fe/Ni metal that accreted to form chondrites at low temperatures. Such chemical trends as can be discerned relate to siderophile or refractory elements that occur within a single component (e.g., siderophile elements contained in the metal, or refractory elements in CAIs). Although nine groups are recognized, there may be many more. However, little ultimate significance attaches to the beloved chondrite groupings; these, like so much else in the solar system, result from stochastic processes. Each appears to have accreted at a distinct region of the nebula. There is, however, no indication of any systematic change in composition with distance from the Sun. A further caveat is that the chondrites available for study represent an unknown sample of the asteroid belt, a view reinforced by the discovery of new meteorites that fall outside the conventional time-hallowed classifications. The Antarctic collection is beginning to reveal how inadequate our present sampling is.

3.7. CHRONOLOGY

The earliest identifiable event that led to the origin of the solar system was the formation of a molecular cloud complex. Within this, denser cores developed, from which a fragment became detached and evolved into the solar nebula. As the Sun began its early violent career, some preexisting grains survived, others grew within the nebula, chondrules formed, the chondrites were assembled, planetesimals began to grow, melting and differentiation occurred, and eventually the solar system took on its recognizable form. Some age limits can be placed even on the earliest of these episodes from the evidence provided from short-lived, now extinct radioactive nuclides. The later events, following the formation of solid bodies, can be dated with increased precision from the decay of the longer-lived ^{238}U, ^{232}Th, ^{87}Rb, and ^{147}Sm nuclides. The record within the meteorites provides our only source of information; it is this aspect that makes meteorites uniquely valuable.

3.7.1. Extinct Radionuclides

There is clear evidence that six short-lived radioactive nuclides were alive at the time of formation of the solar system. These were, in order of increasing half-life, ^{26}Al, ^{53}Mn, ^{107}Pd, ^{129}I, ^{244}Pu, and ^{146}Sm. Table 3.7.1 lists the daughter nuclides, mean lives, and their inferred abundance ratios relative to a

TABLE 3.7.1. Extinct radioactive nuclides.

Parent Nuclide	Daughter Nuclide	Mean Life*	Reference Nuclide	Observed Ratio
^{26}Al	^{26}Mg	1.07×10^{6} yr	^{27}Al	5.0×10^{-5}
^{53}Mn	^{53}Cr	5.3×10^{6} yr	^{55}Mn	4.4×10^{-5}
^{107}Pd	^{107}Ag	9.4×10^{6} yr	^{108}Pd	2.0×10^{-4}
^{129}I	^{129}Xe	2.31×10^{7} yr	^{127}I	1.0×10^{-4}
^{244}Pu	†	1.18×10^{8} yr	^{238}U	7×10^{-3}
^{146}Sm	^{142}Sm	1.49×10^{8} yr	^{144}Sm	1.5×10^{-2}

* Mean life is the reciprocal of the decay constant.
† Fission Xe nuclides at masses 131, 132, 134, 136.

Data from Cameron A. G. W. (1991) in *Protostars and Planets III* (G. Levy et al., eds.), Univ. of Arizona, Tucson, in press, and Wasserburg G. J. (1985) in *Protostars and Planets II* (D. C. Black and M. S. Matthews, eds.), p. 203, Univ. of Arizona, Tucson.

stable reference nuclide [127,128]. The mean lives range from a little over 10^6 yr for ^{26}Al to nearly 150×10^6 yr for ^{146}Sm, and provide interesting constraints both on the sites for the production of these nuclides, and on the relation of the molecular cloud complex to the solar nebula. The production of ^{26}Al and ^{107}Pd most likely occurred in a red giant star within the molecular cloud complex [127]. The other nuclides, ^{53}Mn, ^{129}I, ^{244}Pu, and ^{146}Sm were formed in supernovae just prior to the formation of the molecular cloud complex. Supernovae cannot have occurred within the cloud, since they are sufficiently energetic to have disrupted it. Cameron [127] has estimated the production rates for the short-lived nuclides, their dilution in the interstellar medium, and the decay period needed to reach the abundance levels observed in the solar system at 4560 m.y. ago. The most critical nuclide is ^{53}Mn, since it is produced in a supernova, and hence must predate the formation of the molecular cloud complex. A probable estimate is that ^{53}Mn was produced about 4.5 mean lives or 24 m.y. before the separation of the solar nebula [127]. Since the molecular clouds take of the order of 20 m.y. to form [129], the solar nebula thus seems to have become separated shortly after the cloud was established as a separate entity.

3.7.2. The Beginning of the Solar System or T_0

It is convenient to have a well-established base line to which a chronology can be tied. The BC-AD boundary forms a useful marker for human history, although 753 BC, the date for the foundation of Rome, would have been a more logical choice for Western Civilization [130].

What should serve for the solar system? Should we set T_0 as the time of separation of the molecular cloud, the formation of the keplerian disk, or that of the Sun? None of these significant events is readily measured by current technology. A convenient marker for the solar system is the age of the oldest solid components, formed within the nebula, that can be dated in meteorites. These can be measured to a precision of a few million years. The oldest most reliably dated objects are the refractory inclusions (CAIs) in Allende (Table 3.7.2) and their rounded Pb-Pb age of 4560 m.y. ago is used here as the zero age or T_0 for the solar system. This age records a depletion of Pb relative to U probably in an early, localized, and turbulent heating episode, occurring at some stage between the detachment of a fragment of the molecular cloud and its evolution into the rotating disk we identify as the "solar nebula."

TABLE 3.7.2. $^{207}Pb/^{206}Pb$ ages of Allende refractory inclusions (CAI).

Inclusion	Age (m.y.)
Egg-1A	4.564 ± 5
Egg-3A	4.561 ± 5
Egg-3B	4.544 ± 8
Egg-4	4.557 ± 4
Egg-6	4.565 ± 5
WA-5	4.561 ± 6
Mean	4.559 ± 5

Data from Chen J. H. and Wasserburg G. J. (1981) *Earth Planet. Sci. Lett.*, *52*, 8, Table 3.

3.7.3. Chondrite Ages

A comprehensive survey of H, L, and LL chondrites gives an Rb-Sr age of 4555 m.y. [131]. This agrees with the most recent Pb-Pb age of 4552 ± 3 m.y. for L5 chondrites [132], and with ^{207}Pb-^{206}Pb and ^{147}Sm-^{143}Nd data from Table 3.7.3. There is no significant difference between the various whole-rock ages for H, L, LL, and E chondrites. A rounded value of 4555 m.y. is adopted here as the best estimate of the age of chondrites. The chondrites, which have not been melted since the formation of their parent bodies, thus have ages apparently a few million years younger than those of the CAIs [133].

It should be noted, however, that chondrites are complicated on the one hand by aqueous alteration (Types 1 and 2) and on the other by what is generally referred to as "metamorphism" (Types 4, 5, and 6; see section 3.6). Accordingly, attention is usually focused on Type 3, which appear to be the least altered by either process. Examples of Type 3 chondrites include Tieschitz (H3) and Semarkona (LL3), references to which appear frequently in these pages. There is close agreement between the age given by the whole-rock isochron for H chondrites and the internal mineral isochron age for the unequilibrated H3 chondrite, Tieschitz. The very reduced enstatite chondrites have Rb-Sr ages indistinguishable from those of the H-chondrites [134]. This indicates that the highly reducing conditions that gave rise to these meteorites occurred somewhere in the nebula within a few million years of T_0 and essentially coeval with the chondrite-forming event [135] (Fig. 3.7.1).

The Sm-Nd data for phosphates from chondrites record an age of 4550 ± 45 m.y. The initial $^{143}Nd/^{144}Nd$ ratio is 0.5067 ± 5. These Sm-Nd dates agree with the other chondrite data, within the rather broad error limits (Fig. 3.7.2).

TABLE 3.7.3. Ages of chondritic meteorites (in m.y.).

Chondrite	$^{207}Pb/^{206}Pb$	$^{147}Sm/^{143}Nd$	$^{87}Rb/^{86}Sr$	Ref.
Tieschitz (H)	—	—	4530 ± 60	1
Guarena (H)	—	—	4460 ± 80	2
L5 (mean)	4552 ± 3	—	—	3
Chainpur (LL)	—	—	4517 ± 56	4
St. Severin (LL)	—	4550 ± 330	4510 ± 330	5
Phosphates	4552 ± 4	—	—	6
Chondrite Ave	4555	—	—	7

1. Minster J. F. and Allègre C. J. (1979) *Earth Planet. Sci. Lett., 42,* 333.
2. Wasserburg G. J. et al. (1969) *Earth Planet. Sci. Lett., 7,* 33.
3. Manhés G. et al. (1987) *Terra Cognita, 7,* 377.
4. Minster J. F. and Allègre C. J. (1981) *Earth Planet. Sci. Lett., 56,* 89.
5. Jacobsen S. B. and Wasserburg G. J. (1984) *Earth Planet. Sci. Lett., 67,* 137.
6. Chen J. H. and Wasserburg G. J. (1981) *Earth Planet. Sci. Lett., 52,* 1.
7. Average from 1, 3, and 4.

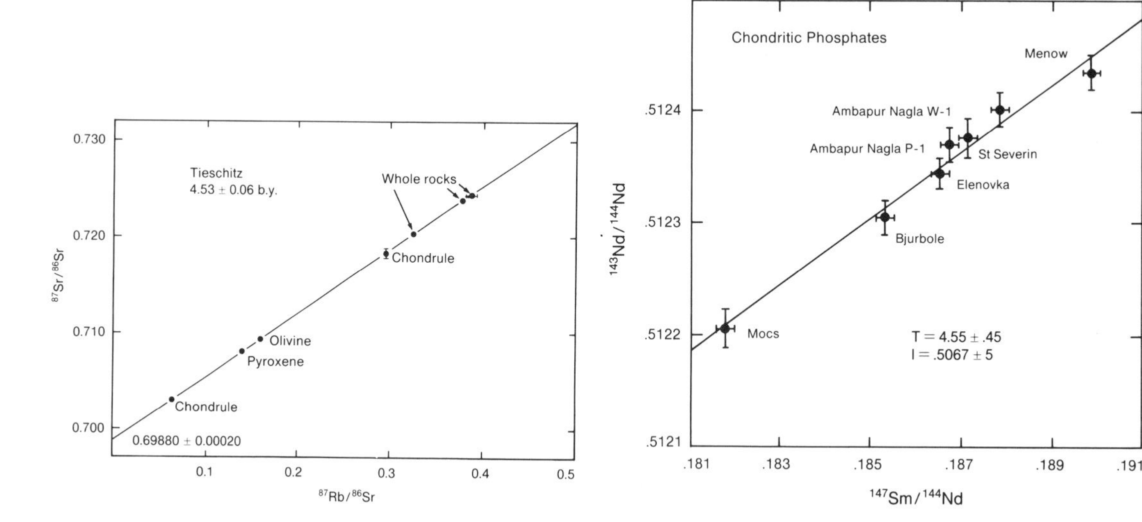

Fig. 3.7.1. *Rb-Sr isochron for the unequilibrated H3 chondrite Tieschitz. After Minster J. F. et al. (1979) Earth Planet. Sci. Lett., 44, 420, Fig. 8.*

Fig. 3.7.2. *Sm-Nd diagram for phosphates from chondrites. After Brannon J. C. et al. (1987) Proc. Lunar Planet. Sci. Conf. 18th, p. 556, Fig. 1.*

3.7.4. Significance of Strontium Initial Isotopic Ratios

Initial Sr ratios are very sensitive indices of early planetary evolution, marking the early separation of volatile Rb from refractory Sr. Table 3.7.4 lists the available precise values for Allende CAIs and achondrites. It should also be noted that there are some unresolved differences among laboratories with respect to the measurement of initial Sr ratios. Several different events seem to have occurred close to the limits of resolution of the techniques, while the systems may also be perturbed by alteration processes.

Low initial ratios are recorded for the Allende CAIs [136]. Presumably the Pb-Pb age of 4560 m.y. also dates this volatile depletion of Rb relative to Sr. This is interpreted as caused by very localized high-temperature events that resulted in the production of the refractory inclusions. As we have seen in section 3.3, the CAI record repeated episodes of evaporation and condensation. Where did these occur? As discussed a little later, they may represent a very turbulent environment during the separation of the nebula from the molecular cloud complex and thus predate the formation of the keplerian disk proper. The reasons for this conclusion depend on the

TABLE 3.7.4. Precise initial $^{87}Sr/^{86}Sr$ ratios.

	Initial $^{87}Sr/^{86}Sr$	Reference
Allende CAI (ALL)	0.69881 ± 2	1
Achondrites		
Angra dos Reis (ADOR)	0.69897 ± 1.5	2
Moore County	0.69897 ± 1.7	2
Basaltic Achondrite Best Initial (BABI)	0.69899 ± 4	3
LEW 86010 (Angrite)	0.69899 ± 1	2
Mean solar system initial	0.69897 ± 1.5	2

1. Podosek F. A. et al. (1991) *Geochim. Cosmochim Acta*, in press.
2. Lugmair G. W. et al. (1989) in *Lunar and Planetary Science XX*, p. 605, Lunar and Planetary Institute, Houston.
3. Papanastassiou D. A. and Wasserburg G. J. (1969) *Earth Planet. Sci. Lett.*, *5*, 361.

evidence from the initial Sr values for the basaltic achondrites, to which we now turn.

It is difficult to measure the Sr initial values in chondrites to a useful degree of precision, so attention has been focused on the basaltic achondrites (Table 3.7.4). The two most significant features of the data from these differentiated meteorites are that they are uniform, and that they have values significantly higher than the Allende CAI. The interpretation based on these data is that it records a large depletion of Rb relative to Sr in the inner nebula. This is interpreted here to be due to early violent T Tauri and FU Orionis stages of the early Sun sweeping out volatile material from the inner nebula. Additional evidence in support of such an event is the widespread depletion of volatile K relative to refractory Rb in the inner planets, so we are looking at a nebulawide rather than local event. Thus, a reasonable case can be made from the Sr initial data for achondrites that they represent a significant event of nebulawide (extending out to 3 AU) depletion of Rb and other volatile elements. On this basis, they mark the depletion of volatile elements in the inner nebula. If this was due to early T Tauri and FU Orionis activity, then we are looking at times within 1 m.y. of the formation of the Sun. The events recorded by the Allende CAI with their very low initial Sr value have to be earlier. A straightforward calculation based on solar Rb/Sr ratios indicates that it takes over 10 m.y. for the $^{87}Sr/^{86}Sr$ ratio to grow from 0.69881 (ALL) to 0.69897 (BABI). This period is longer than the probable lifetime of the nebula, and also contrasts with an apparent 5-m.y. gap between the ages of the CAI and of the chondrites [137]. One resolution of this dilemma is that the production of CAI occurred in turbulent conditions some time between the separation of the fragment of the molecular cloud and its evolution into a disk.

This major volatile depletion event recorded by BABI occurred either before or during chondrule production, and the meteorites were assembled from material already partly depleted in volatile elements. Later, local perturbations due to heating of individual planetesimals added considerable noise level to the isotope data. Such a sequence of events appears from the interpretation of the chondrule data discussed in section 3.5 and the evidence for loss of the volatile elements from the inner nebula discussed in section 2.22.

This interpretation on the timing of volatile loss (when Rb was depleted relative to Sr, and Pb relative to U) is preferred to the possibility of accreting the material cold, and losing Rb during individual heating episodes within the parent bodies.

3.7.5. Early Igneous and Metamorphic Processes

Igneous processes in the solar system began only a few million years after T_0 indicating that melting, differentiation, and "planetary" processes were operating very early. The average crystallization age for these basalts is 4539 ± 4 m.y., which is measurably younger (20 m.y.) than T_0 (Table 3.7.5).

The initial Sr ratio for ADOR [138] is indistinguishable from BABI, although measurably higher than ALL, and so does not record a distinct event. It should be noted here that these measurements are difficult and are being made close to the limits of the present technology. The cumulate eucrites are apparently younger, but perhaps the latter are cooling ages. The noncumulate eucrites appear to have ages that cluster within less than 10 m.y. [139].

The Pb-Pb age for the eucrite, Ibitira, is 4556 ± 6 m.y. compared with an Sm-Nd age of 4460 ± 20 m.y. [140]. These data are interpreted as indicating an

TABLE 3.7.5. Ages of achondritic meteorites (in m.y.).

Achondrite	$^{207}Pb/^{206}Pb$	$^{147}Sm/^{143}Nd$	$^{87}Rb/^{86}Sr$
Angra dos Reis (ADOR)	4551 ± 4 (1)	4564 ± 37 (2)	—
ADOR phosphate	4553 ± 8 (1)	—	—
Juvinas (eucrite)	4539 ± 4 (3)	4500 ± 60 (4)	4500 ± 70 (5)
Ibitira (eucrite)	4556 ± 6 (8) 4560 ± 3 (7)	4460 ± 20 (6)	—
Y75011 (eucrite)	—	4550 ± 140 (9)	—

1. Chen J. H. and Wasserburg G. J. (1981) *Earth Planet. Sci. Lett*, *52*, 1.
2. Jacobsen S. B. and Wasserburg G. J. (1984) *Earth Planet. Sci. Lett.*, *67*, 137.
3. Manhés G. et al. (1984) *Geochim. Cosmochim. Acta*, *48*, 2247.
4. Lugmair G. W. et al. (1975) *Proc. Lunar Sci. Conf. 6th*, p. 1419.
5. Allègre C. J. et al. (1975) *Science*, *187*, 436.
6. Manhés G. et al. (1987) *Meteoritics*, *22*, 453.
7. Prinzhofer D. A. et al. (1989) in *Lunar and Planetary Science XX*, p. 872, Lunar and Planetary Institute, Houston.
8. Chen J. H. and Wasserburg G. J. (1985) in *Lunar and Planetary Science XVI*, p. 119, Lunar and Planetary Institute, Houston.
9. Nyquist L. E. et al. (1986) *J. Geophys. Res.*, *91*, 8137.

extreme depletion of volatile Pb relative to refractory U at 4560 m.y. (close to T_0), followed by igneous fractionation producing a basaltic liquid at 4460 m.y. The Pb-Pb age dates the period of major volatile depletion [141].

3.7.6. Formation Ages and "Metamorphism"

The equilibrated chondrites indicate some kind of internal redistribution for periods up to 100 m.y. following the initial separation. This is generally referred to as "metamorphism" in parent bodies, but there are several problems [137]; however, these are of secondary importance to the major conclusion that separation of volatile from refractory elements apparently occurred within a very short period of time (<5 m.y.), effectively at T_0. Some isotopic mobility occurred within the mineral phases, leading to mineral isochrons that give younger ages and higher initial Sr ratios.

The general explanation is given in Fig. 3.7.3 [137]. The time between separation of the chondrites with a lower Rb/Sr ratio than CI and the isotopic closure of the mineral phases, is generally referred to as the "formation interval." This is a relatively straightforward explanation of the Rb/Sr data, and formation intervals of up to 200 m.y. are obtained. Although these periods are usually ascribed to metamorphism, there is little correlation between the formation

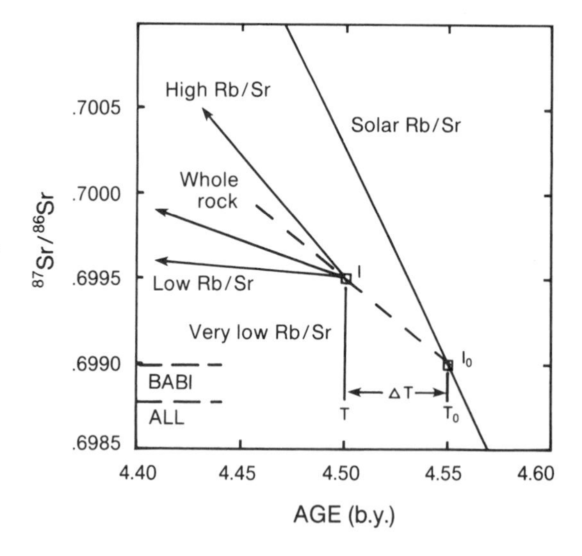

Fig. 3.7.3. *The model evolution of initial $^{87}Sr/^{86}Sr(I)$ as a function of time. The solar nebula has a high Rb/Sr ratio, and I evolves rapidly, passing through the ALL-BABI region about 4.56 b.y. ago. At time T_0, there is a depletion of volatile Rb relative to refractory Sr in the inner nebula; this diminishes with increasing distance from the Sun. Bodies form with lower Rb/Sr and follow a different evolution, shown by the dashed line. Because of heating, the mineral phases do not become isotopically closed until later at time T, when they show a common initial ratio. After Brannon J. C. et al. (1987) Proc. Lunar Planet. Sci. Conf. 18th, p. 561, Fig. 4.*

TABLE. 3.7.6. Formation interval calculations for the Rb-Sr method compared with those from I-Xe.

Meteorite		Initial $^{87}Sr/^{86}Sr$	Whole Rock $^{87}Rb/^{86}Sr$	ΔT (m.y.)	ΔT(I-Xe) (m.y.)
LL4	Soko Banja	0.69959 ± 24	1.563	35 ± 10	3
LL6	St. Severin	0.69907 ± 12	0.134	148 ± 59	8,30
L4	Bjurbole	0.70015 ± 11	0.792	115 ± 9	≡0
L6	Peace River	0.69970 ± 10	0.749	82 ± 9	—
L6	Elenovka	0.69939 ± 30	0.746	55 ± 26	—
H3	Tieschitz	0.69880 ± 20	0.361	5 ± 37	—
H6	Guarena	0.69995 ± 10	0.869	90 ± 8	—
EH5	St. Marks	0.69979 ± 22	0.320	210 ± 45	5

Data from Brannon J. C. et al. (1988) *Proc. Lunar Planet. Sci. Conf. 18th*, p. 561, Table 6.

intervals and the degree of metamorphism. Although Tieschitz (Type 3) has the expected short formation interval, the lowest metamorphic grade, and the lowest initial Sr, the correlation fails for most others. Clearly the relationship is not simple [137]. It should be recalled that other parameters, such as cooling rates, also do not correlate with "metamorphic" grade [142]. A serious problem is that the formation intervals calculated from the precise I-Xe system disagree by very large factors with the dates obtained from the Rb-Sr system [143] (Table 3.7.6). The I-Xe data record intervals typically of 10^7 yr, rather than the 10^8 yr obtained from the Sr data. The interpretation of these data is controversial, and it is not clear what is being measured. This seems to imply that chondrites are recording a complex history of breakup and reassembly; the common factor is the depletion of volatiles close to T_0.

3.7.7. Meteorite Ages and Planetesimals

A final question of importance is the relationship of the meteorite ages to the planetesimal hypothesis. The basic question to be addressed is the apparent absence of meteorites with ages appropriate to the 10^8-yr time span that are generally allowed for the accumulation of the terrestrial planets from precursor planetesimals. As noted above, the meteorite ages record a much shorter sequence of events, closer to T_0.

Does this gap in the age record invalidate the planetesimal hypothesis? Two aspects make this unlikely. First, the timescale for accretion of the terrestrial planets may be overestimated, and could well be closer to 10^7 yr. Second, and contrary to popular belief, the present population of meteorites is not necessarily a sample of the building blocks of the terrestrial planets. This is discussed at length in section 3.14, but the following points may be made here.

1. The asteroid belt is a frozen relic of an early stage of nebular development, representing an arrested step in nebular evolution, due to the early development of Jupiter, which starved the belt of material. Accordingly, no large planetesimals developed, and we have no samples of that stage of the evolution of the planets.

2. The asteroid belt is highly structured and zoned. Meteorites are perturbed into Earth-crossing orbits somewhat by chance. The failure to identify a source within the belt for the ordinary chondrites, at present the most common objects to fall, makes it likely that we have only a random sampling.

3. All the large planet-making objects have been swept up. Since it is likely that the terrestrial, and also the outer, planets were accumulated from narrow feeding zones, it is to be expected that none of this material survives; only the asteroid belt remains, in its state of arrested adolescence.

3.8. COMPOSITION OF THE NEBULA

We are surprisingly well informed about the composition of the primordial solar nebula. Although it was chiefly composed of hydrogen with 27% helium, the main interest here revolves around the nongaseous (although not nonvolatile) constituents. The composition of the Type 1 carbonaceous chondrites (CI) closely resembles that derived from spectra of the solar photosphere for the nongaseous elements (sections 2.12, 2.15, and 3.8).

Although the CI chondrites are not particularly primitive, having undergone aqueous alteration in some kind of asteroidal environment, they are the only class of meteorite to have retained such a close connection to the solar abundance pattern. Further support for this concept, with some minor circularity in reasoning, comes from the predictions of element abundance from theories of element formation. A reasonable level of agreement appears to have been reached on this iterative process.

Excluding those elements that are not well determined in the solar photosphere, significant discrepancies exist between the meteoritic and solar data for six elements. Of these Li, Be, and B are destroyed by nuclear reactions in the Sun. This depletion immediately refutes theories that wish to derive the planets from an already formed Sun (for example, the tidal hypotheses), unless such an event occurred during solar collapse, before or in the very early stages of ignition of the nuclear reactions.

A few remaining elements (Mn, Ge, and Pb) appear to show real differences. Could this be due to fractionation in the meteorites? It could be expected, if that were the case, that other elements of similar geochemical behavior would show like differences (e.g., Bi and Tl should correlate with Pb), but such variations are not observed. The Ge, Mn, and Pb solar data are probably in error. With these exceptions, the CI data are within ±9% of the solar values for the well-determined elements in the solar spectra. The meteoritic values are adopted in this book as the best estimate of the composition of the "solar system," in agreement with the latest compilation [144].

The Fe abundance problem has been resolved as the latest evaluation of the solar spectral data, based on the Fe II lines, agrees with the CI data [145], despite earlier claims that solar abundances were 40% higher [146]. Fe/Si ratios vary considerably between the meteorites, interplanetary dust particles, and Comet Halley with respect to the solar values.

The expectation that Comet Halley might provide the long-awaited sample of "primordial" material does not seem to have been fulfilled. Although only preliminary data are available [147], it appears that while the volatile elements C, N, and O approximate to solar values (H is depleted), the nonvolatile elements in the dust fraction possess Fe/Si and Mg/Si values far below solar.

The solar system abundances, as derived from the CI data [148], do not form a good match for the compositions of the terrestrial planets. The terrestrial planets are more depleted in volatile elements than are the chondrites, as shown by low K/U ratios for Earth, Venus, and Mars compared with the bulk nebular values. This depletion appears to be a general feature of the inner solar nebula. It is considered to result from heating from unknown causes, probably connected with early T Tauri and FU Orionis stages of solar activity, occurring within about 10^6 yr of the arrival of the Sun on the main sequence. An additional heating stage occurs during chondrule formation. Chondrules also inherited the inner nebular depletion in volatile elements, although to a lesser extent than that experienced by the parental material that went to make up the terrestrial planets [149].

3.9. "CONDENSATION SEQUENCE"

The condensation sequence originated with the concept of a hot nebula in that all the constituents were present in CI abundances in gaseous form at $p = 10^{-3}$ atm. Thermodynamic calculations predicted the sequence of minerals condensing from a vapor phase; these bore a general resemblance to the sequence observed in meteorites [150] (Table 3.9.1). This agreement in turn led to the concept that the composition of the planets could be explained by accretion of condensing phases from an initially vaporized nebula, the so-called equilibrium condensation hypothesis [151]. However, the petrography of chondrules does not correspond to a condensation

TABLE 3.9.1. The equilibrium condensation sequence, listing the major minerals that would condense from a gas at 10^{-3} atm and solar system element abundances.

Temperature* (K)	Mineral
1758 (1513)	Corundum, Al_2O_3
1647 (1393)	Perovskite, $CaTiO_3$
1625 (1450)	Melilite, $Ca_2Al_2SiO_7$-$Ca_2MgSi_2O_7$
1513 (1362)	Spinel, $MgAl_2O_4$
1471	Fe, Ni metal
1450	Diopside, $CaMgSi_2O_6$
1444	Forsterite, Mg_2SiO_4
1362	Anorthite, $CaAl_2Si_2O_8$
1349	Enstatite, $MgSiO_3$
<1000	Alkali-bearing feldspar, $(Na,K)AlSi_3O_8$-$CaAl_2Si_2O_8$
<1000	Ferrous olivines, pyroxenes, $(Mg,Fe)_2SiO_4$, $(Mg,Fe)SiO_3$
700	Troilite, FeS
405	Magnetite, Fe_3O_4

* In most cases, temperature at which condensation or reaction begins in a cooling system. At temperature in parentheses, reaction has completely used up the phase and converted it into other minerals.

Data from Wood J. A. (1988) *Annu. Rev. Earth Planet. Sci., 16*, 57, Table 1.

sequence. In fact, it predicts that these obvious igneous minerals would vaporize at temperatures below their melting points.

More recently, the condensation sequence has been recalculated in an attempt to match the nebular region of chondrite formation more closely [152]. Nebular pressures of 10^{-5} atm, considered to be more realistic than the canonical value of 10^{-3} atm, and an enhanced O/H ratio (41 times the cosmic ratio) were used. This is consistent with the formation of minerals in the inner nebula after most of the H and He have gone [153].

The revised condensation sequence is given in Fig. 3.9.1. Under the specified conditions, condensation is thought to occur above 1400 K, a region in which mafic silicate melts are stable. Thus, Fe olivine is stable at 1400-1800 K under high O/H ratios compared with <500 K in a hydrogen-rich nebula. Melt-vapor systems are not easily treated theoretically. The first solid to condense from the gas phase is corundum, followed by hibonite, gehlenite, and spinel (Fig. 3.9.1).

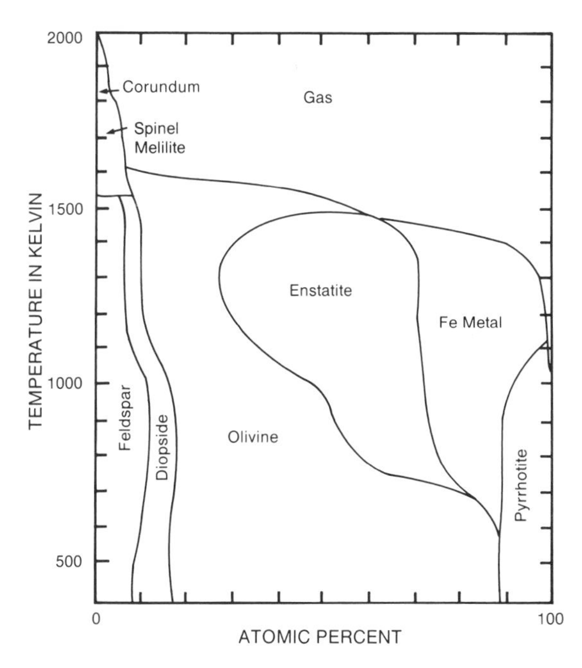

Fig. 3.9.1. *The atomic fractions of Si, Al, Fe, Mg, Ca, Na, and Ti occurring in the various mineral phases at 10^{-5} atm, with the dust/gas ratio enhanced to 100 times the primordial solar nebula ratio. After Wood J. A. and Hashimoto A. (1988) in Lunar and Planetary Science XIX, p. 1293, Fig. 1, Lunar and Planetary Institute, Houston.*

The condensation sequence was interpreted for a number of years as a cooling sequence from an initially vaporized nebula, with equilibrium maintained during cooling. The reality in the nebula appears to be otherwise, with equilibrium conditions rarely being approached. In view of the complexity observed in the Allende inclusions, which record many stages of condensation and evaporation, the condensation sequences observed in meteorites are now thought to record local, rather than nebulawide conditions.The cooling times for chondrules were so short as to preclude formation in a hot portion of the nebula, while the CAIs record many cycles of evaporation and condensation.

3.10. COMETS

"The appearance of comets, followed by these long trains of light, has for a long period terrified mankind, always agitated by extraordinary events of which the causes are unknown. The light of science has dissipated the vain terrors that comets, eclipses, and many other phenomena inspired in the ages of ignorance" [154].

In many discussions of the solar system there has been a tendency to ignore or downplay the importance of comets relative, for example, to meteorites. However, the total mass of comets is substantial [155], totalling perhaps 50 Earth masses of rock and ice, and they represent material from the outer edge of the system. It is also becoming difficult to separate comets from such large outer solar system icy planetesimals as Triton and Pluto, while Chiron has recently joined the family.

Comets as we observe them are characteristic of mature planetary systems. They are not likely to appear in an evolving and growing planetary system, since the following steps have to occur. First, condensation of ices has to occur in the nebula. The next requirement is the formation of large bodies, such as the giant planets, whose gravitational forces will scatter the comets into orbits around the central star. Only at that stage will comets become clearly visible objects as the stellar radiation produces the highly visible coma and tails. Accordingly comets are not to be expected to be observed until the planetary system has reached maturity.

3.10.1. Oort and Kuiper Clouds

About six comets are observed each year from the Earth, so there is a steady supply of these short-lived objects. Although they were originally thought to be derived from the classical Oort Cloud, there now seem to be three sources: Outer and Inner Oort

Clouds, and a closer-in Kuiper Cloud. The Outer Oort Cloud is more or less spherical and extends from about 20,000-50,000 AU. The total number of comets in the classical Oort Cloud is uncertain, but it may contain $1\text{-}2 \times 10^{12}$, representing perhaps 40% of the original population [156].

The Inner Oort Cloud contains possibly about $2 \times 10^{12}\text{-}10^{13}$ comets, located at about 3000-20,000 AU. The Kuiper Cloud is located beyond the orbit of Neptune at about 30-100 AU, with an apparent gap between the Kuiper and Inner Oort Clouds. Both the Inner Oort and Kuiper Clouds are probably flattened disks, in contrast to the more diffuse spherical form of the Outer Oort Cloud [156].

Short-period comets, about 100 of which are known, orbit the Sun with periods of less than about 200 years. Most have orbital periods less than about 15 years. Another 21 have orbital periods between 15 and 200 years. Their orbits lie close to the plane of the ecliptic with low inclination, and are also prograde in the same sense as the planets, except for four. They must come from a source more or less in the plane of the ecliptic.

It has been suggested that the short-period comets are derived from a disk of material at about 35-50 AU, just beyond Neptune (i.e., the Kuiper Cloud). This could represent left-over material from the primordial solar nebula [157] and thus be a better sample of that system than comets from the outer Oort Cloud, which may contain a galactic component. However, the search for "primitive" material is turning out to be a chimera. The amount of close-in material cannot exceed a few Earth masses at most, since the Pioneer 10 spacecraft would have detected the gravitational perturbations of larger masses. A fraction of an Earth mass would suffice as a source for the short-period comets. Perturbations by Neptune could serve to scatter them into Sunward orbits. Possibly Chiron is one of the parent bodies for the short-period comets [158]. The source of these comets in a belt extending out to a few hundred AU or so is consistent with the IRAS observations of dust shells of similar dimensions around stars [159]. The 600 or so long-period comets are derived from the dynamically active outer cloud. Table 3.10.1 gives a set of estimated ranges in composition for the physical parameters for cometary nuclei.

3.10.2. Original Site of Cometary Formation

Perturbations of cometary orbits by passing stars and interactions with interstellar molecular clouds have randomized them, so the original source region for comets has been obscured. Thus, there is little general agreement on the location of the region where comets formed, except that they are solar system objects. No confirmed hyperbolic orbits, which could indicate an origin from outside the solar system, have been found for comets [156]. Carbon monoxide, which vaporizes at significantly lower temperatures than water ice, is abundant in comets. If it is a true parent molecule, and not formed by dissociation of formaldehyde (H_2CO) then the cometary nucleus has not been heated above 25 K. Even more telling is the presence of S_2, which has a sublimation temperature of 20 K. Such temperatures

TABLE 3.10.1. Expected ranges in physical parameters for cometary nuclei.

Parameter	Minimum Value	Expected Value	Maximum Value
Diameter (km)	0.3	5	40
Surface gravity (m/s^2)	10^{-5}	10^{-3}	10^{-2}
Rotation period (d)	0.2	1	14
Mean density (g/cm^3)	0.1	0.5	1.5
Porosity	10%	30%	80%
Density at surface (g/cm^3)	0.005	0.05	1.5
Thickness of mantle (m)	10^{-5}	0.1	1
Albedo (geometric)	0.01	0.03	0.5
Surface temperature (K)	100	150	300
Dust/ice mass ratio	0.1	0.6	1.0
Diameter range of refractory particles (cm)	$\leq 10^{-6}$		≥ 10
Dust composition: silicates		~70%	
carbonaceous compounds		~30%	
Gas production rate (molecules/s)	10^{26}	10^{27}	10^{30}

Data from D. J. Stevenson, personal communication, 1989.

would indicate an original source in the outer reaches of the solar system. This places their formation site at least as far out as Uranus and Neptune, from which regions they have been ejected to the present locations in the Kuiper and Oort Clouds. The original angular momentum of the Oort Cloud is about 3 times that of Uranus and Neptune, and is consistent with formation in the Uranus-Neptune zone, rather than at their present locations [160].

3.10.3. Comet Halley

It was widely anticipated that comets would prove to be pristine samples of the solar nebula. This was based on the somewhat wishful thinking that leads us to attach uniformity or simplicity to unknown regions. Unlike the medieval mind that populated *terra incognita* with monsters, the modern agnostic viewpoint, in rejecting such fantasies, and in the absence of evidence to the contrary, has tended to the opposite extreme and considers these regions (the deep interior of the Earth, the primordial solar nebula, and the interstellar medium) as uniform.

Basic questions are whether comets are solar nebular condensates, small icy planetesimals driven out to the Oort Cloud by planetary or stellar interactions, or are composed of pristine interstellar material; the encounter with Comet Halley was expected to provide such answers.

The nucleus, as measured during the 1986-87 encounter, was $15 \times 7.2 \times 7.2$ km. The gas outflow was very localized in jets, involving less than 15% of the surface area. Enhanced activity observed during March 1986 was consistent with rotation of an active area into sunlight about every 7.4 days [161]. The gas emission measured by the Giotto mission was 20 tons/s, and dust emission was 3-10 tons/s. The total mass of the comet was 10^{11} tons; Halley loses 10^8 tons per solar encounter, or about 1/1000 of its mass. Thus, there is copious emission of both dust and gas even from veterans such as Halley, which has made perhaps 300 solar passes during the 23,000 years it has spent in the inner solar system [162] (Fig. 3.10.1). This leads us to suppose that the mass loss does not fractionate more than a surficial layer; the frozen nucleus should retain its pristine composition. Whether this resembles the material in the primordial solar nebula is another question.

The gas composition was 80% H_2O, 10% CO, 3% CO_2, 2% CH_4, <1.5% NH_3, and 0.1% HCN. The CH_4/CO ratios range from 0.1 to 0.4 and NH_3/N_2 ratios are >0.1-2. These ratios are higher than observed in the interstellar medium, and indicate that the material has been processed somewhere, and did not form by direct accretion from interstellar material. No meaningful values were obtained for the density of Halley, so the density of the nucleus is not well constrained, and it is not known whether the interior is fluffy and underdense, or closer to the expected value for icy objects of about one gm/cm^3 [163].

Although the difficult data analysis has still to be completed, it appears unlikely that Comet Halley will provide the expected unaltered sample of the primordial nebula. Although there is a general similarity with solar and CI abundances, there are some differences that are outside the error limits. Compared to solar abundances, H is strongly depleted as expected. Carbon, O, and probably N appear to have solar values, so that the "icy" or gas component might be primordial. This could imply that Halley is more volatile-rich and closer to expected "primordial" compositions than, for example, CI material. However, the situation for the nonvolatile elements is even less clear. Figure. 3.10.2 shows a comparison with the solar abundances [164,165].

Table 3.10.2 lists abundance data for Halley [166]. Although there is a first-order resemblance to the CI abundances "within a factor of two, which coincides with the accuracy of the ion yields" [167] there are apparently significant differences. The Fe/Si and Mg/Si ratios for Halley are the lowest measured in the solar system, and are the most distinct from the solar ratios for the Earth or any of the chondrite groups (except for the Moon, which is a special case). These elements are contained in the dust component, which must therefore have experienced a similar fractionation to that common to the rest of the solar system [164]. Most grains studied by the *in situ* measurement by Vega 1 were approximately chondritic; no Ca-Al rich grains resembling CAIs were identified [168]. The $^{12}C/^{13}C$ ratio for comet Halley, as measured in the HCN component, is 65, compared to a value of 90 for the solar system and of 43 for "interstellar" carbon [169].

All these observations raise awkward questions. What is the relation of the Halley composition to that of the solar photosphere or the nebula? Interplanetary dust particles (see next section) also show discrepancies, so are not a pristine sample of the "primordial" nebula [170]. The large amounts of CH_4 and NH_3 in Halley have even led some workers to conclude that this is "persuasive evidence that this is not derived from pristine solar nebula or of interstellar origin but that it has been processed . . . in a subnebula of one of the giant planets" [171].

In summary, the long-awaited data from Comet Halley have provided us with a new set of cosmochemical problems. It would be wise to recall,

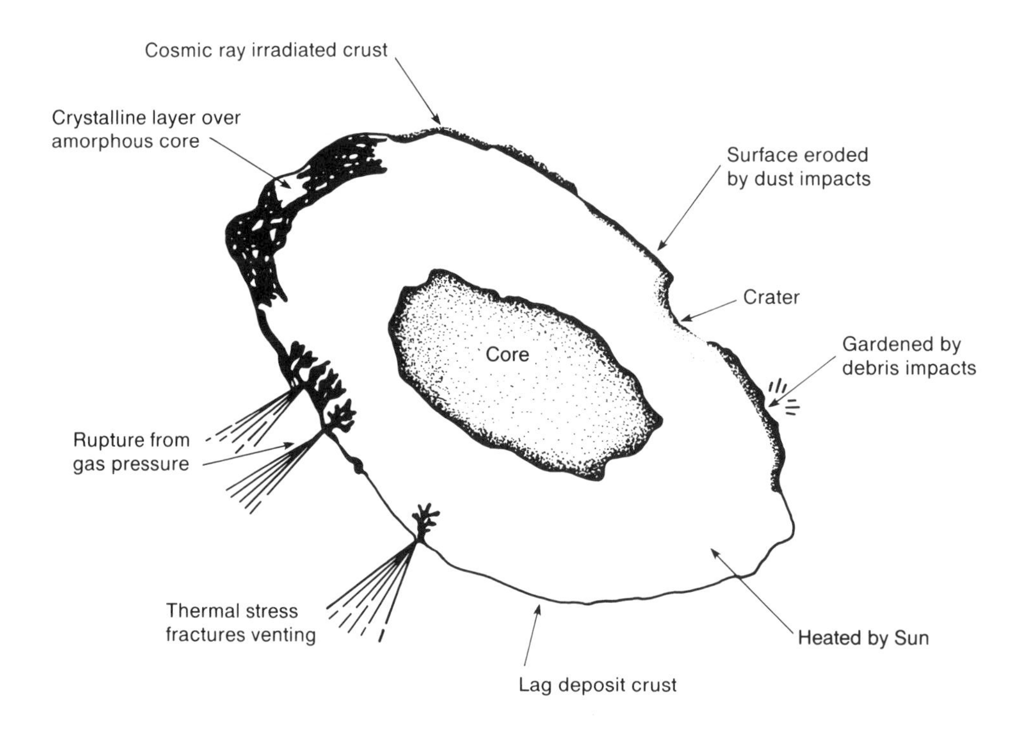

Fig. 3.10.1. *A diagrammatic representation of a typical cometary nucleus, about 10 km in diameter, showing the different processes that will modify its pristine character. After McSween H. Y. and Weissman P. R. (1989) Geochim. Cosmochim. Acta, 53, 3268, Fig. 2.*

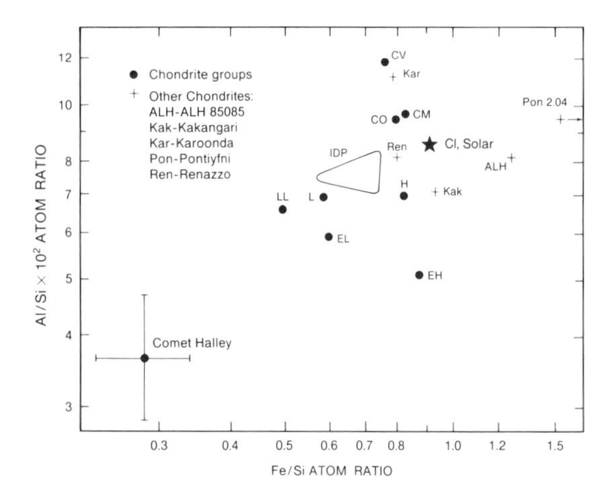

Fig. 3.10.2. *The variation in Al/Si and Fe/Si among the nine recognized chondrite groups, several ungrouped chondrites, interplanetary dust particles (IDP), Comet Halley, and the Sun. There is no systematic change in Fe/Si in the sequence EH-EI, H-L-LL, CO-CV-CM, CI, which is often thought to be the order of distance of formation from the Sun. After Scott E. R. D. and Newsom H. E. (1989) Z. Naturforsch., 44a, 928, Fig. 3.*

TABLE 3.10.2. Element abundances in Comet Halley, Sun, and solar system, normalized to Mg ($\log N_H = 12.00$).

Element	Comet Halley	Sun	Solar System	Halley-Sun
H	9.47	12.00	12.00	-2.53
C	8.64	8.56	8.56	0.08
N	8.05	8.05	8.05	0.00
O	8.99	8.93	8.93	0.06
Na	6.58	6.33	6.31	0.25
Mg	7.58	7.58	7.58	0.00
Al	6.41	6.47	6.48	-0.06
Si	7.85	7.55	7.55	0.30
S	7.44	7.21	7.27	0.23
K	4.88	5.12	5.13	-0.24
Ca	6.38	6.36	6.34	0.02
Ti	5.18	4.99	4.93	0.19
Cr	5.53	5.67	5.68	-0.14
Mn	5.28	5.39	5.53	-0.11
Fe	7.30	7.51	7.51	-0.21
Co	5.06	4.92	4.91	0.14
Ni	6.19	6.25	6.25	-0.06

Data from Anders E. and Grevesse N. (1989) *Geochim. Cosmochim. Acta, 53,* 210, Table 7; Fe in Sun from Holweger H. (1990) *Astron. Astrophys., 232,* 510.

however, how much information we could expect to gain if we had only one meteorite sample available for study, and a recent arrival at that. Accordingly, a new understanding is to be expected from samples returned from comet nuclei, such as the planned Rosetta mission. These may be expected to provide, through laboratory examination, fresh insights into the nature of interstellar dust, a subject fraught with uncertainty at present.

3.10.4. Chiron

This is one of the most isolated bodies in the solar system. It is a dark gray-black object that has an albedo of 0.1, a diameter of 175 km, and occupies a 51-yr orbit with a semimajor axis at 13.7 AU. It is the only recognized body in the immense void between the saturnian and uranian planetary systems, themselves separated by about 10 AU. Recent searches have failed to find any other objects, even though Chiron was 3 magnitudes brighter than the detection limit (mag. 20) [172].

Chiron's closest approach to the Sun is 8.5 AU, inside the orbit of Saturn, while its orbit extends out to 19 AU, almost to Uranus. It shows some brightness variations that could indicate loss of volatiles. Probably it is a comet, since a coma, perhaps due to sublimation of CO_2 ice—it is too far from the Sun for water ice to sublime—has been detected [173]. If so, it is the largest known comet. Its low albedo would suggest that it is coated, like the Comet Halley nucleus, with dark material. Chiron is very similar in size and color to Phoebe, one of the captured saturnian satellites. As noted earlier, the distinction between comets and icy planetesimals is becoming less clear [174].

3.10.5. Comets as Pristine Samples of the Nebula

There has been a general perception that comets are likely to provide the best sample of the original nebula, and indeed might represent nearly unaltered samples of the fragment of the molecular cloud, which was the original source of the nebula. Thus, a sample from inside the nucleus of a comet might be the long-sought primitive sample. This search, like those for other Eldorados, appears to be based on false assumptions. There is apparently as much diversity among comets as in most other natural phenonoma. However, such searches, although they have usually been conducted for the wrong reasons, have mostly produced unexpected benefits and fruitful discoveries.

The internal structure of cometary nuclei is unknown. Some of the nuclei may be composed of heterogeneous assemblages of ice and dust, which could explain the differences observed between apparitions of the same comet. Variable production rates of ammonia and formaldehyde relative to water were observed between different vents of comet Halley, indicative of a heterogeneous composition of the nucleus. A wide variety of processes may affect cometary nuclei. These include impacts, irradiation, and solar heating, and are illustrated in Fig. 3.10.1. Many changes may be induced by such processes. The most dramatic is the formation of a refractory veneer, but others affect the nucleus, including hydration reactions, loss of volatiles, formation of organic compounds, as well as shock effects from collisions. Since the porous (CP) interplanetary dust particles (IDP) are probably derived from comets (see next section), they are likely to have undergone similar processing, and are also not pristine.

The chemical data for Comet Halley discussed above also indicate that the composition for the crucial Fe/Si and Mg/Si ratios, is far removed from those for either the CI meteorites or the solar photosphere, while the nucleus itself is heterogeneous on a local scale. The carbon isotope measurements resemble neither those of the solar system nor the interstellar medium. This information is providing new insights into the complexities of nebular chemistry; it indicates that our previous motivation was misjudged, a common human condition, and that we are likely to obtain unanticipated benefits from future cometary missions. It renders more urgent the need to sample cometary nuclei. Like the seekers of Eldorado, we have stumbled across something else.

3.11. ASTEROIDAL AND COMETARY DUST

A distinction can be made between dust in the interstellar medium and the large amount of dust (visible as the zodiacal light) of solar system origin, which is mainly derived from comets and from collisions in the asteroid belt. Some confusion has resulted fom terming all such extraterrestrial dust as "cosmic" dust.

The particle lifetime in the solar system is controlled by several factors. These include the Poynting-Robertson effect, the ratio of solar gravitational forces to radiation pressure, collisions among particles, and gravitational effects, principally from Jupiter and Saturn.

Considerable insights into the nature of interplanetary material have been gained through the study of particles of extraterrestrial origin collected from a variety of environments. Most of the stratospheric particles that have been analyzed have compositions analogous to CI or CM chondrites, except that they mostly have higher amounts of carbon, being enriched over CI abundances by factors of 2-5 [175], although not so enriched as Halley dust. The C/Si ratios, relative to solar abundances, are Halley dust, 0.5; interplanetary dust particles, 0.15; CI, 0.06; CM, 0.03; and L chondrites, 0.001 [176].

Thus, the interplanetary dust particle composition is closer to solar than chondritic, and indicates a less processed form of material than our sample of the meteorite population. Those particles from deep-sea sediments are larger, but again have CI/CM compositions; other types of meteorites are not represented. Since only 3% of falls are CI or CM it is clear that the ordinary chondrites are not a representative sample of the asteroid belt. This is becoming clearer as we study the Antarctic meteorites.

3.11.1. CP and CS Particles

There are two dominant groups among the stratospheric particles. One group is referred to as chondritic porous (CP) since they are porous on a micrometer scale. The other particles are smooth on a micrometer scale and are distinguished as chondritic smooth (CS). These CS types have a morphology reminiscent of a cluster of grapes, with individual "grapes" about 0.3 μm in diameter. The CS particles look like CI phyllosilicate matrix.

It seems clear that the CS particles are derived from CI or CM meteorites. However, the morphology of the CP particles does not resemble that of any known meteorite group. They appear to be exceptionally well preserved. Curiously, they contain no metal, although this might have been expected if they were nebular condensates. The mineral compositions are not Fe-rich [177]. In this respect, they resemble chondrules; clearly much metal and silicate separation occurred very early, so they might resemble the precursor material of chondrules. The CP particles are perhaps samples of cometary dust, since the Halley dust particles recorded by the Giotto mission appear similar to them [178]. Their fragile nature makes it unlikely that they are derived from asteroids.

3.11.2. Mineralogy

Data for nearly 200 samples of interplanetary dust, in the size range from 4 to 40 μm (mostly 6-15 μm) reveal that the CS particles are mostly hydrated silicates; the CP particles are mainly pyroxene. Most of the particles have "chondritic" compositions. The dust particles tend to be more homogeneous in composition than meteorites. Olivine particles are rare, and most dust particles appear to be either phyllosilicates in CS or pyroxene in CP particles [179].

Distinct divisions can be made between particles that contain hydrous minerals (CS) and those that do not (CP). The hydrated phases are serpentine with basal spacings of 7.2 Å and smectites with 11 Å spacing. Mostly they are compact platey masses.

3.11.3. Deuterium/Hydrogen Ratios

The possible presence of presolar grains among the interplanetary dust particles must also be recognized. The best evidence to date is the very high D/H ratios [180] with enrichments up to 2500 per mil so that they may represent particles from interstellar molecular clouds [181]. Such high values are relatively uncommon in solar system materials, but are common in molecular clouds, apparently due to ion-molecule reactions. Similar enrichments in meteorites such as Semarkona testify to the survival of interstellar material [182].

3.11.4. IRAS Dust Belts

Measurements of infrared emission taken by the IRAS suggest that dust from asteroidal collisions may be a significant source of dust [183]. The IRAS study of the zodiacal light, due to solar reflection of fine dust mainly lying in the plane of the ecliptic, identified narrower bands derived from the asteroid belt. The observations of these bands and dust trails associated with comets indicate that both constitute the major sources for the zodiacal cloud and of stratospheric cosmic dust particles. The respective contributions may be derived from the relative proportions of CP and CS particles [184]. These bands can be linked to specific asteroidal families. Dust from the Themis and Koronis families appears to be responsible for the central bands, which have at least four components, while the adjacent bands come from the Eos family [185]. The dust must be continuously derived by collisional processes between small fragments, since the dust lifetime is only 10^4-10^5 years. Some bands are not associated with known families, so a lot of dust must be produced by erosion of small objects. The importance of these observations is that possibly 50% of the zodiacal dust is derived from the asteroid belt, rather than from comets.

How does this information correlate with the dust collections? It is tempting to correlate the smooth, nonporous particles (CS) with the dust produced by collisions of solid objects in the asteroid belt. In contrast, the porous particles (CP), which constitute about half those analyzed, are plausibly linked with a cometary origin [186].

The larger significance of the dust observations is that 85% of dust, most of the asteroids, and the Halley dust most closely resemble the carbonaceous chondrites, which constitute only 3% of falls. Possibly this dust is a better sample of the asteroid belt than most of the meteorites.

3.11.5. The Terrestrial Record

About 10^4 tons of interplanetary dust accrete to the Earth each year. Particles of 10-μm size accrete at a rate of 1 per m^2 per day, while 100-μm particles arrive at a rate of one per m^2 per year. They are thus ubiquitous, and "cosmic dust is so common that quite literally every footstep a person takes contacts a fragment of cosmic dust" [186].

Collecting the particles is not trivial. Typical impact velocities with the Earth are probably 15 km/s and space collection is handicapped by these high collision velocities and low flux rates. The concentration of particles in the stratosphere is enhanced by a factor of 10^6 over that in space, due to the braking effect of the atmosphere [187,188].

The influx rate does not appear to be constant. The current terrestrial flux of particles less than 10^{-7} g in mass [189] is an order of magnitude higher than is indicated by the crater density on lunar samples. These latter data provide an average rate over the past 10^6-10^7 years [190]. Yet another problem is that dust is lost from the interplanetary medium on timescales of 10^5 years due to Poynting-Robertson effects, so that there must be constant fresh injection [191]. This may help to explain the discrepancy between the lunar and terrestrial influx rates.

Particles larger than 100 μm have been melted and form "cosmic spherules" containing magnetite. Accordingly, they are readily collected from deep-sea sediments and Greenland ice lakes, where they occasionally form placer deposits, with particles up to 1 mm in size. Many are unmelted chondritic grains in the size range of 100-300 μm [192]. Deep-sea spherules are typically either iron or silicate of chondritic composition and were identified as extraterrestrial last century [193]. They are typically about 300 μm in diameter. The iron spheres are magnetite or wüstite with Fe/Ni cores and Pt-group nuggets.

3.11.6. Dust as a Solar System Sample

Dust possesses certain advantages over meteorites as samples. Meteorites must have undergone orbital perturbations to reach Earth-crossing orbits, survive the stresses of atmospheric entry and be big enough to be found.

Are these particles a typical sample of the smaller objects in the solar system? If so, they are mainly of CI and CM composition. This means that the "ordinary" meteorites come from uncommon objects since only a small percentage of the dust particles can be derived from common chondrites or achondrites. If the particles come from comets, then the major dust-producing comets are of CI or CM composition.

The majority of the particles in the 5-1000-μm size range have compositions similar to CI and CM meteorites, which indicates that "cosmic dust" is homogeneous in composition on a scale of micrometers. However, no meteorite is homogeneous at this scale. If planets were formed from this material, they should all be homogeneous. In contrast, it is a major conclusion of this book that the planets are both distinct in composition from one another and that they accreted from a hierachy of differentiated objects of varying sizes.

In summary, it is significant that most cosmic dust particles contain volatile elements [194], minerals stable at low temperatures, and solar flare tracks that would be erased above about 600° C. These all indicate low relative atmospheric entry velocities, consistent with an origin from the asteroid belt [195]. Much of the cosmic dust thus appears to be derived from parent bodies in the asteroid belt, with a significant contribution from cometary sources.

3.12. FRACTIONATED METEORITES AND PARENT BODIES

Achondrites form an interesting group since their compositions most nearly simulate those of rocks on the terrestrial planets. Chondrites, in contrast, appear to have never been modified by residence in large asteroids, so they retain many primitive characteristics that make them especially valuable records of early solar system events [196].

However, the differentiated meteorites also have their tale to tell [197]. The achondrites are of particular importance because they preserve a record of igneous processes at the beginning of solar system history.

3.12.1. Howardites, Eucrites, and Diogenites (HEDs)

The most studied class of the fractionated meteorites are the howardites, eucrites, and diogenites, the so-called HEDs.

The eucrites, the most common achondrites, closely resemble lunar basalts in composition. Most basaltic achondrites have crystallization ages between 4.52 and 4.54 b.y. indicating formation within 10-20 m.y. of T_0 [198]. In contrast, cumulate eucrites (e.g., Moama, Serra de Mage, and Moore County) are younger (4.41-4.46 b.y.). Accordingly, they may not be derived directly from the basaltic eucrites, and represent either a younger igneous event or perhaps very slow cooling in the interior of a large asteroid.

Diogenites are brecciated orthopyroxenites possibly complementary to the eucrites. Howardites are brecciated mixtures of diogenites and eucrites. HEDs probably came from a single differentiated asteroid referred to as the eucrite parent body. This appears to have been originally chondritic in bulk composition, at least for the refractory lithophile elements, since the eucrites have, except for Eu enrichments and depletions, nearly flat chondrite-normalized REE patterns (Fig. 3.12.1).

Although it is often held that asteroidal melting was not universal, insofar as the population of meteorites sampled by the Earth is typical, there is considerable evidence from the asteroid belt that the inner regions are mostly populated with differentiated objects. The eucrite parent body must have either segregated a metal core or been derived from metal-free precursors to account for the very low Ni contents of the eucrites. Most likely a metal core formed, extracting the siderophile elements from the silicate mantle. There must have been a considerable degree of melting of the mantle to allow for metal segregation. Subsequent to this, the mantle must have crystallized. The eucrites then formed by partial melting of this mantle, followed by fractional crystallization. These events mimic much of early lunar history, but were accomplished on timescales very close to T_0 [199]. Eucrites resemble very-low-Ti lunar mare basalts. Partial melting of the interior peridotite of the eucrite parent body produced either volcanic eucrites or plutonic diogenites. Meteoritic impacts mixed the two, forming howardites. No samples of the postulated depleted peridotite mantle have been recovered, leading to the supposition that the body may still be intact.

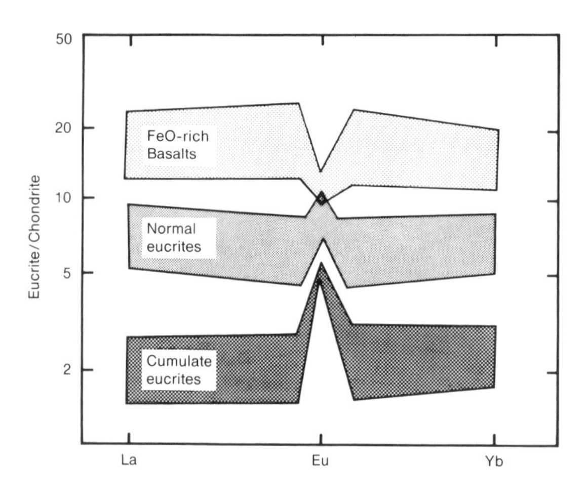

Fig. 3.12.1. *Rare earth element (REE) patterns in eucrites, illustrating the role of fractional crystallization of plagioclase in producing Eu anomalies. After Mittlefehldt D. W. and Lindstrom M. M. (1988) in Lunar and Planetary Science XIX, p. 791, Fig. 1, Lunar and Planetary Institute, Houston.*

The third-largest asteroid, 4 Vesta, is a prime candidate for the eucrite parent body, since the reflectance spectra from the surface matches that measured for eucrites. It may, however, lie too deep within the asteroid belt for samples to readily reach the Earth. Vesta is 550 km in diameter, and remains unique with a distinctive spectrum in the asteroid belt (Fig. 3.12.2). The surface detail of 4 Vesta is, however, consistent with the eucrite-diogenite-peridotite model. Vesta shows a eucritic spectrum on one hemisphere, and a diogenitic spectrum on the other. It also contains a large dunitic area, possibly representing an impact crater into a dunite mantle [200].

The asteroid 3551 (1983 RD) is one of three small Earth-crossing bodies that has a similar spectrum to Vesta, indicating that there may be more than one parent body available to provide a source for the eucrites [201].

3.12.2. Ureilites

"The ureilites are arguably the most bizarre and perplexing of the achondrites" [202].

Ureilites have chemical characteristics that indicate the operation of planetary igneous processes. However, they have primitive oxygen isotope signatures that suggest nebular, rather than planetary conditions. Over 20 examples of these heavily shocked meteorites are known. Consisting mainly of olivine and low-Ca pyroxene (pigeonite) with some interstitial graphite, diamond, Fe-Ni metal, and troilite, ureilites have a coarse grain size that has often led

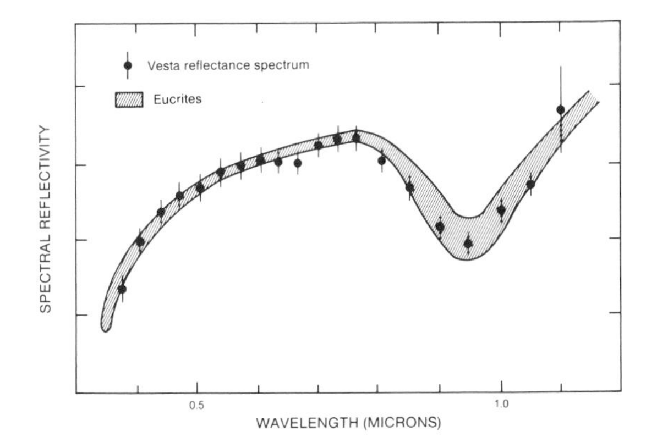

Fig. 3.12.2. *The laboratory spectra of eucrites (band) closely resembles the reflectance spectra of 4 Vesta (points with error bars). After McSween H. Y. (1987) Meteorites and Their Parent Planets, p. 140, Fig. 5.9, Cambridge Univ., New York.*

to their classification as cumulate igneous rocks [203]. However, their oxygen isotope compositions show a wide range (Fig. 3.12.3) similar to the dark inclusions in Allende C3 chondrites. These data are not consistent with derivation of the ureilites from a single differentiated parent body. They appear to owe their isotope characteristics to nebular processes, such as condensation, evaporation, or to metal-silicate separation [204]. Thus, they appear to be among the most primitive of the achondrites.

Some correlations exist between the chemical composition (e.g., FeO, C, and Ir concentrations) of ureilites and their isotopic systematics, indicating that nebular processes determined their composition; the cumulate textures must result from later igneous processes. In this context, there is Sm-Nd isotopic evidence in the ureilites for an event at 3.74 b.y. [205]. However, the ureilite chemistry and textures are consistent with a primitive rather than an igneous origin. The olivine crystals preserve a primary orientation, often attributed to cumulate deposition (hence accounting for the popularity of igneous origins for ureilites), but possibly a relic from the growth of large stacks of olivine platelets from a vapor phase. Most ureilites possess V-shaped REE patterns, indicative of mixing of two distinct sources, the HREE residing in olivine and the LREE in the matrix. Their bulk compositions are widely variable. Ureilites also contain a high abundance of carbon (including diamond) and primitive rare gases depleted in ^{40}Ar. The ureilites appear to be primitive meteorites, subject to some later recrystallization and metasomatism [206].

Models for ureilite origin have spanned a broad spectrum, from nebular condensates through to igneous cumulates. A recent proposal is that they are impact-melted material from CV chondritic parents [207]. This model is based on the observation that the ureilites are derived from the same oxygen isotope reservoir as the CV chondrites. Impact on an isotopically heterogeneous parent body can explain the shock features and most of the features of ureilite chemistry and petrography.

3.12.3. Angra dos Reis (ADOR)

The unique basaltic achondrite, Angra dos Reis (ADOR), which consists mostly of Al-Ti augite, early attracted attention because of its low initial $^{87}Sr/^{86}Sr$ ratio, intermediate between ALL and BABI. However, subsequent work has shown that this ratio is not distinct from the initial ratio of the other basaltic achondrites (Table 3.7.4) [208]. Although commonly interpreted as a cumulate ultramafic igneous rock containing 93% fassaite (Al-Ti pyroxene) [209], it may instead represent a quenched and devitrified magma [210]. The chief interest in this meteorite is that it has a chemical fractionation history as complicated as that of terrestrial or lunar basalts; this implies derivation from an already differentiated

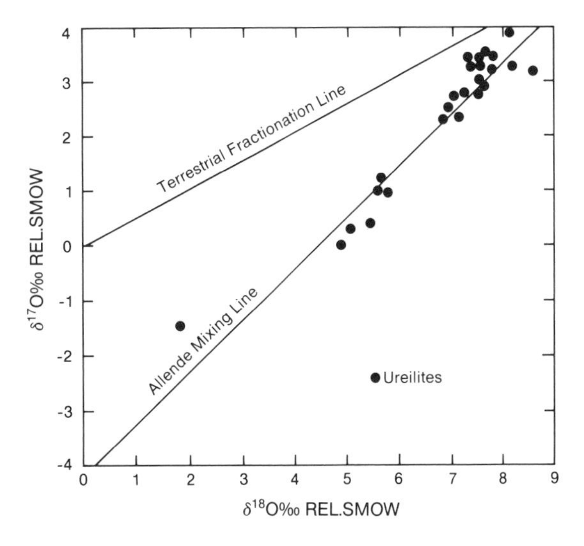

Fig. 3.12.3. *The oxygen isotope composition of ureilites compared with the terrestrial fractionation line and the Allende mixing line. After Clayton R. N. and Mayeda T. K. (1988) in Lunar and Planetary Science XIX, p. 197, Fig. 1, Lunar and Planetary Institute, Houston.*

body. ADOR is depleted in volatile and siderophile elements, probably due to precursor events. The refractory lithophile elements, not affected by these processes, record a complex fractionation history that occurred at times very close to T_0. The Antarctic meteorite LEW 86010 is similar to ADOR, with an initial Sr ratio similar to the other basaltic achondrites (Table 3.7.4) [211].

3.12.4. Aubrites

Aubrites or enstatite achondrites (e.g., Norton County) represent another curious class, consisting mainly of Fe-free enstatite. Sometimes regarded as direct nebular condensates, they are formed from igneous rocks by brecciation during collisional impacts, followed by reassembly. They are depleted in Eu, but since plagioclase is not a liquidus phase, perhaps this element has been removed in CaS (oldhamite) [212].

They have the same oxygen isotope signature as enstatite chondrites, indicating that they were formed in the same part of the nebula, probably under similar reducing conditions, for they contain such highly reduced compounds as oldhamite. Although they exhibit all the characteristics of igneous origin, they were not derived from enstatite chondrites by such processes on the same parent body, but come from a separate parent body [213]. This enables an interesting conclusion to be drawn: Identical oxygen isotope compositions do not necessarily indicate derivation of objects from the same body. Thus, the fact that the Earth and the Moon have the same oxygen isotope composition does not imply a common origin, but merely derivation from the same general part of the solar nebula.

3.12.5. Iron Meteorites

Iron meteorites, perhaps the ultimate in differentiated meteorites, are of much interest in the present context. Melting and solidification occurred very early in the history of the solar system, about 4.6×10^9 yr ago, followed by slow cooling below 900 K [214]. Radiogenic ^{107}Ag derived from ^{107}Pd ($t_{1/2} =$ 6.5 m.y.) is common in iron meteorites, also indicative of a very early origin, while ^{107}Pd was still alive [215].

Iron meteorites were derived from a wide variety of parent bodies. The "shattered planet" hypothesis, untenable for many years, would predict that they were all similar. However, one of their more interesting properties is that they are so different. At least 60 groups and grouplets, distinguished by Ni, Ga, and Ge abundances, have been identified in our terrestrial collections; presumably many more exist in the asteroid belt (Table 3.12.1). This diversity indicates

TABLE 3.12.1. Properties of iron meteorite groups.

Group	Freq. (%)	Bandwidth (mm)	Structure*	Ni (mg/g)	Ga (μg/g)	Ge (μg/g)	Ir (μg/g)
IA	17.0	1.0-3	Om-Ogg	64-87	55-100	190-520	0.6-5.5
IB	1.7	0.01-1.0	D-Om	87-250	11-55	25-190	0.3-2.0
IC	2.1	<3	Anom, Og	61-68	49-55	212-247	0.07-2.1
IIA	8.1	>50	H	53-57	57-62	170-185	2-60
IIB	2.7	5-15	Ogg	57-64	46-59	107-183	0.01-0.9
IIC	1.4	0.06-0.07	Opl	93-115	37-39	88-114	4-11
IID	2.7	0.4-0.8	Of-Om	96-113	70-83	82-98	3.5-18
IIE	2.5	07-2	Anom	75-97	21-28	62-75	1-8
IIF	1.0	0.05-0.21	D-Of	106-140	9-12	99-193	0.8-23
IIIA	24.8	0.9-1.3	Om	71-93	17-23	32-47	0.15-20
IIIB	7.5	0.6-1.3	Om	84-105	16-21	27-46	0.01-0.15
IIIC	1.4	0.2-3	Off-Ogg	62-130	11-92	8-380	0.07-2.1
IIID	1.0	0.01-0.05	D-Off	160-230	1.5-5.2	1.4-4.0	0.02-0.07
IIIE	1.7	1.3-1.6	Og	82-90	17-19	34-37	0.05-6
IIIF	1.0	0.5-1.5	Om-Og	68-85	6.3-7.2	0.7-1.1	0.006-7.9
IVA	8.3	0.25-0.45	Of	74-94	1.6-2.4	0.09-0.14	0.4-4
IVB	2.3	0.006-0.03	D	160-180	0.17-0.27	0.03-0.07	13-38

* Structure abbreviations: H, hexahedrite; Ogg, Og, Om, Of, Off, Opl, coarsest, coarse, medium, fine, finest, and plessitic octahedrites; D, ataxite; Anom, anomalous structure that does not fit into the other categories.

Data from Wasson J. T. (1985) *Meteorites*, Table II-4, Springer-Verlag, New York.

that iron meteorites were derived from many parent bodies, consistent with the planetesimal hypothesis.

Iron meteorites have exposure ages ranging to over 1000 m.y, with some well-developed peaks at 400 (IVA group) and 650 m.y. (IIAB). This is evidence that these iron groups (about 40% of iron meteorites) were produced in a single collision. Others (e.g., IAB and IIAB) were involved in up to four collisions, and the spread of ages indicates numerous collisions in the past billion years. In contrast, chondrites all have exposure ages less than 100 m.y.; less than 10% of irons have such low ages. This difference may simply reflect the relative strength of irons vs. silicates. The irons are also depleted in the volatile siderophile elements and so show the general volatile-element depletion that characterized the inner portions of the solar nebula.

Of more immediate concern here is the inter-element fractionation within the various iron groups. The inverse correlation between Ni and Ir is a typical example (Fig. 3.12.4). This fractionation is a consequence of the redistribution of elements between solid and liquid during fractional crystallization of the metal [216] rather than by partial melting [217]. Extremely large fractionations were achieved. In the IIAB and IIIAB irons, Ir varies by a factor of more than 1000, which contrasts with the very limited variation of Ir (less than a factor of 2) in the metal in chondrites.

The formation of the fractionated asteroids and the segregation of iron cores apparently involved high degrees of partial melting [218], for otherwise the metal is unable to segregate. This implies that core formation on the terrestrial planets requires a high degree of melting. Thus, it appears that the small metal-bearing asteroids were perhaps >50% molten.

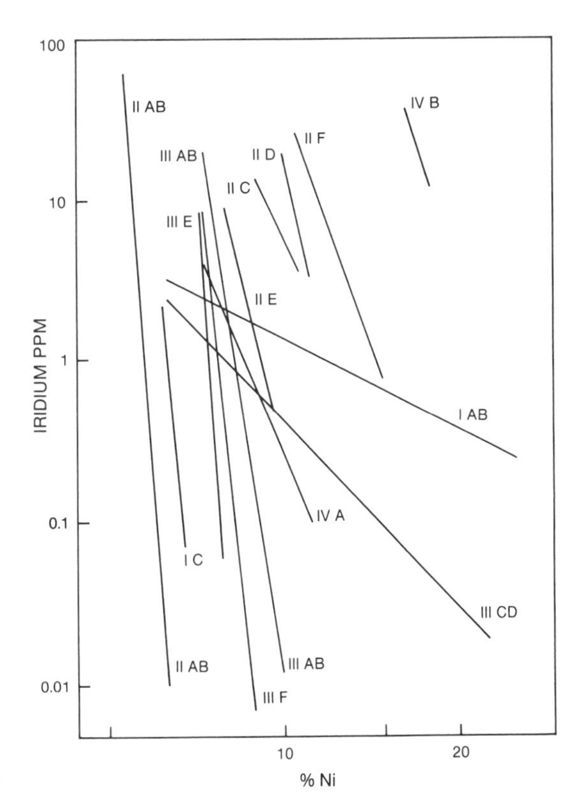

Fig. 3.12.4. *The variation of Ir and Ni among several differing iron meteorite groups. Iridium shows a variation over 5 orders of magnitude. After Wasson J. T. (1985) Meteorites, Fig. II-11, Springer-Verlag, Berlin.*

3.12.6. Cooling Rates

Cooling rates for iron meteorites have been the subject of much controversy. Wasson [219] considers that it is "probable that the reported cooling rates are systemically [sic] low, perhaps by as much as a factor of 10, and that the radii of the parent bodies of the most-slowly-cooled iron meteorites were probably nearer 100 km than 300 km" (Table 3.12.2) [220].

Cooling occurred very slowly (1-100 K/m.y.) through the temperature range 900-400 K [215]. This is generally interpeted as due to burial at depths of 10-200 km, implying rather large asteroids. If the asteroids were coated with fine dust (which has a thermal conductivity 100 times less than solid rock), then these depths might be reduced. Pressure esti-

TABLE 3.12.2. Comparison of published cooling rates for iron meteorites.

Meteorite, Type	Cooling rate in °C/m.y.			
	1	2	3	4
Tazewell, IIID	—	2	500	10
Toluca, IA	1	2	760	25
Bristol, IVA	>10	50	20,000	150-300

1. Wood J. A. (1964) *Icarus, 3,* 429.
2. Goldstein J. I. and Short J. M. (1967) *Geochim. Cosmochim. Acta, 31,* 1733.
3. Narayan C. and Goldstein J. I. (1985) *Geochim. Cosmochim. Acta, 49,* 397.
4. Saukumar V. and Goldstein J. I. (1988) *Geochim. Cosmochim. Acta, 52,* 725.

Data from Saukumar V. and Goldstein J. I. (1988) *Geochim. Cosmochim. Acta, 52,* 725.

mates yield maximum values of a few kilobars, implying parent bodies with diameters of a few hundred kilometers at most.

There is a minor puzzle because there are so many parent bodies. It seems unlikely that over 50 of these bodies would be disrupted relatively recently in solar system history. A resolution of this problem could be that the irons melted and solidified in small (1-10 km) bodies, which were then reaccreted into large planetesimals where they were slowly cooled through 800 K. This model is consistent with the planetesimal hypothesis, but there is a caveat. The accretion of these bodies must have been at low velocities, otherwise the iron core would have been disrupted. This adds another argument for accretion of planetesimals from localized zones, rather than from a random mixture from widely separated zones when high approach velocities would be common.

The iron meteorites provide evidence for low-pressure metal-silicate fractionation before planetary accretion. These differentiated iron meteorites were almost never (except for Group IIE) remixed with silicates. Clearly, little mixing occurred between the iron cores and silicate mantles following disruption of the parent bodies. This is consistent with rather quiet conditions in the asteroid belt following the differentiation of the asteroids.

Possible sources of iron meteorites are the M- and S-type asteroids. Candidates include 6 Hebe, 8 Flora, and 18 Melpomene, all of which are S-type with diameters in the range 50-200 km.

3.12.7. Pallasites and Mesosiderites

Brief mention must also be made of the pallasites and mesosiderites, which constitute some of the most spectacular museum specimens. Containing roughly equal amounts of metal and silicate, they formed by mixing of metal and silicate, perhaps by the disruption of the core-mantle boundary of differentiated asteroids. The mesosiderites are surface-derived breccias, containing many fragments of diverse meteorites. The metal appears similar to chondritic metal, and may have been produced by surface melting in a collisional event. The pallasites, in contrast, are mixtures of typical fractionated metal similar to that of group IIIAB and olivine, probably derived from core-mantle boundaries in differentiated asteroids.

3.12.8. Lunar and Martian Meteorites

The chief significance of lunar and martian meteorites in the present context is the information they provide on the compositions of their parent bodies. Thus, the lunar meteorites (Table 3.12.3) provide useful additional information, since they are from regions of the lunar crust, both highlands and mare [221], not sampled by the Apollo or Luna missions. They are from several sites, as indicated by the cosmic-ray-induced radionuclides and the rare-gas isotope data [222].

It must now be considered well established that the SNC meteorites are derived from Mars [223,224]. The nakhlites (Nakhla, Lafayette, Governador Valadares), which are pyroxenites, and Chassigny, a dunite, have crystallization ages of about 1.3 b.y. [225]. A considerable controversy has arisen over the crystallization ages of the shergottites [226]. They have a "whole-rock" Sm-Nd "isochron" of 1.3 b.y. and Rb-Sr internal "isochrons" of 180 m.y., generally interpreted as due to shock [227]. An alternative interpretation suggests a 360-m.y. Sr-Nd crystallization age [228], but doubt exists as to the reality of this date in view of the inability of ion-microprobe

TABLE 3.12.3. Lunar meteorites.

Discovery	Location/Number	Classification	Weight (g)
1982	Allan Hills 81005	Anorthositic breccia	31
1984	Yamato 791197	Anorthositic breccia	52
1985-7	Yamato 82192/3/86032	Anorthositic breccia	37/27/648
1987*	Yamato 793274	Basaltic breccia	9
1989	MacAlpine Hills 88104/5	Anorthositic breccia	61/662
1989**	Elephant Moraine 87521	Basaltic breccia	31
1990	Asuka 31	Mare gabbro	442
1990**	Yamato 793169	Mare gabbro	6

*Y793274 was classified as a lunar anorthositic regolith breccia in 1987, but was found to be a basalt-rich breccia in 1990.
** EET87521 and Y793169 were originally classified as eucrites.

Data from Lindstrom M. et al. (1990) *Proc. NIPR Symposium on Antarctic Meteorites, 4.* The first lunar meteorite from outside Antarctica (Calcalong Creek, a 19-g breccia) has been recovered from western Australia [Hill D. H. et al. (1991) *Nature, 352,* 614].

studies to locate a proposed LREE enriched carrier, essential to the model. Most likely, the sample has become contaminated with a LREE phase, possibly from cigarette lighter flints [229]!

3.13. THE ASTEROID BELT AS A SOURCE

In this section the significance of the asteroid belt as a source for meteorites is assessed; the main discussion of the origin of the asteroids is given in section 5.11.

The absence of a planet in the asteroid belt, which contains over 4000 numbered bodies as well as countless smaller ones, is probably due to the influence of massive Jupiter, which swept up or ejected many of the bodies. The total mass of the myriad small objects in the belt is less than 5% of the mass of the Moon.

Perhaps the principal questions are (1) the source of meteorites, (2) the nature of the Earth-crossing Apollo and Amor objects, (3) whether the present structure of the asteroid belt resembles that of the primordial belt, and (4) whether the asteroids grew from narrow feeding zones. Since the cratering flux in the inner solar system has been relatively uniform over the past 4 b.y. (since the end of the early heavy bombardment), the asteroid belt appears to have been stable over that period of time.

3.13.1. The Immediate Source: Apollo, Aten, and Amor Objects

The meteorites that fall on the Earth at present are derived from the Earth-crossing Apollo and Aten objects; the Amors are Earth-approaching objects. The Atens have smaller orbits than that of the Earth, but overlap the Earth's orbit near aphelion, while the Apollo objects have orbits greater than Earth's (perihelion less than 1.017 AU). The orbits of the Amor objects, with perihelion greater than 1.017 AU and less than 1.30 AU [230], do not intersect the Earth's orbit. It is estimated that there are 100 Aten objects, 1000 Apollos, and 1000-2000 Amors. These predictions are based on the estimated production rates from the inner asteroid belt Sunward of 2.6 AU, and are in reasonable agreement with observation.

Extinct comets may provide up to 40% of the observed flux of Earth-crossing objects [231]. Extinct comet nuclei may, however, be very difficult to distinguish from primitive C-, P-, and D-class asteroids. Possibly some of the Apollo, Aten, and Amor asteroids belong to this class of objects. Extinct cometary nuclei are predicted to have dark surfaces coated with organic compounds and refractory grains, and low densities [232]. Although S-type asteroids appear to be more common among near-Earth asteroids than in the main belt, this may be an observational artifact [233]. The principal source region is at 2.5 AU, at the 3:1 Jovian resonance, a chaotic zone [234]. The remainder are derived from the innermost asteroidal belt (2.17-2.3 AU).

3.13.2. The Spectrophotometric Paradox

A perennial problem is the apparent absence of a parent body for the ordinary chondrites among the multitude of candidates in the asteroid belt. A few 1-2-km Q-type Earth crossers (1862 Apollo, 1980 WF, and 1981 QA) have the appropriate spectral signature, but none of the studied members of the belt has reflectance spectra that match that measured in terrestrial laboratories for ordinary chondrites. This is sometimes referred to as the "spectrophotometric paradox" [235].

The S-class asteroids in the inner belt have usually been considered the most likely candidates, but their spectra are not compatible with those of the ordinary chondrites [236]. In particular, S asteroids have a larger metal signature and are more olivine-rich than the ordinary chondrites. A wide variety of explanations has been offered [237]. Among the most popular is that some kind of space weathering is overprinting the mineralogical spectra. However, suggestions that the presence of impact melts, glass, shock-induced blackening, breakdown of hydrocarbons, or accumulation of projectile material are responsible have all been shown to be inadequate [238]. The spectrum of Vesta, for example, resembles closely that of the eucrites, and has survived 4500 m.y. of exposure without apparent alteration, while the carbonaceous chondrites are a reasonable match to the C-class asteroids. Even if the few Apollo-Amor Q-class objects are the immediate source, they need to be replenished from the main belt, where Q-class bodies have not been identified at likely resonances in the belt. The problem remains at present a considerable dilemma.

3.14. METEORITES AS PLANETARY BUILDING BLOCKS

Meteorites have long been favorite analogues for planetary compositions. The presence of metallic, sulfide, and silicate phases in meteorites has lent credibility to models of the internal structures of the terrestrial planets involving metallic iron cores overlain by silicate mantles.

In this section, the relevance of the present population of meteorites as building blocks of the solid planets is assessed. The principal question to be addressed is whether any classes of meteorites, either alone or in combination, could be used to construct the terrestrial planets.

The question of the composition of the large outer planet cores, which, according to current wisdom, must be put together from planetesimals, cannot yet be addressed except in the most general terms. One might expect that the cores are composed of primitive rocky material plus ices on account of their distance from the Sun. However, the existence of fractionated meteorites in the asteroid belt, with their evidence of at least local high temperatures, must give pause to such conjectures.

If chondrules form due to the action of nebular flares, then chondrule formation might extend far out in the nebula, although such activity might be expected to diminish with distance from the Sun [239]. Thus, some fractionation of volatile or siderophile elements due to the chondrule-forming event could extend beyond the "snow line" at about 4 AU. This view is reinforced by the failure of the composition of Comet Halley and interplanetary dust particles to match either solar or CI abundances. Therefore, the cores of the giant planets are probably not "primitive" but are fractionated to some degree, and the common habit of assigning CI abundances of heat-producing elements (K, U, and Th) to them, and to outer solar system bodies in general, is probably in error to an uncertain degree.

One popular model forms the terrestrial planets from two components, a reduced component (A) equated with E chondrites, that forms in this scenario inside 1 AU. The other, a metal-free, oxidized, and "primitive" component (B) is equated with CI carbonaceous chondrites, formed in the asteroid belt [240].

There is, however, little evidence for lateral mixing on a scale that mixes material from Mercury to the outer edge of the asteroid belt. The interrelationships between the compositions of the chondritic meteorites and the terrestrial planets reveal few points of similarity. There is little overlap in density between meteorites and planets. Since the silicate chemistry of the inner planets and the chondritic meteorites are broadly similar, the major element chemistry is not particularly diagnostic. "One can call the Earth 'chondritic' in a general way, but it cannot be matched up with a specific chondrite class" [241].

The oxygen isotope data show that, except for the EH and EL groups, there is no overlap with terrestrial values [242]. Likewise the SNC oxygen isotope data, assumed to represent martian values, occupy a distinct field remote from those of the other meteorites. This important observation rules against nebulawide mixing or homogenization during accretion of the planets. The refractory-volatile element ratios are more useful, since K/U ratios are known to some degree of confidence for both the meteorites and the inner planets (except Mercury). They display no overlap between the chondrites and the inner planets. The V-Cr-Mn relationships are mostly related to volatile depletion and are not unique to the Earth or Moon, but are found in CM, CO, and CV chondrites as well. Comparison of refractory element/Si ratios is hampered by uncertainties in the terrestrial data because the upper-lower mantle compositional problem is unresolved. Siderophile trace-element data are complicated by the possibility of a late addition of material to the terrestrial upper mantle. The noble gas data indicate that no known meteorite class has appropriate elemental or isotopic compositions close to those of the inner planets.

Although the composition of the inner planets is chondritic in a general sense, it does not appear possible to build the Earth and the other terrestrial planets out of the building blocks supplied by the currently sampled population of meteorites. There appear to be substantial differences between the asteroid belt and the zone Sunward of about 2 AU, in which the terrestrial planets accumulated. Apart from the oxygen isotopes, the other most significant variation between these regions appears to have been a greater depletion of the volatile elements in the latter.

The meteorites are derived from 2-4 AU. There is really no indication that they were closer to the Sun at an early stage, or that they are typical samples of the planetesimals from which the inner planets accumulated, or that the zonal structure in the asteroid belt is not a primary feature. The meteorites are mostly less strongly depleted in volatile elements, as shown by K/U ratios, than the inner planets. E chondrites, placed closest to the Sun in many scenarios [243], retain their complement of volatiles, but have the lowest refractory/Si ratios.

In conclusion, although we have not identified candidate building blocks for the terrestrial planets from among the present population of meteorites, they provide us with valuable information, on both the state of the early solar nebula, and the nature of the asteroid belt. If there is little identifiable input into the inner planets from the meteorites from the asteroid belt, then this has important implications for planetary accretion models. One corollary is that the terrestrial planets accumulated from rather restricted (perhaps <0.5 AU) zones in the solar nebula. Even if there were lateral mixing during accretion, it did not homogenize the composition of the inner planets nor that of the asteroid belt. Thus, the present zonal structure in the asteroid belt may be an analogue for the structure of the nebula, at least within 4-5 AU of the Sun. This view receives some support from the differences in composition between the inner planets, particularly Earth and Mars, which differ in K/U ratios, Mg/(Mg+Fe) ratios, V-Cr-Mn abundances, and oxygen isotopes, all of which indicate accretion from differing populations of planetesimals. The Earth-Moon relationship is also explicable if little material is derived from outside the terrestrial neighborhood. The most likely conditions for lunar origin by the large impactor hypothesis involve low-velocity collisions of objects about 0.12-0.17 Earth masses.

Although the Moon and the terrestrial mantle differ significantly in composition, the similarity in oxygen isotopes between the Earth and the Moon is consistent with an origin for both bodies in the same portion of the nebula. Models that call upon nebula-wide mixing of planetesimals during planetary accretion receive little support from this study, which seems to require a nebula with some compositional zoning, analogous to that of the present asteroid belt.

NOTES AND REFERENCES

1. This inscription was placed on a 127-kg chondritic LL6 meteorite that fell on November 16, 1492, at 11:30 a.m. in Ensisheim, Alsace, then part of the Holy Roman Empire but now, after a turbulent history, a province of France. A 56-kg piece is still preserved in the Rathaus, Ensisheim, and other specimens are in eight museums.
2. The debate over possible differences in the iron abundance between the CI and solar data has been resolved. The iron abundances in the solar photosphere match those of the CI chondrites; Holweger H. et al. (1990) *Astron. Astrophys., 232,* 510.
3. Kerridge J. F. (1987) *Geochim. Cosmochim. Acta, 51,* 2527.
4. Wannier P. G. (1980) *Annu. Rev. Astron. Astrophys., 18,* 399; Smith D. G. et al. (1982) *Astrophys. J., 263,* 123.
5. Yang J. and Epstein S. (1983) *Geochim. Cosmochim. Acta, 47,* 2199.
6. McKeegan K. D. et al. (1985) *Geochim. Cosmochim. Acta, 49,* 1971; McKeegan K. D. (1987) *Science, 237,* 1468.
7. McNaughton N. J. et al. (1982) *Proc. Lunar Planet. Sci. Conf. 13th,* in *J. Geophys. Res., 87,* A297.
8. A review of isotopic anomalies in major elements Mg, Si, Ca, Ti, Cr, Fe, and Ni in meteorites is given by Clayton R. N. et al. (1988) *Philos. Trans. R. Soc., A325,* 483.
9. Esat T. M. (1988) *Geochim. Cosmochim. Acta, 52,* 1409.
10. See Clayton R. N. et al. (1985) in *Protostars and Planets II* (D. C. Black and M. S. Matthews, eds.), p. 755, Univ. of Arizona, Tucson.
11. Clayton R. N. et al. (1988) *Philos. Trans. R. Soc., A325,* 493.
12. Thiemens M. H. (1988) in *Meteorites and the Early Solar System* (J. F. Kerridge and M. S. Matthews, eds.), p. 899, Univ. of Arizona, Tucson.
13. Wood J. A. and Pellas P. (1991) in *The Sun in Time,* Univ. of Arizona, Tucson, in press.
14. Mahoney W. A. et al. (1984) *Astrophys. J., 286,* 578.
15. Clayton D. D. and Leising M. D. (1987) *Phys. Rep., 144,* 1.
16. For example, El Goresy A. et al. (1978) *Proc. Lunar Planet. Sci. Conf. 9th,* p. 1279; Bischoff A. and Palme H. (1987) *Geochim. Cosmochim. Acta, 51,* 2733.
17. Blum J. D. et al. (1988) *Nature, 331,* 405.
18. For a contrary view, see Palme H. et al. (1989) in *Lunar and Planetary Science XX,* p. 814, Lunar and Planetary Institute, Houston.
19. Wood J. A. (1988) *Annu. Rev. Earth Planet. Sci., 16,* 62.
20. See reviews by Grossman L. (1980) *Annu. Rev. Earth Planet. Sci., 8,* 559; Kornacki A. S. and Wood J. A. (1984) *Proc. Lunar Planet. Sci. Conf. 14th,* in *J. Geophys. Res., 89,* B573; MacPherson G. J. et al. (1988) in *Meteorites and the Early Solar System* (J. F. Kerridge and M. S. Matthews, eds.), p. 746, Univ. of Arizona, Tucson.
21. Mason B. and Taylor S. R. (1982) *Smithson. Contrib. Earth Sci., 25,* 30 pp.
22. Fegley B. Jr. and Palme H. (1985) *Earth Planet. Sci. Lett., 72,* 311.
23. Fegley B. Jr. and Kong D. (1989) in *Lunar and Planetary Science XX,* p. 279, Lunar and Planetary Institute, Houston.
24. Morfill G. (1988) *Icarus, 73,* 1371; Boss A. P. (1988) *Science, 241,* 565.
25. Weidenschilling S. J. (1988) in *Meteorites and the Early Solar System* (J. F. Kerridge and M. S. Matthews, eds.), p. 348, Univ. of Arizona, Tucson.
26. Ireland T. R. (1988) *Geochim. Cosmochim. Acta, 52,* 2827; ibid., 2841.

27. Fegley B. Jr. and Palme H. (1985) *Earth Planet. Sci. Lett., 72,* 311; Rubin A. et al. (1988) in *Meteorites and the Early Solar System* (J. F. Kerridge and M. S. Matthews, eds.), p. 488, Univ. of Arizona, Tucson.
28. Scott E. R. D. et al. (1984) *Geochim. Cosmochim. Acta, 48,* 1741; Scott E. R. D. et al. (1988) in *Meteorites and the Early Solar System* (J. F. Kerridge and M. S. Matthews, eds.), p. 718, Univ. of Arizona, Tucson.
29. For example, Dodd R. T. (1981) *Meteorites: A Petrologic-Chemical Synthesis,* Cambridge Univ., New York, 368 pp.
30. For example, Larimer J. W. and Anders E. (1967) *Geochim. Cosmochim. Acta, 31,* 1239; Grossman L. and Larimer J. W. (1974) *Rev. Geophys. Space Phys.,12,* 71.
31. Kurat G. (1988) *Philos. Trans R. Soc., A325,* 459.
32. Brearley A. et al. (1989) *Geochim. Cosmochim. Acta, 53,* 2081.
33. Ibid., 2091. This supports the conclusions of of Kurat G. (1988) *Philos. Trans. R. Soc., A325,* 459; Kurat G. (1987) *Mitt. Osterr. Miner. Ges., 132,* 9.
34. Grossman J. and Wasson J. T. (1987) *Geochim. Cosmochim. Acta, 51,* 3003.
35. Hutchison R. and Bevan A. W. R. (1983) in *Chondrules and Their Origins* (E. A. King, ed.), p. 162, Lunar and Planetary Institute, Houston; Alexander C. M. et al. (1985) *Meteoritics, 19,* 184; Hutchison R. et al. (1988) *Philos. Trans. R. Soc., A325,* 445; Jones R. H. and Scott E. R. D. (1988) *Proc. Lunar Planet. Sci. Conf. 19th,* p. 523.
36. Nagahara H. et al. (1988) *Nature, 331,* 516.
37. Stephens J. R. and Kothari B. K. (1978) *Moon Planets, 19,* 139; Reitmeijer F. J. M. and McKay D. S. (1986) in *Lunar and Planetary Science XVII,* p. 710, Lunar and Planetary Institute, Houston; Rietmeijer F. J. M. et al. (1986) *Icarus, 66,* 211; Nuth J. A. and Donn B. (1983) *J. Geophys. Res., 88,* A847; Kornacki A. S. and Wood J. A. (1984) *Geochim. Cosmochim. Acta, 48,* 1663.
38. Wood J. A. (1988) *Annu. Rev. Earth Planet. Sci., 16,* 53.
39. Sorby H. C. (1877) *Nature, 15,* 495.
40. See, for example, the lengthy discussion on "definitions and difficulties" of defining chondrules by Grossman J. N. et al. (1988) in *Meteorites and the Early Solar System* (J. F. Kerridge and M. S. Matthews, eds.), pp. 622-623, Univ. of Arizona, Tucson.
41. Prinz M. et al. (1988) *Meteoritics, 23,* 297.
42. Grossman J. N. and Wasson J. T. (1982) *Geochim. Cosmochim. Acta, 46,* 1081; Grossman J. N. and Wasson J. T. (1983) *Geochim. Cosmochim. Acta, 47,* 759; Grossman J. N. and Wasson J. T. (1983) in *Chondrules and Their Origins* (E. A. King, ed.), p. 88, Lunar and Planetary Institute, Houston.
43. Gooding J. L. and Keil K. (1981) *Meteoritics, 16,* 19.
44. Hewins R. (1988) in *Meteorites and the Early Solar System* (J. F. Kerridge and M. S. Matthews, eds.), p. 660, Univ. of Arizona, Tucson.
45. Lofgren G. and Russell W. T. (1986) *Geochim. Cosmochim. Acta, 50,* 1715; Connolly H. C. et al. (1988) in *Lunar and Planetary Science XIX,* p. 205, Lunar and Planetary Institute, Houston.
46. Lofgren G. and Russell W. T. (1986) *Geochim. Cosmochim. Acta, 50,* 1715; Radomsky P. M. and Hewins R. H. (1988) *Meteoritics, 23,* 297.
47. Grossman J. N. et al. (1988) *Earth Planet. Sci. Lett., 91,* 33.
48. Shaw D. M. et al. (1988) *Geochim. Cosmochim. Acta, 52,* 2311. The mean boron abundance for chondrites is 0.55 ppm; Shaw D. M. (1988) in *Lunar and Planetary Science XIX,* p. 1067, Lunar and Planetary Institute, Houston.
49. See Nakamura N. et al. (1989) *Anal. Chem., 61,* 755, for details of the direct loading isotope dilution mass spectrometric analytical method involving determination of REE at 10^{-13}-g levels.
50. Misawa K. and Nakamura N. (1988) *Geochim. Cosmochim. Acta, 52,* 1699.
51. Rubin A. (1986) *Meteoritics, 21,* 283.
52. Gooding J. L. (1983) in *Chondrules and Their Origins* (E. A. King, ed.), p. 61, Lunar and Planetary Institute, Houston.
53. Wilkening L. L. et al. (1984) *Geochim. Cosmochim. Acta, 48,* 1071; Scott E. R. D. and Taylor G. J. (1983) in *Lunar and Planetary Science XIV,* p. 680, Lunar and Planetary Institute, Houston.
54. Weidenschilling S. J. (1988) in *Meteorites and the Early Solar System* (J. F. Kerridge and M. S. Matthews, eds.), p. 348, Univ. of Arizona, Tucson.
55. Grossman J. N. (1982) in *Papers Presented to the Conference on Chondrules and Their Origins,* p. 23, Lunar and Planetary Institute, Houston.
56. Grossman J. N. and Wasson J. T. (1985) *Geochim. Cosmochim. Acta, 49,* 925.
57. Gooding J. L. et al. (1983) *Earth Planet. Sci. Lett., 65,* 209.
58. Kring D. A. and Wood J. A. (1988) *Meteoritics, 23,* 283.
59. Swindle T. D. et al. (1983) in *Chondrules and Their Origins* (E. A. King, ed.), p. 246, Lunar and Planetary Institute, Houston.
60. Gooding J. L. et al. (1983) in *Chondrules and Their Origins* (E. A. King, ed.), p. 210, Lunar and Planetary Institute, Houston.
61. The best review of the subject was given by Taylor G. J. et al. (1983) in *Chondrules and Their Origins* (E. A. King, ed.), p. 262, Lunar and Planetary Institute, Houston.

62. For example Ringwood A. E. (1966) *Rev. Geophys. Space Phys., 4,* 113; Fredriksson K. et al. (1973) *Moon, 7,* 475.
63. One example of an igneous inclusion with a fractionated REE pattern and an H-group O-isotopic signature appears in the Barwell L6 chondrite; Hutchinson R. et al. (1988) *Earth Planet. Sci. Lett., 90,* 105.
64. Dence M. R. (1971) *J. Geophys. Res., 76,* 5552; Grieve R. A. F. and Floran R. J. (1978) *J. Geophys. Res., 83,* 2761.
65. Dodd R. T. (1981) *Meteorites: A Petrologic-Chemical Synthesis,* Cambridge Univ., New York, 368 pp.
66. Wänke H. et al. (1981) in *Lunar and Planetary Science XII,* p. 1139, Lunar and Planetary Institute, Houston; Hutchison R. (1982) *Phys. Earth Planet. Inter., 29,* 199.
67. Rubin A. E. et al. (1982) *Geochim. Cosmochim. Acta, 46,* 1763; Rubin A. E. (1982) *Meteoritics, 17,* 275.
68. Taylor G. J. et. al. (1983) in *Chondrules and Their Origins* (E. A. King, ed.), p. 270, Lunar and Planetary Institute, Houston.
69. Keil K. et al. (1991) *Geochim. Cosmochim. Acta,* in press.
70. Ringwood A. E. (1959) *Geochim. Cosmochim. Acta, 15,* 257; Fredriksson K. and Ringwood A. E. (1963) *Geochim. Cosmochim. Acta, 27,* 639.
71. Dodd R. T. (1981) *Meteorites: A Petrologic-Chemical Synthesis,* Cambridge Univ., New York, 368 pp.
72. Huss G. et al. (1981) *Geochim. Cosmochim. Acta, 45,* 33; Scott E. R. D. et al. (1984) *Geochim. Cosmochim. Acta, 48,* 1741.
73. Wood J. A. and Morfill G. E. (1988) in *Meteorites and the Early Solar System* (J. F. Kerridge and M. S. Matthews, eds.), p. 329, Univ. of Arizona, Tucson.
74. Kracher A. et al. (1984) *Proc. Lunar Planet. Sci. Conf. 14th,* in *J. Geophys. Res., 89,* B559.
75. Larimer J. (1988) *Meteoritics, 23,* 284.
76. Misawa K. and Nakamura N. (1988) *Nature, 34,* 47.
77. Nagahara H. (1983) in *Chondrules and Their Origins* (E. A. King, ed.), p. 211, Lunar and Planetary Institute, Houston; Nagahara H. (1981) *Nature, 292,* 135.
78. See also Rambaldi E. R. (1981) *Nature, 293,* 558.
79. Kracher A. (1984) *Proc. Lunar Planet. Sci. Conf. 14th,* in *J. Geophys. Res., 89,* B559.
80. Rambaldi E. R. and Wasson J. T. (1982) *Geochim. Cosmochim. Acta, 46,* 929.
81. Scott E. R. D. et al. (1984) *Geochim. Cosmochim. Acta, 48,* 1741.
82. Grossman J. N. and Wasson J. T. (1987) *Geochim. Cosmochim. Acta, 51,* 3003.
83. See Grossman J. N. and Wasson J. T. (1987) *Geochim. Cosmochim. Acta, 51,* 3006, Fig. 1.
84. Sears D. W. G. (1988) *Philos. Trans. R. Soc., A325,* 566; Grossman J. N. and Wasson J. T. (1985) *Geochim. Cosmochim. Acta, 49,* 925.
85. Taylor S. R. and Norman M. D. (1990) in *The Origin of the Earth* (H. E. Newsom and J. H. Jones, eds.), p. 29, Oxford Univ., New York.
86. For a general description of solar flares, see Tandberg-Hanssen E. and Emslie A. G. (1988) *The Physics of Solar Flares,* Cambridge Univ., New York, 273 pp.
87. Levy E. H. and Araki S. (1989) *Icarus, 81,* 74, following earlier suggestions of analogous mechanisms by Sonett C. P. (1979) *Geophys. Res. Lett., 6,* 677.
88. Sugiura N. and Strangway D. W. (1983) *Earth Planet. Sci. Lett., 62,* 169; Sugiura N. and Strangway D. W. (1988) in *Meteorites and the Early Solar System* (J. F. Kerridge and M. S. Matthews, eds.), p. 595, Univ. of Arizona, Tucson.
89. See review by Wasson J. T. and Kallemeyn G. W. (1988) *Philos. Trans. R. Soc., A325,* 535.
90. Scott E. R. D. (1988) *Earth Planet. Sci. Lett., 91,* 1.
91. From Sears D. W. G. and Dodd R. T. (1988) in *Meteorites and the Early Solar System* (J. F. Kerridge and M. S. Matthews, eds.), p. 21, Fig. 1.1.10, Univ. of Arizona, Tucson.
92. Wasson J. T. et al. (1988) *Meteoritics, 23,* 309; Kallemeyn G. W. et al. (1989) *Geochim. Cosmochim. Acta, 53,* 2763.
93. Palme H. et al. (1988) in *Meteorites and the Early Solar System* (J. F. Kerridge and M. S. Matthews, eds.), p. 436, Univ. of Arizona, Tucson.
94. Dennison J. E. et al. (1986) *Nature, 319,* 390; Dennison J. E. and Lipschutz M. E. (1987) *Geochim. Cosmochim. Acta, 51,* 741; see also Koeberl C. and Cassidy W. A. (1990) *LPI Tech. Rpt. 90-01,* 102 pp.; Koeberl C. and Cassidy W. A. (1991) *Geochim. Cosmochim. Acta, 55,* 3.
95. Zinner E. et al. (1987) *Nature, 330,* 730.
96. Tang M. et al. (1989) *Nature, 339,* 351.
97. Ibid., 354.
98. Pillinger C. T. (1984) *Geochim. Cosmochim. Acta, 48,* 2739.
99. Larimer J. W. and Wasson J. T. (1988) in *Meteorites and the Early Solar System* (J. F. Kerridge and M. S. Matthews, eds.), p. 417, Univ. of Arizona, Tucson.
100. Ibid, p. 433.

101. Scott E. R. D. et al. (1986) *Proc. Lunar Planet. Sci. Conf. 17th*, in *J. Geophys. Res., 91,* E115; Sears D. W. G. and Weeks K. S. (1986) *Geochim. Cosmochim. Acta, 50,* 2815; Morgan J. W. et al. (1985) *Geochim. Cosmochim. Acta, 49,* 247; Heyse J. W. (1978) *Earth Planet. Sci. Lett., 40,* 365.
102. Fredriksson K. (1983) in *Chondrules and Their Origins* (E. A. King, ed.), p. 44, Lunar and Planetary Institute, Houston.
103. Kallemeyn G. W. et al. (1989) *Geochim. Cosmochim. Acta, 53,* 2763.
104. Tomeoka K. and Buseck P. R. (1988) *Geochim. Cosmochim. Acta, 52,* 1628.
105. Frederiksson K. and Kerridge J. F. (1988) *Meteoritics, 23,* 35. Possibly the many organic compounds, including amino acids, found in carbonaceous chondrites, originated during episodes of aqueous alteration in these meteorites; see Shock E. L. and Schulte M. D. (1990) *Nature, 343,* 728.
106. Tomeoka K. and Buseck P. R. (1988) *Geochim. Cosmochim. Acta, 52,* 1627.
107. Zolensky M. E. et al. (1989) *Icarus, 78,* 411.
108. $\delta D = [(D/H)water - (D/H)SMOW]/[(D/H) SMOW] \times 1000$; SMOW = Standard Mean Ocean Water.
109. Barber D. J. (1985) *Clay Minerals, 20,* 415; Tomeoka K. and Buseck P. R. (1985) *Geochim. Cosmochim. Acta, 49,* 2149.
110. Clayton R. N. and Mayeda T. K. (1984) *Earth Planet. Sci. Lett., 67,* 151.
111. Grimm R. E. and McSween H. Y. (1989) *Icarus, 82,* 249.
112. Kerridge J. F. and Bunch T. (1979) in *Asteroids* (T. Gehrels, ed.), p. 745, Univ. of Arizona, Tucson; Fredriksson K. and Kerridge J. F. (1988) *Meteoritics, 23,* 35; Zolensky M. E. et al. (1989) *Icarus, 78,* 411.
113. Macdougall J. D. et al. (1984) *Nature, 307,* 250.
114. Olsen E. J. et al. (1988) *Geochim. Cosmochim. Acta, 52,* 1615.
115. Keil K. (1988) *Meteoritics, 23,* 278.
116. Bell J. F. (1986) in *Lunar and Planetary Science XVII,* p. 985, Lunar and Planetary Institute, Houston; Gaffey M. J. (1986) *Meteoritics, 21,* 365; Lipschutz M. E. et al. (1989) in *Asteroids II* (R. P. Binzel et al., eds.), p. 740, Univ. of Arizona, Tucson.
117. Pellas P. (1988) *Meteoritics, 23,* 296.
118. Anders E. (1978) in *Asteroids: An Exploration Assessment* (D. Morrison and W. C. Wells, eds.), p. 145, NASA CP-2053.
119. Lipschutz M. E. et al. (1989) in *Asteroids II* (R. P. Binzel et al., eds.), p. 740, Univ. of Arizona, Tucson.
120. Taylor G. J. et al. (1987) *Icarus, 69,* 1.
121. Cooling rates for chondrites are given by Hewins R. H. (1988) in *Meteorites and the Early Solar System* (J. F. Kerridge and M. S. Matthews, eds.), p. 660, Univ. of Arizona, Tucson, and by Lipschutz M. E. et al. (1989) in *Asteroids II* (R. P. Binzel et al., eds.), p. 740, Univ. of Arizona, Tucson; see also Saikumar V. and Goldstein J. I. (1988) *Geochim. Cosmochim. Acta, 52,* 715.
122. Rubin A., Fegley B. Jr., and Brett R. (1988) in *Meteorites and the Early Solar System* (J. F. Kerridge and M. S. Matthews, eds.), p. 488, Univ. of Arizona, Tucson.
123. Ibid., p. 494.
124. For example, Hutchison R. and Bevan A. W. R. (1983) in *Chondrules and Their Origins* (E. A. King, ed.), p. 162, Lunar and Planetary Institute, Houston.
125. Skinner W. R. (1989) in *Lunar and Planetary Science XX,* pp. 1018, 1020, Lunar and Planetary Institute, Houston.
126. Fredriksson K. and Kerridge J. F. (1988) *Meteoritics, 23,* 40.
127. Cameron A. G. W. (1991) *Protostars and Planets III* (G. Levy et al., eds.), Univ. of Arizona, Tucson, in press.
128. Wasserburg G. J. (1985) *Protostars and Planets II* (D. C. Black and M. S. Matthews, eds.), p. 703, Univ. of Arizona, Tucson.
129. Parker E. N. (1968) *Astrophys J., 154,* 895.
130. Dates in the Roman Empire were listed as AUC or *ab urbe condita* ("from the founding of the city") in 753 BC, 1990 AD thus being 2743 AUC.
131. Using the revised decay constant of 1.402×10^{-11}/yr. L chondrites: Minster J. F. and Allègre C. J. (1979) *Meteoritics, 14,* 235; H chondrites: Minster J. F. and Allègre C. J. (1979) *Earth Planet. Sci. Lett., 42,* 333; LL chondrites: Minster J. F. and Allègre C. J. (1981) *Earth Planet. Sci. Lett., 56,* 89.
132. Manhés G. et al. (1987) *Terra Cognita, 7,* 377.
133. Tilton G. R. (1988) in *Meteorites and the Early Solar System* (J. F. Kerridge and M. S. Matthews, eds.), p. 259, Univ. of Arizona, Tucson.
134. Minster J. F. and Allègre C. J. (1981) *Earth Planet. Sci. Lett., 56,* 89.
135. Minster J. F. et al. (1979) *Earth Planet. Sci. Lett., 44,* 420.
136. Podosek F. A. et al. (1991) *Geochim. Cosmochim. Acta, 55,* 1083.
137. Brannon J. C. et al. (1987) *Proc. Lunar Planet. Sci. Conf. 18th,* p. 555.
138. Lugmair G. W. et al. (1989) in *Lunar and Planetary Science XX,* p. 604, Lunar and Planetary Institute, Houston.
139. Tera F. et al. (1989) in *Lunar and Planetary Science XX,* p. 1111, Lunar and Planetary Institute, Houston.

140. Prinzhofer A. et al. (1989) in ***Lunar and Planetary Science XX,*** p. 87, Lunar and Planetary Institute, Houston.
141. Chen J. H. and Wasserburg G. J. (1985) in ***Lunar and Planetary Science XVI,*** p. 119, Lunar and Planetary Institute, Houston.
142. Scott E. R. D. and Rajan R. S. (1981) ***Geochim. Cosmochim. Acta, 45,*** 53.
143. Brannon J. C. et al. (1987) ***Proc. Lunar Planet. Sci. Conf. 18th,*** p. 561.
144. Anders E. and Grevesse N. (1989) ***Geochim. Cosmochim. Acta, 53,*** p. 198, Table 1; E. Anders, personal communications, May 1989 and January 1990.
145. Holweger H. (1988) in ***The Impact of Very High S/N Spectroscopy on Stellar Physics*** (G. Cayrel de Strobel and M. Spite, eds.) p. 411, IAU; Holweger H. et al. (1990) ***Astron. Astrophys., 232,*** 510.
146. Breneman H. H. and Stone E. C. (1985) ***Astrophys. J., 299,*** L57; see also summary of earlier solar photospheric data by Anders E. and Grevesse N. (1989) ***Geochim. Cosmochim. Acta, 53,*** 197.
147. Delsemme A. H. (1988) ***Philos. Trans. R. Soc., A325,*** 509; Jessberger E. K. et al. (1988) ***Nature, 332,*** 691.
148. See Burnett D. S. et al. (1989) ***Geochim. Cosmochim. Acta, 50,*** 471, for a test and discussion of the CI abundance pattern.
149. Wood J. A. and Morfill G. E. (1988) in ***Meteorites and the Early Solar System*** (J. F. Kerridge and M. S. Matthews, eds.), p. 329, Univ. of Arizona, Tucson.
150. Larimer J. W. (1967) ***Geochim. Cosmochim. Acta, 31,*** 1215; Grossman L. (1972) ***Geochim. Cosmochim. Acta, 36,*** 597; Grossman L. and Larimer J. W. (1974) ***Rev. Geophys. Space Phys., 12,*** 71.
151. Lewis J. S. (1973) ***Annu. Rev. Phys. Chem., 24,*** 339; Lewis J. S. (1974) ***Science, 186,*** 440; Lewis J. S. (1974) ***Sci. Am., 250,*** 50.
152. Wood J. A. and Hashimoto A. (1988) in ***Lunar and Planetary Science XIX,*** p. 1292, Lunar and Planetary Institute, Houston; see also Wood J. A. (1988) ***Annu. Rev. Earth Planet. Sci., 16,*** 53.
153. Si, Mg, Fe, Al, Ca, Na, Ni, Ti plus CI proportions of H, C, N, O, and S were increased by a factor of 100 relative to remaining H, C, N, O, and S.
154. Laplace P. S. (1809) ***The System of the World, Vol. 1, Book I*** (J. Pond, trans.), p. 97, R. Phillips, London.
155. See review on the chemistry of comets by Delsemme A. H. (1988) ***Philos. Trans. R. Soc., A325,*** 509.
156. For example, Weissman P. R. (1986) in ***The Galaxy and the Solar System*** (R. Smoluchowski et al., eds.), p. 219, Univ. of Arizona, Tucson; Duncan M. et al. (1987) ***Astron. J., 94,*** 1330; Duncan M. et al. (1988) ***Astrophys. J. Lett., 328,*** L69; Weissman P. R. et al. (1989) ***Geophys. Res. Lett., 16,*** 1241; Torbett M. V. (1989) ***Astron. J., 98,*** 1477. For a contrary view asserting that all comets are derived from the classical spherical Oort Cloud, see Stagg C. R. and Bailey M. E. (1989) ***Mon. Not. R. Astron. Soc., 241,*** 507.
157. For example, Fernandez A. J. (1980) ***Mon. Not. R. Astron. Soc., 192,*** 481.
158. Duncan M. et al. (1988) ***Astrophys. J., 328,*** L69; Hartmann W. K. et al. (1989) in ***Lunar and Planetary Science XX,*** p. 379, Lunar and Planetary Institute, Houston.
159. For example Weissman P. R. (1986) in ***The Galaxy and the Solar System*** (R. Smoluchowski et al., eds.), p. 204, Univ. of Arizona, Tucson.
160. Weissman P. R. (1990) ***Bull. Am. Astron. Soc., 22,*** 1093. This contrasts with earlier high estimates of the angular momentum of the Oort Cloud implying ***in situ*** formation for comets; see Marochnik L. S. et al. (1988) ***Science, 242,*** 547. Their estimates of the cometary population and nucleus masses are probably too high by at least an order of magnitude.
161. McFadden L. A. et al. (1987) ***Astron. Astrophys., 187,*** 333.
162. Jones J. et al. (1989) ***Mon. Not. R. Astron. Soc., 238,*** 179.
163. Peale S. J. (1989) ***Icarus, 82,*** 36.
164. Anders E. and Grevesse N. (1989) ***Geochim. Cosmochim. Acta, 53,*** 197.
165. Using gas-dust ratios from Delsemme A. H. (1988) ***Philos. Trans. R. Soc., A325,*** 509.
166. Anders E. and Grevesse N. (1989) ***Geochim. Cosmochim. Acta, 53,*** 210, Fig. 7.
167. From the PUMA-1 mass spectrometer; Jessberger E. K. et al. (1988) ***Nature, 332,*** 691; Jessberger E. K. et al. (1989) in ***Origin and Evolution of Planetary and Satellite Atmospheres*** (S. K. Atreya et al., eds.), p. 167, Univ. of Arizona, Tucson.
168. Jessberger E. K. et al. (1989) in ***Origin and Evolution of Planetary and Satellite Atmospheres*** (S. K. Atreya et al., eds.), p. 185, Univ. of Arizona, Tucson.
169. Lunine J. I. (1989) in ***The Formation and Evolution of Planetary Systems*** (H. A. Weaver and L. Danly, eds.), pp. 233, 238, Cambridge Univ., New York.
170. Schramm L. S. et al. (1988) in ***Lunar and Planetary Science XIX,*** p. 1033, Lunar and Planetary Institute, Houston.
171. Prinn R. G. and Fegley B. Jr. (1987) ***Annu. Rev. Earth Planet. Sci., 15,*** 171.
172. Kowal C. T. (1989) ***Icarus, 77,*** 122.
173. Meech K. J. and Belton M. J. S. (1989) ***IAU Circular,*** 4770.
174. Hartmann W. K. et al. (1989) in ***Lunar and Planetary Science XX,*** p. 379, Lunar and Planetary Institute, Houston; Hartmann W. K. et al. (1990) ***Icarus, 83,*** 1.

175. Klock W. et al. (1988) *Meteoritics, 23,* 280.
176. Bradley J. P. et al. (1988) *Meteoritics, 23,* 259.
177. Klock W. et al. (1989) in *Lunar and Planetary Science XX,* p. 522, Lunar and Planetary Institute, Houston.
178. Schramm L. S. et al. (1989) *Meteoritics, 24,* 99.
179. Schramm L. S. et al. (1988) in *Lunar and Planetary Science XIX,* p. 1033, Lunar and Planetary Institute, Houston.
180. Zinner E. and McKeegan K. D. (1984) in *Lunar and Planetary Science XV,* p. 961, Lunar and Planetary Institute, Houston; Zinner E. (1988) in *Meteorites and the Early Solar System* (J. F. Kerridge and M. S. Matthews, eds.), p. 956, Univ. of Arizona, Tucson.
181. McKeegan K. D. et al. (1985) *Geochim. Cosmochim. Acta, 49,* 1971.
182. For example, Yang J. and Epstein S. (1983) *Geochim. Cosmochim. Acta, 47,* 2199.
183. Zook H. A. and McKay D. S. (1986) in *Lunar and Planetary Science XVII,* p. 977, Lunar and Planetary Institute, Houston.
184. Dermott S. F. and Nicholson P. (1988) IAU meeting Baltimore; Sykes M. V. (1988) *Astrophys. J. Lett., 334,* L55.
185. All are reliable families; see section 5.11.
186. Brownlee D. E. (1985) *Annu. Rev. Earth Planet. Sci., 13,* 147.
187. A basic reference is McDonnell J. A. M. (1978) *Cosmic Dust,* Wiley, New York, 693 pp. The paper by D. W. Hughes (p. 123) is a standard reference for meteor influx.
188. Brownlee D. E. (1985) *Annu. Rev. Earth Planet. Sci., 13,* 147; Brownlee D. E. (1987) *Philos. Trans. R. Soc. London, A323,* 305.
189. Cour-Palais B. G. (1974) *Proc. Lunar Sci. Conf. 5th,* p. 2451.
190. Hörz F. et al. (1975) *Planet. Space Sci., 23,* 151.
191. Grun E. et al. (1985) *Icarus, 62,* 244.
192. Maurette M. et al. (1987) *Nature, 328,* 699.
193. Murray J. (1876) *Proc. R. Soc. Edinburgh, 9,* 247.
194. Some of the smaller particles have retained volatile components indicated by CI abundances of C, S, Zn, and Br: Van der Stap C. C. A. H. et al. (1986) in *Lunar and Planetary Science XVII,* p. 1013, Lunar and Planetary Institute, Houston.
195. Flynn G. J. (1989) *Icarus, 77,* 287.
196. Kurat G. (1988) *Philos. Trans. R. Soc., A325,* 474.
197. A general treatment is given by McSween H. Y. (1987) *Meteorites and Their Parent Planets,* Cambridge Univ., New York, 237 pp.
198. Carlson R. W. et al. (1988) in *Lunar and Planetary Science XIX,* p. 166, Lunar and Planetary Institute, Houston.
199. Hewins R. H. and Newsom H. E. (1988) in *Meteorites and the Early Solar System* (J. F. Kerridge and M. S. Matthews, eds.), p. 73, Univ. of Arizona, Tucson.
200. Gaffey M. J. (1983) in *Lunar and Planetary Science XIV,* p. 231, Lunar and Planetary Institute, Houston.
201. Bell J. F. (1988) in *Lunar and Planetary Science XIX,* p. 55, Lunar and Planetary Institute, Houston.
202. McSween H. Y. (1987) *Meteorites and Their Parent Planets,* p. 123, Cambridge Univ., New York.
203. Goodrich C. A. et al. (1987) *Geochim. Cosmochim. Acta, 51,* 2255.
204. Clayton R. N. and Mayeda T. K. (1988) in *Lunar and Planetary Science XIX,* p. 197, Lunar and Planetary Institute, Houston.
205. Goodrich C. A. et al. (1988) *Meteoritics, 23,* 269.
206. Kurat G. (1988) *Philos. Trans. R. Soc., A325,* 474.
207. Rubin A. E. (1988) *Meteoritics, 23,* 333.
208. Lugmair G. W. (1989) in *Lunar and Planetary Science XX,* p. 605, Lunar and Planetary Institute, Houston; Shih C.-Y. et al. (1989) in *Lunar and Planetary Science XX,* p. 1004, Lunar and Planetary Institute, Houston; Tilton G. (1988) in *Meteorites and the Early Solar System* (J. F. Kerridge and M. S. Matthews, eds.), p. 259, Univ. of Arizona, Tucson.
209. Prinz M. et al. (1977) *Earth Planet. Sci. Lett., 35,* 317.
210. Treiman A. H. (1988) in *Lunar and Planetary Science XIX,* p. 1203, Lunar and Planetary Institute, Houston.
211. Crozaz G. and Lundberg L. L. (1988) in *Lunar and Planetary Science XIX,* p. 231, Lunar and Planetary Institute, Houston.
212. Taylor G. J. et al. (1988) in *Lunar and Planetary Science XIX,* p. 1185, Lunar and Planetary Institute, Houston.
213. Brett R. and Keil K. (1986) *Earth Planet. Sci. Lett., 81,* 1.
214. Scott E. R. D. (1979) in *Asteroids* (T. Gehrels, ed.), p. 892, Univ. of Arizona, Tucson.
215. Kaiser T. and Wasserburg G. J. (1983) *Geochim. Cosmochim. Acta, 47,* 43.
216. Scott E. R. D. (1979) in *Asteroids* (T. Gehrels, ed.), p. 892, Univ. of Arizona, Tucson.
217. Kelly W. R. and Larimer J. W. (1977) *Geochim. Cosmochim. Acta, 41,* 93.
218. Taylor G. J. (1989) in *Lunar and Planetary Science XX,* p. 1109, Lunar and Planetary Institute, Houston.
219. Wasson J. T. (1985) *Meteorites: Their Record of Early Solar-System History,* p. 92, Freeman, New York.

220. See the recent assessment by Saikumar V. and Goldstein J. I. (1988) *Geochim. Cosmochim. Acta, 52,* 715.
221. Delaney J. S. (1989) *Nature, 342,* 889; Warren P. H. and Kallemeyn G. W. (1989) *Geochim. Cosmochim. Acta, 53,* 3323.
222. Eugster O. (1989) *Science, 245,* 1197.
223. A basic reference for martian meteorites is McSween H. Y. (1985) *Rev. Geophys., 23,* 391.
224. Smith M. R. et al. (1984) *Proc. Lunar Planet. Sci. Conf. 14th,* in *J. Geophys. Res., 89,* B612; McSween H. Y. (1985) *Rev. Geophys., 23,* 391; Laul J. C. et al. (1986) *Geochim. Cosmochim. Acta, 50,* 909; Treiman A. H. (1986) *Geochim. Cosmochim. Acta, 50,* 1061; Treiman A. H. et al. (1986) *Geochim. Cosmochim. Acta, 50,* 1071; Vickery A. M. and Melosh H. J. (1987) *Science, 247,* 738.
225. Nakamura N. et al. (1982) *Geochim. Cosmochim. Acta, 46,* 1555.
226. Shergotty, Zagami, ALHA 77005, and EETA 79001.
227. Nyquist L. E. et al. (1979) *Geochim. Cosmochim. Acta, 43,* 1057; Shih C.-Y. et al. (1982) *Geochim. Cosmochim. Acta, 46,* 232.
228. Jagoutz Z. and Wänke H. (1986) *Geochim. Cosmochim. Acta, 50,* 939.
229. Lundberg L. et al. (1988) *Geochim. Cosmochim. Acta, 52,* 2158.
230. McFadden L. A. et al. (1989) in *Asteroids II* (R. P. Binzel et al., eds.), p. 442, Univ. of Arizona, Tucson; Shoemaker E. M. et al. (1990) in *Global Catastrophes in Earth History,* p. 155, Geol. Soc. Am. Spec. Paper 247.
231. Wetherill G. W. (1988) *Icarus, 76,* 1.
232. Weissman P. R. et al. (1989) in *Asteroids II* (R. P. Binzel et al., eds.), p. 880, Univ. of Arizona, Tucson.
233. Luu J. and Jewitt D. (1989) *Astron. J., 98,* 1905.
234. Wisdom J. (1985) *Icarus, 63,* 272.
235. Wetherill G. W. and Chapman C. R. (1988) in *Meteorites and the Early Solar System* (J. F. Kerridge and M. S. Matthews, eds.), p. 46, Univ. of Arizona, Tucson; Gaffey M. J. et al. (1990), in *Asteroids II* (R. P. Binzel et al., eds.), p. 98, Univ. of Arizona, Tucson; see also Cloutis E. A. et al. (1990) *J. Geophys. Res., 95,* 281.
236. Gaffey M. J. (1986) *Icarus, 66,* 468; Bell J. F. and Keil K. (1988) *Proc. Lunar Planet. Sci. Conf. 18th,* p. 573.
237. For a summary, see Wetherill G. W. and Chapman C. R. (1988) in *Meteorites and the Early Solar System* (J. F. Kerridge and M. S. Matthews, eds.), p. 49, Univ. of Arizona, Tucson.
238. J. F. Bell, personal communication, November 1989.
239. Levy E. H. and Araki S. (1989) *Icarus, 81,* 74; Wood J. A. and Morfill G. E (1988) in *Meteorites and the Early Solar System* (J. F. Kerridge and M. S. Matthews, eds.), p. 329, Univ. of Arizona, Tucson.
240. Wänke H. and Dreibus G. (1988) *Philos. Trans. R. Soc., A325,* 545.
241. Basaltic Volcanism Study Project (1981) *Basaltic Volcanism on the Terrestrial Planets,* p. 648, Pergamon, New York.
242. Dreibus G. et al. (1988) in *Lunar and Planetary Science XIX,* p. 283, Lunar and Planetary Institute, Houston.
243. Smith J. V. (1982) *J. Geol., 90,* 1.

Chapter 4

The Role of Impacts

Illustration, preceding page—*The impact crater Danilova (named for the Russian ballerina). This crater is located in the Lavinia region of Venus (27° S, 339° E) and is about 40 km in diameter. The rough blocky ejecta blanket and central peak of the crater strongly reflect the Magellan radar signal (wavelength 12.5 cm), while the smoother basaltic plains appear darker. The northwest sector of the ejecta blanket is missing, probably a consequence of turbulent interaction between the thick (100 bar) venusian atmosphere and the incoming projectile, which, according to this interpretation, entered from a northwesterly direction. (NASA JPL Magellan P-36711.)*

The Role of Impacts

4

4.1. A RELUCTANT CONVERSION

"A hypervelocity meteorite impact is an extraordinary event—wreaking change instantaneously. Such a process violates every tenet of uniformitarianism. Largely for this reason, hypotheses of impact origin for craters on the Earth and the Moon were vigorously opposed for the better part of the last century. . . . research has now established, beyond doubt, the authenticity of impact as a geological process, but . . . a wide chasm still persists between the views of impact specialists and those of terrestrial geologists" [1].

It might be considered surprising to include a separate chapter on this topic in a book that is principally concerned with examining the origin and evolution of the solar system from a cosmochemical perspective. However, it has become clear that collisions between bodies have played a significant role in the evolution of the solar system. These effects have occurred at all times and stages, beginning with the sticking together of grains in low-velocity collisions in the dusty midplane of the nebula, and continuing with the growth of planetesimals from meter- to kilometer- and eventually to planet-sized bodies. Innumerable collisions occurred during the sweep-up of these smaller bodies into the planets, culminating in the massive final collisions that tilted the planets, spun out planetary subnebulae, formed the Moon, and stripped off the silicate mantle of Mercury. The final stage was the terminal heavy bombardment, whose effects are so dramatically visible on the Moon, but the processes continue at a reduced scale even today; the encounter of the Earth with an Apollo or Aten asteroid would produce a major catastrophe.

The significance of impacts in solar system history has been only slowly appreciated, as the results of recent planetary exploration have become disseminated in the wider scientific community. A good example of the difficulties in attempting to account for the present state of the solar system without considering the effects of impacts can be gained from a recent condensation model for the solar system.

In this model, "the overall characteristics of the Solar System are interpreted as arising from an initial condensation of material at five separate centres (Sun and the four major planets) . . . Each primary resulting body acts as the centre for orbiting secondaries, the terrestrial planets for the Sun and the "icy" satellites for the major planets" [2].

Although this model is able to account for some features of the system, the author concludes that a number of apparently intractable problems remain. Thus, "the rotational orientation of Uranus (and of Pluto) essentially in the ecliptic plane remains unexplained except by supposing a formation from a pathologically strange initial cloud motion. A collision with another body of sufficient magnitude to cause the effect would be difficult to construct on the present arguments . . . The model does not explain the slow retrograde rotation of Venus . . . There is no explanation for the occurrence of pairs of bodies, and particularly of the Earth-Moon system. This remains as a difficult problem" [3].

These comments form a typical example of the difficulties that have pursued solar system theorists for many years. The philosophical problem arises because such theories attempt to build planets by subdividing the gaseous nebula, forming planets like stars, from the top down, as in the giant gaseous protoplanet hypothesis. In this book, the opposite process, of assembling the planets from smaller pieces, from the bottom up, is a preferred approach. This is consistent with the chemical evidence and yields some solutions to the difficulties discussed above, since large collisions are common in the latter stages of planetary accretion.

Curiously, the concept that the asteroid belt represented the fragments of a disrupted planet has always enjoyed wide acceptance. Even when it was realized that their collective mass was much less than that of the Moon, the idea that many asteroids were collisional fragments of somewhat larger parent bodies was never seriously questioned. This intellectual advance seems to have been principally due to the work of Hirayama [4].

Historically, much time was spent on the discussion of the volcanic, or internal, vs. the impact origin of lunar craters, for the dominating influence of impacts in early solar system history has only recently been realized. The history of such attempts, and of the distinction between volcanic and impact cratering processes, is a fascinating chapter in science, replete with misconceptions, misidentifications, and faulty conclusions. However, it is not the purpose of this book to retrace that interesting but arduous path, and the field is now mainly the province of the historians of science [5].

Neither is there space here to devote attention to the remarkable and instructive controversy over the origin of tektites, except to make the following comment. The source of tektites has been demonstrated beyond reasonable doubt as being due to melted terrestrial (usually sedimentary) rock splashed during meteorite impact [6]. The whole argument over a lunar vs. terrestrial origin of tektites was an interesting example of the inability of protagonists for a lunar origin to recognize the decisive geochemical evidence in favor of a terrestrial origin.

The history of the study of the lunar surface is another cautionary tale. Again, full justice cannot be done here to this fascinating topic, although some comments appear at other places in the text. The final resolution of many of the various controversies is well documented by Wilhelms [7], whose evaluation of the evidence should be carefully studied as a lesson in the correct approach to the interpretation of planetary landscapes.

The early studies of the surface of the Moon provided a great impetus to crater studies, particularly after it was realized that a stratigraphic sequence could be constructed from the relative numbers of craters on each surface unit [8]. Thus, the cratering record on the planets and satellites can in principle provide much information about the early history of these bodies. However, conclusions on the interpretation of the solar system cratering record vary widely. "Crater counting and the interpretation of size-frequency distribution plots has developed into as much an art as a science" [9,10].

Even coauthors of papers in this subject may not agree. Thus, in a recent discussion on the interpretation of cratering statistics, "the authors disagree between themselves about the crater population on the lightly cratered regions of the Moon (maria) and Mars (younger plains). One of us (G. Neukum) believes the size-frequency distribution is similar to the highlands population, while the other (R. Strom) believes it is significantly different" [11].

One ultimate objective of establishing the cratering histories on planets and satellites throughout the solar system is the interplanetary correlation of geological time. However, it has proven very difficult to correlate between bodies. Cratering fluxes appear to vary widely in different parts of the system, and there does not appear to be that prerequisite, a uniform solar-system-wide flux of impactors. Only sample return missions to the various bodies followed by isotopic dating in terrestrial laboratories will answer this question [12].

Cratering has now become a discipline in its own right, complete with conferences and textbooks [13]. In this work, it is possible only to touch lightly on the large amount of information that can be gained from a study of crater morphology, and the effects and relative ages of resurfacing processes, whether volcanic or due to impact ejecta blankets. Another significant objective is to assess the nature of the impacting bodies, whether comets, asteroids, debris in circumplanetary orbits, or projectiles left over following accretion. All the above features provide useful insights, but perhaps the most significant in the present context is the difference that the study of impact processes, with their essentially random and stochastic nature, has made to our philosophical outlook on the origin of the solar system. Before this objective is reached, it is necessary to explore the somewhat controversial topic of interplanetary correlations.

4.2. SURFACE HISTORIES OF THE PLANETS AND SATELLITES

The best-documented study is that of the Moon, where the oldest landscape is dominated by impact events. The stratigraphy of the Moon is a key to much else, since "the lunar pre-Nectarian System provides our closest look at the early solar system" [14]. Apart from the Moon, from which dated samples are available, and, of course, the Earth, the cratering record provides our only method of obtaining relative ages

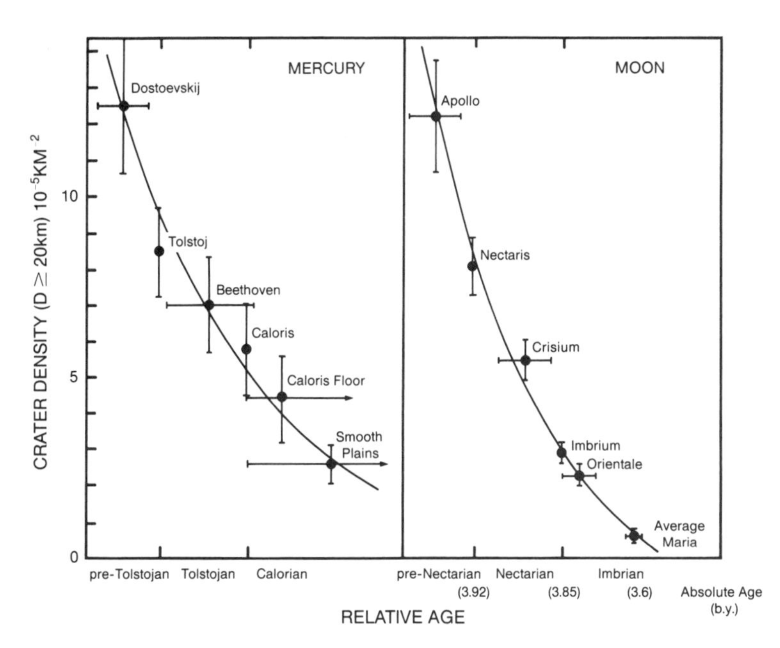

Fig. 4.2.1. *The crater density of geological units on Mercury and the Moon as a function of stratigraphic age. Symbols with no horizontal bars mark the commencement of stratigraphic systems. [After Spudis P. D. and Guest J. E. (1988) in Mercury (F. Vilas et al., eds.), p. 162, Fig. 26, Univ. of Arizona, Tucson.]*

for planetary surfaces. The stratigraphic succession of units can be obtained for a single planet or satellite from the density of primary craters. However, the correlation of terrains between planets or satellites can only be carried out if we know the relative cratering flux for the various bodies. Even on the Moon, we are heavily dependent on this information.

The interpretation of the cratering record is itself often beset with controversy and it requires an exceptionally astute observer to get the record straight, avoiding the many traps such as counting of secondary craters. A good evaluation is provided by Wilhelms [15], which includes the following comment: "Most of the pitfalls [of crater dating] can be avoided by common sense and examination of a surface from the viewpoint of the sequence of events through which it was shaped" [16].

A reading of the recent extensive literature on the cratering history of bodies from Mercury to the outer reaches of the solar system indicates how little this sage advice has been heeded; many workers seem content to defend a particular bias or preconception. However, it is possible to discern some progress through the morass of conflicting interpretations.

From the time of Galileo, the cratered surface of the Moon was the subject of much interest, but the full implications for the early history of the solar system have been realized only very recently. The most significant result to appear from our study by spacecraft of the remoter reaches of the solar system is the evidence for a history of impacts of staggering proportions. Every old surface was battered by impacts of all dimensions. The better preserved the surface, the more evidence is retained. Grains from the lunar surface contain micrometer-sized pits, and at the other end of the scale, the nearside of the Moon is dominated by the Procellarum Basin, 3200 km in diameter. Craters and basins exist on the Moon at all scales between [17].

A second major conclusion is that most of the large impacts on the Moon were early, with a cut-off point of about 3860 m.y. This date comes from the age of the Imbrium impact [18]. Extrapolation of this date to mark the end of the early heavy bombardment to other bodies is often made. The termination of heavy cratering is almost certainly very early, but not necessarily synchronous throughout the solar system. It may be possible to extend this evidence from the Moon to correlate the cratering history of the lunar surface with the old heavily cratered surfaces on Mercury (Fig. 4.2.1) and possibly Mars in the inner solar system, but it appears to this reviewer that there is no way to extend the correlation from the inner to the outer solar system. Even the extrapolation from the Moon to the inner planets depends on correlating the populations of impacting objects. This process is fraught with uncertainty in a topic where stochastic processes dominate. These problems will be addressed both in this chapter and in the succeeding ones on planets and satellites.

A basic question on which much energy has been expended is whether the observed planetary and satellite surfaces are saturated with craters. Saturation is reached when the formation of new craters or basins obliterates old craters. Since many more small craters are formed than large ones, surfaces become saturated with small craters, and the number of these craters is in a steady state or equilibrium condition. If crater saturation has not been reached for a particular crater diameter, then counting of all craters above that size (the so-called production function) can form the basis for an estimate of the age of the surface. The problem is not without controversy [19]. Like so many polarized debates in the natural sciences, workers are commonly seeing different aspects of the same problem, and arguing at cross purposes or on nonintersecting lines [20].

The diameter of the largest crater to which saturation applies is conventionally given as C_s. The increase in C_s with age is illustrated in Table 4.2.1 for the well-documented lunar case. The formation of large ringed basins not only destroys smaller craters within the excavation cavity, but covers most craters within about one basin radius outside the basin with an ejecta blanket. This possibility of resurfacing during the bombardment has usually been underestimated. The end result is that fresh surfaces are effectively produced on which a new generation of craters can accumulate. In addition to ejecta blankets, volcanic activity, including both lava flows and ash showers, and the little understood processes involving low-melting-point ices on the outer satellites, may generate fresh surfaces at any stage of the cratering history. Thus, the crater distribution that we see on the present surfaces may well represent a production population, but it is only recording the final generation of impacts. These craters, however, have formed on a surface that has been saturated and resurfaced many times since the original formation of the solid crust. The significance of saturation cratering is that the older record is removed and we can date surfaces by crater counting only back to the point when they were saturated with craters or resurfaced by impact basin ejecta or by volcanic processes. Perhaps the most extreme examples are those outer solar system satellites that may have been broken up by large impacts and reassembled, although this is subject to much debate (see Chapter 6). The most dramatic examples are the satellites such as Mimas and especially Miranda, which may have been fragmented by collisions and reassembled.

TABLE 4.2.1. Increase in C_s, the diameter of the largest crater to which saturation applies, with age, for lunar craters.

Lunar Stratigraphy	C_s (m)
Eratosthenian System	<100 (mare)
Upper Imbrian Series	80-300 (mare)
Lower Imbrian Series	320-860 (basin)
Nectarian System	800-4000 (basin)
Pre-Nectarian System	>4000 (basin)

Data from Wilhelms D. E. (1987) *USGS Prof. Paper 1348*, p. 130, Table 7.3.

The evidence for saturation in the lunar highlands has been summarized by Hartmann [21]. Crater and basin saturation were mostly achieved at a very early stage. The arguments for this appear overwhelming to this reviewer, and are typified by the following comment: "The predominant view has been that the highlands crater populations are in some sense saturated on all three bodies [Moon, Mars, and Mercury] and differ due to different endogenic processes... Everyone agrees that the more lightly cratered terrains on all three bodies represent a production function due to comets and asteroids in the inner solar system during recent aeons" [22]. Even previous opponents of this notion agree "that the heavily cratered surfaces in the solar system may be relatively close to saturation density, but whether or not they have actually reached equilibrium is uncertain" [23].

The concept of "saturation equilibrium" is based on the assumption that crater densities fluctuate around a particular level. Apart from the destruction of older craters within the new basin or crater rims the ejecta blankets create new surfaces, completely or partially burying older craters, depending on their depth and the thickness of the ejecta blanket [24]. A test of this hypothesis is available from a comparison between the cratered surface of Phobos and that of the most heavily cratered sectors of the lunar highlands. Both show similar crater densities. This is interpreted to indicate that both surfaces have reached some sort of "saturation equilibrium," that accordingly represents a "universal... [saturation equilibrium] level" [24,25].

4.3. EFFECTS OF CRATERING

In this section, attention is focused mainly on the effects of large collisions in the origin and evolution of planets and satellites. Collisions between bodies affect material at all levels, from crushing to mild shock to total melting and vaporization; some of these aspects will be dealt with in subsequent sections.

The smallest recorded craters are the "zap pits" found on lunar samples, which range down to 0.01 μm (Fig. 4.3.1). The presence of large craters on planetary surfaces provides direct evidence of the previous existence of planetesimals and forms one of the cornerstones leading to the concept that the planets accreted from a hierarchy of objects, rather than from dust. A consequence of this hypothesis is that the early solar system teemed with large bodies and that planetary accretion was far from orderly. Not only were primary bodies present, but many secondary objects, produced as by-products of the large collisions, contributed to the rain of projectiles.

Only a brief summary of the mechanics of cratering is given here, since a detailed discussion has recently appeared in textbook form. In that work [26; see also 27], separate chapters are devoted to stress waves in solids, contact and compression stages, excavation stages, and the modification stages of cratering mechanics. Additional chapters deal with ejecta deposits and scaling, as well as the problems of crater morphology, multiring basins, cratered landscapes, atmospheric interactions, and planetary evolution. The interested reader is referred to this comprehensive study for a full treatment of the mathematics of cratering mechanics.

Impact events occur on timescales measured in fractions of a second to minutes. For convenience, several separate stages can be identified as happening during the very short timescales of large impact events. It should be emphasized that these stages grade into one another, and that the division is merely a device to improve our understanding of the complex, although orderly sequence of processes that occur (Fig. 4.3.2). These are:

1. Penetration of the impactor and transfer of the bulk of its kinetic energy into the target through a shock wave. Initial shock wave velocities are about 10 km/s, so this contact and compression phase occurs on timescales of less than a second, except for the largest craters. The shock wave compresses and accelerates the target material downward and outward.

2. Rarefaction of this shock wave at the impactor's and, more importantly, at the target's free surfaces, leading to decompression of the volume traversed by the compressive shock wave.

3. Deflection of material by the rarefaction wave to more upward and outward flow, and excavation of the material cavity. Initial ejection velocities may be as high as a few kilometers per second, but velocities attenuate rapidly, and typical ejection velocities are in the 100-m/s range. A cavity, known as the transient cavity, is produced by a combination of excavation and downward displacement of target mate-

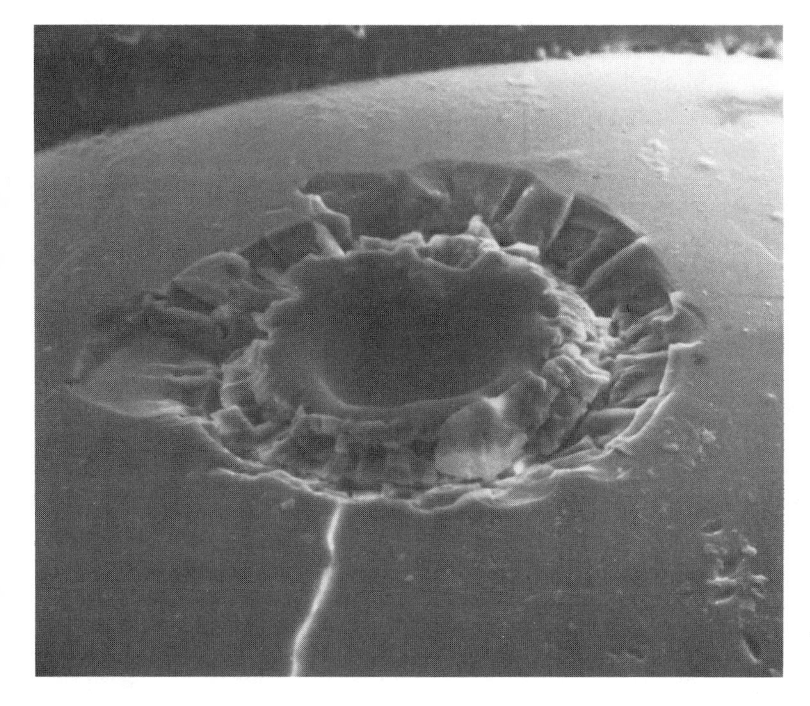

Fig. 4.3.1. *Zap pit. An SEM photo of a high-velocity impact pit on an Apollo 11 glass sphere. The central pit is 30 μm in diameter. (Courtesy D. S. McKay, NASA JSC; NASA S70-18264).*

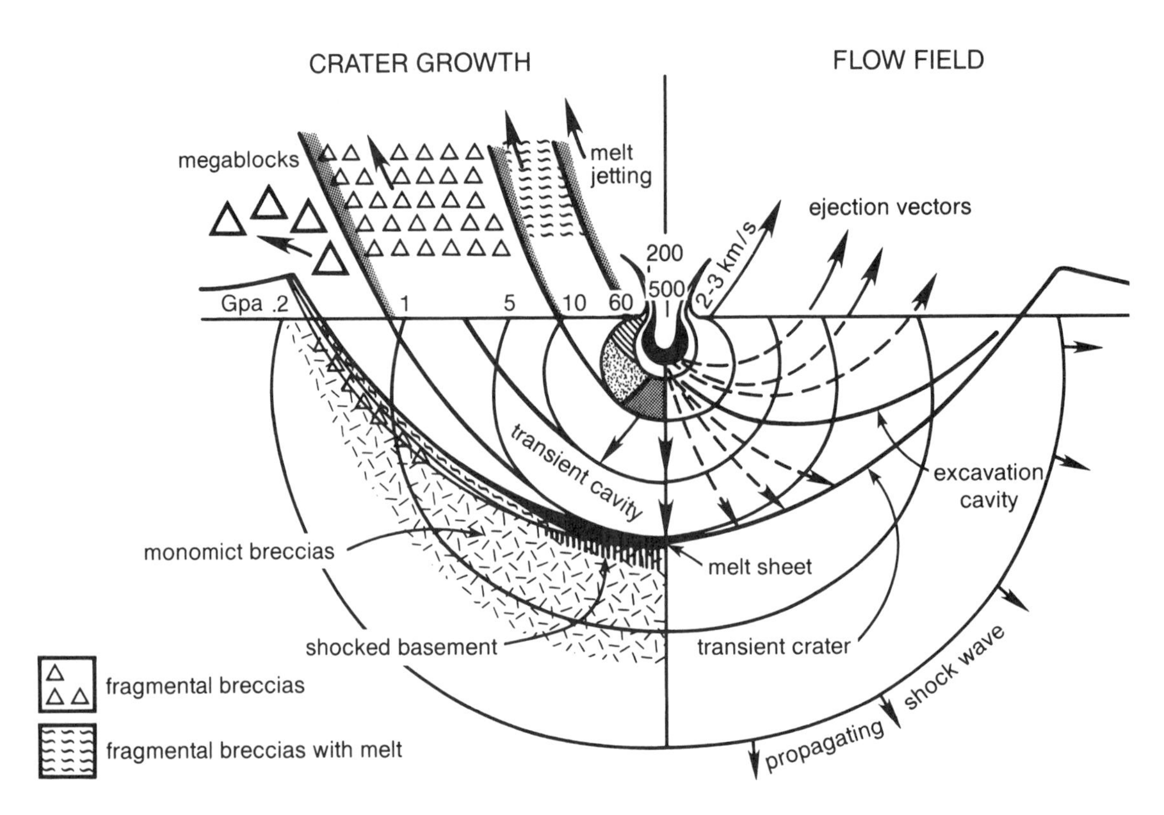

Fig. 4.3.2. *Schematic cross section through a growing impact crater before crater modification takes place. Various regimes of the melt zone are shown and the formation of various types of breccias are depicted on the left side. The right side shows the flow field of particle motion. After Stöffler D. (1981) LPI Tech. Rpt. 81-01, p. 132.*

rials. Ejected material covers the area around the growing crater. This excavation stage takes from seconds to minutes to complete, depending on the size of the crater.

4. Modification of the transient crater cavity due to gravitational forces and to the relaxation of the compressed target materials. Effects include slumping of crater walls, formation of central peaks in larger craters, and production of the great concentric rings of mountains forming the spectacular multiring basins, the largest observed structures produced by impact.

5. On some surfaces, isostatic readjustment may subdue the crater, so that eventually only a palimpsest may remain as mute evidence of a catastrophic event.

The total time for a typical 10-km-diameter crater to form is less than a few minutes. Peak pressures in the shock wave may reach 5000-10,000 kbar (500-1000 GPa) at typical impact velocities. The main significance of this is the effect on the material involved. Under such conditions, the projectile and some of the target material will be vaporized. Rock engulfed by the 700-kbar isobar will be melted, while certain individual minerals will melt at 400-600-kbar pressures. Framework silicates (e.g., feldspars) form solid-state glasses such as maskelynite at lower pressures (300-400 kbar). Fracturing, mechanical deformation, and microscopic planar deformation features occur at pressures below 250 kbar. The significance for planetary formation is that much of the materials that were eventually incorporated into the terrestrial planets went through such processes many times.

The depth of excavation of very large craters and basins is constrained from the evidence from the lunar highland crust, which is from 60 to over 100 km thick. No unequivocal mantle samples appear to have been excavated from beneath this crust, and even lower crustal samples are rare in returned lunar samples, only a few indicating derivation from depths of 50 km [28]. Thus, even the giant basin impacts, forming ringed basins over 1000 km in diameter, have not dug into the lunar mantle.

Another possible effect of impact cratering is impact-induced devolatilization of hydrated silicates and hydrogen isotope fractionation. This is likely to be important for small (Mars-sized) bodies, on which

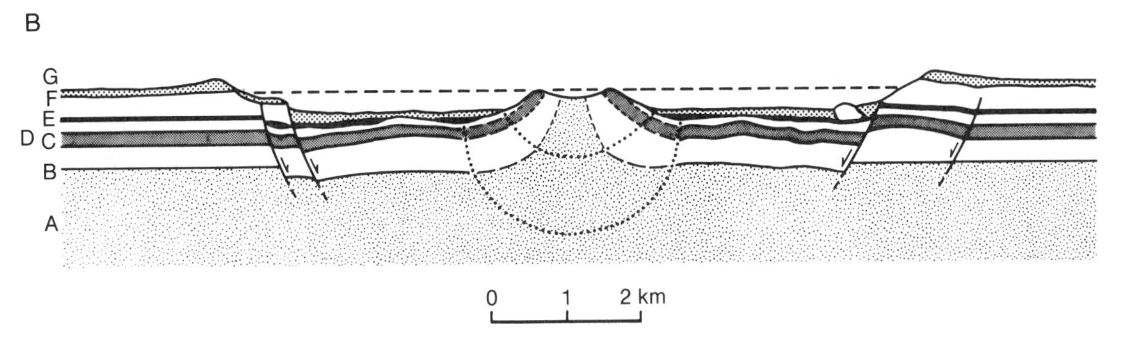

Fig. 4.3.3. *A schematic diagram of a large complex crater with slumped walls and an uplifted central peak. The lowest unit, A, has been displaced upward by about 1 km. Units A and B dip vertically in the central peak. Unit G is ejecta on the crater rim and forms breccia within the crater. Impact melt is not shown. The smaller semicircle marks the penetration limit of the impacting projectile, while the larger semicircle outlines the zone of total disruption. (Courtesy R. J. Pike, USGS).*

the gas and solid isotopic reservoirs can be isolated by burial or by gas escape [29,30].

Several distinct classes of crater are recognized in order of increasing size: simple, complex-immature, complex-mature, central-peak crater (or protobasin) (Fig. 4.3.3), peak-ring basin (or two-ring basin) and multiring basin (Fig. 1.11.2) [31].

The transition from one class to the next is controlled by three principal factors: gravity, strength of target materials, and impact energy. Of these, gravity is the most important. This is shown by the abrupt transition from simple to complex crater forms on the terrestrial planets (Fig. 4.3.4). The change is a major function of gravity, except on Mars, where the transition occurs at smaller crater diameters. This is interpreted as due to the presence of subcrustal volatiles such as water, a deduction supported by the fluidized ejecta deposits commonly observed surrounding martian craters. The role of the impactor type appears to be less significant.

There are significant differences between the characteristics of martian craters at low and high latitudes. Those at low latitudes have sharp rims, and the smaller craters are crisp and bowl shaped. In contrast, at higher latitudes (>30°) the crater rim crests are commonly rounded, ejecta blankets are close to rims, there are many debris flows within craters, and the terrain is softer and more rolling. These latter characteristics are all suggestive of the presence of ice in the subsurface down to depths of about 2 km [32].

Two unique features of martian impact craters are consistent with the presence of ice. Craters with pitted central peaks, even down to small craters, are

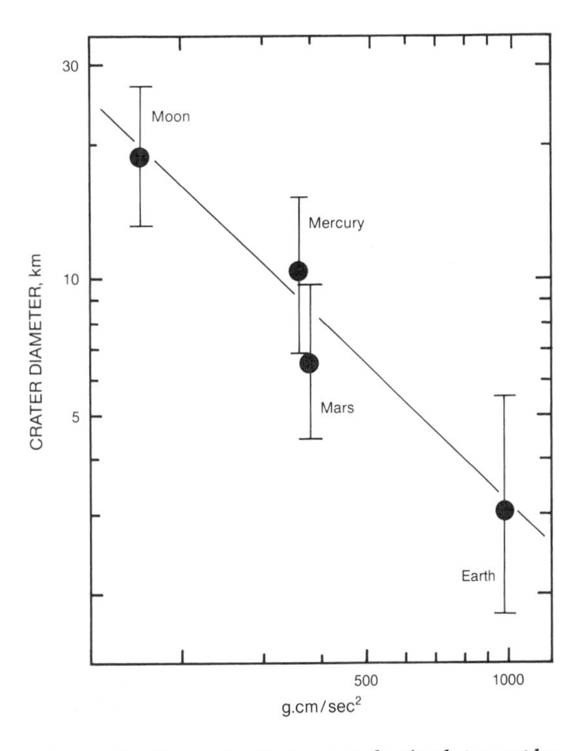

Fig. 4.3.4. *The crater diameter at the simple-to-complex crater transition on the Moon, Mercury, Mars, and the Earth is a simple function of gravity. The transition at Mars is at a lower diameter than for the other bodies, probably due to the effect of subsurface volatiles enhancing the change to complex crater form. After Pike R. J. (1988) in Mercury (F. Vilas et al., eds.), p. 231, Fig. 24, Univ. of Arizona, Tucson.*

common, although very rare on the Moon or Mercury. The lobate ejecta blankets look like mud flows. Target rheology has some influence for the smallest craters. Once the impacts are large enough to form basins, gravity and target strength are overwhelmed. Hydrodynamic effects predominate over these parameters, and wave phenomena produce the dramatic rings.

The controversy over relaxation of craters and basins on icy surfaces began with early predictions that icy satellites would be unable to maintain topographic relief over long periods [33]. However, many basins and craters were subsequently found on the icy satellites. Craters persist on icy surfaces for periods comparable with the age of the solar system, although some crater relaxation has occurred on the saturnian and uranian satellites [34]. Simple craters appear to be 30-40% shallower on the icy satellites of Saturn and Uranus than those on the silicate surfaces of the inner planets. This effect must be due principally to the nature of the surface. The spectacular development of central peaks, for example, on Mimas, indicates that floor rebound is more readily achieved on icy surfaces due to their lower viscosity. Incidentally, this provides additional evidence in favor of fluidized models for central peak, peak ring, and ringed basin development. The change from simple to complex crater morphology occurs at smaller diameters on icy surfaces, again consistent with the expected differences compared to silicate surfaces. Viscous relaxation occurs only for large structures. Thus, the spectacular crater Herschel (140 km in diameter and 10.5 km deep) on Mimas is unrelaxed, but "Tirawa" (350 km in diameter, 7.5 km deep) on Rhea has relaxed about 7 km [35]. Relaxation of craters differs when the crater diameter is a significant fraction of the radius of the satellite. It appears that most viscous relaxation of icy craters occurs at an early stage in the history of the basin, when the stresses are high, and that subsequently there is little relaxation over most of geological history [36].

The effects of cratering on icy satellites are not well understood, which is not particularly surprising since we are only beginning to obtain a thorough understanding of cratering on the rocky planets [37]. Theoretical considerations, based on ice viscosities, should indicate significant relaxation. This seems to occur only for large craters (e.g., >60-km diameter on Ganymede and Callisto), whereas the effect might be expected for craters as small as 20 km in diameter. Clearly some other factor is operating, perhaps a "stiffening" of the ice by a rocky component [38].

4.4. IMPACTS AND PLANETARY OBLIQUITIES

A list of planetary obliquities is given in Table 4.4.1. They constitute the best evidence for the collision of the planets with massive bodies, up to Earth-sized, in the early solar system. It is difficult to account for the large variations in the tilts of the planets by other mechanisms. Although one might expect, on the basis of the planetesimal hypothesis, that the terrestrial planets might be tilted, what is remarkable is that three of the four giant planets show significant tilts relative to the plane of the ecliptic. Saturn is tilted at 26.7°, Neptune at 29°8′, and Uranus at 97°, while even Jupiter has a 3.1° tilt. The best model solution for a large impact capable of tilting Uranus on its side calls for an Earth-sized impactor [39]. One major constraint is that very little mass finishes up in orbit as a result of the large collision; the uranian rings and satellite systems are very small, consisting of only about 1.05×10^{-4} of the mass of Uranus.

In the planetesimal hypothesis for the evolution of the solar system, such massive impacts are a universal phenomenon. They occur at all stages, be-

TABLE 4.4.1. Planetary obliquities and orbital inclinations.

	Inclination of Equator to Orbit	Inclination of Planetary Orbit to Plane of the Ecliptic	Eccentricity
Mercury	0°	7°00′	0.206
Venus	~177°	3°24′	0.007
Earth	24°25′	Defines ecliptic	0.017
Mars	25°2′	1°51′	0.093
Jupiter	3°1′	1°18′	0.049
Saturn	26°7′	2°29′	0.053
Uranus	97°0′	0°46′	0.046
Neptune	29°8′	1°47′	0.012

Data from *Proc. Lunar Planet Sci. Conf. 9th*, vol. 3, frontispiece.

ginning with the accretion of grains, meter-, and kilometer-sized bodies in the midplane of the nebula, and finishing with the large collisions responsible for knocking Uranus on its side, for tipping Venus over and causing it to rotate backwards, for depleting the silicate mantle of Mercury, and for forming the Earth's moon. The important observation in the context of the present work is that impacts are ubiquitous. Most of the planets are displaced from the plane of the ecliptic by varying degrees. The random tilts of the planets appear as a natural consequence of formation of the planets by accretion of a hierarchy of planetesimals. In this process, the largest collisions occur toward the end of the accretionary events. Pluto and its anomalously large satellite, Charon, are not only in an eccentric and highly inclined orbit, but are in synchronous rotation in a plane that is close to 90° to the plane of the ecliptic. Such a curious situation is a likely result of a massive collision.

No general laplacian theory accounts for the observed obliquities, nor indeed predicts that the planets should be significantly tilted. It must be judged fortunate that Laplace did not have access to too much detail. Otherwise he might not have perceived the broad overview that resulted in the concept of the solar nebula.

4.5. FORMATION OF PLANETARY NEBULAR DISKS

The formation of the regular satellites (i.e., those in prograde orbits in the equatorial plane) of the giant planets raises several interesting questions. Their orbital characteristics are too regular to suppose that they represent captured planetesimals. In contrast, the many satellites in nonequatorial or retrograde orbits are more reasonably explained as captured objects. The regular satellites have often been regarded as analogues for the formation of the solar system. Accordingly, they might be expected to provide some insights into the formation of the planets themselves. A basic question in this context is whether they formed from a disk left over following accretion of the parent planet or was the disk spun out or knocked out following the formation of the planet?

Here the concept is explored that the regular satellites form from planetary subnebulae as second-order features rather than being a natural outcome of giant planet formation. There are some good, although not compelling, reasons that can be advanced in support of this viewpoint. In the giant gaseous protoplanet hypothesis, in which the planets condense from fragments of the nebula, residual disks of solar nebula composition might be expected to form. The preferred model for the formation of the giant planets starts with the accretion of planetesimals into rocky cores. When these reach about 10 Earth masses, gravitational collapse of H and He rapidly completes the planet. Whether an extended disk remains as a remnant of the accretion process, whether one is spun out from the growing planet, or whether a disk forms from material knocked out by collisions, is uncertain. What is clear is that the process or processes are different for each of the giant planets. This makes the operation of some regular inevitable Laplacian theory for the formation of planetary systems less likely than the probability that chaotic events dominate the end result. If the disks from which the regular satellites of Jupiter, Saturn, and Uranus form are spun out by collisions, the analogy often drawn between the planetary system and the regular satellite systems is in fact not very close.

In the collisional model, there is an obvious difficulty concerning the regular satellites. They are in equatorial orbits. A massive collision, analogous to that which produced the terrestrial moon, might be expected to produce a nonequatorial orbit as is the case for our satellite. The example of the uranian system, however, is instructive. Both the rings and satellites are in the equatorial plane, which is tilted at 90° to the plane of the ecliptic. Various scenarios are possible. Uranus presumably formed in a normal orientation, accreting first a 10-15-Earth-mass core that then collected ices and whatever gas was available in the later stages of nebular evolution. As noted earlier, it is likely that these events took appreciably longer [40] than the formation of Jupiter. In this model, the dispersion of the nebula was well under way before the cores of Uranus and Neptune had grown large enough to cause gravitational collapse of gases and ices. If the rings and satellites had formed when Uranus was oriented normal to the plane of the ecliptic, it is unlikely that they would reorient themselves so perfectly. Thus, it is possible that the uranian satellite-forming disk resulted as a by-product of the collisional event. This ring and satellite-forming event must have occurred quite late in the accretionary history of Uranus. The reason for this is that enough time (10^7-10^8 years) is needed for the requisite Earth-sized body to form before it could collide with Uranus and knock it on its side. Thus, it is reasonable to suppose that the uranian rings and satellites formed subsequently to the collisional event that produced the present obliquity. On this basis, it is assumed that the material from which the rings and satellites formed was placed in equatorial orbit either by the major collisional event or by another large impact.

From the preceding discussion, it appears that satellites do not necessarily represent miniature solar systems, nor are they an inevitable consequence of planetary formation. Like rings, they are decorative, but not an essential accompaniment to the formation of the solar system (see section 6.2 for a further discussion of subplanetary nebulae).

4.6. ADDITION AND REMOVAL OF PLANETARY ATMOSPHERES

There are no traces of primordial atmospheres on the terrestrial planets. The most likely candidate in the solar system to have retained one is Titan. The planetesimal hypothesis does not make many predictions about the state of the primitive atmosphere of the Earth, except that accretion most likely took place in a gas-free environment. It is a thesis of the models discussed here that the inner planets accreted on timescales of 10^7-10^8 yr, measurably longer than the lifetime of the gaseous solar nebula (10^6 yr). Atmospheric gases are accordingly derived from what is brought in by later planetesimals or comets, not from the primitive nebula. In this model, the terrestrial atmosphere was entirely secondary, rather than being in any way representative of primary condensation or accretion from the solar nebula (see section 5.10). Accordingly, it is perhaps not surprising that the atmospheres of the inner planets are so different among themselves, both in elemental and isotopic abundances, and from our ideas about the composition of the primordial solar nebula.

However, another major factor enters to complicate the problem further. Degassing during accretion may have produced a primitive atmosphere. Such an atmosphere, however, is likely to have been removed during later large impact events, such as that responsible for the backward rotation of Venus or that which most likely was responsible for the formation of the terrestrial moon. Similar events may be postulated for Mars, since the present tenuous atmosphere of that planet is entirely secondary. "The most likely history is that the evolving planetary embryos suffered several giant impacts each, and lost their primordial atmospheres repeatedly during their accretionary phase" [41].

The prospect that the rare gas content of the present atmospheres of the terrestrial planets can be related back to that of the nebula in any meaningful way must remain as one of the casualties of our improved understanding of the effects of large impacts. Such events are likely to remove atmospheres by shock waves formed during the passage of the impactor through the atmosphere, or by entraining atmospheric gases in ejecta shocked to escape velocities [42]. These mechanisms are likely to remove only rather small amounts, comparable to the volume of atmosphere encountered by the impacting object or the ejected material. However, in large impacts, the vapor cloud produced by the impact interacts with the atmosphere. Much of the kinetic energy of the projectile is transferred to the vaporized material. If the impact is large enough that the expanding vapor cloud exceeds the planetary escape velocity, and if the mass of impact-generated vapor exceeds the mass of the atmosphere, all the atmosphere above a tangent plane may be removed [43].

This removal process is clearly more efficient for smaller bodies, an illustration of the law that the poor get poorer. Thus, the impact of a 3-km-diameter object on Mars would remove all the atmosphere above a plane tangent to the surface, but a 13-km-diameter planetesimal would be required to accomplish the same effect on the Earth, and a 70-km-diameter body would be needed to remove such a slice of the present atmosphere of Venus. Clearly, such processes would be effective before the close of the period of heavy bombardment about 3800 m.y. ago. Since that time, the atmospheres of the terrestrial planets have been stable against massive removal. The giant impact probably responsible for the formation of the Moon is expected to have been of sufficient magnitude to have caused the removal of any preexisting terrestrial atmosphere, as well as causing melting of the mantle [44]. The entire martian atmosphere could be removed completely by such processes in a period of 100 m.y. during the period of the early heavy bombardment.

If all the atmospheres of the inner planets are secondary, and have been subject to random removal by impacts, then it will be difficult to discern any information about the initial nebula from them [45]. Instead, the evolution of each planetary atmosphere is unique. While this is interesting as a scientific problem in its own right, the present atmospheric composition will only indirectly, or perhaps not at all, provide insights into the origin of the individual planets.

Relevant to this discussion is the role of comets as potential suppliers of water and other volatile compounds (sections 3.10 and 5.10). Scenarios in which the terrestrial planets accrete from "wet" CI-type planetesimals do not seem very likely. The terrestrial planets are assumed to have accreted from rather dry planetesimals derived from the inner nebula after the volatiles had been driven out to a "snow line" in the vicinity of Jupiter by early intense solar activity. Some volatiles may have been trapped in planetesimals that had grown before the early intense solar activity depleted the inner nebula in gas, dust, and small particles. Such planetesimals may be the principal source of water (perhaps locked in hydrated minerals) for the terrestrial planets, but it is not clear whether this source is likely to be adequate to account for the terrestrial volatile budget. Another possibility is that most of the water in the terrestrial hydrosphere was accreted as a late veneer from comets, which, because of their eccentric orbits, are not so restricted by the "Jupiter barrier," which inhibits the transfer of bodies from the outer to the inner solar system [46].

The potential amount of water that might be added to the Earth can be calculated from the estimated impact flux on the Moon for the period 4.4-3.8 b.y. There is some disagreement about this [47] and the assumptions about the flux rate are critically dependent on whether there was a late lunar "cataclysm." If the assumption is made that 10% of the impactors are comets with 50% water, then it is possible to accrete the mass of water in the terrestrial oceans solely during the final stages of the heavy bombardment. Whether this is a reasonable assumption, however, may be open to question. The bodies responsible for the late bombardment may be debris from a late collisional event with a large object in the inner solar system, and have nothing to do with comets.

Comets impact the inner planets at much higher velocities than planetesimals that originate in the inner solar system, so they are more likely to erode an atmosphere than deposit one [48]. The total absence of water on the Moon, even at parts per billion levels, following 4400 m.y. of impacts of comets with 50% H_2O, suggests that on smaller bodies at least, not even a drop is retained. Even though the Moon is small, with a low escape velocity, one might have expected some of the heavier volatile elements to be present. The only evidence is the very minor alteration observed in the notorious rusty rock, 66095, from the lunar highlands [49].

The high velocities of cometary impacts, on the contrary, may make them prime eroding agents of planetary atmospheres, as noted above in the discussion of the role of the expanding vapor clouds resulting from the impact in removing massive amounts of existing atmospheres [50]. CI chondrites are an unlikely source in the volatile-depleted inner solar system, and would deposit primordial abundances of the other volatile elements. Comets possess the advantage of being 50% water ice compared with 17% for CI chondrites. A recent estimate is that the Earth acquired a net gain of 0.3-1.0 oceanic masses during the period between 4.4 and 3.8 aeons ago. Venus is estimated to have gained a similar inventory, but Mars only received a layer of water 10-100 m deep, since it was more susceptible to impact erosion [51]. All these conjectures depend heavily on the assumed influx, but this in its turn, is beset with uncertainty, as we shall see in the next sections.

4.7. LUNAR CRATERING AND ORIGIN

A major insight from the study of the Moon was the evidence for massive impact cratering early in the history of the solar system. One of the most striking features of the lunar surface is the evidence of meteorite impacts at all scales, from large basins, hundreds to over a thousand kilometers in diameter, with concentric rings of mountains, down to micrometer-sized pits caused by micrometeorites impacting on surficial grains. Although the larger craters were long thought to result from volcanic activity, their origin due to the impact of asteroids, comets, and meteorites was unequivocally established only about 25 years ago, following much controversy over impact vs. volcanic interpretations of the face of the Moon [52,53].

In contrast to the Earth, the major structural features of the lunar crust were formed by these events, accounting for the prevalence of circular landforms (Fig. 4.7.1). These range from the mountain rings to the circular basins filled with dark basaltic lava. The complexity of overlapping multiring basins and ejecta blankets produces gridlike patterns sometimes mistakenly attributed to thermal cooling or internal tectonic processes.

The record preserved on the lunar surface indicates that, following lunar accretion and crustal solidification at about 4440 m.y. ago, a population of large objects (up to a hundred kilometers in diameter) struck the Moon over the next 500 m.y (Table 4.7.1). These impacts formed at least 80 ringed basins with diameters greater than 300 km on the solid crust (Fig. 1.11.2). Smaller colliding bodies created an additional 1000 craters in the range 30-300 km in diameter, and there is a host of smaller-sized craters [54]. This bombardment declined steeply in intensity after the formation of the final ringed basins (Imbrium and Orientale) at about 3850 m.y. ago. Similar heavily cratered surfaces on Mercury, Mars, and the satellites of the giant planets show that these catastrophic events were not unique to the Moon. Since

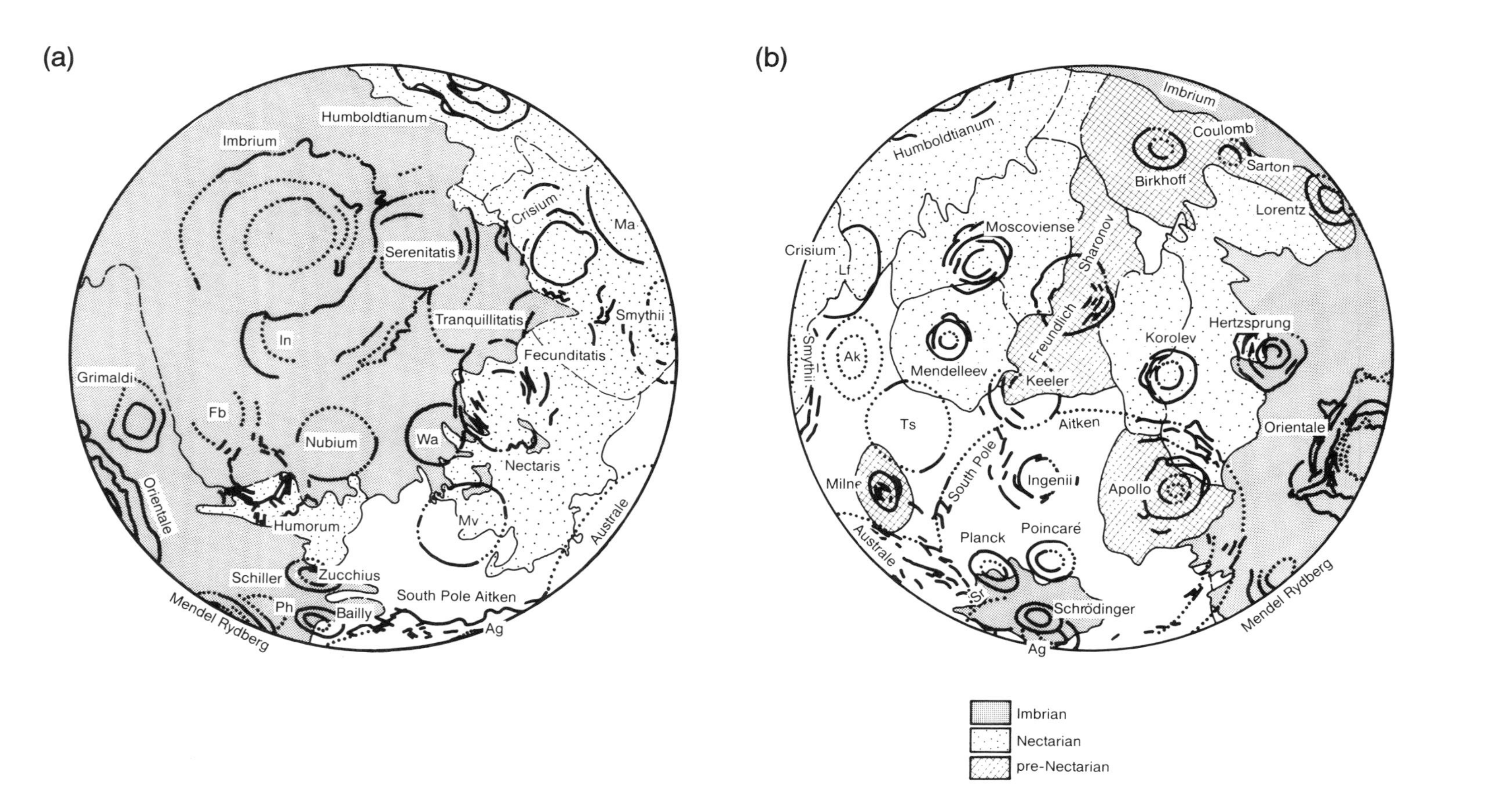

Fig. 4.7.1. *The distribution of multiring basins on the Moon: **(a)** lunar nearside; **(b)** lunar farside. (Courtesy D. E. Wilhelms, USGS.)*

TABLE 4.7.1. Estimates of ages of geological provinces on the Moon.

Geologic Province	Crater Density Relative to Average Lunar Mare	Estimated Crater Retention Age (aeons) Minimum	Best Estimate	Maximum	Regolith Thickness (m)
Tycho	0.1	0.09	0.3	0.6	—
Aristarchus	0.2	0.5	0.9	1.3	—
Copernicus	0.3	0.8	1.5	2.2	—
Mare Crisium	0.5	2.1	2.6	3.5?	—
Mare Imbrium	0.50	2.2	2.6	3.5	—
Whole Mare Serenitatis	0.65	2.8	3.1	3.7	—
S. Oceanus Procellarum	0.75	3.0	3.3	3.7	3
Mare Fecunditatis	0.93	3.2	3.4	3.8	—
Average of front-side maria	1.00	3.3	3.5	3.8	5
Mare Orientale	1.10	3.3	3.5	3.8	—
Mare Tranquillitatis	1.30	3.5	3.6	3.9	5
Mare Humorum	1.50	3.6	3.7	3.9	—
Mare Nubium	2.5	3.8	3.9	4.0	—
Orientale Basin and ejecta	2.5	3.8	3.9	4.0	—
Fra Mauro	3.0	3.9	4.0	4.1	8
Fill in Schrödinger Basin	5.1	4.0	4.1	4.2	—
Fill in Mendeleev Crater	9	4.1	4.2	4.2	—
Front-side highlands	10	4.1	4.2	4.2	10-15
Heavily cratered highlands and basins	32	4.3	4.4	4.5	megaregolith
"Pure" highlands	36	4.3	4.4	4.5	megaregolith

From Taylor S. R. (1982) *Planetary Science: A Lunar Perspective*, p. 102, Table 3.3, Lunar and Planetary Institute, Houston.

it is probable that the Earth also suffered this ordeal, this helps to explain the absence of rocks older than 3.9 b.y. on this planet. The impact of all these bodies at high velocities resulted in a zone of fractured and brecciated rubble on the Moon, termed the megaregolith, of uncertain depth.

Meteorite impacts on the Moon, since the termination of the great bombardment about 3850 m.y. ago, have occurred at a much slower rate, roughly consistent with the recent observed flux on the Earth (see section 4.15). The youngest such major event on the Moon was the formation, about 100 m.y. ago, of the crater Tycho, 85 km in diameter, probably due to the impact of an object about 3-5 km in diameter. Material ejected during this impact forms the bright rays that extend across the visible face and are a spectacular feature of the full Moon, particularly when viewed through binoculars. This large impact on the Moon is not far removed in time from the similar catastrophe probably responsible for the Cretaceous-Tertiary boundary event on the Earth.

The origin of the Moon is discussed more fully in sections 6.17 and 6.18, but some comments are appropriate here. The Earth-Moon system has an anomalously high angular momentum compared to other planets [55]. This excess angular momentum cannot arise though multiple small impacts, which could contribute an excess of probably not more than 10%. However, one large impact could account for the observed excess. A *minimum* mass of that of Mars is required for such an impactor to accomplish this [56]. The lunar orbit is nonequatorial, being inclined at 5.1° to the plane of the ecliptic, while the axial plane of the Earth is inclined at 24.4°. Both these facts are explicable in the context of giant impacts, but appear as unsolved problems in other versions of early Earth-Moon history.

Pre-Apollo theories for the origin of the Moon all failed for various reasons following the sample return. Capture of an already-formed Moon from elsewhere is improbable on dynamic grounds. Formation of the Moon from a ring of silicate debris produced by disruption of incoming planetesimals, with their iron cores accreting to the Earth, has been shown to be unlikely, but perhaps not impossible. Even totally molten planetesimals may not be disrupted by tidal forces during grazing collisions [57], but further work may be needed [58].

Fission hypotheses, deriving the Moon from the terrestrial mantle (popular since they provided a low-density Moon, but testable following access to lunar samples) have failed to account for the significant chemical differences. None of these theories accounted for the strange lunar orbit and for the high angular momentum of the Earth-Moon system, a rock on which all hypotheses foundered.

The single-impact theory resolves many of the problems associated with the origin of the Moon and has become nearly a consensus [59]. The physics of such a massive impact have been studied by several groups. The optimum conditions are for the impactor to have a mass 0.14 that of the Earth mass and an impact velocity of 5 km/s [60]. This body, as well as the Earth, is assumed to have differentiated by the time of the collision into a metallic core and silicate mantle. In one particular simulation, the collision disrupts the impactor, much of which goes into orbit about the Earth [61] (Fig. 4.7.2). The gravitational torques, due to the asymmetrical shape of the Earth following the impact, are sufficient to accelerate material into orbit. Another effect that promotes material into orbit is the acceleration away from the Earth due to expanding gases from the vaporized part of the impactor. Following the impact the metal core of the impactor separates from the silicate mantle, and the mantle material is accelerated and the core decelerated relative to the Earth. The metallic core accretes to the Earth within about four hours. Whether the remaining material forms several moonlets or coalesces rapidly into the Moon is not known. These two models, however, have interesting consequences. In the first case, considerable cooling will occur before the Moon forms, so a half-molten Moon might form. In the second case, a totally molten Moon could result.

Both scenarios account for the presence of a deep magma ocean. Realistic minimal depths for the magma ocean based on the requirements to form the 100-km-thick aluminous lunar highland crust are about 500 km or about half lunar volume. Although early estimates of the allowable change in radius due to cooling placed geophysical restrictions on the depth of the magma ocean to 200 km [62], revisions allow a depth of 640 km [63], in accordance with the minimal geochemical estimates. Whether the whole Moon was melted is still uncertain.

Only a small amount of material from the Earth's mantle eventually ends up in the Moon. The models all indicate an amount less than 16%. The lower FeO content of the upper terrestrial mantle (8%), compared to that of the bulk Moon (13%), also suggests that the contribution from the Earth's mantle may be small.

The past history of the lunar orbit could shed some additional light on lunar origins, but attempts to determine from tidal calculations whether the Moon was once much closer to the Earth involve many uncertainties [64]. This is due to the changing distribution of shallow-shelf seas, so the paleotidal regime of the Earth is not a serious constraint on lunar origins at times so far distant as 4450 m.y. ago. However, surprisingly little variation from present Earth-Moon distances is evident throughout the geological record. Ancient tidal deposits, for example in the 3.3-b.y.-old Moodies Group, South Africa, appear to be of similar magnitude to present day examples [65]. In a younger example, the Elatina Proterozoic sediments (about 680 m.y. ago) show banding consistent with a Moon-Earth distance only about 4% less than the present value [66].

The existence of another planet-sized body, with its own geochemical history, involved in the origin of the Moon adds some complexity to the problem. However, a limited number of reasonable endmember scenarios can explain most of the geochemical problems associated with the composition of the Earth and the Moon. The siderophile abundances in the impactor's mantle and the size and composition of the impactor's core can be derived by considering the constraints imposed by the siderophile element depletion patterns in the Earth and Moon. We adopt the following conditions for the impactor. It was 0.14 Earth masses and differentiated into a metallic core and silicate mantle. It formed in the same part of the solar nebula as the Earth (between Venus and Mars), in order to account for the low relative impact velocity (5 km/s) and the similarity in oxygen isotopes.

The impact event was sufficiently energetic to vaporize much of the material that went to make up the Moon. This naturally explains such unique geochemical features as the bone-dry nature of the Moon and the extreme depletion of very volatile elements. Bismuth and Tl are depleted, for example, by factors of 200 relative to cosmic abundances. However, if the impactor formed in the region of the terrestrial planets, then it probably had already been somewhat depleted in volatile elements, in common with the rest of the inner solar system. This first-stage depletion must have occurred very early, as is shown by the very primitive $^{87}Sr/^{86}Sr$ ratios observed in lunar highland samples (LUNI values are close to BABI; section 3.7). It seems safe to conclude that the impactor mantle was not composed of material of CI composition, but had undergone a similar volatile-element depletion to that general in the inner solar system.

A final consequence of the giant impact model for lunar origin is that the event is energetic enough to melt the terrestrial mantle. Such melting appears to be an inevitable consequence of the accretion of large planetesimals. In contrast to the infall of fine dust where the energy is largely lost by radiation from the planetary surface, the thermal energy of large

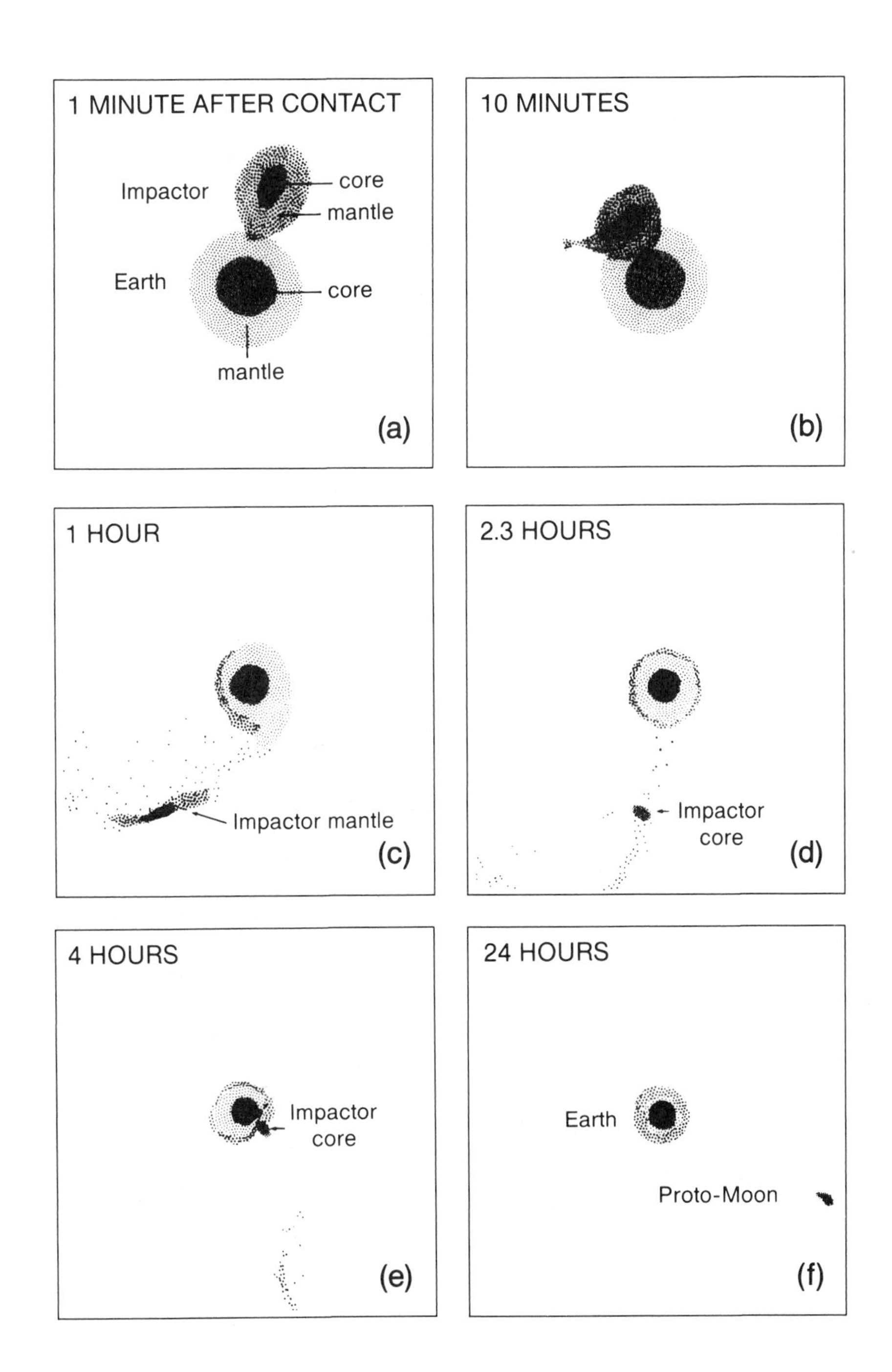

Fig. 4.7.2. *Computer simulation of the formation of the Moon by the oblique collision of a 0.14-Earth-mass body with the Earth at a velocity of 5 km/s. Both the Earth and the impactor have differentiated into a metallic core and a silicate mantle. Time following the impact is shown in the boxes. After the collision (a,b), the impactor is spread out in space (c), but the debris clumps together through gravitational attraction. The iron core of the impactor separates from the silicate mantle (d) and accretes to the Earth (e) about four hours after the initial encounter. About 24 hr following the impact, a silicate lump of about lunar mass is in orbit around the Earth (f). This material comes principally from the silicate mantle of the impactor. (Courtesy A. G. W. Cameron and W. Benz, Smithsonian Astrophysical Institute.)*

bodies is buried deeply within the planet. The consequences of a terrestrial magma ocean are discussed in Chapter 5.

4.8. MERCURY

In many ways, the discussions over the impact history of Mercury epitomize the problem of interpretation of the cratering record. Most of the problems that beset the lunar record have surfaced again in Mercury. Table 4.8.1 lists crater densities on differing mercurian geological units [67,68] (see also section 5.3).

Earlier views that Mercury had fewer large basins than the Moon have proven erroneous; mapping has revealed that they are at least as plentiful as on the Moon. If the age of the Caloris Basin is placed at 3860 m.y. by analogy with Imbrium, then the heavily cratered mercurian highlands have an age of about 4200 m.y. The "volcanic" fill of Caloris cannot be distinguished from the age of the basin; possibly the fill is impact generated rather than volcanic [69]. An ancient pre-Tolstojan system of large basins, which apparently covered the mercurian surface, was mostly erased by the formation of the "intercrater plains," which obliterated craters less than about 500 km in diameter. Whether these plains formed by volcanic resurfacing or represent ejecta blankets is currently in dispute and unlikely to be resolved until future missions to Mercury. This debate, along with that over the origin of the younger "smooth plains," bears a considerable resemblance to the arguments over the lunar Cayley Plains, eventually shown to be basin ejecta, rather than of volcanic origin.

This reviewer is impressed by the lack of albedo contrast on Mercury, in contrast to the naked-eye distinction between the lunar basaltic maria and the highlands. Since basaltic volcanism is the general and voluminous feature of the inner planets, if the intercrater and smooth plains on Mercury are volcanic, then they are unique. Silicic volcanism, which might produce low-albedo deposits, seems an unlikely candidate on Mercury and may indeed be a mainly terrestrial phenomenon [70,71]. It seems unwise to use petrological models derived from other bodies to interpret the surface composition of Mercury, in view of the probable violent early history of Mercury. If much of the silicate mantle were removed by a giant impact, it is not clear that the composition of the reassembled debris would have been similar to that of the original mantle composition, or indeed to that of the mantles of the other terrestrial planets. Accordingly, magmas derived from the mercurian mantle may indeed be unique.

TABLE 4.8.1. Crater densities on some mercurian geological units.

Unit	Area ($10^3 km^2$)	Number ($D \geq 20$ km)	Density ($10^{-5} km^{-2}$)
Smooth plains	494	12	2.4 ± 0.7
Smooth plains	528	15	2.8 ± 0.7
Smooth plains	1022	27	2.6 ± 0.5
Caloris floor	280	11	3.9 ± 1.2
Bach Basin	120	5	4.2 ± 1.9
Caloris Basin	360	21	5.8 ± 1.3
Beethoven Basin	440	31	7.0 ± 1.3
Tolstoj Basin	413	35	8.5 ± 1.4
Dostoevskij Basin	360	45	12.5 ± 1.9
Surikov Basin	120	19	15.8 ± 3.6

Data from Spudis P. D. and Guest J. E. (1988) in *Mercury* (F. Vilas et al., eds.), p. 154, Univ. of Arizona, Tucson.

The later smooth plains are likewise the subject of controversy. The case here for a volcanic origin is stronger [72], but if so, the lavas are distinct from conventional basalts. Perhaps the mantle of Mercury is more refractory than that of the Moon, Venus, Mars, or the Earth. If "Mercury underwent large-scale volcanic resurfacing after the Caloris basin impact" [72], then this indicates that the mercurian mantle is distinct in composition to that of the other terrestrial planets; as noted above, this is a conceivable effect resulting from a catastrophic early collision.

4.8.1. Crater Populations

The general conclusion from crater counting studies is that the crater size-frequency distribution is similar to that of Mars or the lunar highlands, but that Mercury is deficient in craters smaller than 50 km. This difference is apparently due to the formation of the intercrater plains late in the cratering history. The crater population on Mercury differs from that of Callisto and Ganymede. This is interpreted to indicate that the late heavy bombardment in the inner solar system differed from that at Jupiter. If so, cratering on the rocky planets was restricted mostly to objects left over in the inner solar system following planetary accretion, rather than to a wide-ranging flux of bodies or comets from the outer regions of the system. Jupiter and Saturn are barriers to both the exchange of rocky material from the inner solar system and icy planetesimals from the outer solar system. Although planetesimals may be pumped up to become Jupiter crossers, most will be ejected from the solar system on timescales of the order of a million years [73].

One view is that there is a general similarity in the size-frequency distribution of crater populations for Mercury, the Moon, and Mars [74]. This, coupled with the differences observed for the satellites of the outer planets, is interpreted to indicate that a common set of bodies in heliocentric orbits in the inner solar system was responsible for the late heavy bombardment, although the impact velocities were higher at Mercury than for the Moon.

An alternative explanation is that there was a set of Mercury impactors ("vulcanoids") in orbit between the Sun and Mercury that could have caused the late heavy bombardment on Mercury [75]. No such objects greater than 50 km in diameter have been detected, but they would have been expected to be swept up within 10^9 years. Vulcanoids would impact Mercury with much lower velocities than bodies in heliocentric orbit and might have been responsible for forming the early, now nearly erased, basins.

The similarity in the size-frequency distribution of the inner solar system cratering record argues, as noted above, for a common Sun-orbiting population of impactors. This conclusion has the advantage that the date established from the lunar cratering record for the end of the heavy bombardment, marked by the formation of the Imbrium and Orientale basins at about 3850 m.y., can be extended to Mercury. However, this interpretation is debatable [76].

Whether a large population of impactors can be stored within the inner solar system for up to 500 m.y. after the main stage of planetary accretion is a major problem, that will reappear in this text. An alternative view is that a few larger (>lunar sized) bodies persisted as "loose cannons" in the inner solar system. The debris resulting from their destruction at random intervals may have provided most of the observable cratering on the Moon, Mars, and Mercury.

4.8.2. Origin

The high density of Mercury has had a fatal attraction for solar system modelers and inventors of grand unified theories for the origin of the solar system. This is because it anchors one end of a sequence extending from the high-density inner planets out to the low-density outer solar system bodies. Just as the apparent match between the density of the Moon and that of the Earth's mantle provided a tantalizing and ironic clue to lunar origin and misled workers from George Darwin (1845-1912) onwards [77], so the density of Mercury was another trap [78]. Of the various explanations that have been offered, high-temperature evaporation can be ruled out (see section 5.3). The current model involving a large impact accounts not only for the loss of the silicate mantle, leaving the metal core and some of the mantle to reaccrete, but it also accounts for the strange orbit of Mercury with respect to the plane of the ecliptic. The size of the body required to cause major fragmentation, disruption, and stripping of the mantle of Mercury is about 0.20 Mercury masses with an impact velocity of 20 km/s [79] (Fig. 4.8.1). It is an outcome of these calculations that bodies in the zone of growth for both Mercury and Mars will be subjected to higher-velocity collisions than those of Earth and Venus [80]. Accordingly, accretion of these smaller planets is likely to be more violent than for the larger terrestrial planets [81].

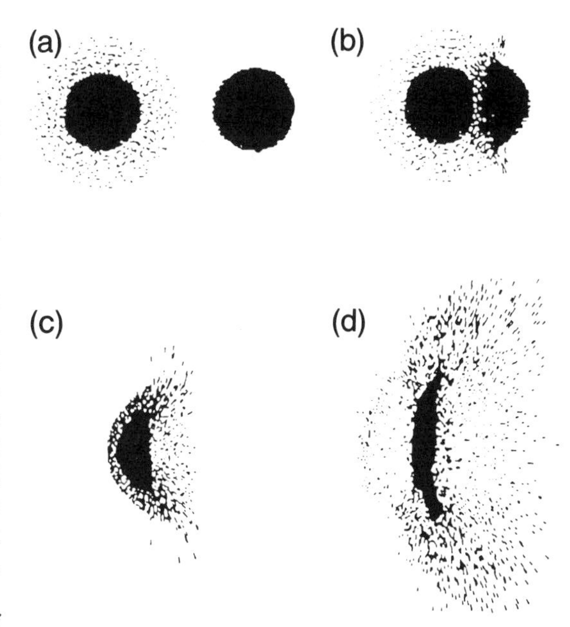

Fig. 4.8.1. *A computer simulation of a giant impact on Mercury at (a) 1.0 min, (b) 2.3 min, (c) 7.7 min, and (d) 41.7 min from impact. (Courtesy A. G. W. Cameron and W. Benz, Smithsonian Astrophysical Institute.)*

4.9. THE EARLY INTENSE BOMBARDMENT

The most spectacular landforms discovered by spacecraft are probably the giant impact basins such as Orientale. However, like the martian landscapes of Vallis Marineris and Olympus Mons, which are so impressive when photographed from orbit, they would be less striking on the ground [82]. Early workers, notably Gilbert, Baldwin, and Urey, all drew attention to the circular Imbrium structure on the Moon and postulated its origin as due to collision with a large asteroid. The resulting discovery of the Orientale Basin [83] from the Lunar Orbiter photographic survey provided lunar mappers with a nearly pristine example of a ringed impact basin. Rapid progress followed as it was realized that the lunar surface was effectively saturated with large (>300 km in diameter) basins [84].

These provided the first direct evidence that the massive bombardment was an order of magnitude more intense than even the long-visible cratered surface of the Moon would suggest. Missions to both Mars and Mercury revealed similar landscapes, while the surveys of the outer satellites showed the ubiquitous extent of cratering throughout the solar system. Radar mapping of Venus showed that even that cloud-shrouded atmosphere (with a surface pressure of 100 bar) provides no protection from large impacting bodies. The venusian surface, however, is less than 1 b.y. old, and there is no sign of an ancient crust nor any evidence of ringed basins preserved on this relatively young surface.

The best estimate of early flux rates comes from the lunar surface [84]. Is this rate similar throughout the solar system? Is the Moon a special case (see section 4.11 on the lunar cataclysm)? Did it encounter a ring of planetesimals in Earth orbit, the tail of the accretionary bombardment to the Earth, or was it struck by a population of late impactors? The problem of storing a local population of basin-forming impactors in close proximity to the Earth-Moon system for half a billion years is considerable [85] and the dated late lunar bombardment may represent an essentially stochastic event. The general similarity of the early cratering history for the Moon, Mars, and Mercury shows that the heavy cratering was not restricted to the Earth-Moon system, but was widespread throughout the inner solar system, although perhaps not coeval.

There are reasonably well-documented differences in cratering rates between the inner solar system, and those recorded on the jovian, saturnian, and uranian satellites. The rates established from their surfaces, in turn, appear to differ among themselves, particularly for the saturnian system. The apparently differing fluxes in different parts of the solar system makes any attempt to establish a systemwide chronology based on cratering a doubtful proposition.

4.10. LARGE IMPACT BASINS

Large basins are of two types: peak-ring (or two-ring) and multiring basins. Although frequently considered to be closely related, they are probably formed by different mechanisms. Peak-ring basins appear to be formed in the same way as central peaks, except on a larger scale. The peak-rings are symmetrical in cross section and form inside the true crater rim. In contrast, the concentric rings in multiring basins are asymmetric in cross section and some form outside the main crater rim (Tables 4.10.1 and 4.10.2). Many hypotheses have been advanced to account for the large ringed basins, and "the origin of multiple rings, both internal and external, has nearly as many interpretations as there are investigators" [86].

Multiring basins were recognized comparatively late in planetary science, only coming to general notice just before the Apollo missions due to the superb photographs of the Orientale basin obtained by Lunar Orbiter IV in May 1967. By this time, impact theory was sufficiently advanced to provide the only rational explanation; thus we have been spared having to consider internal origins for these structures on the Moon, planets, and satellites. Regarding the origin of these structures on Earth, some skepticism has remained among geologists. The two terrestrial examples of large impact structures mostly cited as being due to internal processes are Sudbury (Canada) and Vredefort (South Africa). The discovery of the high-pressure phase, stishovite, in pseudotachylite veins in the 140-km-diameter Vredefort structure [87] should finally convince even the most hardened proponents of an internal origin of the reality of an impact origin [88,89].

No large basins have been identified on Venus, the present surface being probably too young to preserve any record. The large basins on the Moon and other planets formed by about 4 b.y. ago; most subsequent impact events were on too small a scale to form these features. There are fewer preserved martian basins (Table 4.10.3) because of the high level of geological activity on Mars that has caused major resurfacing. A list of basins on the satellites of the giant planets is given in Table 4.10.4.

The details of the formation of two-ring and multiring basins remains somewhat controversial. Many hypotheses have been advanced to account for these superb landscape forms. These include "megaterrace" and "collapse," "nested crater," "rock tsunami," "elastic plate," and "oscillating peak" models. Most

TABLE 4.10.1a. Lunar basins, listed in order of increasing size and complexity.

Basin	Center (lat., long.)		Major Rings	
			Diameter (km)	Radius (km)
Bailly	67°S,	68°W	300	—
			150	—
Schrödinger	75°S,	134°E	320	—
			155	—
Schiller-Zucchius	56°S,	44.5°W	325	—
			165	—
Planck	57.5°S,	135.5°E	325	—
			175	—
Mendeleev	6°N,	141°E	330	—
			140	—
Birkhoff	59°N,	147°W	330	—
			150	—
Poincaré	57.5°S,	162°E	340	—
			175	—
Lorentz	34°N,	97°W	360	—
			185	—
Grimaldi	5°S,	68°W	430	—
			230	—
Korolev	4.5°S,	157°W	440	—
			220	—
Moscoviense	26°N,	147°E	—	275
			445	—
			210	—
Apollo	36°S,	151°W	505	—
			250	—
Coulomb-Sarton	52°N,	123°E	530(?)	—
			400	—
			180	—
Ingenii	34°S,	163°E	560(?)	—
			325	—
Hertzsprung	1.5°N,	128.5°W	570	—
			—	205
			265	—
			140	—
Freundlich-Sharonov	18.5°N,	175°E	600	—
			?	?
			?	?
Humboldtianum	61°N,	84°E	—	600
			600	—
			275	—
Mendel-Rydberg	50°S,	94°W	630(?)	—
			460	—
			—	155
			200	—
Serenitatis	27°N,	19°E	—	450
			740	—
			420	—
Keeler-Heaviside	10°S,	162°E	780(?)	—
			—	270
			—	170
Humorum	24°S,	39.5°W	—	?
			820(?)	—
			—	350
			—	280
			440	—
			325	—
Smythii	2°S,	87°E	—	480
			840	—
			—	330
			360	—
Nectaris	16°S,	34°E	860	—
			600	—
			450	—
			350	—
Australe	51.5°S,	94.5°E	880	—
			—	275
			—	95
Orientale	20°S,	95°W	—	650
			930	—
			620	—
			480	—
			320	—
Crisium	17.5°N,	58.5°E	1060	—
			—	450
			—	380
			635	—
			500	—
			380	
Imbrium	33°N,	18°W	—	900
			1160	—
			670	—
South Pole-Aitken	56°S,	180°	2500	—
			—	900

Data from Wilhelms D. E. (1987) *USGS Prof. Paper 1348*, p. 64.

TABLE 4.10.1b. Probable lunar basins, in order of increasing size.

Basin	Center (lat., long)		Apparent Diameter (km)
Pingré-Hausen	56°S,	82°W	300
Sikorsky-Rittenhouse	68.5°S,	111°E	310
Amundsen-Ganswindt	81°S,	120°E	355
Werner-Airy	24°S,	12°E	500
Balmer-Kapteyn	15.5°S,	69°E	550
Flamsteed-Billy	7°-8°S,	45°W	570
Marginis	20°N,	84°E	580
Al-Khwarizmi/King	1°N,	112°E	590
Insularum	9°N,	18°W	600
Grissom-White	44°S,	161°W	600
Lomonosov-Fleming	19°N,	105°E	620
Mutus-Vlacq	51.5°S,	21°E	690
Nubium	21°S,	15°W	690
Tsiolkovskiy-Stark	15°S,	128°E	700
Tranquillitatis	7°N,	40°E	800
Fecunditatis	4°S,	52°E	990
Procellarum	25°N,	15°W	3200

Data from Wilhelms D. E. (1987) *USGS Prof. Paper 1348,* Table 4.2.

TABLE 4.10.2. Multiring basins on Mercury.

Basin	Center	Age	Ring Diameters (km)*
Caloris	30°,195°	C	(630), 900, *1340*, (2050), (2070), (3700)
Tolstoj	-16°,164°	T	(260), 330, *510*, (720)
Van Eyck	44°,159°	T	150, *285*, (450), 520
Shakespeare	49°,151°	pT	(200), *420*, 680
Sobkou	34°,132°	pT	490, *850*, 1420
Brahms-Zola	59°,172°	pT	340, *620*, 840, (1080)
Hiroshige-Mahler	-16°,23°	pT	150, *355*, (700)
Mena-Theophanes	-1°,129°	pT	260, 475, *770*, 1200
Tir	6°,168°	pT	380, 660, 950, *1250*
Andal-Coleridge	-43°,49°	pT	(420), 700, 1030, *1300*, 1750
Matisse-Repin	-24°,75°	pT	410, *850*, 1250, (1550), (1990)
Vincente-Yakovlev	-52°,162°	pT	360, *725*, 950, 1250, (1700)
Eitoku-Milton	-23°,171°	pT	280, 590, 850, *1180*
Probable basins			
Borealis	73°,53°	pT	860, *1530*, (2230)
Derzhavin-Sor Juana	51°,27°	pT	*560*, 740, 890
Budh	17°,151°	pT	580, *850*, 1140
Ibsen-Petrarch	-31°,30°	pT	425, *640*, 930, 1175
Hawthorne-Riemenschneider	-56°,105°	pT	270, *500*, 780, 1050
Possible basins			
(Gluck-Holbein)	35°,19°	pT	240, *500*, 950
(Chong-Gauguin)	57°,106°	pT	220, 350, 580, *940*
(Donne-Moliere)	4°,10°	pT	375, 700, (825), *1060*, 1500
(Bartok-Ives)	-33°,115°	pT	480, 790, *1175*, (1500)
(Sadi-Scopas)	-82.5°,44°	pT	360, 600, *930*, (1310)

* Italicized diameters correspond to physiographically most prominent ring (basin rim); diameters in parentheses reflect uncertain measurement due to discontinuous rings.

Age: C = Calorian; T = Tolstojan; pT = pre-Tolstojan. Data from Spudis P. D. and Guest J. E. (1988) in *Mercury* (F. Vilas et al., eds.), p. 139, Table 2, Univ. of Arizona, Tucson.

TABLE 4.10.3. Multiring basins on Mars.

Basin	Center	Ring Diameters* (km)
Elysium	33°N, 201°W	800?, 1540, 2000, 3600, **4970**
Utopia	48°N, 240°W	3300, **4715**
North Tharsis	11°N, 97.5°W	1455, 2330, 3650, **4500**
Chryse	22°N, 46.5°W	891, 1534, 2596, **3600**, 4600
Scopulus	5°N, 278°W	1900, **2700**
Daedalia	14.5°S, 127°W	1475, **2540**, 3960
Hellas	43°S, 291°W	1350?, **2295**, 4200
Memnonia-A	22°S, 165.5°W	1593, **2065**
Acidalia	60°N, 30°W	1000, 1500, **1950**
Near Arcadia	31.7°N, 166.5°W	880, 1463, **1925**
Isidis	13°N, 272.5°W	1100, **1900**
Argyre	50°S, 42°W	540, 1140, **1850**
Sirenum	44°S, 166.5°W	500, 710, 1000, **1548**
South Hesperia	32°S, 255°W	900, **1255**
Cassini-A	13.7°N, 323.7°W	354, 653, 928, **1204**
Al Qahira	20°S, 190°W	141?, 353, 715, **1034**
South of Hephaestus	10°N, 233°W	500, **1000**
Memnonia-B	30°S, 180°W	180, 340, **1000**
Al Qahira-A	13.2°S, 183.5°W	335?, 530, 731, **994**
Ladon	18°S, 29°W	270, 470, 580, **975**, 1300, 1700
South Polar	83°S, 267°W	**850**
West Tempe	56°N, 78°W	**830?**
Overlapped by Newcomb	22°S, 4°W	380, **800**
Amazonis	6°N, 168°W	**800**
South of Renaudot	38°N, 297°W	**600**
Holden	25°S, 32°W	260, **580**
Mangala	0°N, 147°W	300, **570**
Overlapped by Schiaparelli	5°S, 346.5°W	140, **560**
Cassini	24°N, 328°W	321, **547**, 930
Southeast of Hellas	58°S, 273°W	225, **500**

* The principal outer ring is shown in bold type.

Data from Schultz R. A. and Frey H. V. (1990) *J. Geophys. Res., 95,* 14177, Table 1.

TABLE 4.10.4. Multiring basins on the satellites of the outer planets.

Basin	Outer Ring Diameter (km)
Ganymede	
Gilgamesh	550
Osiris	500
Western equatorial basin	185
(also four poorly defined Valhalla-type basins)	
Callisto	
Valhalla	4000
Asgard	1640
Lat. 53, long. 36	920
Lat. 45, long. 138	500
Grimr	180
Alfr	163
Loni	123
Lat. 41, long. 262	71

Data from Passey A. R. and Shoemaker E. M. (1982) in *Satellites of Jupiter* (D. Morrison, ed.), p. 379, Univ. of Arizona, Tucson.

of these models encounter difficulties, but those that embody wave effects in which the target has minimal strength are more successful in explaining the structures. As in many other topics, there has been dispute over some of the basic parameters. A fundamental question is over both the number and spacing of the rings. Such information is crucial to any understanding of the processes of ringed basin formation.

The large multiring basins form the major structural features of the crusts of the Moon, Mars, and Mercury, although their complex overlapping structural patterns have often been confused with some sort of "grid" due to internal thermal or tectonic processes. On the terrestrial planets, up to seven concentric rings may form in the bigger basins. The ring spacing averages $\sqrt{2.0} \pm 0.3$ D, this value having been statistically established for nearly 500 rings constituting over 100 two-ring and multiring basins [90]. This spacing remains the same for the Moon, Mars, and Mercury, and is thus independent of gravity, target strength, and velocity of the impactor. It must represent a primary feature of large basin formation, and is consistent with a fluidized wave mechanism as responsible for the multiring basin structures (Fig. 4.10.1).

Several features of the target may affect the formation of these basins. These include the structure and topography prior to impact. The somewhat distorted shape of Mare Imbrium is probably due to its off-center superposition on the much older and larger Procellarum Basin (Fig. 4.10.2). Crustal layering and lithospheric thickness may also modify the ring spacing. A major conclusion is that all such structures are relatively shallow, most basin ejecta being derived from crustal rather than mantle levels. This is evident from the large multiring basins on the Moon, which, although well over 1000 km in diameter, have not brought up material from beneath the 60-100-km-thick lunar highland crust [91].

A basic question is whether the large basins were formed by collapse into a deep central cavity, with the rings developing by slumping around ring faults, or whether the structure developed essentially instantaneously in a low-strength, fluid-like medium, with the rings representing frozen ripples and the central peak resulting from rebound. Modeling from small, simple bowl-shaped craters is not relevant to these large, relatively shallow multiring basins, of which the lunar Orientale Basin is the type example. A caveat may be added here that Orientale is the youngest of the major lunar basins and was thus emplaced into the coldest and thickest lunar lithosphere. It may thus possess some characteristics different from those of older large multiring basins.

The evidence from large terrestrial basins is in favor of rapid formation in materials of low strength. Conventional structural collapse along outer ring faults plays only a minor role. Central peak rebound is most dramatically shown in the lunar crater Anaxagoras, where the peak has not formed symmetrically, but has apparently collapsed sideways in the course of formation (Fig. 4.10.3).

On Mars, multiring basins show a regular variation with increasing diameter. Basins with diameters less than 1850 km resemble the lunar multiring basin, Orientale. Basins with diameters in the range 1850-3600 km (Argyre type) have a rugged mountainous annulus, possibly the result of impact into a relatively weak lithosphere. Larger Chryse-type basins with diameters exceeding 3600 km have very shallow topography. They have many concentric rings, scarps, and massifs. Spherical target geometry may be the controlling factor for these structures [92].

Basin-forming impacts on icy crusts produce multiring structures (Fig. 4.10.4), while impacts into planets with thick atmospheres produce asymmetric ejecta blankets (Fig. 4.10.5).

4.10.1. Basin Ages

The only direct evidence for large basin ages comes from the dated lunar samples. The stratigraphical evidence that the basins are older than the mare basalts that fill them provides younger limits of about 3850 m.y. for the youngest basins (Imbrium and Orientale). Considerable interest was generated by the isotopic dating of the returned lunar samples interpreted as ejecta from Imbrium (Apollo 14, 16) or Serenitatis (Apollo 17). Most dates clustered around 3850-3950 m.y. The youngest large basin on the Moon is Mare Orientale, which clearly postdates the Imbrium collision. However, Orientale ejecta is buried by the basaltic lavas of Oceanus Procellarum and those of other maria. This constrains the age of Orientale to be older than the ages of the mare surfaces.

One of the oldest basalts collected from a mare surface is 10003. Ironically, this was almost the first sample picked up by the Apollo 11 astronauts. Ages of 3860 m.y. (Ar-Ar) or 3760 m.y. (Rb-Sr) have been determined. Mare basalt sample 10062 has ages of

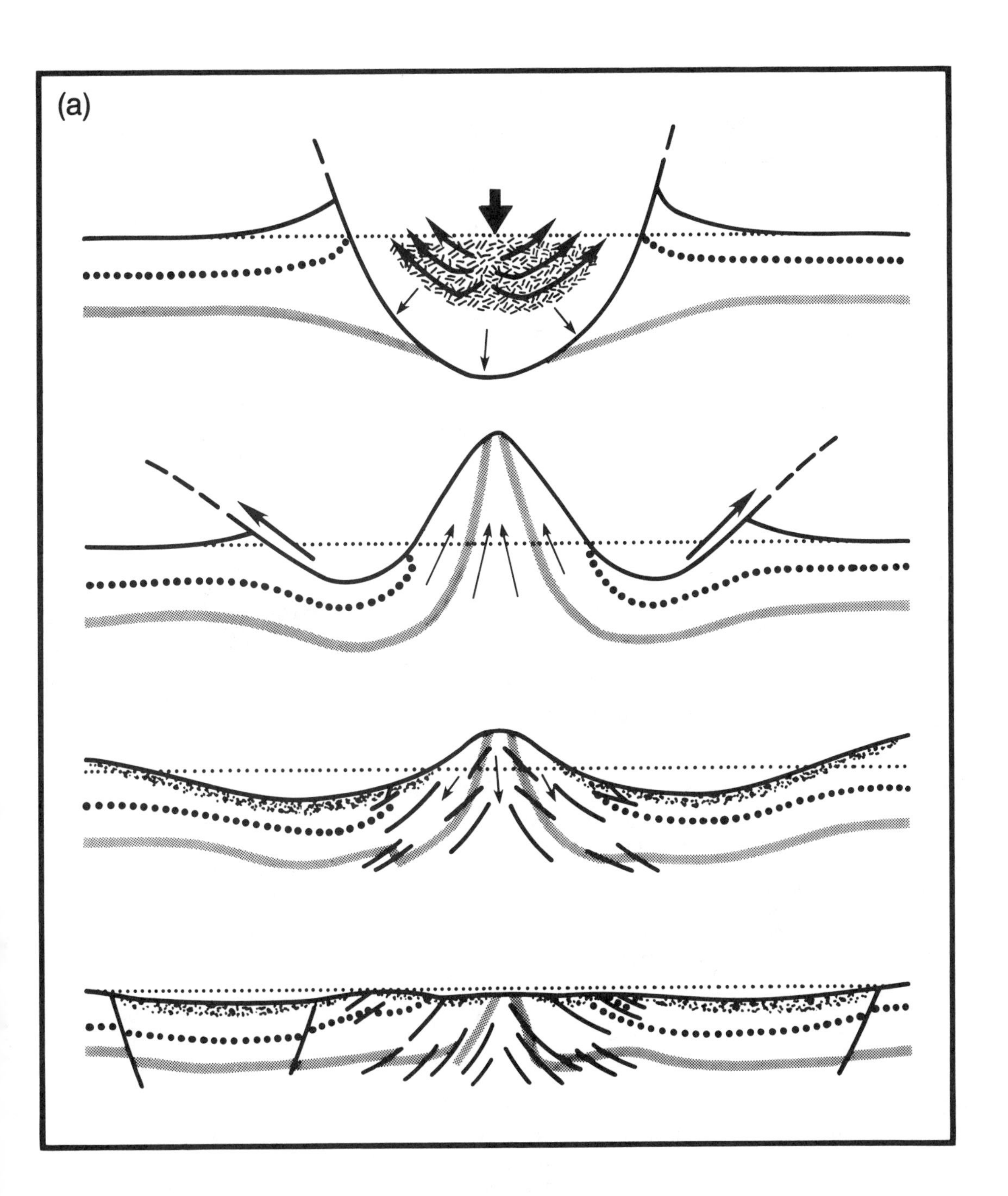

Fig. 4.10.1. *(a) Model for the formation of ring structures. At the top, a transient cavity forms by excavation and compression. In the next stage, rebound of the compressed base forms a central peak. In the third stage, collapse of the central peak occurs. Major excavation has ceased by this stage. At the bottom, rim and uplift have reached equilibrium height controlled by rock strength and gravity. Excess volume in the initial uplift, which cannot be accommodated by reverse faulting, forms a ring. (Courtesy R. A. F. Grieve, Geological Survey of Canada.)*

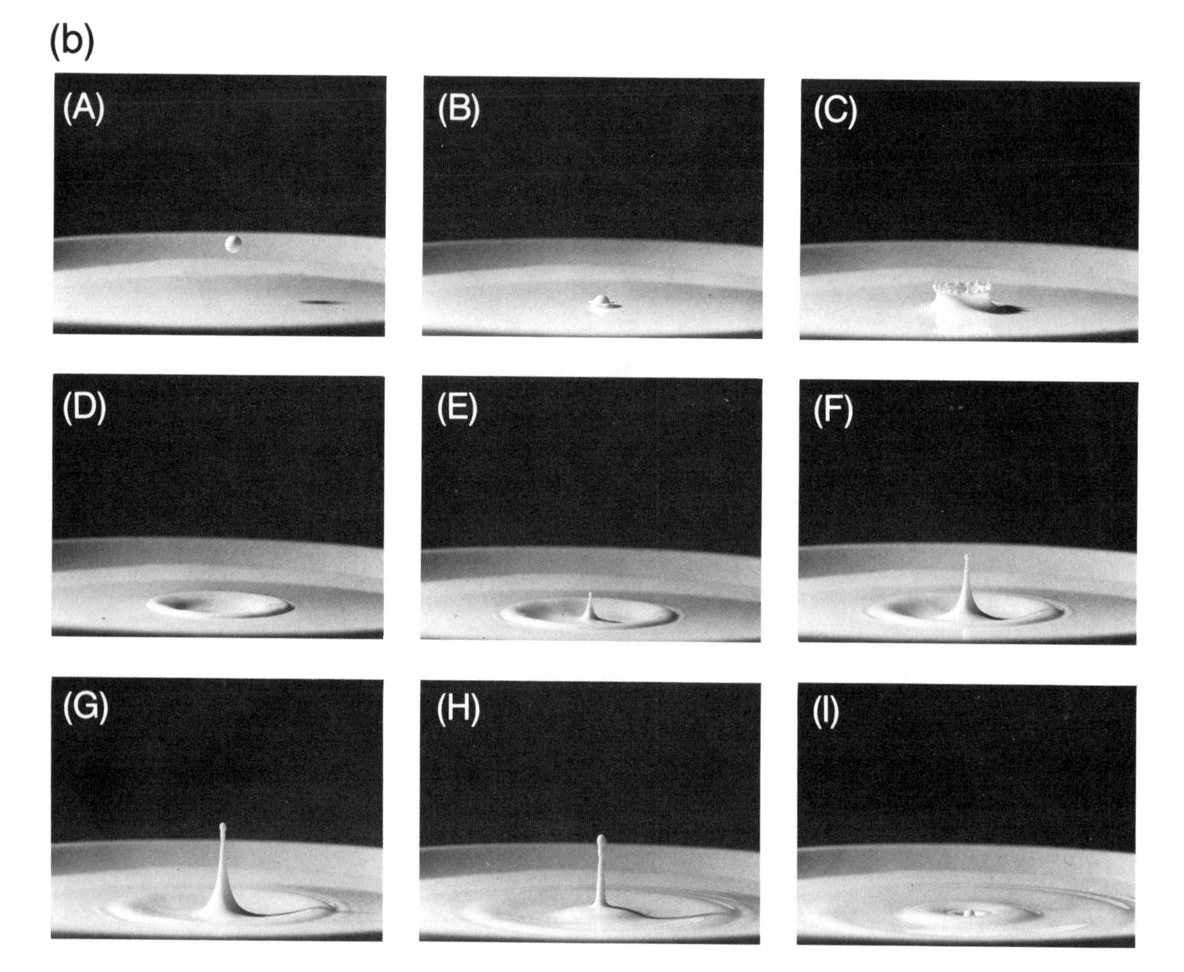

Fig. 4.10.1. *(continued)* ***(b)*** *The hydrodynamic theory of the formation of central peaks and rings is shown by this sequence of photographs of the impact of a milk droplet into a pool of 50/50 milk/cream: (A) before impact; (B) impact; (C) initial crater; (D) primary tsunami wave; (E) central peak rebound commences; (E–G) growth of central peak; (H) collapse of central peak begins and a second concentric ring forms; and (I) end of central peak collapse, with additional ring formation. (Photo sequence courtesy R. B. Baldwin, Oliver Machinery Corp. Photographs by Gene Wentworth of Honeywell Photograph Products.)*

3880 m.y. (Sm-Nd) or 3920 m.y. (Rb-Sr) and is apparently a little older than 10003 [93]. Of course, these samples were from crater ejecta blankets and have been excavated from depth. Although they do not represent material exactly from the present surface, which clearly postdates the Imbrium and Orientale collisions, they are probably from within a few meters of the mare surface [94]. The many dated basalts from surfaces that postdate stratigraphically the last of the great collisions have provided ages sufficiently close to 3800 m.y. that they provide a younger limit to the end of the massive bombardment.

The relative ages of the lunar basins have been well established by the work of Wilhelms [84], but the assignment of absolute ages remains uncertain, and there is no consensus. Imbrium may be as young as 3750-3770 m.y. [95], although most estimates place it at 3860 m.y., based on the Apollo 14 dates [84]. Nectaris is probably younger than 3960 m.y. if the ages of 3960-4150 m.y. for the melt breccias at Apollo 16 site represent pre-Nectarian rocks. The basic problem is that all the Apollo sites are dominated by ejecta from Imbrium, although ejecta from Serenitatis and Nectaris are present at the Apollo 16

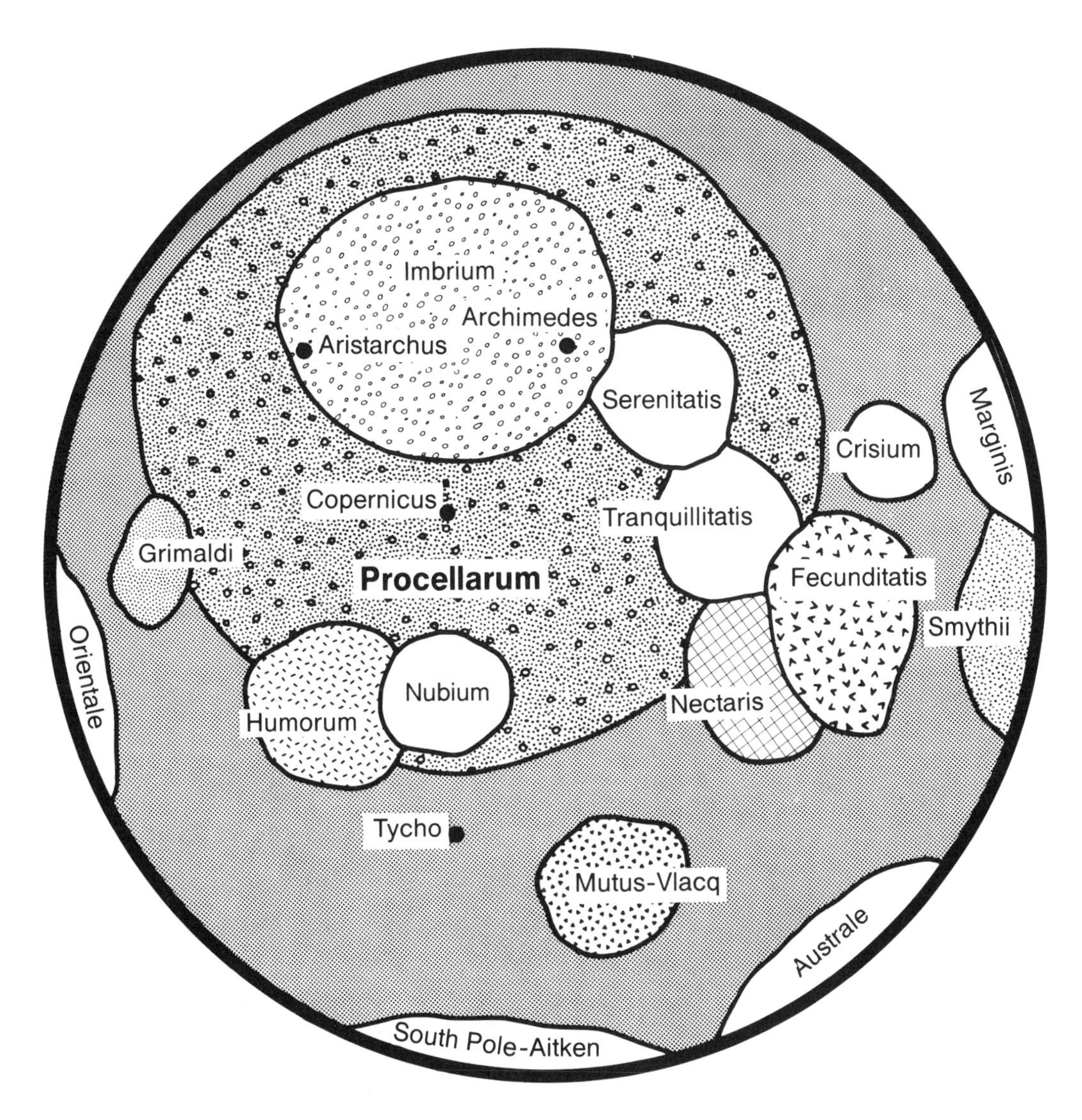

Fig. 4.10.2. *The giant Procellarum impact basin on the nearside of the Moon. After Wilhelms D. E. (1987) USGS Prof. Paper 1348.*

and 17 sites. The very old Procellarum Basin on the nearside has also removed much preexisting crust, perhaps explaining the paucity of old ages on the nearside.

4.10.2. Lunar Swirls

Among the curious features on the lunar surface probably related to impact are swirls. These are features of medium albedo, with a swirl-like pattern. They are found mainly on the eastern limb and on the farside of the Moon. They have no apparent relief, and are associated with the uppermost surface of the lunar regolith. They are mostly concentrated in the antipodal regions of large young impact basins (Orientale, Imbrium, Serenetatis, Crisium) associated with pitted and furrowed terrain. Such terrain is considered to result from focused seismic activity due to the giant basin-forming impacts. Only the youngest examples are expected to survive. Alternative explanations associated with cometary impacts would be expected to provide a random distribution, and do not account for the antipodal correlation with the young basins [96]. The chaotic terrain on Mer-

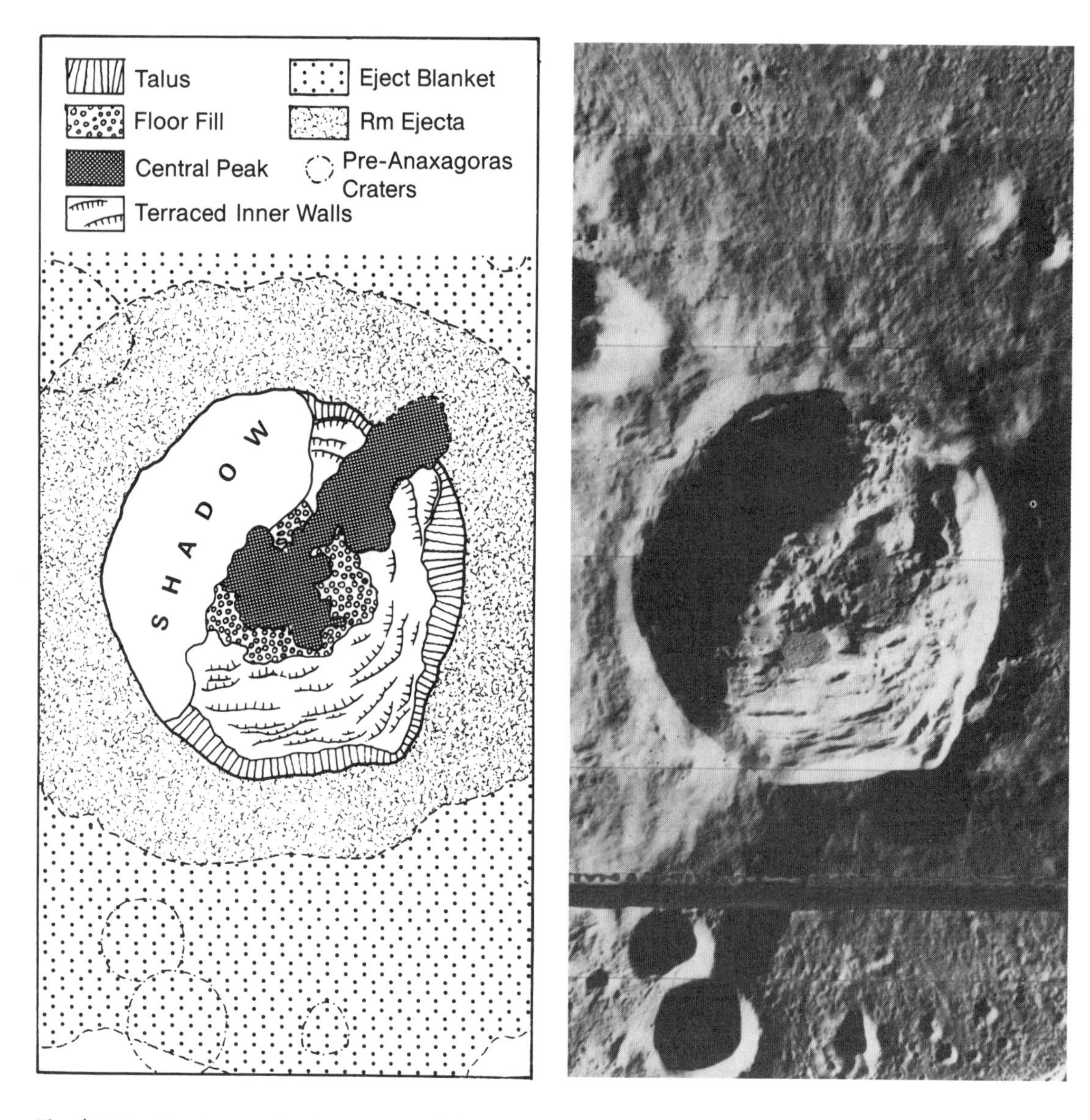

Fig. 4.10.3. *The lunar crater Anaxagoras (51 km in diameter) showing an asymmetric central peak. The geological sketch map (courtesy J. B. Murray, University of London) indicates that much of the central peak material overlies the terraced crater walls, so that the central peak seems to have collapsed sideways during formation.*

cury, antipodal to the large Caloris Basin, may also be due to such effects.

The swirls show a correlation with lunar magnetic anomalies ("magcons"). Reiner Gamma is the type example. How these are related to basin formation is unclear. Why are the swirls lighter in color? The low albedo of the swirls themselves may be related to deflection of the solar wind by the magnetic anomalies, resulting in albedo contrasts due to differing intensities of the ion bombardment of the surface. This process causes darkening of the surface with time. However, the swirls must have originated about 4 b.y. ago, so the lighter color of these areas has persisted over most of solar system history.

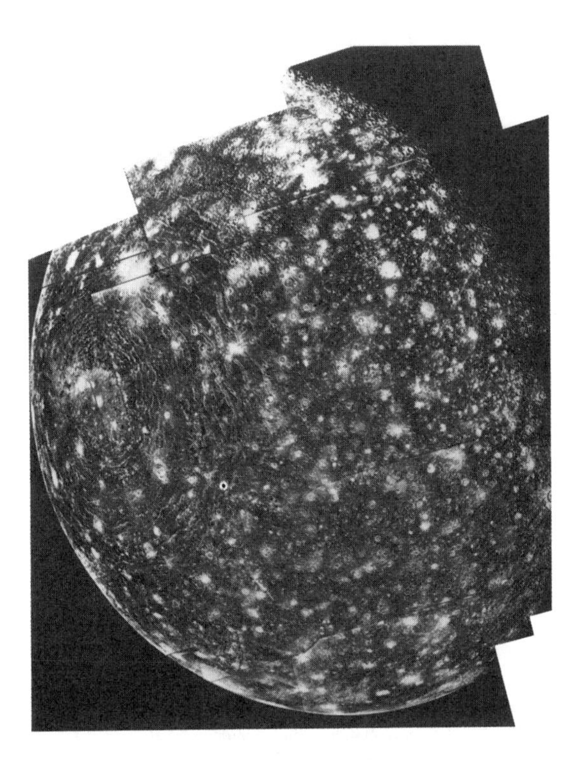

Fig. 4.10.4. *The multiring Valhalla Basin, 4000 km in diameter, on Callisto, centered at 10° latitude and 55° longitude (NASA photo).*

4.11. LUNAR CATACLYSMS?

The concentration of ages of returned lunar highland samples between about 3850 and 3950 m.y. led to the concept of the "Lunar Cataclysm." This was perceived as a late spike in the cratering flux, following a pause after the initial heavy bombardment of planetesimals during accretion at about 4500 m.y. There has been an extensive debate on the reality of this observation [97]. The problem is exacerbated by the localized nature of the Apollo and Luna samples, although the lunar meteorites from several random sites show similar clustering of ages [97]. There is clear isotopic evidence that parts of the feldspathic highland crust crystallized by 4440 m.y., and that the interior, the source regions of the mare basalts, was solid by 4400 m.y. The crystallization in the interior of the final remnants of the "magma ocean" was completed by 4350 m.y. [98]. The presently observable craters postdate these events, and so do not record the tail of lunar accretion, but were formed by later impacts that added little mass. Accordingly, the Moon was essentially formed by 4440 m.y. and little was added by the later bombardment whose visible effects are so dramatic.

The stratigraphic record of the lunar basin cratering history (Table 4.10.1), involving over 40 basins, extends back to the giant 3200-km-diameter Procellarum Basin. The date of this basin is uncertain, but is unlikely to be more than 4200 m.y. on the basis

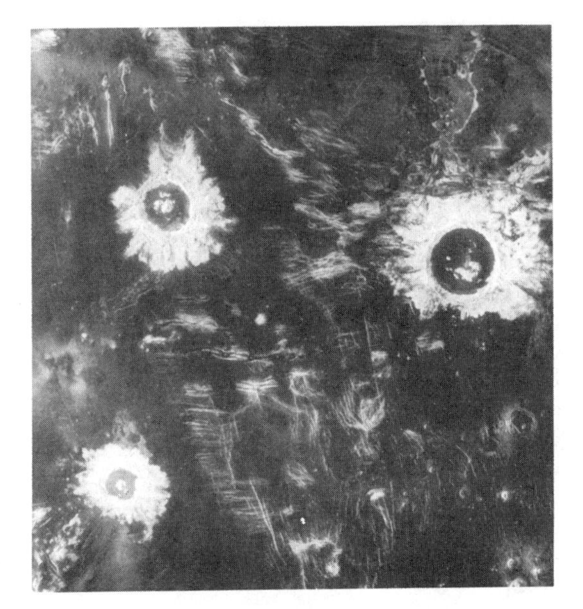

Fig. 4.10.5. *A "crater farm" on Venus. The mosaic, centered at 27° S latitude and 339° E longitude, is in the Lavinia region of Venus. Three large impact craters, with diameters ranging from 37 to 50 km, are located in a region of fractured basaltic plains. The craters exhibit typical features of meteorite impact craters, such as terraced inner walls and central peaks. The asymmetric ejecta blankets are probably due to interaction with the turbulent atmosphere resulting from the oblique passage of the incoming projectile. The crater floors are flooded with basaltic lavas. Many volcanic domes, 1–12 km in diameter, occur in the southeastern corner of the mosaic. The resolution of the data is 120 m. (NASA Magellan P-36711, courtesy Jet Propulsion Laboratory.) [See also Boyce J. M. et al. (1991) Science, 252, 288.]*

of flux estimates. It was already excavated into a solid lunar crust, that had an unknowable previous cratering history.

The arguments for and arguments against the "cataclysm" include:

1. No pre-Nectarian basins were sampled by the Apollo or Luna missions. All sites are dominated by Imbrium ejecta, leading to a biased young sampling.
2. Age dates are likely to be continually reset during an ongoing bombardment. However, mare basalt fragments in the lunar breccias with dates extending back to 4230 m.y. appear not to have had their ages reset.
3. There are a number of reliable pre-4000-m.y. ages, with a range up to 4150 m.y. [99]. However, many of these dates are from feldspathic fragments rather than impact glasses [97], so they may not necessarily be dating impact events.
4. At the Apollo 14 and 16 sites impact melt lithologies are abundant, comprising up to 70% in some samples, and are more common than would be predicted by a late spike [99].
5. Serenitatis is not a particularly old basin [100], though it was long considered to be so on account of its degraded appearance. This is due mainly to its proximity to Imbrium. The Imbrium impact destroyed its western sector. Serenitatis is probably a double basin formed by the simultaneous impacts of two bodies [101].
6. A very complex history is recorded by the differences in crater frequency on the ejecta blankets of the large impact basins [84].
7. There are at least 30 pre-Nectarian lunar basins. The large number of very degraded basins, both on the Moon and Mercury, argues for a continuing flux, and is consistent with the sweep-up of the accretionary tail of the swarm of planetesimals.
8. Among the unresolved questions with a late "cataclysm" is the storage problem; it is dynamically difficult to keep a large population of impactors in the Earth-Moon vicinity for 600 m.y. The contrary interpretation that the massive lunar bombardment represents the tail of accretion seems equally erroneous. Simple backward extrapolation of the cratering flux around 3800-3900 m.y. produces much too large a Moon. In the most probable scenario for lunar origin, the Moon was assembled very quickly prior to 4440 m.y. following the giant impact, in periods ranging from 24 hr to 100 yr (section 6.18).
9. The Pb isotope data form a well-defined array betwen 4460 and 3860 m.y., suggesting a large-scale remobilization of Pb at the younger time [102] (Fig. 4.11.1). The intercept ages do not scatter between these two ages as would be expected from a continuing resetting due to large collisions. The alternative view is that all these samples were influenced by the Imbrium collision.

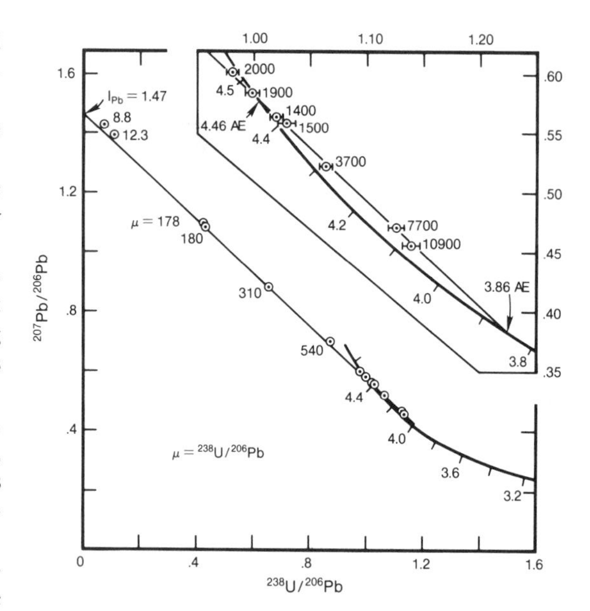

Fig. 4.11.1. *U-Pb evolution diagram for lunar highland breccias. Inset shows data points near the concordia curve. Most of the data points lie on a line intersecting concordia at 4460 m.y. and 3860 m.y. The upper intersection represents the time of early lunar differentiation and crust formation. The lower intersection corresponds to the time of intense bombardment, which has reset the ages. Samples whose data points do not lie precisely on the line may have impact ages slightly different from 3860 m.y. The bombardment affected almost all sampled lunar highland rocks, causing severe U/Pb fractionation in the breccias. This is reflected in the linear distribution of the data points, and the strong correlation of μ values with position of the data points. The measured value of* $\mu = {}^{238}U/{}^{204}Pb$ *is given for each sample [102]. (Courtesy G.J. Wasserburg.)*

10. Stratigraphic evidence for an extended sequence of collisions, so elegantly described by Wilhelms [100], may of course occur within a quite limited time frame. Structures such as Orientale are emplaced within minutes. It seems possible to accommodate the entire lunar basin sequence from Procellarum to Orientale within 200 m.y.
11. There could be a clustering of large impacts due to the break-up of a large planetesimal; there is evidence of such gaps in the lunar cratering record [102].

In summary, the view is taken here that that the accretion of the Moon was essentially completed by 4440 m.y., and that the record is consistent with a major spike of basin-forming collisions between about 4000 and 3850 m.y. (Fig. 4.11.2). Whether impacts spread over 200 m.y. or so constitute a "cataclysm" may be left to experts in semantics; the term is now too thoroughly entrenched in the lunar literature to be changed.

This cataclysm raises a formidable question. What was the source of the bodies responsible for the production of 40 lunar basins over 300 km in diameter, 1000 craters between 30 and 300 km in diameter, and over 10,000 craters, the smallest of which would resemble Canon Diablo or Wolf Creek? Storage of individual objects is a major dynamical problem. The planetesimal hypothesis suggests that the largest objects are the last to be swept up. The giant lunar-forming impact forms an example. If a lunar or larger body remained as a "loose cannon" in the system until it collided with the Earth or Venus at perhaps 4000 or 4100 m.y. ago, the debris from this collision might produce the observed sequence of small and large craters and basins on the Moon, down to the final sweep-up at about 3850 m.y., producing Imbrium and Orientale. The implications of such a stochastic event are considerable. Most of the observed cratering record is dominated by essentially random events in this scenario. Correlations based on alleged similarities in cratering populations even within the inner solar system become uncertain.

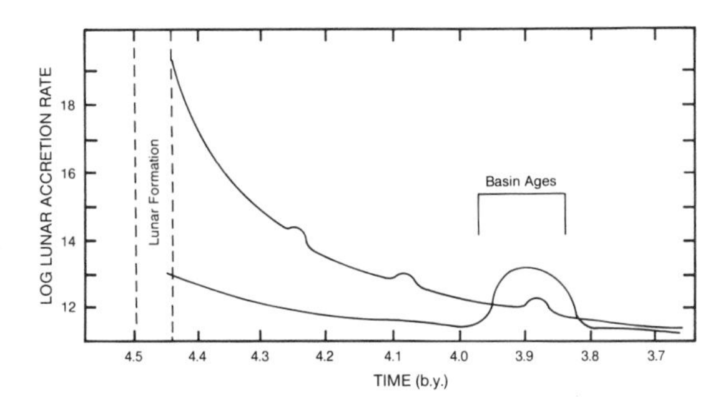

Fig. 4.11.2. *Two versions of the lunar bombardment history. The upper line represents a continuously declining flux, the tail of planetary accretion, with some superimposed spikes. The lower line assumes the Moon was fully accreted by 4450 m.y. and that little material was added after that date. In this interpretation there is a sudden spike or cataclysm around 4000–3850 m.y. due to debris from the breakup of a large body. Adapted from Ryder G. (1989) in Lunar and Planetary Science XX, p. 934, Lunar and Planetary Institute, Houston; Ryder G. (1990) Eos Trans. AGU, 71, 313.*

4.12. EARLY CRATERING FLUX ON THE TERRESTRIAL PLANETS

The heavily cratered terrains of the Moon, Mercury, and Mars have similar crater size/frequency distributions. Mars and Mercury lack the large number of craters with diameters less than about 50 km that are common on the Moon. This is probably due to the formation of intercrater plains on Mercury, either by volcanic processes or ejecta blankets. On Mars, the operation of igneous processes and erosion has erased some of the record.

Analyses of cratering curves for Mars, the Moon, and Mercury, and consideration of the orbits for the impacting objects suggest that the bodies that caused the late heavy bombardment were derived from within the inner solar system [103]. This observation lends further support to the notion that there was little interaction or mixing outside the somewhat localized feeding zones for each planet or between the inner and outer parts of the solar system. The late heavy bombardment does not appear to have been caused by bodies coming from the outer solar system, but as noted above, may well result from the stochastic events within the inner solar system.

4.13. EARLY CRATERING FLUX IN THE OUTER SOLAR SYSTEM

"... the bombardment history of the solar system remains a subject of almost schismatic controversy..." [104].

Most of the conclusions in this and the following sections on cratering and impactor populations in the outer solar system must be regarded as tentative. The evidence favors bombardment by objects from local sources rather than from a solar-system-wide population of bodies. This reinforces many other lines of evidence for the growth of planets and satellites from localized sources within the nebula, rather than from a widely dispersed set of precursor planetesimals.

4.13.1. Jovian System

From a study of impact craters on Ganymede and Callisto there appear to be two separate cratering populations in the jovian system (see also section 6.9). The first, responsible for most of the early small craters, is restricted to the jovian system and characterized by low impact velocities; the second, which appears throughout the outer solar system, appears to have been a high-velocity population, responsible for most of the large craters, and has been dominant since the time of formation of the light grooved terrain on Ganymede. This interpretation, however, like so much else in the cratering record, is controversial [105].

Ganymede is of interest because it has a similar surface gravity to the Moon (respectively 143 and 162 cm/s^2). The largest impact structure on Ganymede is the Gilgamesh Basin, 534 km in diameter. Ganymede has an icy surface, which enables comparisons with impacts on rocky surfaces. Preliminary work has suggested that ejecta deposits on icy surfaces were larger by a factor of 3 compared to those on the Moon [106] and craters in icy surfaces appear to be about 30% shallower than those on the Moon [104].

4.13.2. Saturnian System

The cratering history recorded by the saturnian satellites (see also section 6.10) appears to be different from that of the jovian system [107]. The best imaged moon is Rhea, the largest, except Titan, of the saturnian moons. The surface of Rhea appears to be saturated with craters up to about 32 km diameter, but the density of craters with diameters greater than 64 km does not approach the saturation equilibrium density. The surfaces of Mimas and Iapetus, likewise, are not saturated with large craters; hence, the crater density probably represents a production function. This observation makes it likely that the classical saturnian satellites are original aggregates, rather than being subjected to many episodes of disruption and reaccretion [108].

4.13.3. Uranian System

The crater populations on the uranian satellites (see also section 6.12) appear, in turn, to differ from those of both the jovian and saturnian systems [109], although this question is not entirely resolved [104]. The cratered terrain of Miranda is the oldest surface represented, followed by the surfaces of Umbriel, Oberon, Titania, and Ariel. The resurfaced coronae of Miranda are the youngest in the uranian system, having about 20 times fewer craters of 10 km diameter than the old cratered terrain. Despite their primitive appearance, Titania, Umbriel, and Oberon were resurfaced at an early stage and have had a turbulent early history. There appear to be two different populations of impactors internal to the uranian system; there is little sign of any craters ascribable to an external population of impactors, or any similarities to the saturnian or jovian system impactors. There is evidence for a progressive loss of larger impactors, which could be expected of a population undergoing collisional evolution.

The local sources for most of the impacting objects make it impossible to relate the cratering history and the cratered surfaces on the uranian satellites to, for example, the dated surfaces on the Moon. Accordingly, it is likely that only the most general comments about relative ages of surfaces in differing satellite systems can be made.

In summary, there are separate populations of cratering objects in the inner solar system, and in the jovian, saturnian, and uranian systems. Cratering in the inner solar system appears to be due to rocky planetesimals or asteroids. In contrast, cratering in the outer solar system is due to either icy planetesimals or comets; the difference between these is probably semantic. This is strong evidence against significant mixing throughout the solar system in the later stages of planetary accretion. Thus, sequential accretion of late-stage volatile-rich material scattered from the ice-rich outer solar system as a source for the water and other volatile components of the terrestrial planets [110] may be on a relatively small scale, and not be statistically apparent in the cratering record. As noted earlier, comets might supply the terrestrial budget of water, but [111] the high velocity of cometary impacts makes them a prime candidate for removing, as well as adding, atmospheres [112].

4.14. THE IMPACTOR POPULATION IN THE EARLY SOLAR SYSTEM

If a single population of impacting objects were responsible for the production of craters throughout the solar system, then it would be possible to establish an interplanetary correlation of geological time. This could be related to absolute ages by reference to the lunar record. This desirable objective could be reached even if two populations of objects existed. The first would be responsible for the heavy bombardment extending down, as shown in the lunar record, to about 3850 m.y. A second set of smaller-sized bodies would be responsible for the cratering record on the lunar maria and elsewhere since that time. It might be possible in the future to reach this goal if we had a better understanding of the orbital characteristics of both satellites and comets [104] and some sample return missions.

However, an examination of the cratering record on the Moon, Mercury, Mars, the Galilean satellites of Jupiter, and the saturnian and uranian satellites reveals that no common suite of asteroids, comets, or other bodies can explain the visible cratering distribution on these bodies. As with most aspects of cratering studies, considerable controversy has surrounded this question. Some workers have favored a single population, but this view, permissible when information on cratered surfaces was restricted to the inner solar system, has become increasingly difficult to sustain as detailed photographs of the satellites of the outer planets have become available.

The existence of a single large population of impactors crossing between the inner and outer solar system would have various effects, which do not seem to have occurred. The asteroid belt is still well zoned and is a relic of very early conditions, argued elsewhere to originate in a heating episode identified with early violent solar activity. This dates the origin of the still-preserved zonal nature of the belt from the earliest period.

Rather than finding that one population was responsible, it seems possible to separate out various classes of impacting bodies. The first is the population responsible for the heavy cratering of the lunar highlands, Mercury, and Mars. This population is of great interest. The bodies that impacted the Moon down to 3850 m.y. were not an accretionary tail; the Moon was assembled very quickly before 4440 m.y. [113] in a manner that was unique and did not resemble that of the terrestrial planets. It is entirely possible that the observed lunar bombardment resulted from debris from a massive planetesimal collision around 4000 m.y. and so has little relevance even to the cratering record on Mars or Mercury, which may have been due largely to similar but unrelated events.

Whether or not a single population was responsible for the heavy cratering within the inner solar system, the cratered surfaces on the satellites of the outer solar system seem to have been produced by distinct populations. Those on Callisto and Ganymede are different from those of the saturnian satellites. These constitute two populations among themselves, an older one that is similar to the inner planetary crater populations and a younger one that has some similarities to those on the Galilean satellites of Jupiter. Many of these craters appear to be due to impacts of local objects in planetocentric orbits. The uranian satellites may also show two different cratering populations of which the youngest resembles that of the young terrestrial planets cratering record, but this is controversial. There is debate over the crater counts on these poorly imaged objects, and only one population may be responsible [104].

The implications of this confusing record are that the accretion of the planets was largely a local affair, that mixing between the inner and outer reaches of the solar system was minimal and was perhaps localized even within the inner solar system, but that the giant planets may have had a common impacting flux, except for the rather well-defined population II in the saturnian system. Fortunately there are relatively few craters formed by internal geological processes that would add further confusion to the record. Volcanic calderas are readily enough distinguished from impact craters, despite much earlier confusion and wishful thinking [114].

Planetesimals, asteroids, and comets produce primary craters and basins. Secondary craters result from the impact of ejecta, generally at relatively low velocities, and may often be distinguished by shape. Nevertheless, such secondary craters have caused much confusion, although they are useful in dating ejecta blankets of large basins [115].

Initially, it was possible to extrapolate cratering on the Moon to Mercury and Mars since, with plausible assumptions, the cratering records on these three bodies could be ascribed to a common suite of inner solar system planetesimals, the tail end of the accretionary swarm. This conclusion was reinforced by the difficulty of storing such objects in terrestrial orbit for 600 m.y. down to the cut-off point of the heavy bombardment at 3850 m.y. However, this appealing and simple picture is in doubt if the lunar record is dominated by a late cataclysm, a type example of a stochastic event.

Difficulties arose immediately from the cratering record on the Galilean satellites, Ganymede, and Callisto. The latter is the most heavily cratered object in the solar system, yet is saturated with smaller craters, with a paucity of craters greater than 60 km in diameter. Although attempts have been made to explain this as due to relaxation of large basins and craters on an icy surface, the observed surface shows a very uniform distribution of smaller craters [116]. The consensus is that a distinct population from that of the inner solar system was responsible for cratering the jovian satellites.

Two populations of objects seem required to account for the cratering on the saturnian satellites, an early, mostly erased Population I, represented by some remnant large craters. Most of the cratering is due to the so-called Population II objects; the consensus is that these were Saturn-orbiting objects, rather than bodies in heliocentric orbit.

Thus, there appear to be several distinct populations within the solar system, with the possibility of additional circumplanetary impactors in the uranian and neptunian systems. The prognosis is accordingly veering away from a single heliocentric swarm of impactors, however useful such a population would be in establishing interplanetary correlations, to a heterogeneous set of bodies being responsible for the cratering of planetary and satellite surfaces.

Bodies in orbits around the outer planets form another distinct population. It is generally supposed that sweep-up of these objects occurred on short timescales. There might be survivors left over after accretion, or debris resulting from ejecta from large impacts on small bodies, or perhaps resulting from the disintegration of a satellite. Since ring systems are common and seem to be relatively short-lived, such catastrophic breakups may be frequent, and circumplanetary debris common.

The differing populations of impactors in separate portions of the solar system do not help geochemists in their inquiry into the composition of the planets. If there were a common population, these might represent primordial compositions from a common source, so that the planets might be assembled from uniform materials. This elegant possibility seems remote; even Comet Halley turned out not to be pristine, and the planets and satellites are so dissimilar that they might well have resided in different planetary systems without exciting comment.

4.15. CRATERING FLUX SINCE THE HEAVY BOMBARDMENT

There are five possible reservoirs to supply impactors at present. These are the asteroid belt, the Kuiper and Inner and Outer Oort Clouds, and the Trojan asteroids located at the Lagrangian points on Jupiter's orbit; the latter are probably available only as a source of impactors for the Jovian satellites.

4.15.1. Earth and Moon

Since the decline of the heavy bombardment, the cratering flux appears to have remained relatively low. The present flux rate for the Earth is $5.4 \pm 2.7 \times 10^{-15}$ per km^2 per yr for craters greater than 20 km in diameter, based on the terrestrial cratering record [117,118]. Thus, about every 2-3 m.y. an asteroid large enough to form a 20-km-diameter crater hits the Earth [119]. These rates are higher than those derived from the dated Apollo 12 and 15 regions, which are based on craters with diameters greater than 4 km. This has led to the suggestion that the recent rate over the past 120 m.y., as estimated from the terrestrial cratering record, is a factor of 2-3 times greater than the average for the past 3.5 b.y. This estimate is compatible with rates estimated from Earth-crossing asteroids [120].

It is generally assumed that there has been an increase in the cratering flux in the past few hundred million years, during the Phanerozoic, possibly due to an increased comet flux [120]. This effect can also be seen in the terrestrial Proterozoic cratering record (E. M. Shoemaker, personal communication, June 1990). There is some disagreement about this, with claims that the present rate is not distinguishable from the average based on the lunar record [121]. It is also possible that the apparent increase in the flux rate is spurious, and is the result of incorrect scaling laws between the Earth and Moon [122].

Although the entire present cratering flux responsible for terrestrial craters larger than 10 km might be obtained from the asteroid belt, more realistic models indicate that only one-third of the flux can be ascribed to Apollo objects from the belt, with an additional one-third due to Apollo-Amor extinct comet nuclei, and one-third from active long- and short-period comets [123,124,125].

Considerable interest has been aroused by the alleged 26-m.y. periodicity in the cratering flux, which has been tied into proposed extinctions on the terrestrial fossil record. The terrestrial cratering record, however, does not contain hard evidence of such periodicity [126], and the proposal appears to be yet another example of false correlations and spurious periodicities (see section 4.16.4). The presence of either planet X or the Nemesis star, which could induce periodic cometary showers, has not been confirmed by IRAS searches [127]. Thus periodic showers of comets seem unlikely. The suggestion that a class of small comets at about 1 AU [128] was providing a large flux of volatiles to the Earth (so that a volume of water equal to that of the oceans would be added in about 5 m.y.) has been shown to be erroneous [129]. In summary, the flux of comets to the Earth is very variable and dependent on so many factors that only order-of-magnitude estimates of terrestrial cratering due to comets are possible [130].

The present micrometeorite flux on the Moon is of interest since it represents a hazard to potential settlers on the Moon. Table 4.15.1 gives an estimate of the crater sizes and flux to be expected [131].

4.15.2. Present Accretion Rates

Estimates of the present accretion rate of extraterrestrial material to the Earth vary widely, from a

TABLE 4.15.1. Micrometeorite flux on the Moon.

Craters/m^2/yr	Crater Diameter (μm)
30,000	≥0.1
1,200	≥1.0
300	≥10
0.6	≥100
0.001	≥1000

Data from Fechtig H. et al. (1974) *Proc. Lunar Sci. Conf. 5th*, p. 2463.

TABLE 4.15.2. Various estimates of the terrestrial global accretion rates for extraterrestrial material.

Method Used in Estimate	Accretion Rate (tons yr^{-1})
Magnetic spherules from Pacific clay	90
^{26}Al in Greenland ice	<100,000
^{26}Al in Pacific sediments	1.3×10^6
^{53}Mn in Antarctic ice	100,000
Iron in particles collected from stratosphere	<90,000
Ni in Pacific sediments	$(1\text{-}2) \times 10^6$
	400,000
Ir in Mn nodules	20,000
	70,000
Ir in Pacific sediments	110,000
	78,000
Ir in Antarctic ice	400,000
Ir in Antarctic and Greenland ice	10,000-20,000
Ir in South Pole aerosol	6,000-11,000
Co in Antarctic atmosphere	600
^{3}He/^{4}He ratio in Pacific sediments	2,000
Satellite, visual, and radio-meteor studies	16,000

Data from Tuncel G. and Zoller W. H. (1987) *Nature*, *329*, 705, Table 2.

low of 90 tons per year to over 10^6 tons per year. A list of these estimates is given in Table 4.15.2. The flux of material is obviously inherently difficult to measure. Accumulation rates of Ir in deep-sea sediments provide an annual amount of 77,000 ± 25,000 tons per year. This rate appears to have been relatively constant during the past 67 m.y. since the end of the Cretaceous [132]. Much lower estimates are calculated from the photographic meteor and satellite data [133] and from the Ir content of South Pole aerosols [134]. These rates are not necessarily in disagreement, despite the wide difference between the estimates from the sediment and the atmospheric data. The satellite and aerosol measurements are effectively recording the instantaneous background flux of fine dust; the South Pole Ir measurements on aerosols were for a period of one year. The marine sediments, in contrast, are integrating both the dust influx and the more sporadic infall of larger bodies over periods of many million years.

4.15.3. Venus

The impact craters on the surface of Venus all appear to be relatively fresh (Fig. 4.10.5), and their density indicates a young (<1 b.y.) surface [135]. The crater population belongs to the post-heavy-bombardment population [136]. The crater densities are similar to those on Olympus Mons and the Tharsis plains on Mars, if the cratering flux for Venus, further removed from the asteroid belt source, is half that of Mars [137]. No craters below 5 km in diameter have been detected on Venus, a tribute to the filtering effect of the thick venusian atmosphere, while multiple craters are common, presumably due to projectile breakup in the atmosphere. Asymmetric ejecta blankets (Fig. 4.10.5) are most likely due to interaction of the ejecta with a highly turbulent atmosphere resulting from the passage of the projectile. Another striking contrast to the lunar craters is the absence of the beautiful far-flung ray systems around the venusian craters. This again must be a consequence of the presence of the thick atmosphere, which has effectively hindered the spray of fine particles from the impact.

4.15.4. Mars

The meteorite flux at Mars is estimated [138] at about 1.33 times that of the Earth due to the closer proximity of Mars to the asteroid belt. The micrometeorite flux is smaller because of the likely decrease in dust with increasing heliocentric distance. The low gravity on Mars, combined with enough atmosphere to provide some deceleration of meteorites, could make Mars a prime site for survival of micrometeorites [139]. The total accreting mass of micrometeorites on Mars is only about 0.09 that of the Earth because of the smaller planetary cross section. However, because of the lower entry velocities and the thin martian atmosphere, micrometeorites should survive entry and be much more abundant than on the Earth. It has been calculated that a 10-gm soil sample in the martian regolith could contain 5000 micrometeorites greater than 100 μm and 10 greater than 1000 μm in diameter [139].

4.16. EFFECTS ON BIOLOGICAL EVOLUTION

Homo sapiens is a notably egocentric animal, more so even than *Felis domesticus*, so it may come as a disappointment to some readers that this text deals mainly with the nonbiological aspects of the solar system and that this section is placed in a chapter dealing essentially with chance events. Among the reasons for this are that life in the solar system is confined to this planet. Although the Earth at present (and hopefully in the future) provides an excellent habitat, this satisfactory situation, like evolution, is to a large degree accidental. The Earth, like the rest of the system, is the product of many chance events; further comments on this theme can be found throughout the text and in the epilogue. Finally, I am conscious of the difficulty in making judgments outside my areas of experience.

Nevertheless, some comments are in order, particularly since impact events may have played a crucial role, both in providing a suitable initial environment and also in changing the course of evolution. Impact events may thus have had a controlling influence on the origin and development of life.

Estimates of the frustration of the development of life by massive impacts that would result in a fresh beginning indicate that such globally sterilizing events were common before 4000 m.y. [140]. However, the subject is beset with uncertainty, and other estimates conclude that an upper limit for the existence of conditions for the survival of life may have existed as early as 4440 m.y., shortly after the Moon-forming impact, although this requires a deep-water environment. In summary, estimates for the origin of life converge around 4000 m.y. [141].

4.16.1. Development of Life

What is surprising is not only that life originated, but that it developed so rapidly [142]. Carbon isotopic evidence in the Isua metasediments is consistent with the existence of microbial life 3.8 b.y. ago [143]. Stromatolites (bacterial fossils) were clearly in existence by 3560 m.y. ago in the Warrawoona Group in Western Australia [144]. Since conditions became less hazardous following the termination of the massive bombardment in the inner solar system about 3.8 b.y. ago, the stromatolite evidence indicates that evolution proceeded very rapidly. Tidal pools are a favored location and the chance formation of the Moon provided this useful environment, for, without our satellite, tides raised by the Sun would be minute. The 24.4° tilt to the plane of the ecliptic, another likely outcome of the lunar-forming collision or a similar catastrophe, provides the seasons, a great stimulus to evolution, while the 24-hr rotation period, also favorable to the development of life, is a consequence of such chance collisional events.

The early experiments of Miller and Urey [145] were based on the concept of an early terrestrial reducing atmosphere of methane, ammonia, or hydrogen. Although the state of the early atmosphere is unknown, it was probably not strongly reducing, although a recent survey harks back to the concept of an early reducing atmosphere [146]. However, the (secondary) atmosphere was probably close to neutral (CO_2, N_2), not reducing as commonly supposed previously. The effect of the large collisions was probably to remove any early atmosphere, whatever its composition. There is considerable agreement that by about 4 b.y. ago the early atmosphere consisted of CO_2, water vapor, and nitrogen, and it appears to this reviewer that life originated under these conditions in Darwin's "warm little pond." Perhaps abiotic organic compounds were supplied from meteorites, arriving intact by "soft-landing" of small fragments decelerated by the atmosphere, or perhaps in cometary dust [147]. However there was probably little cratonic or shallow shelf crust available for the development of life. This has directed attention to the possibility that life perhaps originated in a more sheltered, deep hydrothermal system. Deep marine hydrothermal environments became protected from the intense bombardment by about 4.2 b.y. [148].

What seems to be necessary is the presence of liquid water and a water-sediment interface, perhaps provided by, for example, clay mineral surfaces. In such an environment, the complex organic compounds produced in the atmosphere by lightning and solar UV radiation, in addition to those brought in by meteorites, may achieve the requisite synthesis.

However, the path must have been difficult since massive impacts seem to have continued well into the Archean. Laterally extensive thick layers of altered spherules, ascribed to impact, have been found in 3.2-3.5-b.y.-old Fig Tree Group and Warrawoona Group sediments. Although heavily altered, these spherule beds contain high concentrations (over an order of magnitude more than in adjacent sediments) of Ir, Au, Pt, and Os in approximately CI proportions [149]. These beds of spherules in early Archean sedimentary sequences from South Africa and Australia apparently represent quenched liquid droplets from massive impacts (Fig. 4.16.1). They are possibly derived from condensation of vapor clouds; there is no sign of any associated breccia or other impact-derived debris. The spherule beds occur toward the top of coarse detrital beds up to 3 m thick, occurring in otherwise quiet depositional environments. The interpretation is that these coarse sediments represent tsunami deposits caused by large oceanic im-

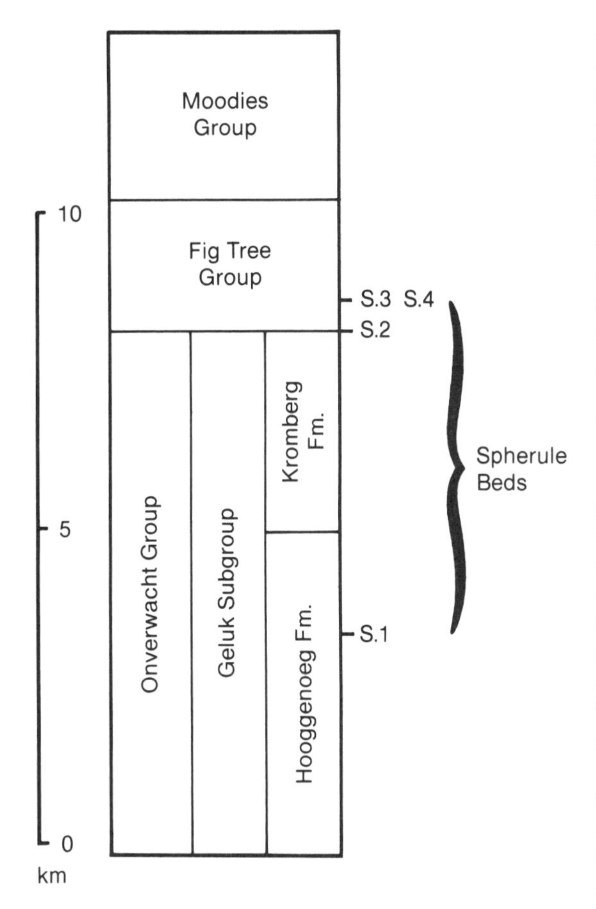

Fig. 4.16.1. *The position of spherule beds in the stratigraphic succession of the upper part of the Swaziland Supergroup, Barberton Greenstone belt, South Africa. After Lowe D. R. et al. (1989) Science, 245, 959, Fig. 1.*

pacts, and that the spherules represent a globewide fallout. It is important that the origin of these spherules be understood [150].

4.16.2. Life on Mars?

"Unanticipated chemistry and uncertain contamination have been the bane of 'exobiology'" [151].

The Viking Lander biochemical experiments attempted to discover whether life was present on Mars. The enigmatic results (e.g., substantial amounts of O_2 gas were evolved when the soil samples were humidified) have been generally attributed to the presence of a strongly oxidizing component in the soil. Other oxygen species in the soil possibly responsible for the release of oxygen include inorganic nitrates [152]. These generally negative results were reinforced by the fact that the Viking GC-MS instrument did not identify any organic compounds at parts per billion levels. This contrasts with the lunar situation, where a few parts per million of organic compounds were present. Since these must also be added to Mars by meteoritic or cometary infall, this is an apparent paradox, and there must be an efficient way of destroying organic molecules on Mars. Probably any such organic material is destroyed by the influx of UV radiation (2000-3000 Å) coupled with the presence of oxidizing species in the soil [153].

4.16.3. Bacteria in Comets?

Some workers have interpreted the infrared spectrum of Halley as evidence for the presence of bacteria. Such an identification would have profound implications for the origin of life, since such bacteria could arrive "ready-made" on the Earth [154]. However, the identification is not unique since almost all organic molecules ($>3 \times 10^6$ are known) show some similar features. It is not difficult to find a "compound with a C-H stretch band that fits this resonant feature as well as do bacteria... if mixtures are included, then the number is essentially infinite... Hoyle and Wickramasinghe rest their case on the fact that the particular shape is unique to living matter (*E. coli*). Unfortunately it is not, as myriads of simple compounds and an infinite number of mixtures possess similar features" [155].

4.16.4. Periodic Extinctions

Considerable interest was aroused by the suggestion that mass extinctions in the fossil record showed a regular pattern with a period of 26-30 m.y. [156]. This suggestion generated considerable literature dealing with the possibility of periodic comet showers from the Oort Cloud, triggered by an undiscovered dark companion, which came to be called the Nemesis star, of the Sun. There is considerable doubt both about the existence of this star and of the periodicity of the extinction record [157]. It seems that incorrect statistical treatments have been used by the proponents of a regular periodicity. "The cyclical pattern of extinction that can currently be observed in the data is indistinguishable from what is expected by chance" [158] and "in both cratering and extinction records, the statistical evidence for periodicity is very weak" [159]. The lack of periodicity is confirmed by Grieve [160], who has made the most detailed study of the terrestrial cratering record. Several further comments are appropriate on this topic. Thus, "there has been a long and dismal history in astronomy of spurious periodicities which have been claimed in many types of data" [161]. "Hidden periodicities used to be discovered as easily as witches in medieval times, but even strong faith must be fortified by a statistical test" [162]. "The most likely situation is that the alleged periodicity in the extinction record is spurious, and that any relationship between impacts and extinctions likely involves single, large, random impacts of comets or asteroids" [163].

4.16.5. Cretaceous-Tertiary Boundary Event

The K-T boundary extinction event is strongly correlated with the following phenomena: Ir and other siderophile element enrichments, derived from the impacting body; shocked quartz and feldspar, and stishovite produced by shock-induced high pressures; spherules; soot from wildfires ignited by the event; and anomalously high $^{87}Sr/^{86}Sr$ ratios (reflecting increased weathering rates due to nitric acid rain formed from nitrogen in the atmosphere during the impact). All these observations strongly support an impact event [164]. Multiple impacts are a possibility, although some trace-element evidence suggests a single source [165]. The worldwide distribution of shocked quartz, for which pressures of about 90 kbar are required, is particularly strong evidence for impact. Peak pressures during volcanic eruptions are unlikely to exceed a few kilobars (the pressure at the base of the crust is 8-10 kbar), much less than the 90 kbar required to produce shock lamellae in quartz. The claims that feldspars and biotites from the volcanic Toba Tuff, Sumatra, show similar shocked features [166] have been shown to be invalid [167]. The site of the impact continues to be elusive, although it is probably close to continental North America, a conclusion indicated by the presence of clasts in the K-T boundary clay in North America, its greater thickness, and the larger size of shocked quartz grains, compared to other localities around the world [168]. The Manson structure in Iowa, a buried 35-km-diameter impact structure of the right age [169], is too small to be responsible for the global effects of the collision. Energy of the order of 10^{23} J, 2 orders of magnitude greater than the Manson event, is required [170]. The evidence for giant-impact-induced wave deposits in southern U.S. marine sections indicates a source in the Caribbean [171]. If the impact were on a shallow marine shelf with carbonate sediments, enough CO_2 could be volatilized into the atmosphere to produce a temporary greenhouse effect, which could raise the mean surface temperature by about 10 K [172]. Such an effect might last for 10^5 yr, explaining the protracted extent of the paleontological extinction record [173]. As noted elsewhere, this event cleared the way for mammalian evolution to proceed unhindered by the giant reptiles, thus finally allowing this present enquiry into past events.

4.17. LARGE COLLISIONS AND PLANETARY ACCRETION: A SUMMARY

"Giant impacts appear to be a likely and normal consequence of planetary accretion, and not a low probability, ad hoc solution to various problems" [174].

Several major effects can be identified as due to very large collisions in the final stages of accretion of the planets:

1. Large collisions provide sufficient energy to melt the terrestrial planets. Core-mantle separation under these conditions will effectively be instantaneous compared to timescales for accretion.

2. The obliquities observed among some of the planets are so large as to demand an origin by massive impact. Collisions with the next largest body in the accretional sequence can account for these observations. Planetary tilts accordingly provide *prima facie* evidence for the accumulation of the planets by accretion from a hierarchy of planetesimals. No hypothesis involving condensation from a laplacian nebula in an orderly manner, or by formation of giant gaseous protoplanets can account for the rather untidy situation in which the planets now find themselves. It would be extraordinary if all planets were tilted to the same degree, just as much as if all planets were identical. If the planets showed zero obliquities, it would be possible to entertain an orderly origin for them. They bear, however, not only on their battered faces, but also from their varying tilts, silent witness to the trauma surrounding their accretion.

3. Collision with an object about 0.2 mercurian masses accounts for the high iron/silicate ratio in Mercury by removing its silicate mantle.

4. Collision with the Earth of an 0.15-Earth-mass object can account for the origin, angular momentum and the strange chemistry of the Moon.

5. Cataclysmic impacts will remove early planetary atmospheres, so present atmospheric compositions

are not necessarily related to, or provide information about primordial atmospheres.

It is sometimes objected that the probability of a collision at just the right velocity, angle, and mass to produce the Moon, or to strip off the silicate mantle of Mercury, is very low. This is true, but of course other collisions involving different angles, velocities, and other parameters might produce equally "anomalous" effects. However, the most convincing argument is the absence of "normal" planets. All bear some signature of having experienced unusual processes. If large impacts of planetesimals are a characteristic feature of the final stages of planetary accretion, then the details of the individual impacts become to some extent free parameters. Those that probably produced the Moon and the high density of Mercury fall well within the predicted range of such events.

NOTES AND REFERENCES

1. Marvin U. B. (1990) in *Global Catastrophes in Earth History,* GSA Spec. Paper 247, p. 147.
2. Cole G. H. A. (1988) in *The Physics of the Planets* (S. K. Runcorn, ed.), p. 385, Wiley, New York.
3. Ibid., pp. 398-399.
4. Hirayama K. (1918) *Proc. Phys.-Math. Soc. Japan, Series 3, 2,* 236.
5. See Marvin U. B. (1990) in *Global Catastrophes in Earth History,* GSA Spec. Paper 247, p. 147. Also Hoyt W. G. (1987) *Coon Mountain Controversies, Meteor Crater and the Development of Impact Theory,* Univ. of Arizona, Tucson, 442 pp.
6. Taylor S. R. (1973) *Earth-Sci. Rev., 9,* 101; Shaw H. F. and Wasserburg G. J. (1982) *Earth Planet. Sci. Lett., 60,* 155.
7. Wilhelms D. E. (1986) *USGS Prof. Paper 1348,* 302 pp.
8. Shoemaker E. M. and Hackman R. J. (1962) in *The Moon,* p. 289, Academic, New York.
9. Weissman P. R. (1989) in *Origin and Evolution of Planetary and Satellite Atmospheres* (S. K. Atreya et al., eds.), p. 234, Univ. of Arizona, Tucson.
10. An excellent and readable review of cratering is given by Chapman C. R. and McKinnon W. B. (1986) in *Satellites* (J. A. Burns and M. S. Matthews, eds.), p. 492, Univ. of Arizona, Tucson.
11. Strom R. G. and Neukum G. (1988) in *Mercury* (F. Vilas et al., eds.), p. 345, Univ. of Arizona, Tucson.
12. The most objective summary of this topic is given by Chapman C. R. and McKinnon W. B. (1986) in *Satellites* (J. A. Burns and M. S. Matthews, eds.), p. 492, Univ. of Arizona, Tucson.
13. Roddy D. J. et al., eds. (1977) *Impact and Explosion Cratering,* Pergamon, New York, 1299 pp.; Grieve R. A. F. (1987) *Annu. Rev. Earth Planet. Sci., 15,* 245; Melosh H. J. (1989) *Impact Cratering: A Geologic Process,* Oxford Univ., New York, 245 pp.
14. Wilhelms D. E. (1986) *USGS Prof. Paper 1348,* p. 139.
15. Ibid., pp. 129-136.
16. Ibid., p. 133.
17. Although the existence of the Procellarum Basin, the largest such structure on the Moon, is somewhat controversial, it explains a number of otherwise puzzling features of the lunar nearside, such as the thin crust, the prevalence of mare basalts, the wide extent and unusual distribution of the lavas forming Oceanus Procellarum, and the nonsymmetrical shape of the Imbrium Basin. The present author agrees with the interpretation of Wilhelms D. E. (1987, *USGS Prof. Paper 1348,* p. 145) that, although now much degraded, Procellarum is the oldest and largest of the observable lunar basins.
18. The date of the Imbrium collision depends on the interpretation of the ages of samples from Apollo 14 and 15 missions. The age of the black and white breccia 15455, probably derived from the impact, is 3860 ± 40 m.y. This agrees with the Pb isotope data [Oberli F. et al. (1978) in *Lunar and Planetary Science IX,* p. 832, Lunar and Planetary Institute, Houston; Oberli F. et al. (1979) in *Lunar and Planetary Science X,* p. 940, Lunar and Planetary Institute, Houston], and is accepted here. Wilhelms D. E. (1987, *USGS Prof. Paper 1348,* p. 224) opts for 3850 m.y. (within the error limits) as the preferred date. There is some argument, not accepted here, for a younger age of 3.75-3.77 b.y. [Deutsch A. and Stöffler D. (1987) *Geochim. Cosmochim. Acta, 51,* 1951].
19. See Hartmann W. K. (1984) *Icarus, 60,* 56; Hartmann W. K. (1988) in *Lunar and Planetary Science XIX,* p. 449, Lunar and Planetary Institute, Houston; Woronow A. (1977) *J. Geophys. Res., 82,* 2447; Woronow A. (1987) *Icarus, 34,* 76; Strom R. G. and Woronow A. (1982) in *Lunar and Planetary Science XIII,* p. 782, Lunar and Planetary Institute, Houston, for various opinions on this topic.
20. An excellent and reasoned discussion of the controversial topic of crater saturation is given by Chapman C. R. and McKinnon W. B. (1986) in *Satellites* (J. A. Burns and M. S. Matthews, eds.), p. 492, Univ. of Arizona, Tucson; see also Weissman P. R. (1989) in *Origin and Evolution of Planetary and Satellite Atmospheres* (S. K. Atreya

et al., eds.), p. 234, Univ. of Arizona, Tucson, while the reader can also be entertained by an extended and sensible discussion of the "silly but furious battles" over the question in Melosh H. J. (1989) *Impact Cratering: A Geologic Process,* pp. 191-196, Oxford Univ., New York.

21. Hartmann W. K. (1984) *Icarus, 60,* 56.
22. Chapman C. R. and McKinnon W. B. (1986) in *Satellites* (J. A. Burns and M. S. Matthews, eds.), p. 551, Univ. of Arizona, Tucson.
23. Strom R. G. and Neukum G. (1988) in *Mercury* (F. Vilas et al., eds.) p. 340, Univ. of Arizona, Tucson.
24. Hartmann W. K. (1989) in *Lunar and Planetary Science XX,* p. 377, Lunar and Planetary Institute, Houston.
25. There is considerable dispute over the level at which saturation equilibrium is reached. The equivalence of crater densities of the surfaces of Phobos and of the heavily cratered lunar highlands lends credence to the estimates by Hartmann [24], since there is now rather general agreement that the most heavily cratered sectors of the lunar highlands are saturated with craters [22]. See, however, Woronow A. et al. (1982) in *Satellites of Jupiter* (D. Morrison, ed.), p. 237, Univ. of Arizona, Tucson.
26. Melosh H. J. (1989) *Impact Cratering: A Geologic Process,* Oxford Univ., New York, 245 pp.
27. Another excellent survey is given by Chapman C. R. and McKinnon W. B. (1986) in *Satellites* (J. A. Burns and M. S. Matthews, eds.), p. 494, Univ. of Arizona, Tucson. This discusses scaling, impact melting and vaporization, ejecta blankets, rays, crater morphology, the problems of collisional breakup and reaccretion of satellites, and cratering statistics.
28. Marvin U. B. et al. (1989) *Science, 243,* 925.
29. Tyburczy J. A. et al. (1988) in *Lunar and Planetary Science XIX,* p. 1209, Lunar and Planetary Institute, Houston.
30. See Pike R. J. (1988) in *Mercury* (F. Vilas et al., eds.), p. 165, Univ. of Arizona, Tucson.
31. Ibid.; Melosh H. J. (1989) *Impact Cratering: A Geologic Process,* Chapters 8 and 9, Oxford Univ., New York. The surface gravity value for Mercury is 370 cm/s^2 compared to the lunar value of 162 cm/s^2. This difference accounts for the smaller ejecta blankets on Mercury compared to the Moon for the same sized crater. Gault D. E. et al. (1975) *J. Geophys. Res., 80,* 2444.
32. Carr M. H. (1986) *Icarus, 68,* 187.
33. For example, Johnson T. V. and McGetchin T. R. (1973) *Icarus, 18,* 612.
34. Moore J. M. and McKinnon W. B. (1991) in *Uranus,* Univ. of Arizona, Tucson, in press; Schenk P. M. (1991) in *Uranus,* Univ. of Arizona, Tucson, in press.
35. Schenk P. M., op. cit. Crater relaxation on icy satellites has also been studied by Thomas P. J. and Squyres S. W. (1988) *J. Geophys. Res., 93,* 14919.
36. Hillgren V. J. and Melosh H. J. (1989) *Geophys. Res. Lett., 16,* 1339.
37. Chapman C. R. and McKinnon W. B. (1986) in *Satellites* (J. A. Burns and M. S. Matthews, eds.), p. 525, Univ. of Arizona, Tucson.
38. Croft S. K. (1988) in *Lunar and Planetary Science XIX,* p. 219, Lunar and Planetary Institute, Houston.
39. Benz W. and Cameron A. G. W. (1989) *Bull. Am. Astron. Soc., 21,* 916.
40. Perhaps 10 m.y., so that by the latter stages of the formation of Uranus, the gas was long gone.
41. Weissman P. R. (1989) in *Origin and Evolution of Planetary and Satellite Atmospheres* (S. K. Atreya et al., eds.), p. 242, Univ. of Arizona, Tucson.
42. Cameron A. G. W. (1983) *Icarus, 56,* 195; Ahrens T. J. and O'Keefe J. D. (1987) *Inter. J. Impact Eng., 5,* 13; Walker J. G. C. (1987) *Icarus, 68,* 87.
43. Melosh H. J. and Vickery A. M. (1989) *Nature, 338,* 487.
44. Benz W. and Cameron A. G. W. (1990) in *Origin of the Earth* (H. E. Newsom and J. H. Jones, eds.), p. 61, Oxford Univ., New York.
45. See O'Keefe J. D. and Ahrens T. J. (1988) in *Lunar and Planetary Science XIX,* p. 887, Lunar and Planetary Institute, Houston.
46. Chyba C. F. (1987) *Nature, 330,* 632; Chyba C. F. (1990) *Nature, 343,* 129.
47. See estimates respectively in Hartmann W. K. et al. (1981) *Basaltic Volcanism on the Terrestrial Planets,* Table 8.2, Pergamon, New York, and in Wetherill G. W. (1977) *Proc. Lunar Planet. Sci. Conf. 8th,* p. 1; McKinnon W. B. (1989) *Nature, 338,* 465; Maher K. A. and Stevenson D. J. (1988) *Nature, 331,* 612; Hartmann W. K. (1980) in *The Lunar Highlands Crust,* Pergamon, New York, p. 155; Wilhelms D. E. (1987) *USGS Prof. Paper 1348,* 302 pp.
48. McKinnon W. B. (1989) *Nature, 338,* 465.
49. Garrison J. R. and Taylor L. A. (1980) in *The Lunar Highlands Crust,* p. 395, Pergamon, New York.
50. Melosh H. J. and Vickery A. M. (1989) *Nature, 338,* 487; McKinnon W. B. (1989) *Nature, 338,* 465.
51. Chyba C. F. (1989) *Bull. Am. Astron. Soc., 21,* 915; Chyba C. F. (1990) *Nature, 343,* 129.
52. For a statement typical of that era, see Green J. (1965) *Annu. New York Acad. Sci., 123,* 385, as well as many other curiously dated papers in "Geologic Problems in Lunar Research" (1965) *Annu. New York Acad. Sci., 123,* 367-1257.

53. See comments on the history of the acceptance of an impact origin for lunar craters by Taylor S. R. (1982) *Planetary Science: A Lunar Perspective,* p. 63, Lunar and Planetary Institute, Houston. It is extraordinary to record that a book published in 1988 still regards the problem as unresolved; see Prior F. W. (1988) *The Moon Observer's Handbook,* Cambridge Univ., New York, 309 pp.
54. Wilhelms D. E. (1986) *USGS Prof. Paper 1348,* 302 pp.
55. This crucial aspect of the problem of lunar origin was pointed out by Cameron A. G. W. and Ward W. R. (1976) in *Lunar Science VII,* p. 120, The Lunar Science Institute, Houston.
56. Benz W. et al. (1986) *Icarus, 66,* 515; Benz W. et al. (1987) *Icarus, 71,* 30.
57. Mizuno H. and Boss A. P. (1984) *Icarus, 63,* 109; Boss A. P. and Benz W. (1989) *Bull. Am. Astron. Soc., 21,* 915.
58. W. B. McKinnon, Washington Univ., personal communication, March 1990.
59. For example, Hartmann W. K. et al., eds. (1986) *The Origin of the Moon,* Lunar and Planetary Institute, Houston, 781 pp.; Newsom H. E. and Taylor S. R. (1989) *Nature, 338,* 29.
60. Benz W. and Cameron A. G. W. (1988) in *Lunar and Planetary Science XIX,* p. 61, Lunar and Planetary Institute, Houston.
61. For example, Benz W. and Cameron A. G. W. (1990) in *Origin of the Earth* (H. E. Newsom and J. H. Jones, eds.), p. 61, Oxford Univ., New York.
62. Solomon S. C. and Chaiken J. (1976) *Proc. Lunar Planet. Sci. Conf. 7th,* p. 3229.
63. Kirk R. L. et al. (1988) in *Lunar and Planetary Science XIX,* p. 605, Lunar and Planetary Institute, Houston.
64. For example, Hansen K. S. (1982) *Rev. Geophys. Space Phys., 20,* 457.
65. Eriksson K. A. (1979) *Precambrian Res., 8,* 153.
66. Sonett C. P. et al. (1988) *Nature, 335,* 806.
67. The cratering history of Mercury is given in detail in F. Vilas et al., eds. (1988) *Mercury,* Univ. of Arizona, Tucson, by Spudis P. D. and Guest J. E., p. 118; Pike R. J., p. 165; Schultz P., p. 274; and Strom R. G. and Neukum G., p. 336; with additional comments by Melosh H. J. and McKinnon W. B., p. 374.
68. From Spudis P. D. and Guest J. E. (1988) in *Mercury* (F. Vilas et al., eds.), p. 138, Univ. of Arizona, Tucson.
69. Strom R. G. and Neukum G. (1988) in *Mercury* (F. Vilas et al., eds.), p. 366, Univ. of Arizona, Tucson.
70. See Francis P. W. (1987) in *Lunar and Planetary Science XVIII,* pp. 300-301, Lunar and Planetary Institute, Houston.
71. Campbell I. H. and Taylor S. R. (1983) *Geophys. Res. Lett., 10,* 1061.
72. Spudis P. D. and Guest J. E. (1988) in *Mercury* (F. Vilas et al., eds.), p. 155, Univ. of Arizona, Tucson.
73. Weissman P. R. (1989) in *Origin and Evolution of Planetary and Satellite Atmospheres* (S. K. Atreya et al., eds.), p. 230, Univ. of Arizona, Tucson.
74. Strom R. G. and Neukum G. (1988) in *Mercury* (F. Vilas et al., eds.), p. 352, Univ. of Arizona, Tucson.
75. Leake M. A. et al. (1987) *Icarus, 71,* 350.
76. Chapman C. R. (1988) in *Mercury* (F. Vilas et al., eds.), p. 16, Univ. of Arizona, Tucson.
77. Darwin G. H. (1879) *Philos. Trans. R. Soc., 170,* 442.
78. Even Kant was misled; see section 1.5.2.
79. Cameron A. G. W. et al. (1988) in *Mercury* (F. Vilas et al., eds.), p. 692, Univ. of Arizona, Tucson; Benz W. et al. (1988) *Icarus, 74,* 516.
80. Wetherill G. W. (1988) in *Mercury* (F. Vilas et al., eds.), p. 670, Univ. of Arizona, Tucson.
81. Ibid.
82. The far rim of Vallis Marineris would be over the martian horizon, so to an observer on the martian surface, this vast structure would probably be less impressive than is the Grand Canyon to a terrestrial viewer.
83. First proposed by Hartmann W. K. and Kuiper G. P. (1962) *Comm. Univ. Arizona Lunar Planet. Lab., 1,* 51.
84. Wilhelms D. E. (1987) *USGS Prof. Paper 1348,* 302 pp.
85. Wetherill G. W. (1975) *Proc. Lunar Sci. Conf. 6th,* p. 1539.
86. Melosh H. J. (1989) *Impact Cratering: A Geologic Process,* p. 131, Oxford Univ., New York. See the perceptive discussions in Wilhelms D. E. (1986) *USGS Prof. Paper 1348,* pp. 77-81, and in Pike R. J. and Spudis P. D. (1987) *Earth Moon Planets, 39,* 129.
87. McHone J. F. and Nieman R. A. (1988) *Meteoritics, 23,* 289.
88. Hoyt W. G. (1987) *Coon Mountain Controversies, Meteor Crater and the Development of Impact Theory,* Univ. of Arizona, Tucson, 442 pp.
89. It is curious to record this example of extreme conservatism in the geological profession, which 150 years ago was the leader in changing both the perceived age of the Earth from the biblical figure of 6000 years by several orders of magnitude, and our understanding of the relationhip of man in the animal kingdom.
90. Pike R. J. and Spudis P. D. (1987) *Earth Moon Planets, 39,* 129; but see also Melosh H. J. (1989) *Impact Cratering: A Geologic Process,* Oxford Univ., New York, pp. 169-173, and Fig. 9.5 for a less sanguine view of the statistical basis for the claimed $\sqrt{2}$ spacing.

91. Baldwin R. B. (1981) in *Multi-ring Basins,* p. 275, Lunar and Planetary Institute, Houston.
92. Schultz R. A. and Frey H. V. (1990) *J. Geophys. Res., 95,* 14175.
93. Taylor S. R. (1982) *Planetary Science: A Lunar Perspective,* Table 6.6, Lunar and Planetary Institute, Houston.
94. Beaty D. W. and Albee A. L. (1978) *Proc. Lunar Planet. Sci. Conf. 9th,* p. 359; Beaty D. W. et al. (1979) *Proc. Lunar Planet. Sci. Conf. 10th,* p. 41.
95. Deutsch A. and Stöffler D. (1987) *Geochim. Cosmochim. Acta, 51,* 1951.
96. Hood L. L. and Williams C. R. (1988) in *Lunar and Planetary Science XIX,* p. 503, Lunar and Planetary Institute, Houston; Hood L. L. and Williams C. R. (1989) *Proc. Lunar Planet. Sci. Conf. 19th,* p. 99.
97. Ryder G. (1989) in *Lunar and Planetary Science XX,* p. 934, Lunar and Planetary Institute, Houston; Ryder G. (1990) *Eos Trans. AGU, 71,* 313.
98. Carlson R. W. and Lugmair G. W. (1988) *Earth Planet. Sci. Lett., 90,* 119; see also summary by Taylor S. R. (1982) *Planetary Science: A Lunar Perspective,* p. 231, Lunar and Planetary Institute, Houston.
99. Maurer P. P. et al. (1978) *Geochim. Cosmochim. Acta, 42,* 1687; Stöffler D. et al. (1985) *Proc. Lunar Planet. Sci. Conf. 15th,* p. 449; Stöffler D. et al. (1989) *Meteoritics, 24,* 328; see, however, Ryder G. (1989) *Meteoritics, 24,* 322.
100. Wilhelms D. E. (1987) *USGS Prof. Paper 1348,* pp. 173, 180.
101. The age of Serenitatis is estimated by Wilhelms D. E. (1987) *USGS Prof. Paper 1348,* p. 178, as 3.86 b.y.
102. Oberli F. et al. (1978) in *Lunar and Planetary Science IX,* p. 832, Lunar and Planetary Institute, Houston; Oberli F. et al. (1979) in *Lunar and Planetary Science X,* p. 940, Lunar and Planetary Institute, Houston.
103. Strom R. G. (1988) in *Lunar and Planetary Science XIX,* p. 1141, Lunar and Planetary Institute, Houston.
104. McKinnon W. B. et al. (1991) in *Uranus,* Univ. of Arizona, Tucson, in press.
105. Croft S. K. and Duxbury E. D. (1988) in *Lunar and Planetary Science XIX,* p. 227, Lunar and Planetary Institute, Houston; Croft S. K. (1988) in *Lunar and Planetary Science XIX,* p. 221, Lunar and Planetary Institute, Houston.
106. Smith B. A. et al. (1979) *Science, 206,* 927.
107. Lissauer J. J. et al. (1988) in *Lunar and Planetary Science XIX,* p. 683, Lunar and Planetary Institute, Houston; Lissauer J. J. et al. (1988) *J. Geophys. Res., 93,* 13776.
108. For example, Smith B. A. et al. (1982) *Science, 215,* 504.
109. Strom R. G. (1987) *Icarus, 70,* 517; Croft S. K. (1988) in *Lunar and Planetary Science XIX,* p. 223, Lunar and Planetary Institute, Houston; Croft S. K. (1991) *Uranus,* Univ. of Arizona, Tucson, in press.
110. For example, Dreibus G. and Wänke H. (1987) in *Lunar and Planetary Science XVIII,* p. 248, Lunar and Planetary Institute, Houston.
111. Chyba C. F. (1987) *Nature, 330,* 632.
112. Melosh H. J. and Vickery A. M. (1989) *Nature, 338,* 487.
113. Carlson R. W. and Lugmair G. W. (1988) *Earth Planet. Sci. Lett., 90,* 119.
114. See Pike R. J. (1980) *USGS Prof. Paper 1046-C* for a sober and skeptical assessment of this problem.
115. A full discussion of crater morphology is given by Pike R. J. (1980) *USGS Prof. Paper 1046-C*; see also Pike R. J. (1988) in *Mercury* (F. Vilas et al., eds.), p. 165, Univ. of Arizona, Tucson. A useful discussion is also given by Wilhelms D. E. (1986) *USGS Prof. Paper 1348,* 302 pp.
116. See Chapman C. R. and McKinnon W. B. (1986) in *Satellites* (J. A. Burns and M. S. Matthews, eds.), p. 561, Fig. 21, Univ. of Arizona, Tucson, for a clear diagram illustrating this point.
117. From Grieve R. A. F. (1984) *Proc. Lunar Planet. Sci. Conf. 14th,* in *J. Geophys. Res., 89,* B403.
118. This is consistent with a value of $4.1 \pm 2 \times 10^{-15}$ per km^2 per year for the production of craters greater than 20 km in diameter estimated by Shoemaker E. M. et al. (1991) in *Global Catastrophes in Earth History,* GSA Spec. Paper 247, p. 155.
119. A list of terrestrial impact craters is given by Grieve R. A. F. (1991) *Meteoritics, 26,* 175.
120. Shoemaker E. M. (1979) in *Asteroids* (T. Gehrels, ed.), p. 253, Univ. of Arizona, Tucson.
121. Neukum G. (1988) in *Lunar and Planetary Science XIX,* p. 850, Lunar and Planetary Institute, Houston.
122. Wetherill G. W. (1989) *Meteoritics, 24,* 15.
123. Ibid.; Shoemaker E. M. et al. (1991) in *Global Catastrophes in Earth History,* GSA Spec. Paper 247, p. 155.
124. A near-Earth miss with Hermes, about 1 km in diameter, occurred in 1937. Another near miss at about the same distance, ~700,000 km, with asteroid 1989FC, of uncertain size between 100 m and 1 km in diameter, occurred on March 23, 1989.
125. Hut P. et al. (1987) *Nature, 329,* 118. Major cometary showers, involving 10^9 comets greater than 3 km in diameter, might occur every 300-500 m.y. and should each result in about 20 impacts on the Earth. Minor showers, involving about 10^8 comets, might occur about every 30-50 m.y. and result in about two terrestrial impacts.
126. Grieve R. A. F. et al. (1986) *Proc. Lunar Planet. Sci. Conf. 18th,* p. 375.
127. Tremaine S. (1986) in *The Galaxy and the Solar System* (R. Smoluchowski et al., eds.), p. 409, Univ. of Arizona, Tucson.
128. For example, Frank L. A. et al. (1986) *Geophys. Res. Lett., 13,* 307; Donahue T. M. et al. (1987) *Nature, 330,* 548.
129. Hall D. T. and Shemansky D. E. (1988) *Nature, 335,* 417.

130. Weissman P. R. (1989) in *Origin and Evolution of Planetary and Satellite Atmospheres* (S. K. Atreya et al., eds.), p. 258, Univ. of Arizona, Tucson.
131. Fechtig H. et al. (1974) *Proc. Lunar Sci. Conf. 5th*, p. 2463.
132. Kyte F. (1986) in *Lunar and Planetary Science XVII*, p. 452, Lunar and Planetary Institute, Houston.
133. 16,000 tons/year; Hughes D. W. (1978) in *Cosmic Dust* (J. A. M. McDonnell, ed.), p. 123, Wiley, New York.
134. 6000-11,000 tons/year; Tuncel G. and Zoller W. H. (1987) *Nature, 329*, 703.
135. Basilevsky A. T. et al. (1987) *J. Geophys. Res., 92*, 12,869.
136. Barlow N. G. (1988) in *Lunar and Planetary Science XIX*, p. 33, Lunar and Planetary Institute, Houston.
137. Hartmann W. K. et al. (1981) *Basaltic Volcanism on the Terrestrial Planets*, p. 1080, Pergamon, New York.
138. Ibid., p. 1049.
139. Flynn G. J. and McKay D. S. (1988) *LPI Tech. Rpt. 88-07*, p. 77.
140. Mayer K. A. and Stevenson D. J. (1988) *Nature, 331*, 612; see, however, Oberbeck V. R. and Fogelman G. (1989) *Nature, 339*, 434, for some corrections to the caculations in the previous paper.
141. Sleep N. H. et al. (1989) *Nature, 342*, 139.
142. A useful and up-to-date review on the origin of life is given by Joyce G. F. (1989) *Nature, 338*, 217.
143. Schidlowski M. et al. (1979) *Geochim. Cosmochim. Acta, 43*, 189; Schidlowski M. et al. (1988) *Nature, 333*, 313.
144. Buick R. et al. (1981) *Alcheringa, 5*, 161.
145. Miller S. L. (1953) *Science, 117*, 528; Miller S. L. and Urey H. C. (1959) *Science, 130*, 245.
146. Dyson F. (1987) *The Origin of Life*, Cambridge Univ., New York.
147. Anders E. (1989) *Nature, 342*, 255.
148. Mayer K. A. and Stevenson D. J. (1988) *Nature, 331*, 612; Oberbeck V. R. and Fogelman G. (1989) *Nature, 339*, 434.
149. Lowe D. R. and Byerly G. R. (1986) *Geology, 14*, 83; Kyte F. T. et al. (1988) *Meteoritics, 23*, 284.
150. Lowe D. R. and Byerly G. R. (1988) in *Lunar and Planetary Science XIX*, p. 693, Lunar and Planetary Institute, Houston; Byerly G. R. and Lowe D. R. (1988) in *Lunar and Planetary Science XIX*, p. 152, Lunar and Planetary Institute, Houston; Lowe D. R. et al. (1989) *Science, 245*, 959.
151. Chyba C. F. (1990) *Nature, 348*, 114.
152. Plumb R. C. et al. (1989) *Nature, 338*, 633.
153. Oro J. (1988) *LPI Tech. Rpt. 88-07*, p. 134.
154. Wickramasinghe D. T. (1987) *Philos. Trans. R. Soc., A323*, p. 377, Fig. D1.
155. Kroto H. W. (1987) *Philos. Trans. R. Soc., A323*, p. 378.
156. Raup D. M. and Sepkoski J. J. (1984) *Proc. Nat. Acad. Sci. USA, 81*, 801.
157. Those readers wishing for a succinct summary of the large and confusing literature on this topic should consult Tremaine S. (1986) in *The Galaxy and the Solar System* (R. Smoluchowski et al., eds.), p. 409, Univ. of Arizona, Tucson.
158. Noma E. and Glass A. L. (1987) *Geol. Mag., 124*, 322.
159. Tremaine S. (1986) "Is there evidence for a solar companion star?" in *The Galaxy and the Solar System* (R. Smoluchowski et al., eds.), p. 409, Univ. of Arizona, Tucson.
160. Grieve R. A. F. et al. (1986) *Proc. Lunar Planet. Sci. Conf. 18th*, p. 375.
161. Tremaine S. (1986) in *The Galaxy and the Solar System* (R. Smoluchowski et al., eds.), p. 413, Univ. of Arizona, Tucson.
162. Feller W. (1971) *An Introduction to Probability Theory and its Applications, Vol. 2*, 2nd edition, p. 76, Wiley, New York.
163. Weissman P. R. (1989) in *Origin and Evolution of Planetary and Satellite Atmospheres* (S. K. Atreya et al., eds.) p. 261, Univ. of Arizona, Tucson.
164. de Silva S. L. and Sharpton V. L. (1988) in *Lunar and Planetary Science XIX*, p. 273, Lunar and Planetary Institute, Houston.
165. Gilmour I. and Anders E. (1988) in *Lunar and Planetary Science XIX*, p. 389, Lunar and Planetary Institute, Houston.
166. Carter N. L. et al. (1986) *Geology, 14*, 380.
167. de Silva S. L. and Sharpton V. L., op. cit.; Sharpton V. L. and Schuraytz B. C. (1989) *Geology, 17*, 1040.
168. Bohor B. F. (1988) *Meteoritics, 23*, 258.
169. Kunk M. J. et al. (1989) *Science, 244*, 1565.
170. Grieve R. A. F. (1989) *Nature, 340*, 428.
171. Hildebrand A. R. and Boynton W. V. (1988) *Meteoritics, 23*, 274.
172. O'Keefe J. D. and Ahrens T. J. (1988) in *Lunar and Planetary Science XIX*, p. 885, Lunar and Planetary Institute, Houston.
173. Berner R. A. et al. (1983) *Am. J. Sci., 283*, 641.
174. Weissman P. R. (1989) in *Origin and Evolution of Planetary and Satellite Atmospheres* (S. K. Atreya et al., eds.) p. 241, Univ. of Arizona, Tucson.

Chapter 5

The Planets

Illustration, preceding page—*Neptune and Triton as seen by Voyager. The outer limit of the planetary system is illustrated by this view of Neptune and Triton. The rim of Neptune (radius 24,764 km) is in the foreground. The thin crescent of the satellite Triton (radius 1352 km) is visible in the distance. (NASA JPL Voyager P-34761.)*

The Planets

5

5.1. A DIFFICULT TASK

"Estimating the compositions of planets has proved to be one of the most difficult tasks in cosmochemistry. Even determining the composition of the earth is a challenge, because planetary differentiation has ensured that there is no place on or within it where we can collect a sample that has the bulk composition of the whole planet" [1].

The planets did not accrete from fine dust, but rather from a hierarchy of planetesimals that included bodies the size of the smaller terrestrial planets in the final stages. The terrestrial planets are all distinct in composition, density, or in oxygen isotopes. They thus preserve some memory of variations or compositional differences either originally present in or induced by fractionation within the nebula. The process of planetary accretion has blurred, but not homogenized these zones, just as collisions have smoothed, but not removed the variations in the asteroid belt. A major assessment is that the planets were assembled from rather localized regions of the solar nebula. These, however, were not individual and separate for each planet, and considerable mixing, as well as mass loss from large collisions, has occurred. It is also clear that there is not a simple variation in composition with heliocentric distance. Probably there was some radial migration of planetesimals. However, there appears to have been little input to the Earth from the asteroid belt, for the present population of meteorites, at least, are not representative building blocks of this planet (section 3.14). The planets are not identical in composition as strict application of such accumulation models might suggest, but they display many differences in composition, reflecting the operation of stochastic processes.

Planetesimal accretion was clearly efficient. Apart from the asteroid belt, the solar system is remarkably clean and there is an absence of stranded small bodies. There appears to be no sign of the "vulcanoids," possible impactors in the region of Mercury (section 4.8). The immense gap, 10 AU wide, between Saturn and Uranus appears to be populated only by tiny Chiron, no other objects up to 3 magnitudes fainter having been detected in this void [2].

Although increasing knowledge of the terrestrial planets revealed many differences in composition among them, the giant planets might at least have been thought to be undifferentiated fragments of the solar nebula, similar in composition to the Sun. Such a proposition would simplify all models of planetary evolution. However, as we gained more information about them, they all turned out, like their terrestrial relatives, to be compositionally distinct.

In this chapter, I have selected the evidence, from the wealth of information available for the planets, relevant to the major issues discussed in this book. Thus, there is a bias toward geochemical observations and considerable variability in treatment of the individual planets. My endeavor is to explain what is observed, with an emphasis on new data. Thus, there is some discussion of the surface features of Venus and Mars since these convey useful compositional information, but in a book of this length, which deals with the whole solar system, a high degree of selectivity is inevitable. Although there are individual sections on each of the terrestrial planets, discussions of their cores, mantles, crusts, atmospheres, and hydrospheres occur in several other sections.

5.2. GIANT AND TERRESTRIAL PLANETS

Although the solar system is customarily divided into inner rocky and outer gaseous planets, there are really three basic divisions of planetary bodies: terrestrial or inner planets, gas giants (Jupiter and Saturn), and ice giants (Uranus and Neptune). The principal compositional differences are listed in Table 5.2.1. There is little reason to suppose that this diversity would be repeated in another solar system, since it seems to depend on a number of chance events, from the initial size of the molecular cloud fragment to the appearance of life on the Earth. In this book my approach is to try to describe what is actually present, rather than to attempt to fit the observations into some universal theory of planetary formation. Too many chance events intervene.

The major event in the solar system was the early appearance of Jupiter. Why is it so large? Probably because it formed at a critical location just past the "snow line" [3]. This enabled it to form a core of

TABLE 5.2.1a. The terrestrial planets.

Orbital Elements

	Mean Distance from Sun (AU)*	Mean Distance from Sun (10^6 km)	Orbital Period (day)	Mean Orbital Velocity (km/s)	Eccentricity	Inclination to Ecliptic (deg)	Obliquity
Mercury	0.387	57.9	87.969	47.89	0.2056	7.00	0°
Venus	0.723332	108.2	224.701	35.05	0.006787	3.39	~177°
Earth	1.000	149.6	365.256	29.79	0.0167	—	24°25′
Mars	1.524	227.9	686.980	24.13	0.0934	1.85	25°2′

*1 AU = 149,597,870 km

Physical Properties

	Mercury	Venus	Earth	Mars
Mass (Earth = 1)	0.0559	0.81503	1.0000	0.1074
Mass (kg)	3.303×10^{23}	4.871×10^{24}	5.976×10^{24}	6.421×10^{23}
Equatorial radius (km)	2439	6051.3	6378	3398
Equatorial radius (Earth = 1)	0.382	0.949	1.000	0.532
Mean density (g/cm^3)	5.42	5.24	5.52	3.94
Zero pressure density (g/cm^3)	5.3	3.95	4.0	3.75
Equatorial surface gravity (m/s^2)	3.78	8.87	9.78	3.72
Equatorial escape velocity (km/s)	4.3	10.4	11.2	5.0
Sidereal rotation period	58.65 days	243.01 days	23.9345 hr	24.6229 hr

Data from Colin L. (1983) in *Venus* (D. M. Hunten et al., eds.), p. 17, Univ. of Arizona, Tucson.

TABLE 5.2.1b. Data for the giant planets.

Parameters		Jupiter	Saturn	Uranus	Neptune
Semimajor axis (mean distance from Sun)	AU	5.203	9.523	19.164	29.987
	10^6km	778.4	1424.6	2866.9	4486.0
Eccentricity		0.049	0.053	0.046	0.0099
Inclination to ecliptic		1°18′	2°29′	0°46′	1°47′
Mean orbital velocity km/s		13.06	9.65	6.80	5.43
Sidereal period of revolution		11.86 yr	29.46 yr	84.01 yr	164.1 yr
Aphelion distance 10^6 km		816.5	1500.1	2998.8	4539.8
Perihelion distance 10^6 km		740.3	1349.1	2735.0	4432.2
Mass (Earth = 1)		317.9	95.2	14.5	17.2
Equatorial radius (km)		71,600	60,000	25,559	24,764
Mean density g/cm^3		1.326	0.686	1.267	1.640
Equatorial surface gravity (cm s^{-2})		2288	905	830	1115
Rotation period, synodic (hr)		9.925	10.23	17.24	15.6
Inclination of equator to orbit		3°06′	26°42′	97°0′	29°8′

Data from *Science* (1986) *233*, 39; *Science* (1989) *246*, 1466; and other sources.

10-15 Earth masses, which then accreted much of the gas being swept out of the inner solar system during the T Tauri and FU Orionis stages of early violent solar activity. In such a model, while Jupiter forms early and accretes the H and He left over after the formation of the Sun, the terrestrial planets form much later in a gas-free environment, from the residual planetesimals in the inner solar system.

The solar system is not a failed double-star system, and the formation of a planetary system is a consequence of the mass, angular momentum, and other physical properties of the fragment that became detached from the molecular cloud. There does appear to be a real gap between Jupiter-sized objects and the smallest stars, which may explain the great scarcity or even absence of brown dwarfs (see section 2.10.3).

In addition to the compositional and density differences between the inner and outer planets, there is also the striking difference in the distribution of mass. There is no equivalent to the Titius-Bode rule of planetary spacing for the distribution of mass in the solar system, indicating that the famous rule is a secondary consequence of planetary dynamics (see section 1.6).

5.3. MERCURY

Although there is a common perception that Mercury is so close to the Sun that it is difficult to observe, it is often clearly visible near the horizon, at least in the southern hemisphere. Like the Moon, the real significance of the anomalous nature of Mercury has only recently been appreciated; both these objects have played the role of red herrings in our attempts to understand the solar system [4]. The high density of Mercury was a particular trap, since it could be related to the proximity of that planet to the Sun, and thus appear to fit into a grand overall scheme of decreasing temperature and density with distance.

The spin and orbital configuration of Mercury are strange. It might have been supposed for a small planet (Mercury's mass is only about 5.5% of that of the Earth), closest of all to the massive Sun, that these properties would be the most regular in the solar system. However, the orbital eccentricity (0.2) and the inclination of the orbit to the plane of the eclipitic (7.2°) are larger than those of the other planets, being exceeded only by those of Pluto! Possibly the orbital plane is linked to the equatorial plane of the Sun, which has a 7° inclination to the plane of the ecliptic, a fact of obscure significance that is rarely commented upon [5].

5.3.1. Density

The best estimate for the mean density of Mercury is 5.43 ± 0.01 g/cm^3 [6]. The uncompressed density of about 5.3 g/cm^3 is the highest for any planet. This contrasts with the much lower values of about 3.75 g/cm^3 for Mars and about 4.0 g/cm^3 for the uncompressed densities of Venus and the Earth. At present, the moment of inertia is unknown. This property could distinguish between a homogeneous planet ($I/MR^2 = 0.4$) and one with an iron core ($I/MR^2 = 0.34$). However, the surface appearance, high albedo, and spectral signature (somewhat reminiscent of the lunar highlands), combined with the presence of a magnetic field, all suggest that Mercury is a well-differentiated planet, with an iron core about 0.75 of the planetary radius.

5.3.2. Magnetic Field

The magnetic field, although weaker than that of the Earth, is a dipole field of internal origin and is most likely caused by an active dynamo in an outer fluid core, which is the same origin as the terrestrial field. The dipole moment is about 3×10^{22} oersted cm^2, and the equatorial surface field is 0.0035 oersted, compared with 0.31 oersted for the Earth [7]. Thus, part of the core is probably still liquid, raising interesting geochemical considerations. A pure iron core would have frozen long ago, so the most likely candidate is an FeS core. This raises questions as yet unanswered about the state of both the venusian and martian cores, particularly since the latter is generally considered to be rich in FeS (see section 5.6).

The presence of the volatile element sulfur as a constituent of the planet closest to the Sun has important implications for models of planetary accretion. If Mercury contains a substantial (2-3%) sulfur content, then this removes much of the rationale for a heliocentric zoning of nebular composition. Models in which Mercury accretes from high-temperature components only are no longer viable. If the innermost planet has a substantial volatile component (although FeS is the probable source of the sulfur), there is little basis for condensation models of planetary accumulation based on heliocentric distance. "The consensus that Mercury is a refractory-rich planet has been so widespread that the possibility that Mercury could be volatile-rich has not been considered seriously, even though such a model is entirely consistent with the observed mean density of the planet" [8].

5.3.3. Surface Composition

The question of the composition of the surface, as interpreted from remote sensing data, is particularly frustrating. Figure 5.3.1 shows the comparison between the integrated spectrum for the mercurian surface and that of a laboratory spectrum for Apollo 16 lunar highlands samples. All that can be said is that the mercurian data are consistent with a regolith surface, possibly containing Fe or Ti-bearing agglutinates. Spectral data from thermal infrared spectra exclude both ultramafic and granitic compositions, and appear to be compatible with a feldspathic crust similar to that of the lunar highlands [9]. The presence of the sodium cloud around Mercury is consistent with the surficial presence of sodium-containing minerals such as plagioclase or perhaps pyroxene [10]. It seems less likely that comets, which approach Mercury at high velocities, are responsible for coating the surface with alkali and other elements. The lunar surface certainly provides no supporting evidence for such a veneer.

A small amount of further information can be gleaned about composition from the surface morphology (Fig. 5.3.2). There are two types of plains units on Mercury, the early intercrater plains of pre-Tolstojan age and the later "smooth" plains of Calorian age (Table 5.3.1 gives a stratigraphic scheme for Mercury). The origin of both sets of plains is disputed between those who equate them with the lunar Cayley-type plains (debris-sheets from impacts) [11] and those who propose that they are volcanic plains analogous to the lunar maria, and hence are formed by rather fluid basaltic lava flows [12]. It is interesting to note parallels with the history of the interpretation of the lunar surface in this controversy.

If the intercrater plains are volcanic, the mercurian lavas must possess the same low viscosity as the lunar basalts; this assumption remains to be tested, but it would imply a similarity in mantle sources, that, given the unique nature of lunar evolution, seems surprising. Their albedo (Table 5.3.2) is twice as high as that of the lunar maria, a fact not consistent with the presence of low-viscosity iron-rich lavas. The wrinkle ridges on the floor of the Caloris Basin, cited in support of an analogy with lunar mare basalts, appear to be very large. Only more data will solve this problem, but it is clear that if these features are volcanic, then their petrogenesis is distinct from that of the lunar mare basalts. The lunar experience in

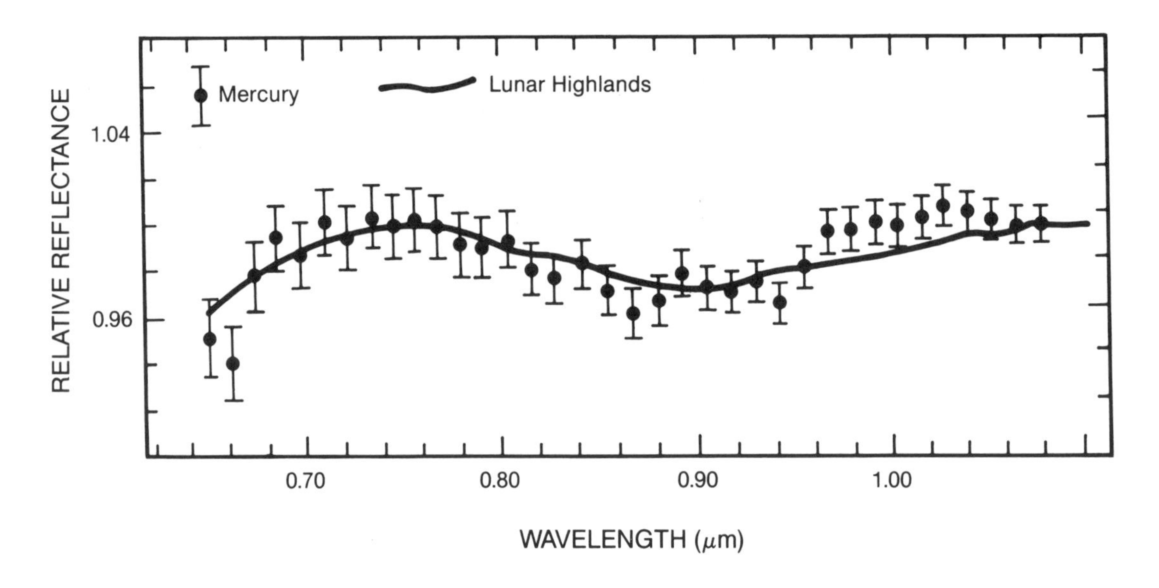

Fig. 5.3.1. *The reflectance spectrum of the surface of Mercury, covering both smooth and intercrater plains, shows a close resemblance to a laboratory spectrum of an Apollo 16 lunar highland soil sample. After Vilas F. (1988) in Mercury (F. Vilas et al., eds.), p. 71, Fig. 6, Univ. of Arizona, Tucson.*

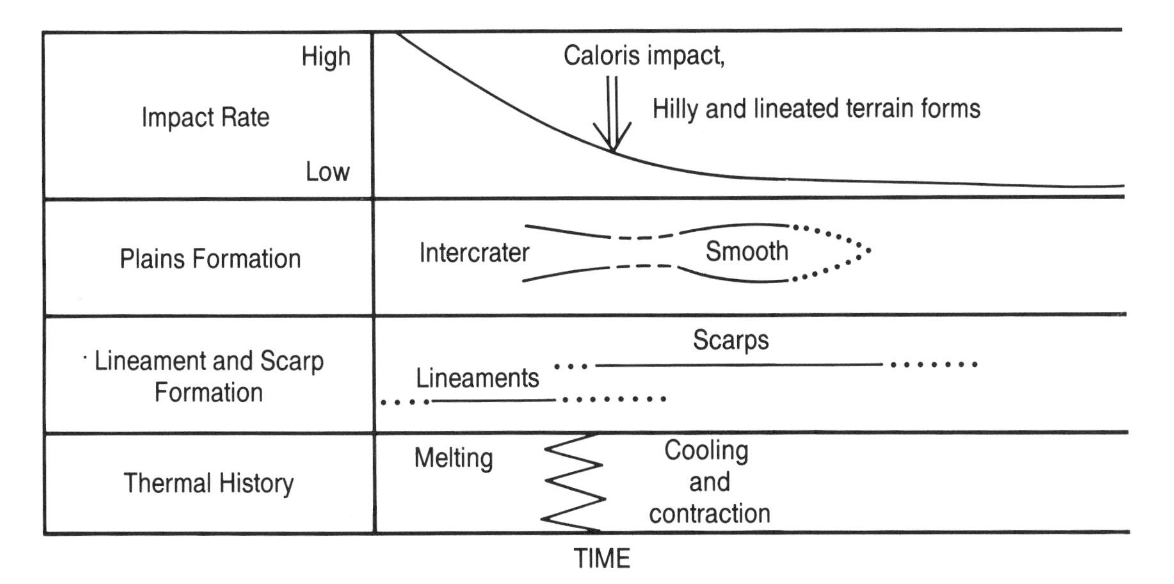

Fig. 5.3.2. *The major tectonic and thermal events in mercurian history. After Melosh H. J. and McKinnon W. B. (1988) in Mercury (F. Vilas et al., eds.), p. 399, Fig. 12, Univ. of Arizona, Tucson.*

TABLE 5.3.1. The mercurian stratigraphic sequence.

System	Major Units	Approx. Age of Base of System*
Kuiperian	Crater materials	1.0 b.y.
Mansurian	Crater materials	3.0-3.5 b.y.
Calorian	Caloris Group; plains, crater, small-basin materials	3.9 b.y.
Tolstojan	Goya Formation; crater, small-basin, plains materials	3.9-4.0 b.y.
Pre-Tolstojan	Intercrater plains, multiring basin, crater materials	pre-4.0 b.y.

* Approximate ages based on the assumption of a lunar-type impact flux.

Data from Spudis P. D. and Guest J. E. (1988) in *Mercury* (F. Vilas et al., eds.), p. 138, Table 1, Univ. of Arizona, Tucson.

TABLE 5.3.2. Comparison of lunar and mercurian albedos.

Terrain	Albedo (%)
Lunar maria	6-7
Mercury Caloris smooth plains	12-13
Lunar highlands	10-11
Mercury highlands (intercrater plains)	16-18
Lunar bright rayed craters	15-16
Mercury bright rayed craters	36-41

Data from Strom R. G. (1988) *Mercury: The Elusive Planet*, p. 94, Smithsonian, Washington, DC.

trying to interpret surface composition from photographic evidence alone should induce caution. Thus, following the study of the lunar Ranger and Orbiter photographs, before the lunar sample return, it was remarked that "a surface cannot be characterized by its portrait . . . the heightened resolution of the pictures did not resolve the arguments. The moon remained inscrutable at all scales" [13].

5.3.4. Evidence for Planetary Contraction

One of the most informative features about Mercury is the evidence that it has undergone a slight contraction in planetary radius at a very early stage of its history (a summary of the tectonic history is given in Fig. 5.3.2). There may have been an early period of mild expansion associated with melting and tidal despinning of the planet. The mercurian grid pattern bears some resemblance to the "lunar grid." However, such patterns on the better-resolved lunar surface are due mainly to lineaments produced during basin-forming impacts rather than due to any underlying tectonic cause, and the mercurian grid pattern may be due to the same cause.

The famous lobate scarps (Fig. 5.3.3), which average about 1 km in height, indicate that a contraction in planetary radius of about 2-4 km has occurred.

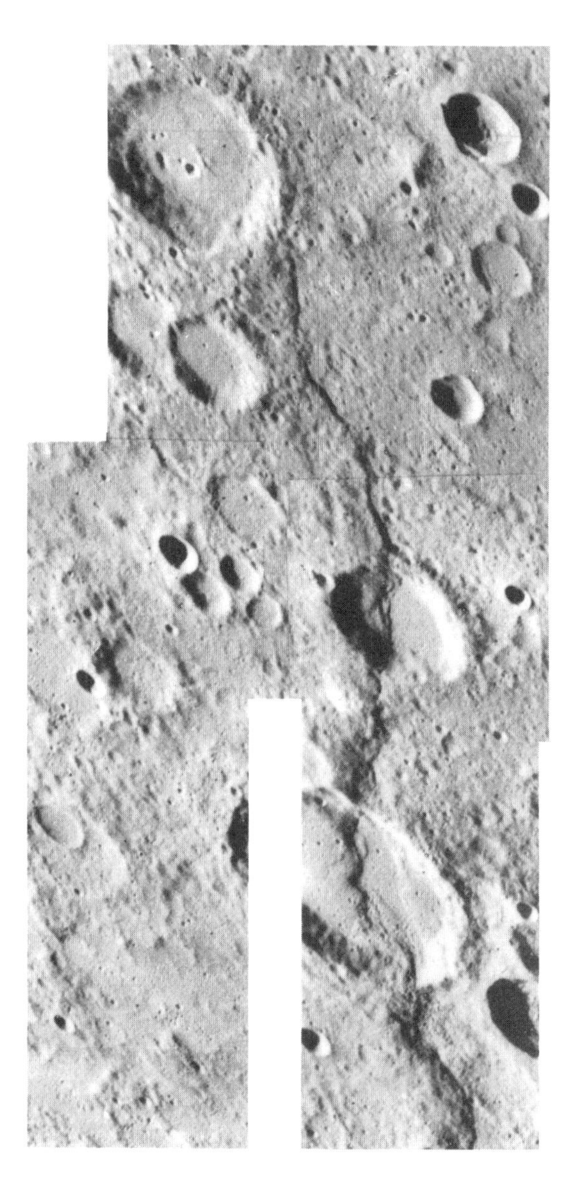

Fig. 5.3.3. *Discovery Scarp, which is one of the largest of the lobate scarps on Mercury, up to 2 km high and 500 km long. (NASA Atlas of Mercury 11-27)*

If the core were totally solidified, a decrease in radius of about 17 km would be expected [14]. Accordingly, much of the contraction is due to the cooling and solidification of the mantle and crust. This contraction occurred toward the end of the massive bombardment and must have occurred earlier than 4000 m.y. by analogy with the dated lunar highland crust. The scarps cut the older, degraded craters and the intercrater plains; younger, fresher craters cut the scarps and some of the smooth plains appear to be younger.

Following the initial contraction, the radius of Mercury has been unchanged for at least 4000 m.y. This fact is fatal for those "expanding Earth" hypotheses that call for a massive increase in the terrestrial radius by a factor of 2 in the last few hundred million years [15].

5.3.5. Composition

Two recent surveys summarize the various models that have been advanced for the composition of Mercury [16]. None of the proposed models are very satisfactory. For example, the "preferred" model for the rocky mantle composition has a range of 10% for silica, a factor of 2 variation for such critical elements as Al, Ca, and Ti, while Na varies by a factor of 5, and Fe by an order of magnitude [17]! These limits are so large that they provide no effective geochemical constraints, since they cover nearly any permissible composition. The combination of high density, magnetic field, and silicate surface indicates that the planet is differentiated, suggesting that Mercury has a large iron core. As noted above, this may well contain a substantial FeS component to account for its probable liquid condition and for the magnetic field. It is also reasonable to suppose that the surface and mantle are composed of silicate that has undergone disruption and heating to an unknown degree by the large impact that most probably stripped off over half the mantle. The presence of Na and K ions in the atmosphere, as well as the albedo and general appearance, suggests something akin to the lunar highland feldspathic crust. If the smooth plains are volcanic, analogous to the lunar maria, then their high albedo suggests a differing composition to the iron-rich lavas on the Moon.

The driving force behind previous attempts to account for Mercury has been to fit the high density of the planet into some preferred overall solar system scheme, such as equilibrium condensation. It has become clear that none of these proposed models work, and the high density is conveniently accommodated by the large-impact hypothesis, which makes Mercury unique. Until we have some more data on the surface composition, it seems pointless to add yet another compositional model on the basis of so little evidence for a planet that has been subjected to such a traumatic event. It is of the utmost significance to secure some more compositional data to see whether they are consistent with a giant collision and planetary disruption. For example, was there preferential recondensation of refractory elements, or was the reassembly isochemical. Thus, the K/U ratio, easily measured by spacecraft, would provide the crucial information on volatile/refractory element ratios. The presence of alkali ions in the tenuous atmosphere suggests that some of the volatile element inventory survived. How much is a question of great interest, but realistic geochemical models for the composition of Mercury will have to wait upon further missions.

5.3.6. Origin

Early attempts to explain the composition of Mercury, and in particular its high iron abundance, were seduced by its proximity to the Sun, since it apparently formed one end member of a sequence extending out to the low-density outer planets. This variation appeared to be consistent with a radial decrease in temperature outward from the Sun. Equilibrium condensation models predicted iron-rich material at 0.3 AU. In an alternative model, low-density silicates were preferentially lost to the Sun by gas drag in a gaseous nebula; such processes appear less likely if the terrestrial planets accreted on timescales of 10^7-10^8 yr, by which time the gaseous nebula was long gone.

The two current hypotheses that explain the high density of Mercury are (1) high-temperature evaporation of the silicate mantle [18] and (2) collisional removal of the mantle [19].

In the high-temperature vaporization scenario, loss of about 80% of the silicate is required, which results in the nearly complete depletion of the alkali elements. The recent discovery of the sodium cloud around Mercury [20], coupled with the evidence from reflectance spectra of a lunar-highland-like plagioclase(?) crust [21], seems at variance with the requirements for nearly total loss of alkalis in evaporation models. Of course, sodium and potassium might be added to the surface by comets or from CI meteorite impacts. As noted earlier, the high velocities of accretion at Mercury and the general absence of such exotic material on the lunar surface make this less likely. Removal of much of the silicate mantle during a collision with a smaller body can account at present for the exceedingly sparse geochemical observations on Mercury. Current estimates [22] place the impactor at 0.20 Mercury masses, with an impact velocity of 20 km/s. The giant collisional explanation for the high density of Mercury places core formation very early, in line with most current thinking about this problem for the terrestrial planets, and emphasizes the stochastic nature of planetary and satellite formation [23]. In this scenario, the inital mass of Mercury would be about twice its present value. The disrupted silicate material, with a grain size less than a centimeter, is either swept away by the Poynting-Robertson effect (solar radiation pressure) or possibly accreted onto proto-Venus or proto-Earth if the orbit of proto-Mercury was crossing the orbits of these much larger bodies.

Impact energies on smaller bodies may exceed those needed to cause planetary disruption by an order of magnitude. Mercury may thus be considered as a survivor, the only one left of several such sized bodies that formerly populated the inner solar system. In contrast, massive collisions on Earth-sized planets are likely to lead to such events as the formation of the Moon; impact energies are less than 20% of that needed for disruption.

5.4. VENUS

The brilliance of this planet, rising in the morning or setting in the evening skies, has been admired since antiquity. More recently it has attracted special interest as an apparent twin of the Earth. The uncompressed density is 3.95 g/cm^3 as compared with 4.0 g/cm^3 for the Earth. Major differences from the Earth include the lack of a satellite, the slow retrograde rotation, and the thick atmosphere with its high noble gas content.

Owing to the slow rotation period (243 days), no value is available for the moment of inertia, but the similarity in density to the Earth and the obvious presence of surface lavas from a differentiated silicate mantle imply a similar internal structure of core and mantle for this planet [24]. The absence of a magnetic field is usually attributed to the slightly smaller size of Venus, which is responsible for the probable absence of a solid inner core. It is the freezing of the Earth's inner core that is thought to drive the terrestrial core dynamo and so produce the magnetic field. This difference between the Earth and Venus is a good illustration that significant differences in planets may arise from small changes in planetary dimensions.

5.4.1. Crustal Composition

Erosion appears to be minimal on Venus, so there is little likelihood of finding a sample such as the martian soil or terrestrial loess that could provide an overall average of the venusian surface. Instead, the available data are limited to spot samples. The early gamma ray data from Venera 8 (Table 5.4.1) indicated high K (4%), U (2.2 ppm), and Th (6.5 ppm)

TABLE 5.4.1a. Major-element composition of Mars, Venus, and Moon surface material.

	Mars		Venus			Moon
	(1)	(2)	(3)	(4)	(5)	(6)
SiO_2	55.01	50.5	46.8	49.8	51.5	45
TiO_2	0.78	0.89	1.66	1.28	0.23	0.56
Al_2O_3	9.1	7.74	16.4	18.3	18	24.6
FeO	19.7	20.2	9.7	9.0	8.7	6.6
MgO	7.5	9.13	11.7	8.3	13	6.8
CaO	7.1	9.82	7.4	10.5	8.5	15.9
Na_2O	—	1.50	(2.1)	(2.5)	—	0.45
K_2O	<0.6	0.19	4.2	0.2	0.1	0.075
Σ	100	100	100	100	100	100

1. Viking lander, Chryse Planitia; Clark B. C. et al. (1982) *J. Geophys. Res., 87,* 10059.
2. Shergotty meteorite; Dreibus G. and Wänke H. (1985) *Meteoritics, 20,* 367.
3. Venera 13; Surkov Yu. A. (1984) *Proc. Lunar Planet. Sci. Conf. 14th,* in *J. Geophys. Res., 89,* B393.
4. Venera 14; Surkov Yu. A. (1984) *Proc. Lunar Planet. Sci. Conf. 14th,* in *J. Geophys. Res., 89,* B393.
5. Vega 2; Surkov Yu. A. et al. (1987) *Proc. Lunar Planet. Sci. Conf. 17th,* in *J. Geophys. Res., 92,* E537.
6. Lunar highland crust; Taylor S. R. (1982) *Planetary Science: A Lunar Perspective,* Lunar and Planetary Institute, Houston; Newsom H. E. and Taylor S. R. (1989) *Nature, 338,* 29.

TABLE 5.4.1b. Potassium, uranium, and thorium in Venus surface rocks.

		Venera			Vega	
		8	9	10	1	2
K	(%)	4.0	0.47	0.30	0.45	0.40
U	(ppm)	2.2	0.60	0.46	0.64	0.68
Th	(ppm)	6.5	3.65	0.70	1.5	2.0

Data from Surkov Yu. A. et al. (1987) *Proc. Lunar Planet. Sci. Conf. 17th,* in *J. Geophys. Res., 92,* E537.

values. These are typical of abundance levels observed in terrestrial granites, and immediately led to the speculation that granite was present on the surface of Venus. The existence of small continent-sized areas such as Aphrodite Terra and Ishtar Terra raised the possibility that these high-standing areas might thus result from the presence of low-density rocks analogous to the terrestrial continents. This would imply an evolution for Venus resembling that of the Earth.

The Venera 9 and 10 gamma ray data, in contrast to the earlier values, gave low values for K, U, and Th, consistent with the levels seen in terrestrial basaltic rocks. The Venera 13 and 14 XRF experiment resolved this question of the petrological nature of the surface by providing data on the major-element composition [25]. The Venera 13 analysis resembles terrestrial alkali basalt, while the Venera 14 data are close in composition to terrestrial midocean ridge basalts. (The high S and Cl values must represent surficial volcanic additions, in the absence of an oceanic sink for these elements. Mars shows similar high surficial values for these elements for the same reason). The high K_2O value of 4% in the alkali basaltic composition measured by Venera 13 indicates that the high K, U, and Th values in the Venera 8 gamma ray data [26] may also be due to alkali basalt or perhaps a rhyolitic "pancake," but does not indicate the presence of large areas of granite, analogous to the terrestrial continental crust.

All these analyses are from near Beta Regio, or from low-lying areas thought to be basaltic on topographic grounds. However, the Vega 2 XRF experiment provided major-element data from the eastern flank of the high-standing Aphrodite Terra. This analysis also turned out to be basaltic (Table 5.4.1), with a very low value for K_2O confirmed by the gamma ray data [27]. Gamma ray data from Vega 1 showed similar low values for K, U, and Th (Table 5.4.1).

The crustal composition appears to be dominantly basaltic over the considerable range in elevations observed. Thus, the high-standing regions of Aphrodite Terra and Ishtar Terra are probably due to tectonic rather than compositional controls [28]. This is generally supported by the topographical information, which appears to indicate extensional deformation in equatorial regions (e.g., Aphrodite Terra) [29] and compressional deformation in northern high latitudes (Atalanta Planitia and Ishtar Terra) [30]. Pyroclastic deposits appear to be rare, probably due to the difficulty of maintaining fire fountain eruptions in the thick venusian atmosphere. The interpretation of the surface as basaltic is reinforced by the presence of thousands of small shield volcanoes, typically 1-10 km in diameter with slopes of about 5°, which occur on the volcanic plains. Over 21,000 have been identified. They most closely resemble terrestrial oceanic floor seamounts in size range and density [31].

In summary, the venusian crust appears to be dominated by basaltic lavas and the presence of wide areas of more fractionated rocks is minimal. Thus, the production of extensive regions of granite appears unlikely on Venus, and this familiar rock type may be restricted to the Earth [32]. A small number of "pancake" features, about 20 km in diameter, which resemble terrestrial rhyolite domes, have been observed. These are probably formed by the extrusion of viscous lavas of rhyolitic composition, formed as residual melts during crystal fractionation of basaltic magma chambers. Whether Venera 8 landed on such a "pancake" is conjectural [33]. It is interesting that the heat-producing elements K, U, and Th are concentrated near the planetary surface on Venus in a manner different from that observed terrestrially, where they are concentrated in the siliceous continents.

5.4.2. Age of the Surface

The surface of Venus appears to be relatively young based on crater counting studies, with estimates averaging about 400 m.y. [34]. This indicates that the observable crust is not the result of some primordial melting event such as is responsible for the lunar highlands, but is secondary (see section 5.9 for a discussion of the distinctions between primary and secondary crusts).

The age of the venusian surface is nearly an order of magnitude older that that of the terrestrial oceanic crust. If the venusian crust is analogous to the terrestrial oceanic crust, the resurfacing rate appears to be an order of magnitude less on Venus than on the Earth, despite similar planetary size and similar abundances of radioactive elements.

5.4.3. Heat Loss, Rates of Volcanism, and Crustal Thickness

It is assumed that the planetary budget of K, U, and Th is similar to that of the Earth. The K/U ratios measured by the U.S.S.R. Venera and Vega missions are similar to terrestrial values (about 10^3).

Most of the heat loss from the Earth takes place through the formation of oceanic lithosphere at the midocean ridges, where about 18 km^3 of lava are produced per year. Another 2 km^3 of volcanic rocks are produced at island arcs or in hot-spot intraplate volcanism [35]. In contrast, the production rate for lavas on Venus is about 1 km^3 per year, which can account only for a trivial amount of the planetary heat loss.

Like Venus, the Moon, Mercury, and Mars lose most of their heat through lithospheric conduction [36]. Crustal growth on Venus is probably mainly accomplished by hot-spot volcanism and vertical accretion, with minor recycling, in contrast to the dominant horizontal growth and recycling of the terrestrial basaltic crust. The crustal thickness appears to be of the order of 10-30 km [36], so that its volume is of the same order as that of the present terrestrial basaltic oceanic crust. The basalt-to-dense-eclogite transition, which would facilitate sinking and recycling of the crust, occurs at a depth of 65 km on Venus. Such a transition will occur only in areas of overthickened crust and is here judged to be a minor process. If little subduction or recycling is occurring, then Venus has produced over geological time an order of magnitude less basalt that has the Earth.

5.4.4. Tectonics

The crust of Venus differs fundamentally from that of the Earth, both in the absence of a bimodal topography, and in the strong positive correlation between gravity and topography [37]. Peak gravity values are 65-70 mgal. The volcanic region, Beta Regio, is nearly 10 km high. The major mountain ranges are in Ishtar Terra. The Maxwell Montes rise 11 km above the mean venusian datum. Isostatic support of the topography on long geological timescales seems unlikely. The correlation of gravity and topography indicates that the compensation level is very deep (100-1000 km) implying dynamic support [36,37].

There are many examples of compressional tectonics, such as the banded terrain of Ishtar Terra. Lakshmi Planum (centered about 65°N, 335°) at the western end of Ishtar Terra is a unique highland plain about 3 km above the mean planetary radius. The plateau is surrounded by compressional features, and appears to result from tectonic convergence, resulting in crustal thicking [38,39].

Coronae are large (150-1000 km diameter) circular features formed of concentric rings of grooves and ridges. They may be the surface expression of hot spots and of mantle upwelling [36]. Other unique features of the venusian surface include the closely packed sets of grooves and ridges or tesserae, which appear to result from compression [38,39,40]. Most of the surface features can be explained as resulting from mantle plumes. Upwellings and downwellings appear to account for the major surface features [36]. Compressional forces acting on a smaller scale than on the Earth account for the widespread evidence of ridges [39].

In summary, the surface of Venus appears to be dominated by basaltic volcanism, and lateral tectonics, so that the high-standing areas are crumpled up basaltic volcanic crust. Ironically, the ridged terrain bears some resemblance to the tectonic structure, but not to the composition, of the terrestrial granitic continents.

5.4.5. Composition

In discussing the composition of Venus, we are on somewhat more secure ground than is the case for Mercury, where much depends on uncertain analogies. The crust of Venus is clearly young and basaltic, and is thus being currently derived from the mantle. The compositions of the lavas are broadly similar to the familiar terrestrial alkali-rich basalts, leading to the supposition that the mantle of Venus is similar to that of the Earth. The probable absence of large surface regions of more silicic rocks is readily attributable to the tectonic differences between the dry one-plate crust of Venus and the wet multiplate subduction environment on Earth. Although geochemical processes of partial melting in the mantle will produce mostly basaltic magmas, these melts will deplete the mantle in incompatible elements. Thus, the heat-producing elements K, U, and Th, will be concentrated over time in the planetary crust. This is a rather common feature in planetary evolution and these near-surface concentrations do not of course imply any enrichment in the bulk planet. The K/U ratios measured on the venusian surface overlap with the terrestrial surficial values, indicating that the depletion of the volatile, relative to the refractory elements, is about the same on both planets.

The density of Venus (5.24 g/cm^3) is about 5% less than that of the Earth (5.514 g/cm^3), and the planet is 320 km smaller in radius. The density difference is mostly due to the lower internal pressures. After correcting for these pressure differences, the uncompressed density of Venus is within about 1% of that of the Earth, consistent perhaps with a slightly smaller core mass, but not requiring any real difference in bulk composition. The often used, but in many respects erroneous, view of Venus as a twin of the Earth, appears to be a reasonable statement of the internal bulk planetary composition. Accordingly, until we acquire some more data, the bulk-Earth primitive mantle and core composition can be used as a guide for the composition of Venus.

Despite the probable similarity in size, density, and composition, there are major differences between Venus and the Earth. Venus rotates very slowly (243 days), it has no satellite, and possesses a thick atmosphere (100 bar, mostly CO_2), which contains about 80 times as much of the nonradiogenic argon isotopes (^{36}Ar, ^{38}Ar) as the Earth. Venus has no detectable magnetic field, the surface temperature is 470°C, and the only detectable water is the atmospheric content of 100 ppm.

Venus lacks the bimodal distribution of topography that reflects the division of the Earth's crust into thick, low-density, sialic continental crust and dense, thin, basaltic oceanic crust. In contrast to the Earth, there is a strong positive correlation on Venus between gravity and topography.

Thus there are remarkable secondary differences between the two planets superimposed on the primary similarities. What are the causes of the differences? The absence of a satellite, the rotation rate, and the atmospheric differences are probably the consequence of a differing collisional history from the Earth [41,42]. The other effects (high surface temperature, absence of a magnetic field, and atmospheric composition) follow from this differing history. The differences seen in surface tectonics (the absence of subduction, the rate of volcanism, and crustal growth) are due principally to the differences in the surficial concentration of water, the atmospheric composition, and surface temperatures. It is sobering to contemplate the differences between those apparent twins, Earth and Venus, in terms of providing a suitable environment for life.

5.5. EARTH

The Earth represents a particularly difficult subject to incorporate in a book dealing with the solar system. What does one say about this planet, unique even by the standards of the solar system? The amount of information available is so large that any attempt to summarize it risks reducing the topic to a trivial discussion. It seems better to assume that any reader of this book is well aquainted with plate tectonics, geological history, and geophysics, or is able to find such information, rather than attempt to condense the material to a trivial level or a recitation of truisms [43].

Another basic dilemma is whether a study of the Earth informs us about the origin and evolution of the solar system? We seem to be placed on one of the least informative locations for this purpose, as indicated, for example, by the difficulty in recognizing that impact cratering is a ubiquitous and important planetary process. Our experience with lunar geology and geochemistry, with their subtle but crucial distinctions from our hard-won terrestrial experience, should warn us of the hazards of trying to extrapolate from unique terrestrial conditions. Possibly it is better to remind ourselves about some areas of which we have little knowledge, for it is ironic that we understand the composition and evolution of the Moon better than that of the Earth [44].

Several other factors conspire to cause problems. The collapse of models relating compositional variations to heliocentric distance make extrapolations from Mercury through Venus, Earth to Mars questionable. Can one deduce general principles from a unique planet accompanied by a unique satellite? In many ways the Earth epitomizes the problem of accounting for the solar system when only one example of each exists.

5.5.1. Composition

The bulk composition of the Earth is not easy to establish. There are two major regions in the Earth about which we have little definite evidence. Those are the lower continental crust and the lower mantle. Both are important for geochemical mass balance calculations, but represent unknowns in attempting to establish the overall composition of the Earth (Table 5.5.1 gives estimates for the composition of these regions). The composition of the upper crust is reasonably well constrained, although there is still uncertainty about the lower crust. This is not a trivial problem, for although the continental crust is less than half of one percent of the mass of the Earth, it contains a significant proportion of the incompatible elements (Table 5.5.1). The composition of the upper mantle is also relatively well understood, estimates being derived from volcanic rocks and xenoliths derived from the upper 200 km or so. This is given in Table 5.5.1, along with the composition of the "primitive" mantle (= present mantle and crust). This composition has CI ratios for most of the major elements, in the absence of any more definitive basis for making an estimate (see Fig. 5.5.1).

5.5.2. Chondritic Interelement Ratios and Terrestrial Mg/Si Values

The compositions given in Table 5.5.1 are discussed in detail in the references given in the table footnotes, but some general observations may be made. The refractory elements (e.g., Ca, Al, REE, Th, U, Sr, etc.) do not appear to separate in cosmochemical processes, except under the extreme conditions of the formation of the refractory CAIs (section 3.3), so that their ratios are unchanged in the Earth relative to the CI values. An interlocking set of ratios (K/U, K/Rb, Rb/Sr, Sr/Eu, Al/Sr, K/Tl, Rb/Cs, K/La, Ca/La) provides a consistency check on the values for the refractory and volatile element abundances. However, the upper mantle composition has an Mg/Si ratio of 1.14, in contrast to the CI Mg/Si ratio of 0.90 (Fig. 5.5.2). This observation has been the focus of a continuing controversy. The unknown composition of the lower mantle is a decisive parameter. If it has the same composition as the upper mantle, then the Earth has a substantially higher Mg/Si ratio compared to the CI and solar abundances. On the other hand, if the bulk Earth has CI ratios, then the lower mantle is richer in Si, unless some Si is partitioned into the core or has been lost during some high-temperature event. There is some seismic evidence in support of a more Si-rich lower mantle [45], but the question is currently unresolved.

A chemical difference between the upper and lower mantle would mean that substantial fractionation has taken place and that the lower mantle is a silica-rich cumulate or residuum caused by olivine flotation and perovskite sinking [46]. It has even been suggested by proponents of uniform upper and lower mantle compositions that the terrestrial upper mantle Mg/Si ratio of 1.14, rather than the CI ratio of 0.90, is in fact the solar ratio, and that the CI meteorites are enriched in Si, volatilized from the inner nebula, so that the Earth, and not the CI meteorites, reflects the primordial nebular ratio [47]! If this is so, the

TABLE 5.5.1a. Element abundances in primitive terrestrial mantle (present mantle plus crust).

	%
SiO_2	49.9
TiO_2	0.16
Al_2O_3	3.64
FeO	8.0
MgO	35.1
CaO	2.89
Na_2O	0.34
K_2O	0.02
Σ	100.1

Li	0.83 ppm	Rb	0.55 ppm	Eu	131 ppb
Be	60 ppb	Sr	17.8 ppm	Gd	459 ppb
B	0.6 ppm	Y	3.4 ppm	Tb	87 ppb
Na	2500 ppm	Zr	8.3 ppm	Dy	572 ppb
Mg	21.2 %	Nb	0.56 ppm	Ho	128 ppb
Al	1.93 %	Mo	59 ppb	Er	374 ppb
Si	23.3 %	Ru	4.3 ppb	Tm	54 ppb
K	180 ppm	Rh	1.7 ppb	Yb	372 ppb
Ca	2.07 %	Pd	3.9 ppb	Lu	57 ppb
Sc	13 ppm	Ag	19 ppb	Hf	0.27 ppm
Ti	960 ppm	Cd	40 ppb	Ta	0.04 ppm
V	128 ppm	In	18 ppb	W	16 ppb
Cr	3000 ppm	Sn	<1 ppm	Re	0.25 ppb
Mn	1000 ppm	Sb	25 ppb	Os	3.8 ppb
Fe	6.22 %	Te	22 ppb	Ir	3.2 ppb
Co	100 ppm	Cs	18 ppb	Pt	8.7 ppb
Ni	2000 ppm	Ba	5.1 ppm	Au	1.3 ppb
Cu	28 ppm	La	551 ppb	Tl	6 ppb
Zn	50 ppm	Ce	1436 ppb	Pb	120 ppb
Ga	3 ppm	Pr	206 ppb	Bi	10 ppb
Ge	1.2 ppm	Nd	1067 ppb	Th	64 ppb
As	0.10 ppm	Sm	347 ppb	U	18 ppb
Se	41 ppb				

Data from Taylor S. R. and McLennan S. M. (1985) *The Continental Crust: Its Composition and Evolution,* Table 11.3, Blackwell, Oxford.

TABLE 5.5.1b. Major-element composition (wt%) of the terrestrial upper mantle.

	1	2
SiO_2	45	45.1
TiO_2	0.15	0.2
Al_2O_3	3.3	3.3
FeO	8.0	8.0
MnO	0.13	0.15
MgO	39.8	38.1
NiO	0.25	—
Cr_2O_3	0.44	0.4
CaO	2.6	3.1
Na_2O	0.34	0.4
K_2O	0.02	0.03
Σ	100.0	98.8

1. Data from Taylor S. R. and McLennan S. M. (1985) *The Continental Crust: Its Composition and Evolution,* Table 11.2, Blackwell, Oxford.
2. Pyrolite; data from Ringwood A. E. (1975) *Composition and Petrology of the Earth's Mantle,* McGraw Hill, New York.

TABLE 5.5.1c. Chemical composition of the upper portion of the terrestrial continental crust.

	%		Norm		
SiO_2	66.0	Q	15.7		
TiO_2	0.5	Or	20.1		
Al_2O_3	15.2	Ab	13.6		
FeO	4.5	Di	6.1		
MgO	2.2	Hy	9.9		
CaO	4.2	Il	0.95		
Na_2O	3.9				
K_2O	3.4				
	99.9				

Li	20 ppm	Ni	20 ppm	In	50 ppb	Tm	0.33 ppm
Be	3 ppm	Cu	25 ppm	Sn	5.5 ppm	Yb	2.2 ppm
B	15 ppm	Zn	71 ppm	Sb	0.2 ppm	Lu	0.32 ppm
Na	2.89%	Ga	17 ppm	Cs	3.7 ppm	Hf	5.8 ppm
Mg	1.33%	Ge	1.6 ppm	Ba	550 ppm	Ta	2.2 ppm
Al	8.04%	As	1.5 ppm	La	30 ppm	W	2.0 ppm
Si	30.8%	Se	0.05 ppm	Ce	64 ppm	Re	0.5 ppb
K	2.80%	Rb	112 ppm	Pr	7.1 ppm	Ir	0.02 ppb
Ca	3.00%	Sr	350 ppm	Nd	26 ppm	Au	1.8 ppb
Sc	11 ppm	Y	22 ppm	Sm	4.5 ppm	Tl	750 ppb
Ti	3000 ppm	Zr	190 ppm	Eu	0.88 ppm	Pb	20 ppm
V	60 ppm	Nb	25 ppm	Gd	3.8 ppm	Bi	127 ppb
Cr	35 ppm	Mo	1.5 ppm	Tb	0.64 ppm	Th	10.7 ppm
Mn	600 ppm	Pd	0.5 ppb	Dy	3.5 ppm	U	2.8 ppm
Fe	3.50%	Ag	50 ppb	Ho	0.80 ppm		
Co	10 ppm	Cd	98 ppb	Er	2.3 ppm		

Data from Taylor S. R. and McLennan S. M. (1985) *The Continental Crust: Its Composition and Evolution*, Table 2.15, Blackwell, Oxford.

TABLE 5.5.1d. Composition of the bulk continental crust.

	%		Norm		
SiO_2	57.3	Q	6.6		
TiO_2	0.9	Or	6.5		
Al_2O_3	15.9	Ab	26.2		
FeO	9.1	An	26.2		
MgO	5.3	Di	8.7		
CaO	7.4	Hy	24.1		
Na_2O	3.1	Il	1.7		
K_2O	1.1				
Σ	100.1				

Li	13 ppm	Ni	105 ppm	In	50 ppb	Tm	0.32. ppm
Be	1.5 ppm	Cu	75 ppm	Sn	2.5 ppm	Yb	2.2 ppm
B	10 ppm	Zn	80 ppm	Sb	0.2 ppm	Lu	0.30 ppm
Na	2.30%	Ga	18 ppm	Cs	1.0 ppm	Hf	3.0 ppm
Mg	3.20%	Ge	1.6 ppm	Ba	250 ppm	Ta	1.0 ppm
Al	8.41%	As	1.0 ppm	La	16 ppm	W	1.0 ppm
Si	26.77%	Se	0.05 ppm	Ce	33 ppm	Re	0.5 ppb
K	0.91%	Rb	32 ppm	Pr	3.9 ppm	Ir	0.1 ppb
Ca	5.29%	Sr	260 ppm	Nd	16 ppm	Au	3.0 ppb
Sc	30 ppm	Y	20 ppm	Sm	3.5 ppm	Tl	360 ppm
Ti	5400 ppm	Zr	100 ppm	Eu	1.1 ppm	Pb	8.0 ppm
V	230 ppm	Nb	11 ppm	Gd	3.3 ppm	Bi	60 ppb
Cr	185 ppm	Mo	1.0 ppm	Tb	0.60 ppm	Th	3.5 ppm
Mn	1400 ppm	Pd	1.0 ppb	Dy	3.7 ppm	U	0.91 ppm
Fe	7.07%	Ag	80 ppb	Ho	0.78 ppm		
Co	29 ppm	Cd	98 ppb	Er	2.2 ppm		

Data from Taylor S. R. and McLennan S. M. (1985) *The Continental Crust: Its Composition and Evolution*, Table 3.5, Blackwell, Oxford.

TABLE 5.5.1e. Proposed composition of the lower continental crust.

	%	Norm	
SiO_2	54.4	Q	3.7
TiO_2	1.0	Or	2.0
Al_2O_3	16.1	Ab	23.7
FeO	10.6	An	30.4
MgO	6.3	Di	9.8
CaO	8.5	Hy	28.6
Na_2O	2.8	Il	1.9
K_2O	0.34		
Σ	100.0	Mg/Mg + Fe = 0.51	

Li	11 ppm	Ni	135 ppm	In	50 ppb	Tm	0.32 ppm
Be	1.0 ppm	Cu	90 ppm	Sn	1.5 ppm	Yb	2.2 ppm
B	8.3 ppm	Zn	83 ppm	Sb	0.2 ppm	Lu	0.29 ppm
Na	2.08%	Ga	18 ppm	Cs	0.1 ppm	Hf	2.1 ppm
Mg	3.80%	Ge	1.6 ppm	Ba	150 ppm	Ta	0.6 ppm
Al	8.52%	As	0.8 ppm	La	11 ppm	W	0.7 ppm
Si	25.42%	Se	0.05 ppm	Ce	23 ppm	Re	0.5 ppb
K	0.28%	Rb	5.3 ppm	Pr	2.8 ppm	Ir	0.13 ppb
Ca	6.07%	Sr	230 ppm	Nd	12.7 ppm	Au	3.4 ppb
Sc	36 ppm	Y	19 ppm	Sm	3.17 ppm	Tl	230 ppb
Ti	0.60%	Zr	70 ppm	Eu	1.17 ppm	Pb	4.0 ppm
V	285 ppm	Nb	6 ppm	Gd	3.13 ppm	Bi	38 ppb
Cr	235 ppm	Mo	0.8 ppm	Tb	0.59 ppm	Th	1.06 ppm
Mn	1670 ppm	Pd	1 ppb	Dy	3.6 ppm	U	0.28 ppm
Fe	8.24%	Ag	90 ppb	Ho	0.77 ppm		
Co	35 ppm	Cd	98 ppb	Er	2.2 ppm		

Data from Taylor S. R. and McLennan S. M. (1985) *The Continental Crust: Its Composition and Evolution*, Table 4.4. Blackwell, Oxford.

TABLE 5.5.1f. The composition of the terrestrial oceanic crust.

	%	Norm	
SiO_2	49.5	Or	0.89
TiO_2	1.5	Ab	23.7
Al_2O_3	16.0	An	30.7
FeO	10.5	Di	21.0
MgO	7.7	Hy	5.9
CaO	11.3	Ol	14.6
Na_2O	2.8	Il	2.6
K_2O	0.15		
Σ	99.5		

Li	10 ppm	Cu	86 ppm	In	72 ppb	Tm	0.54 ppm
Be	0.5 ppm	Zn	85 ppm	Sn	1.4 ppm	Yb	5.1 ppm
B	4 ppm	Ga	17 ppm	Sb	17 ppb	Lu	0.56 ppm
Na	2.08%	Ge	1.5 ppm	Te	3 ppb	Hf	2.5 ppm
Mg	4.64%	As	1.0 ppm	Cs	30 ppb	Ta	0.3 ppm
Al	8.47%	Se	160 ppb	Ba	25 ppm	W	0.5 ppm
Si	23.1%	Rb	2.2 ppm	La	3.7 ppm	Re	0.9 ppb
K	1250 ppm	Sr	130 ppm	Ce	11.5 ppm	Os	<0.004 ppb
Ca	8.08%	Y	32 ppm	Pr	1.8 ppm	Ir	0.02 ppb
Sc	38 ppm	Zr	80 ppm	Nd	10.0 ppm	Pt	2.3 ppb
Ti	0.90%	Nb	2.2 ppm	Sm	3.3 ppm	Au	0.23 ppb
V	250 ppm	Mo	1.0 ppm	Eu	1.3 ppm	Hg	20 ppb
Cr	270 ppm	Ru	1.0 ppb	Gd	4.6 ppm	Tl	12 ppb
Mn	1000 ppm	Rh	0.2 ppb	Tb	0.87 ppm	Pb	0.8 ppm
Fe	8.16%	Pd	<0.2 ppb	Dy	5.7 ppm	Bi	7 ppb
Co	47 ppm	Ag	26 ppb	Ho	1.3 ppm	Th	0.22 ppm
Ni	135 ppm	Cd	130 ppb	Er	3.7 ppm	U	0.10 ppm

Data from Taylor S. R. and McLennan S. M. (1985) *The Continental Crust: Its Composition and Evolution*, Table 11.6, Blackwell, Oxford.

TABLE 5.5.1g. Bulk composition of the Earth.

	wt%	
SiO_2	34	Mantle and crust Silicates and oxides*
TiO_2	0.1	
Al_2O_3	2.5	
FeO	5.5	
MgO	24	
CaO	2.0	
Na_2O	0.2	
Fe	28.5	Metallic core
Ni	1.7	
S or O	1.5	
	100.0	

* Also K_2O = 140 ppm.

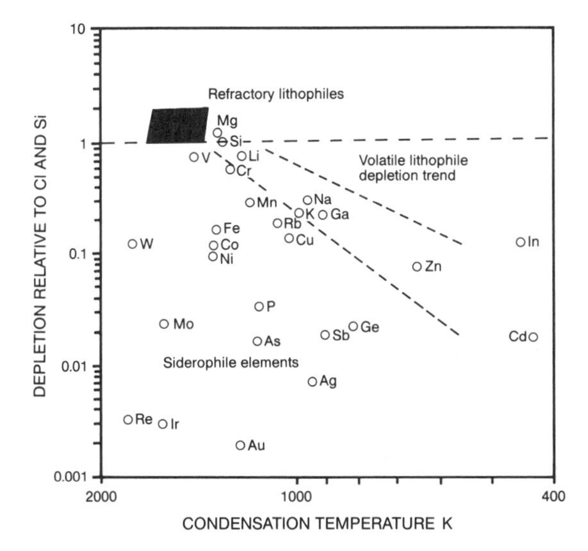

Fig. 5.5.1. *The abundance of elements in the silicate mantle of the Earth, relative to primordial CI abundances, normalized to Si, showing the general depletion of volatile lithophile elements. Courtesy H. E. Newsom, Univ. of New Mexico.*

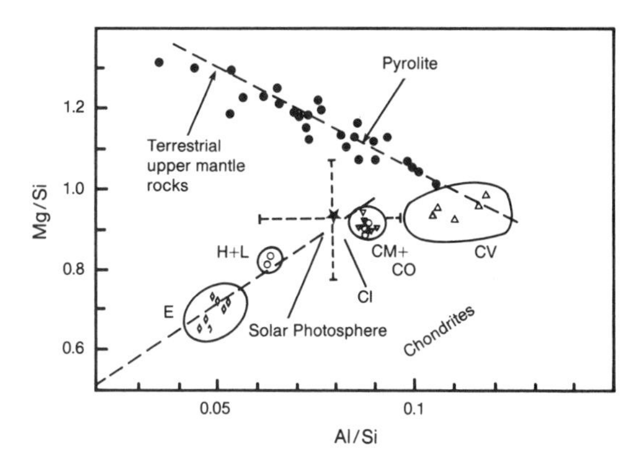

Fig. 5.5.2. *The variation in Mg/Si and Al/Si ratios in terrestrial upper mantle rocks and in meteorites. The composition of the primitive upper mantle lies either at "pyrolite" or further to the right along the terrestrial line. It has a higher Mg/Si ratio than that of the primitive solar nebula, given either by the CI or solar values, and of all chondrites except the CV group. The contrasting variation in Mg/Si and Al/Si among the chondritic groups is commonly ascribed to "cosmochemical" fractionation, but merely reflects the accretion of differing proportions of mineral phases during growth of the chondrite parent bodies in different parts of the nebula, and has no wider significance. After Palme H. (1990) Nature, 343, 23.*

close match between the CI and solar abundances for almost all elements (Fig. 2.15.2) becomes a curiosity.

However, it is not necessary to seek so extreme a solution to the dilemma. There is substantial variation in the Mg/Si ratio and other refractory element ratios among the various classes of meteorites, as discussed in section 3.6. Figure 5.5.2 shows this variation, which has commonly been attributed to unspecified cosmochemical fractionation processes. However, it is merely the result of random mixing of the various components that happened to be accreted. If the lower mantle turns out to have the same composition as the upper mantle, it will merely reinforce the concept that substantial deviations occurred between the compositions of the planetesimals that accreted to form the Earth and the present population of meteorites [48]. A high Mg/Si ratio for the bulk Earth has substantial implications for mixing (or rather lack of mixing) of planetesimals in the inner nebula, so the Earth must have accreted from a localized suite and there was no mixing between 1 and 2.5-4 AU during the accretion of the Earth. For the present, it seems preferable to retain the CI interelement ratios until the question is resolved. Figure 5.5.3 shows the mineral composition of the terrestrial mantle as a function of pressure. The question of core composition is addressed in more detail in section 5.8.

5.5.3. Accretion and Mantle Melting

A significant question in the present context is the relative sizes of the planetesimals that, in the current preferred scenario, accreted to form the terrestrial planets. In this hypothesis, large precursor objects were relatively common. Probably 50-75% of present Earth mass was originally present in large planetesimals of lunar or larger mass. These bodies may al-

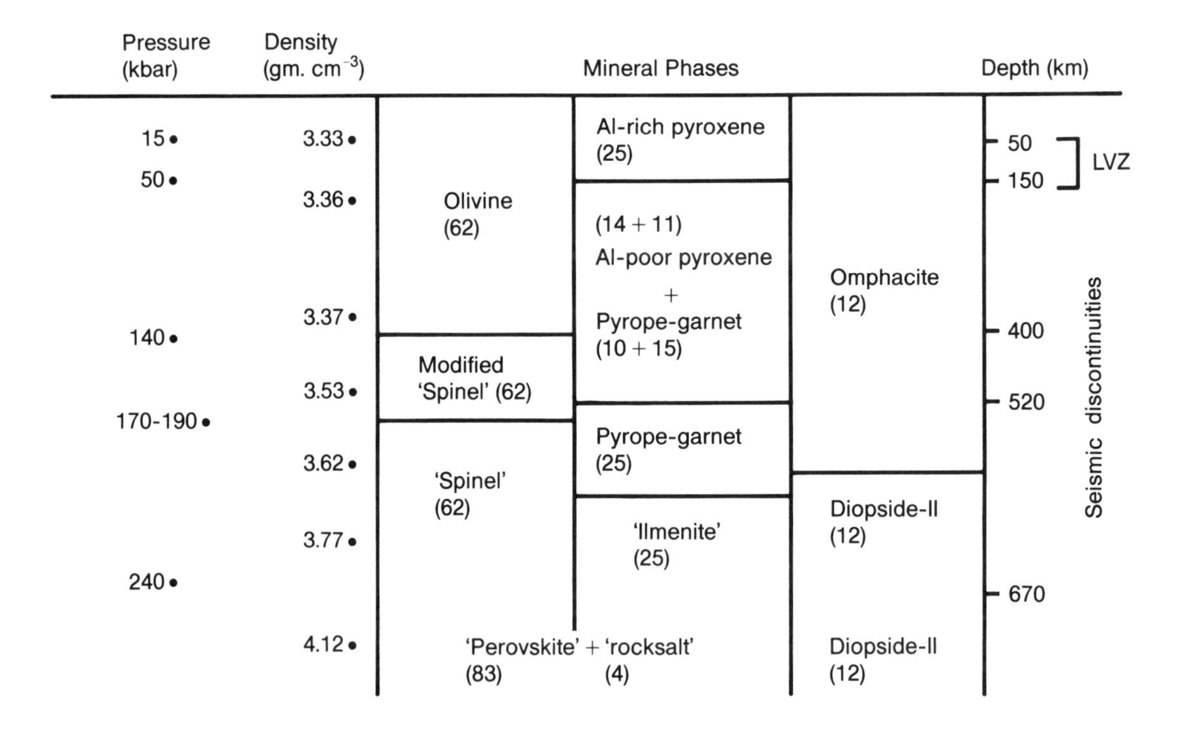

Fig. 5.5.3. *Mineral phases as a function of pressure in the mantle, corresponding to olivine and pyroxene components in the upper mantle. The weight percentage of each mineral is given by the numbers in parentheses. After Taylor S. R. and McLennan S. M. (1985) The Continental Crust: Its Composition and Evolution, Blackwell, Oxford, p. 262, Fig. 11.4. This source gives mineral formulae and coordination numbers for the cations involved.*

ready have been melted and formed metallic cores and silicate mantles. The addition of such differentiated, possibly molten, bodies to the Earth adds a complicating factor to our present limited understanding of terrestrial core-mantle relationships, since much of the core and mantle arrived, in this view, prepackaged.

This creates difficulties for models that attempt to build the Earth from two components ("A" and "B," usually identified with reduced E and oxidized CI chondrites respectively [49]). Such models reduce the iron in the oxidized CI by reaction with water, and dispose of the resulting hydrogen by escape. However, if large planetesimals are accreted, their volatiles are buried, along with the accretional energy, deep within the growing planet, making escape of volatiles more difficult. This general problem has bedeviled long-standing attempts to build planets from oxidized carbonaceous chondrites; how are the excess volatiles resulting from the reduction of oxidized iron to metal disposed of? The problem is not trivial, since the models predict 15% of oxidized component "B" in the Earth and 40% in Mars. The problem is compounded by the fact that the E chondrites, usually identified with component "A," are not depleted in volatiles. In summary, the attempts to build the terrestrial planets from these two components seem to raise more difficulties than they solve.

When did the accretion of the largest planetesimals occur? Most probably during the closing stages of the formation of the planets about 4500 m.y. ago, with a residuum of 100-km-sized and smaller objects being swept up in the next 500 m.y. The sweep-up of these bodies was mostly completed by about 3800 m.y. ago. This bombardment would have disrupted any early crust. Minimal estimates indicate that, besides innumerable smaller craters, at least 200 basins with diameters exceeding 1000 km formed in this interval. If 10% of these objects were comets from the outer reaches of the solar system, they could have delivered sufficient water to provide the observed oceans. The high encounter velocities of comets, however, by stripping off atmospheres, may take away more than they bring (see sections 3.10 and 5.10).

If there is little identifiable input into the inner planets from the meteorites now coming from the asteroid belt, then this has important implications for planetary accretion models. One corollary is that there was little lateral mixing in the nebula and that the terrestrial planets accumulated from rather narrow (perhaps <0.5 AU) concentric zones in the solar nebula. The present asteroid belt has a zoned structure, ranging from apparently differentiated objects in the inner belt to apparently primitive objects that dominate in the outer reaches. Whatever heating mechanism was responsible for the differentiation, this appears to have been controlled by heliocentric distance. The asteroid belt, although the zone has been broadened through time by collisions, may be an analogue for the structure of the nebula. Temperature, and hence fractionation, is expected to increase Sunward of the inner edge of the belt. This view receives some support from the differences in composition between the inner planets, particularly Earth and Mars, which differ in K/U ratios, Mg/(Mg + Fe) ratios, V-Cr-Mn abundances, and oxygen isotopes, all of which indicate accretion from somewhat differing populations of planetesimals. It is argued above that the terrestrial depletion in K and other volatile elements was an inherent feature of the incoming planetesimals and was due to a prior loss of volatile elements throughout the inner nebula.

A consequence of the planetesimal hypothesis is that planetary melting occurs due to the stored energy from the large impacts; in scenarios of accretion from fine dust, the heat from gravitational infall is mostly reradiated to space. Melting of planets during accretion is widely held to have occurred [50] once they reach about 10% of Earth mass [51] and "deep-seated gobal melting seems unavoidable one way or the other" [52]. This would be assisted if many of the incoming planetesimals were hot or molten.

The crystallization of a completely molten terrestrial mantle is not, however, clearly understood. Thus, it might be expected that fractionation of majorite or perovskite should disturb the chondritic relative ratios that are observed in the upper mantle. The siderophile element abundances in the upper mantle (Fig. 5.5.4) appear to be in chondritic relative proportions. Ni/Co and Ir/Au are about chondritic; fractionation of olivine should disturb these ratios, since both Ni and Ir preferentially enter olivine [53].

A similar scenario of late accretion of volatile-rich planetesimals is often invoked to account for the volatile (e.g., H_2O) inventory of the Earth. Such late veneers of CI-type material, although popular, fail to satisfy rare gas isotopic constraints [54]. A late cometary influx could be an equally viable source [55]. However, late infall of planetesimals to planets that are essentially complete is like adding icing to a cake: the decoration may give little insight about the composition of the interior. It is also worth recalling that the Moon shows no evidence of a late veneer, although it should be clearly evident.

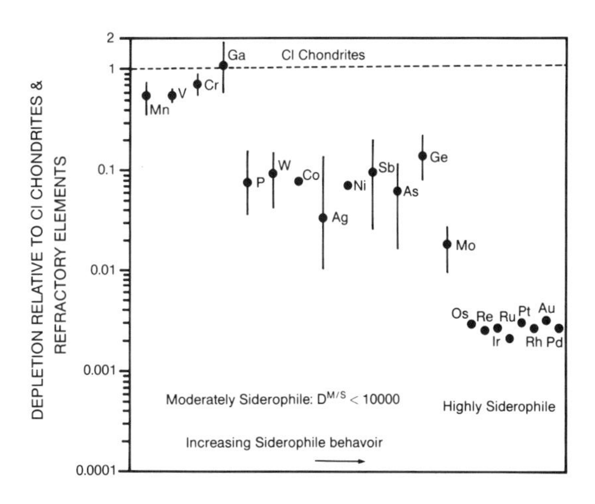

Fig. 5.5.4. *The depletion of the siderophile elements in the terrestrial upper mantle relative to CI chondrites and normalized to the refractory elements. The abundances of the volatile siderophile elements have been corrected for the general depletion of the volatile elements in the Earth. The resulting patterns show that the moderately and highly siderophile elements are depleted to differing degrees, but both are subparallel to the CI levels. The significance of these patterns remains an area of controversy. Courtesy H. E. Newson, Univ. of New Mexico.*

5.6. MARS

The unique fascination that Mars has exercised on the human imagination, illustrated by its identification with the God of War by the Romans, by Lowell's canals, and by H. G. Wells' invasion of the Earth by martians, seems to be due to the recognition that the planet is potentially inhabitable, and that the inhabitants of the red planet might share some of the less desirable of human characteristics.

5.6.1. Crustal Composition

> *"It is the superposition of rock units that provides the most significant reading in the geologic record of a planet. Yet, on Mars, the pages are partially glued together" [56].*

North of a boundary inclined at about 28° to the equator, the martian surface consists of volcanic plains and large volcanoes, all probably basaltic [57]. In contrast, the southern hemisphere of Mars is broadly composed of an Ancient Cratered Terrain,

TABLE 5.6.1. Martian surface composition measured by the Phobos mission, compared to earlier measurements.

Element	Phobos 2 (%)	Mars 5 (%)	Viking 1 (%)	Viking 2 (%)
O	48 ± 5	44 ± 5	50.1 ± 4.3	50.4
Mg	6 ± 3	—	5.0 ± 2.5	—
Al	5 ± 2	5 ± 2	3.0 ± 0.9	—
Si	19 ± 4	14 ± 3	20.9 ± 2.5	20.0
S	—	—	3.1 ± 0.5	2.6
Cl	—	—	0.7 ± 0.3	0.6
Ti	1 ± 0.5	—	0.51 ± 0.2	0.61
K	0.3 ± 0.1	0.3 ± 0.1	0.25	0.25
Ca	6 ± 3	—	4.0 ± 0.8	3.6
Fe	9 ± 3	14 ± 4	12.7 ± 2.0	14.2
U	$(0.5 \pm 0.1) \times 10^{-4}$	$(0.6 \pm 0.1) \times 10^{-4}$	—	—
Th	$(1.9 \pm 0.6) \times 10^{-4}$	$(2.1 \pm 0.5) \times 10^{-4}$	—	—

Data from Surkov Yu. A. et al. (1989) *Nature, 341*, 597.

older than about 4000 m.y. based both on crater counting and the lunar analogy. Its composition is unknown, but there are a number of convergent lines of evidence that indicate that it is unlikely to be acidic or very different in composition from the basaltic plains. The Viking Lander XRF data at the two sites 4000 km apart were both similar and basaltic in composition. Although both landers were in the northern hemisphere, it is argued that the fine material analyzed by the Viking landers (Table 5.6.1) represents a planetwide dust average (analogous to loess) and could be expected to provide some kind of average sample of the surface. The composition contains no constituent suggestive that the average crust is more silicic than basalt [58]. The U.S.S.R. Phobos mission obtained gamma ray measurements of K, U, and Th, in addition to major-element analyses of a broad region encompassing Lunar Planum and Xanthe Terra [59] (Table 5.6.1). The values of 0.5 ppm U and 2 ppm Th are consistent with basaltic rocks.

It might be expected that the Ancient Cratered Terrain has contributed significantly to the composition of the fine material analyzed by the Viking landers. The terrain is heavily cratered and so should possess a high proportion of dust and finely comminuted debris. This material is likely to be a significant component in the planetwide dust storms. Accordingly, the composition of the martian fine material might even be biased toward that of the Ancient Cratered Terrain unless dust derived from that terrain was locked up early in polar layered deposits. In summary, the Ancient Cratered Terrain is probably basaltic in composition. Thus, the martian crust appears to be dominated by rocks with low silica contents, a conclusion reinforced by the compositions of the SNC meteorites, and which is discussed later. The ages of the volcanic plains, based on crater counting, extend over much of geologic time and so they form an example of a secondary [60] crust, composed of basalt derived by partial melting from the martian mantle.

The Viking data indicate a high surficial iron content, consistent both with that expected on the "red" planet and with remote sensing studies, which yield a strong surface signature of Fe^{3+}, indicative of the presence of hematite (α-Fe_2O_3) on the surface. The surficial light and dark units revealed by remote sensing do not correlate with the surface geology. This lack of correlation between the light and dark terrains and the surface geology probably indicates that the remote sensing techniques are looking at a thin coating of windblown dust [61].

Pyroxene, presumably indicating unweathered igneous rocks, has been detected on the dark areas. No carbonate has been detected on the surface of Mars. Scapolite has been suggested to be widespread on the martian surface as a sink for the CO_2 in a 2-bar primordial martian atmosphere. Possibly scapolite is a weathering product, but this seems unlikely from our knowledge that the terrestrial synthesis of the mineral requires pressures of a few kilobars and temperatures greater than 500°C. Moreover, the identification of scapolite is doubtful. The critical absorption feature at 2.36 μm may be due to atmospheric absorption, and the other spectral features may be due to bicarbonate or bisulfate, both of which are expected to be common on the martian surface [62].

5.6.2. Dust

The problem of the upper crustal composition of Mars is simplified by the ubiquitous presence of a martian analogue of terrestrial loess. Global dust storms deposit a thin coating of about 45 μm per storm in the equatorial regions. Dust is removed from the dark regions and deposited in the bright areas where the thickness is of the order of a meter, insufficient to bury rocks. The dust particles are small, 2-40 μm. The thickness and rates of accumulation suggest that the deposits undergo cyclical erosion and deposition, possibly related to changes in the martian orbit. The source of the dust is ancient, since there is little evidence of recent erosional activity.

The dust deposits inform us of three things: (1) they are the cause of the albedo markings on Mars; (2) they are very thin and continously reworked; (3) the dust should provide an excellent integrated global average composition, except that it may be biased in favor of the older heavily cratered terrain, rather than the younger, less-eroded basaltic lavas.

The frequency of martian dust storms is high at present, and they occur nearly every martian year. However, it is not clear whether the present rate is abnormally high and whether dust storms arise due to some aperiodic forcing function [63]. The dust arises mainly from Hellas or Solis Planum, and is moved primarily from the southern to the northern hemisphere, the region sampled by Viking. This constitutes an additional reason for supposing that the Ancient Cratered Terrain is mainly basaltic and resembles neither the terrestrial granitic nor the lunar anorthositic crusts.

It appears that erosion over most of the martian surface has been very slow, consistent with the unweathered nature of the rocks at the Viking Lander sites. The conclusion is that eolian redistribution of sediments occurs at a much greater rate than erosion and production of fresh debris; accordingly, much of the sedimentary debris that now covers the surface was probably produced early in martian history and has been reworked ever since. Early impact cratering, producing much breccia and dust, was conceivably an important source of this material. This also informs us that the dust is probably biased toward the composition of the ancient crust.

5.6.3. SNC Meteorites

It is clear that the SNC (Shergotty, Nakhla, Chassigny) meteorites [64] come from a geochemically evolved planet. The presence of a trapped atmospheric component [65] is probably the best evidence for a martian origin for the SNC meteorites. Possibly their most curious feature is that they are cumulate rocks. Of the 10 separate lithologies, 8 are cumulates. The other two show signs of crystal fractionation. The shergottites bear some analogies to terrestrial komatiites. Shergotty and Zagami, although closer to normal basalts, are enriched in pyroxene. Since they must be derived from near the martian surface, this may mean that shallow differentiated igneous intrusions are common. Indeed, most of the SNC meteorites are relatively fine grained, with glassy mesostasis, indicative of shallow emplacement and rapid cooling. Even such extreme cases as the Nakhlites (mainly augite) and Chassigny (dominately olivine) must be crustal material. Nakhla has an Sm-Nd crystallization age of 1.26 ± 0.07 b.y. and is derived from ancient planetary mantle that was depleted in light REE, consistent with a martian origin [66].

The problems of ejecting material from the martian surface are considerable and have led to the proposal that all the SNC meteorites were derived from a single large (about 100 km diameter) crater about 200 m.y. ago. The ejected blocks were large enough to provide shielding from cosmic radiation until collisional breakup a few million years ago, so that the cosmic ray exposure ages are young. Such a model accounts for the dynamic problems of ejection from Mars, and would also explain the cumulate nature of the SNC meteorites, consistent with derivation from a shallow subsurface [67].

Clearly the igneous petrology and geochemistry of our samples of the surface rocks of Mars are very complex in detail. They appear to have been derived about 1300 m.y. ago from a mantle source that melted very early in martian history. This makes the establishment of bulk planetary compositions more difficult since the mantle at 1.3 b.y. may have been depleted in volatile elements by much earlier events.

The composition of the shergottites is very similar to that measured by the Viking Lander. Thus, the chemistry of these random samples of the martian crust reinforces the information from the Viking and Phobos missions. It does not encourage the speculation that Mars possesses a low-density granitic crust analogous to the terrestrial continental crust or to the distinctly different lunar highlands. A significant feature is that the SNC meteorites as a class are heavily depleted in chalcophile elements. This observation is generally taken to indicate that sulfur has been extracted from the martian mantle and forms a major component in the martian core.

Arguments based on the SNC data for the amount of water and halogens in Mars are questionable [68] because of the evidence for a period of early degassing. If there is a period of strong early degassing, much of the water content of the original mantle will escape. Igneous rocks subsequently derived from the mantle come from a much drier environment. This constraint also applies to the halogens, so that the ratios of, for example, Br to the refractory element La, in SNC meteorites will not necessarily yield the primitive volatile content of the planet.

This random handful of samples from a geologically complex planet presents us with the same difficulties of interpretation as a similar suite of rocks from the Earth. It is interesting to speculate how much we would have learnt about the Moon if the lunar Antarctic meteorites were our sole samples! The feldspathic nature of the highland crust might have been established, but the scarcity of mare samples would have been puzzling in the absence of a view of the lunar farside. It would have been argued that one cannot work out the history of a planetary body from a handful of samples. However, the Apollo and Luna missions did provide an adequate sampling of the Moon since the Moon is well mapped and has had a relatively simple history. Accordingly, only a few years' study was needed to establish the broad history of lunar geological and geochemical evolution. Mars is somewhat more complex.

5.6.4. Tharsis

The Tharsis bulge is the dominating structural feature on the martian surface. It is about 10 km high at the center and 8000 km across and occupies about one-quarter of the martian surface. The largest volcanoes are superimposed on it. The Valles Marineris system originates near the center of the Tharsis plateau and extends eastward. Tharsis is a unique feature among the terrestrial planets and accordingly calls for comment. A radial fracture pattern is centered on the plateau, although there appear to be distinct tectonic provinces within the overall structure [69]. Most of the large features due to catastrophic flooding are concentrated around the edges of the Tharsis bulge. A large positive gravity anomaly, up to 500 mgal, exists across the bulge [70]. Detailed geophysical data, particularly for elevation and gravity, are lacking; only low-resolution data are available [71].

The origin of the Tharsis plateau is enigmatic. A basic problem is to explain the permanent nature of the bulge. It has existed for too long a period to be supported by mantle convection processes. One model constructs the plateau by outpouring of volcanics, forming a lava plateau [72]. This model is supported by the concentration of the large volcanoes on the Tharsis bulge. The second hypothesis suggests that the bulge originated by uplift due to isostatic compensation for density differences in the mantle [73]. This explanation accords with the presence of some ancient cratered terrain on the bulge and the development of large-scale tensional fracturing around the plateau culminating in the great chasm of Valles Marineris. In this model, preferred here, the buoyancy required for uplift is provided by the low-density mantle residuum resulting from partial melting that produces iron-rich magmas. Most of these are required to intrude the upper mantle and crust, rather than to overplate the crust. This situation is analogous to the presence of low-density keels beneath the terrestrial continents.

The Tharsis Bulge on one side of the planet affects the moment of inertia and thus strongly influences the present obliquity of Mars. The obliquity of Mars averages 24.4° ± 12.7° [74]. At present, it is 25.2° and increasing, so the similarity to the obliquity of the Earth is a coincidence. The obliquity cycles through maxima and minima on timescales of 10^6 years. Possibly this effect is what drives the periodicity exhibited in the polar layered deposits [75].

However, extrapolations of martian obliquity history beyond a few million years are "uncertain. Projections into the more distant past are merely conjectural" [76] due to factors such as the axial precession rate, which is poorly known. A further problem is that the planetary motions may undergo chaotic behavior, making all such predictions somewhat uncertain [77]. Before the formation of the Tharsis Bulge, martian obliquity could have been more extreme, ranging perhaps from 25° to a maximum of 43°, although this conclusion is subject to the uncertainties noted above. The highest value would tilt the polar ice cap toward the Sun, facilitating melting. Possibly this mechanism could produce enough water water vapor in the atmosphere to induce episodic rainfall.

5.6.5. Moment of Inertia

The radius of Mars is 3389.92 ± 0.04 km [78]. Neither the global topography and gravity field nor the basic parameter of the coefficient of the moment of inertia of Mars are well known. The commonly accepted value for I/MR^2 is 0.365 [79], but a much lower value of 0.345 is possible, and values anywhere between 0.345 and 0.365 are allowable from the gravitational field. Cosmochemical models have generally chosen I/MR^2 to lie between 0.360 and 0.375, but do not provide an independent check [80].

The lack of a well-determined value for this crucial parameter makes it impossible to place close constraints on the composition of Mars and no such models are unique. If the moment of inertia values turn out to be low, the martian mantle must be low in iron content. However, most of the geochemical parameters indicate a high iron content for the mantle. If the value is close to 0.365, a wide range of mantle compositions are possible [80].

Geochemists would prefer a high iron content for the mantle, on account of the high surface iron contents measured by the Viking and Phobos missions, the red surface color due to hematite, the apparently low viscosity lavas on the martian surface, and from the compositions of the SNC meteorites. However, the geodetic constraints neither support nor disprove this supposition.

5.6.6. Bulk Composition

Mars is much smaller than the Earth and has a lower density (3.934). The uncompressed density is only 3.75 g/cm^3, so it is distinct in composition from the Earth and Venus. The crustal composition is dominated by iron-rich basalts, and there is no evidence, as discussed above, to indicate the presence of any more siliceous compositions, the Ancient Cratered Terrain being apparently similar in composition to the northern plains. The SNC meteorites present further confirmatory evidence of a basic or ultrabasic crust, since they must have been excavated from shallow depths. The moment of inertia indicates that the planet is differentiated, with a denser core. Unfortunately, as noted above, the uncertainty in that parameter makes the size of the core indeterminate.

Figure 5.6.1 shows the possible core radius vs. density relationships for Mars. If Mars has a CI composition and is completely differentiated, then the core mass is about 21% of Mars and the core radius is 50% of the planetary radius, but the fit is not exact, and none of the other meteorite classes are good matches either. It is generally supposed that the martian core contains a substantial amount of FeS, principally on account of the great depletion of the SNC meteorites in chalcophile elements.

A wide variety of models has been proposed for the bulk or mantle composition of Mars. These are mostly based on theoretical models of nebular composition, such as equilibrium condensation, and are now of historical interest [81]. Table 5.6.2 gives probably the most reliable geochemical estimate of the bulk composition of Mars, based on the SNC meteorite data [82]. Figure 5.6.2 compares the abundances in Mars with those in the Earth. It makes the reasonable assumption that the refractory element ratios, Fe/Si and Ni/Si, are present in CI ratios. A total sulfur abundance of 3.1%, assuming all sulfur is in the core, yields a core mass of 21.7%. This value depends rather heavily on the assumed amount of sulfur, and the apparent agreement is probably for-

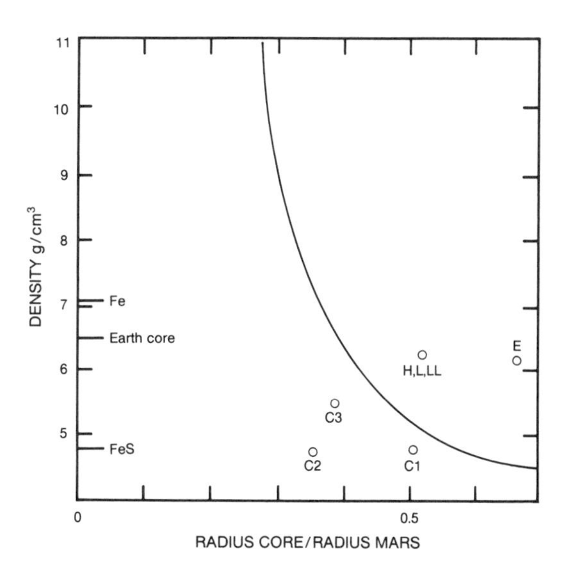

Fig. 5.6.1. *Variations in core radius vs. core density for models of Mars. The data for the meteoritic compositions are plotted on the basis that all Fe, Ni, and FeS is in the core. After Anderson D. L. (1989) Theory of the Earth, Blackwell, Oxford, p. 21.*

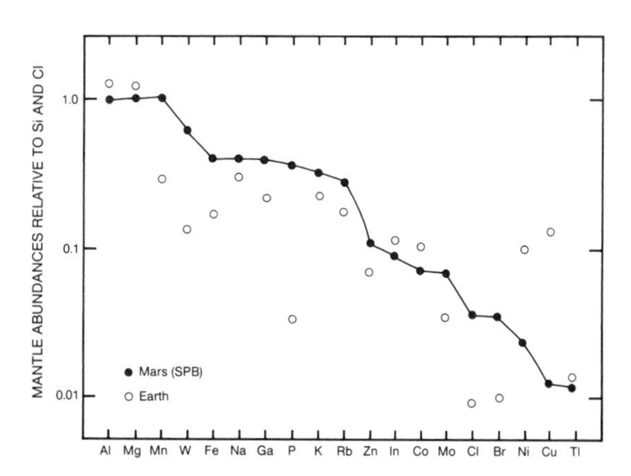

Fig. 5.6.2. *The abundances of elements in the martian mantle (from SNC data) compared with those in the terrestrial mantle. The strong depletion of the chalcophile elements in the martian mantle is ascribed to removal of sulfide phases into a core. After Laul J. C. et al. (1986) Geochim. Cosmochim. Acta, 50, 924, Fig. 14.*

TABLE 5.6.2. The bulk composition of Mars based on the assumption that it is the parent body of the SNC meteorites [86].

Element	Abundance	Element	Abundance
Mantle + Crust			
MgO	30.2%	In	14 ppb
Al_2O_3	3.02%	Tl	3.6 ppb
CaO	2.45%	Cl	44 ppm
SiO_2	44.4%	Br	165 ppb
TiO_2	0.14%	I	37 ppb
FeO	17.9%	La	475 ppb
Na_2O	0.50%	Th	56 ppb
P_2O_5	0.16%	U	16 ppb
Cr_2O_3	0.76%		
MnO	0.46%	Core	
K	315 ppm	Fe	77.8%
Rb	1.12 ppm	Ni	7.6%
Zn	74 ppm	Co	0.36%
Ga	6.6 ppm	S	14.24%
		Core Mass	21.7%

tuitous. The mantle in this model is iron-rich, in accordance with the surficial evidence. Although this composition can be represented by a mixture of 60% reduced component "A" and 40% component "B" [83], this arithmetical exercise produces its own set of problems, particularly with respect to the oxygen isotopes and the removal of excess volatiles.

The K/U ratios and Rb-Sr systematics of the SNC meteorites indicate that Mars is more volatile-rich than the Earth, by a factor of about 2. Thus, K/U ratios are closer to 2×10^3 rather than about 10^3 for the Earth (Fig. 5.6.3). (Although Mars is volatile rich, it has a low abundance of the noble gases, probably due to early atmospheric removal by collisions.) Additional evidence that Mars is richer than the Earth in volatile elements can be inferred from the lead isotope signatures of the SNC meteorites. SNC meteorites have $^{206}Pb/^{204}Pb$ values of 12-15 [84], while the values for the Earth are 19-20, and 150-500 for the Moon [85]. The lead in the SNC meteorites evolved in an environment of $^{238}U/^{204}Pb$ (μ) = 5 that is much lower than that for the Earth (μ = 8), the Moon (μ = 20-300), and the eucrites (μ = 100-150).

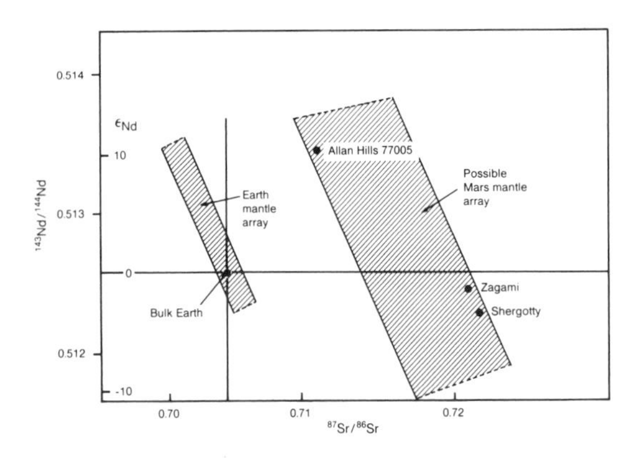

Fig. 5.6.3. *The difference between the martian and terrestrial mantles on the Nd-Sr isotopic diagram is consistent with a higher volatile element content in Mars. The resulting higher Rb/Sr ratio leads to increased $^{87}Sr/^{86}Sr$ values and a displacement of the "martian mantle array" to the right.*

Thus, a reasonably consistent picture of the bulk composition of Mars is beginning to emerge. Compared to the Earth and Venus, it is more volatile rich, by a factor of about 2 [86], but it is still much depleted in comparison with the primordial solar nebula and solar values. Its core probably has a substantial component of FeS that could act as a sink for the chalcophile elements that are so highly depleted in the SNC meteorites. This model for the composition of Mars should be regarded as an approximation, but consistent with our present knowledge. Earlier models that predicted a volatile-depleted Mars [87] were based on very early U.S.S.R. measurements of surface K/U ratios that resembled the lunar values, and which are probably in error.

5.6.7. Accretion and Evolution

Mars accreted from a different population of planetesimals than the Earth or Venus. This interpretation

TABLE 5.6.3. Principal events in martian history.

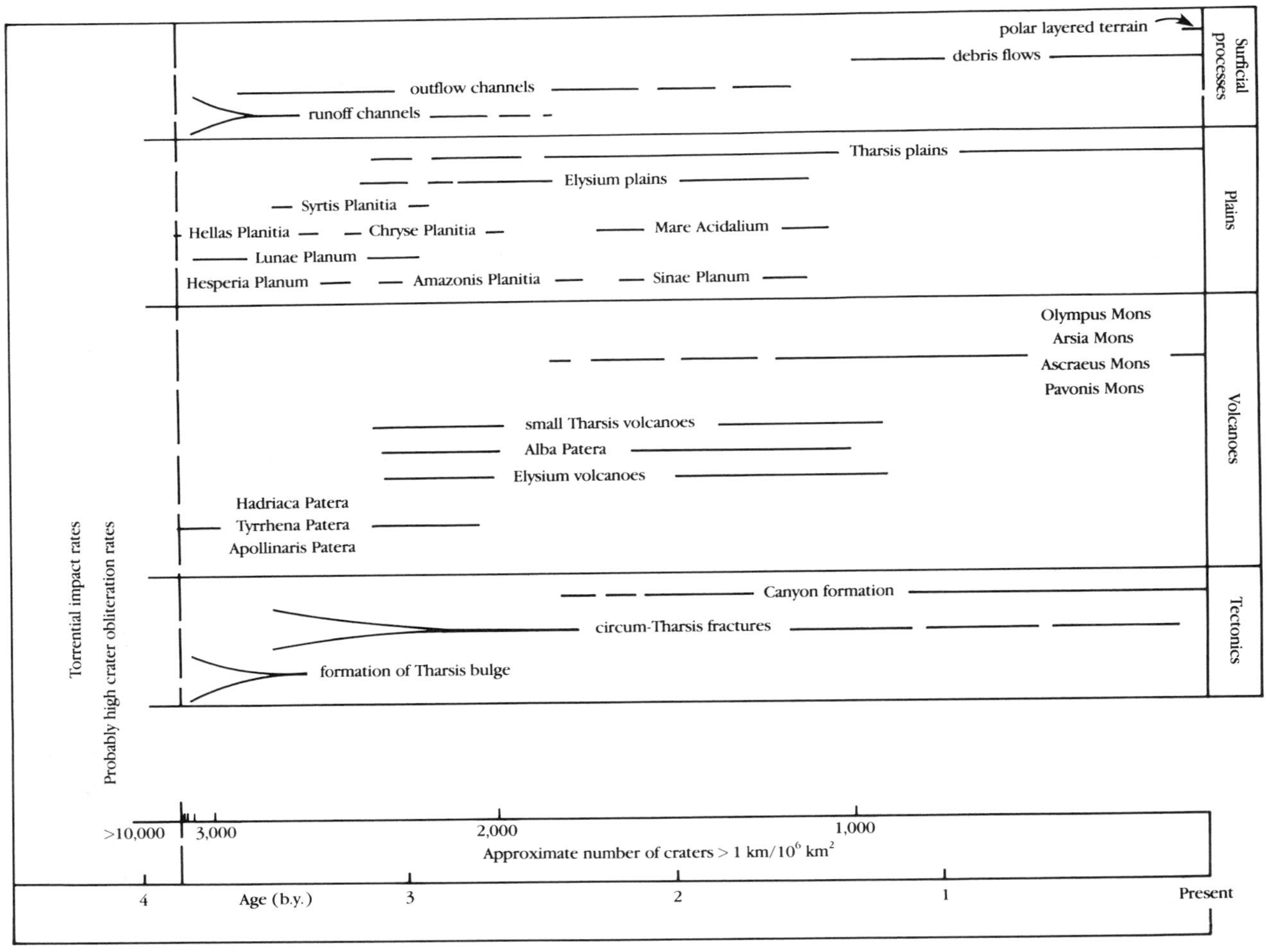

Surficial processes
polar layered terrain
debris flows
outflow channels
runoff channels

Plains
Tharsis plains
Elysium plains
Syrtis Planitia
Hellas Planitia
Chryse Planitia
Mare Acidalium
Lunae Planum
Hesperia Planum
Amazonis Planitia
Sinae Planum

Volcanoes
Olympus Mons
Arsia Mons
Ascraeus Mons
Pavonis Mons
small Tharsis volcanoes
Alba Patera
Elysium volcanoes
Hadriaca Patera
Tyrrhena Patera
Apollinaris Patera

Tectonics
Canyon formation
circum-Tharsis fractures
formation of Tharsis bulge

Torrential impact rates
Probably high crater obliteration rates

>10,000 3,000 2,000 1,000
Approximate number of craters > 1 km/10^6 km^2

4 Age (b.y.) 3 2 1 Present

After Carr M. H. (1981) *The Surface of Mars*, Fig 16.1, Yale Univ., New Haven.

is supported by the differing oxygen isotope compositions of the SNC meteorites compared to the terrestrial value. Impact velocities at Mars (and Mercury) are high (>10 km/s) according to the Wetherill model. For this reason, one does not expect to see moons developing as a consequence of large impacts at either planet. Like Mercury, Mars is a survivor. Other martian-sized objects formed within the inner solar system, but were swept up by Venus or the Earth, one providing us with our unique satellite.

A possible scenario of the evolution of Mars begins with melting of the planet as an inevitable consequence of planetesimal accretion. Core formation occurs early and a transient magma ocean forms. During this period the mantle was depleted in chalcophile elements, presumably scavenged by FeS into the core. Large volumes of basaltic crust form by partial melting due to the high heat flux following the solidification of the magma ocean. This thick early crust was subjected to heavy cratering, and survives as the Ancient Cratered Terrain (see Table 5.6.3). Recycling of the early basaltic crust is difficult since the eclogite stability field is not reached near the surface because of the low pressures on Mars. This model predicts that large volumes of basaltic crust will form early in planetary history, so the bulk of the Ancient Cratered Terrain should be basaltic.

The north-south crustal dichotomy is probably due to an early global convective pattern, perhaps aided by massive impacts; the greater thickness of the lunar crust on the farside of the Moon is usually attributed to a similar cause. No plate tectonics appear to have operated on Mars, so it forms another example of a one-plate planet. The absence of subduction has also inhibited the development of more acidic rocks, and the view is taken here that granites and similar evolved rocks are mostly restricted to the Earth [88]. There is thus a clear contrast between the growth of the continental crust of the Earth and the ancient crust of Mars.

5.7. COMPOSITIONAL DIFFERENCES AMONG THE TERRESTRIAL PLANETS AND METEORITES

Meteorites have long been favorite planetary building materials, since the presence of metal, sulfide, and silicate phases provided a small-scale analogue for models of the interiors of planets. In this section, some of the differences between the terrestrial planets and meteorites are examined.

5.7.1. Density

The densities of the inner planets are given in Table 5.2.1. This lists both actual and uncompressed densities. The latter enable comparison with chondritic densities. The uncompressed densities are uncertain within ±10-20% because of assumptions about mineral phases and the extrapolations to megabar pressures and high temperatures for Venus and the Earth. The chondrite data are for those groups (EH, EL, IAB, H, L, LL, CV, and CO) that do not contain hydrated silicates. The uncompressed density for Mars falls within the range for chondrites, but the values for the Earth and Venus barely overlap the meteorite field, while that for Mercury lies far outside the meteorite data. The fact that the meteoritic and planetary data for this fundamental property do not match is a significant constraint on the nature of the material that accreted to form the inner planets.

5.7.2. Major-Element Chemistry

The FeO/(FeO + MgO) ratio is not known for the Earth with great precision since there is a major uncertainty on account of both the metallic core and the unknown Fe and Mg contents of the lower mantle. With these uncertainties in mind, one can compare the FeO/(FeO + MgO) ratios for various chondrite classes with those of the planets. None of the meteorite classes match that of the Earth, which falls between the H group and that of the silicate inclusions in the IAB irons. Comparisons with other major elements suffer from the lack of knowledge about the composition of the lower mantle. As noted earlier, the bulk Earth ratios (e.g., Mg/Si) are commonly assumed to be chondritic, but this begs the question. As discussed above, the bulk Earth Mg/Si ratio may well be higher than the overall solar nebular value, and reflect terrestrial accretion from a local set of heterogeneous planetesimals.

5.7.3. Volatile Element Depletion

From a geochemical point of view, one of the most significant observations is the depletion in volatile elements that appears to be common to all four terrestrial planets. This depletion is well shown on a plot of K (a moderately volatile element) vs. U (a refractory element) (see Fig. 2.22.1).

This is one of the few geochemical measurements available for Earth, Venus, and Mars, as well as for the meteorites, and it provides crucial information (see section 2.22). The most general observation is that the terrestrial planets are depleted in K relative to the meteorites. Measurements of K/U ratios for Venus [89] indicate an overlap with those for the terrestrial surface. The K/U ratios for the SNC meteorites appear to be somewhat higher (1.5×10^4), consistent with a higher volatile content for that planet, but still a factor of 4 lower than CI values. This conclusion is supported by the Nd and Sr isotopic systematics for both planets, assuming the SNC connection with Mars (Fig. 5.6.3). The common classes of chondrites (E, H, L, LL, and CI) all have high K/U ratios and so are unsuitable building blocks for the inner planets, although the C2 and C3 classes have lower ratios, approaching the values for Mars. The Rb/Sr isotopic systematics indicate that the Earth is also depleted in volatile Rb relative to refractory Sr. No samples derived from the mantle bear any evidence of having been in an Rb/Sr environment as high as that of CI since T_0. Since Rb has closely similar properties to K, it is unlikely that either K or Rb is present in the mantle in CI concentrations. Thus, the presence of a CI component in the Earth, required in the two-component models [90], should be apparent from Sr isotopic systematics, but is not obvious.

5.7.4. Eucrites and the Moon

The only meteorites that have very low K/U ratios are the eucrites, which have ratios less than those of the Earth and overlap those of the Moon. The K/U ratios for lunar samples are very low (about 2500), consistent with a second stage of volatile loss during lunar formation. The Moon appears to have a unique chemical composition among the satellites and to be so far removed in composition from that of the primordial nebula and of most meteorites, except for the eucrites, that it does not seem profitable to discuss the possibility that any of the present population of meteorites were its building blocks. It would be a very difficult task to assemble a circumterrestrial ring composed exclusively of eucrite parent body planetesimals from which the bone-dry, refractory-rich, volatile-poor Moon might be formed. This, however, raises the question of how the eucrites lost their volatile elements. This apparently happened close to T_0, since the eucrite ages are about 4.55 b.y.

Thus, at least one asteroid went through a sequence of volatile and siderophile element depletion similar to that experienced by the Moon. Possibly a large collisional event is required to account not only for the Moon but also for the eucrite parent body. The canonical view is that asteroids with surfaces of achondritic composition are rare [91], so the processes that produced the eucrite parent body appear to have been uncommon in the early solar nebula, at least in the region of the asteroid belt. However, such processes producing "igneous" planetesimals may have been more common in the inner nebula.

5.7.5. Vanadium, Chromium, and Manganese

Relative to CI abundances, moderately volatile Cr and Mn are depleted with respect to more refractory V in CAI inclusions and refractory chondrules from ordinary and enstatite chondrites [92], in the CO, CM, and CV classes of chondrites [93], and in the Earth and Moon [94]. In contrast, the eucrites have effectively chondritic ratios, while the SNC meteorites (Mars?) have CI abundances of Mn, but are depleted in both V and Cr to the same extent. The SNC values are consistent with the separation of a sulfide-rich core in Mars [95]. It is clear that the V-Cr-Mn patterns are not related to planetary size since similar patterns occur both in the Earth and in some meteorite parent bodies. The depletions in the order V-Cr-Mn observed in the Earth, Moon, and various meteorites listed above are in the order of increasing volatility of the elements, and are the reverse of those expected on the basis of their siderophile characteristics. These observations indicate that the Earth-Moon abundances are not unique, but probably reflect the compositions of the accreting planetesimals both to the Earth and to the Mars-sized impactor, whose mantle was most likely the source of the protolunar material.

5.8. CORE-MANTLE RELATIONSHIPS

Formation of metal-rich cores in the terrestrial planets is expected to occur catastrophically once melting temperatures are reached and metallic iron in the incoming planetesimals melts and sinks rapidly through the silicate mantle. It is generally accepted that metal was present in the accreting planetesimals rather than being produced by reduction following accretion [96], so the metal-silicate equilibrium was established in a low-pressure environment external to the Earth. The scenario in which the terrestrial planets accrete from planetesimals that were already mostly differentiated into metallic, silicate, and sulfide phases implies little further reaction between metal and silicate once these bodies accreted to the Earth.

However, a massive event such as the collision of the Earth with a Mars-sized body may change this scenario by raising the temperature of the terrestrial mantle and core to several thousand Kelvin, thus altering established low-temperature metal-silicate partitioning, as discussed in section 5.8.1.

5.8.1. The Terrestrial Siderophile Element Paradox

The upper mantle abundances in the Earth of Re, Au, Ni, Co, and the platinum group elements (PGE = Ru, Rh, Pd, Os, Ir, Pt) are too high for the present upper mantle to have been in equilibrium with the core. The distribution of the PGE appears to be rather uniform, and they are present in approximately CI proportions [97].

Four hypotheses [98,99] have been advanced to explain this pattern: (1) inefficient core formation, (2) equilibrium between Fe-S-O rather than Fe-Ni, (3) late meteoritic bombardment following core formation, or (4) addition of the impactor core during the lunar-forming event.

In addition to the melting induced, the giant impact origin for the Moon has implications for the siderophile budget of the Earth. This depends on the fate of the metal core from the impactor. Most of the metal core ends up in the Earth, with the metal penetrating the mantle and ending up wrapped about the Earth's core [99]. Such an event would not disturb siderophile abundance patterns already present in the Earth's mantle. However, a significant amount of material from the impactor's core, enriched in siderophile elements, will probably be vaporized and redistributed into the mantle. Therefore, a complicated geochemical scenario for the Earth must be constructed. This requires additional metal segregation to occur in the Earth's mantle to deplete the highly siderophile elements below their present abundances, followed by the accretion of the late veneer to establish the abundances of the highly siderophile elements; alternatively the late veneer was caused by material added from the impactor's core (Fig. 5.5.4; see also Fig. 6.17.1).

There are other possible consequences of the giant Moon-forming impact. Very high temperatures are reached. The mantle is melted in such an event, with minimum temperatures in the range 3000-3500 K throughout much of the mantle, reaching 4500 K or possibly higher at the core-mantle boundary. If equilibrium partitioning of elements between metal and silicate occurs at such temperatures, then extrapolation [100] from the measured distribution coefficients at 1500-1600 K predicts that the siderophile element patterns so calculated do indeed match those in the mantle (Fig. 5.4.4; see also Fig. 6.17.1). The study by Rama Murthy [100] may have resolved the long-standing terrestrial mantle siderophile element dilemma.

5.8.2. Sulfur and Oxygen as Core Components

The metallic core of the Earth contains about 10% of a light element, either oxygen or sulfur. If high-pressure core-mantle equilibrium was not attained in the early Earth, then it seems unlikely that oxygen entered the core, since this requires megabar pressures, as is the case for potassium. Sulfur then becomes the most viable candidate for the light element in the Earth's core.

It is frequently asserted that sulfur is an unlikely candidate for the light element in the Earth's core, on account of its relative volatility. Although sulfur is more volatile than, for example, potassium, which is depleted in the Earth relative to CI, the argument fails to recognize that most of the sulfur accreting to the Earth in planetesimals is stabilized by combination with iron as troilite (FeS). It is also curious that sulfur shows no correlation with metamorphic grade in chondritic meteorites, being no more depleted in H6 than in H3 chondrites. Accordingly, there is no reason to suppose that it was strongly depleted in the material accreting to the Earth. This remains, however, as another current area of controversy. From the long-standing comparison with meteorites, cores have been supposed to be formed variously of Fe-Ni metal or Fe-FeS mixtures. Less commonly suggested are magnetite cores. Other proposals in the past for high-density forms of silicates, metallic hydrogen, and so forth have all been discredited and are not discussed further here.

Oxygen may enter the core only at megabar pressures, whereas FeS is available in the incoming planetesimals. Accordingly, the presence of oxygen in the core depends on the establishment of high-pressure equilibrium during accretion. If the metal-silicate equilibrium was established under low pressures in precursor planetesimals, then oxygen is less likely to be the light element in the core.

5.8.3. Potassium as a Core Constituent

Another element occasionally suggested as a core component is potassium. This suggestion arose in an attempt to account for the depletion of potassium relative, for example, to uranium, as shown by low K/U ratios in the outer parts of the Earth. The question arose whether the depletion of potassium was planetwide, or whether it might be hidden in the core. The central pressure in Venus (r =

6070 km) is 2.8-2.9 Mbar, only slightly less than that in the Earth's core (3.6 Mbar), so K might be hidden in the core of that planet. However, the central pressure in Mars ($r = 3390$ km) is only 400 Kbar, which is insufficient to allow K to enter a martian core. Thus, the depletion of K in Mars must be due to some other process. Elements of the atomic weight of potassium cannot be lost from the terrestrial planets, even at elevated temperatures once they have reached their present size. If they were boiled off in some manner during accretion, the K/U ratio should vary with planet size. This does not apppear to be the case. In addition to K, many other volatile elements are depleted in the Earth relative to CI abundances. It is unlikely that they are all hidden in a metallic core.

It has been argued earlier that the depletion of K (and other volatile elements) relative to refractory elements such as uranium is due to volatile loss in the planetesimals or the nebula prior to accretion of the planets. Thus, it was a property of the incoming precursor planetesimals from which the inner planets accumulated.

5.8.4. Terrestrial Planetary Cores

As discussed in section 5.3, the mercurian core possibly has a large FeS component, enabling it to remain liquid, and so be responsible for the presently observed field. It is curious that Mercury, but not Mars, possesses a detectable magnetic field, despite the common perception of an FeS core for Mars. The very large depletions in chalcophile elements observed in the SNC meteorites are usually taken as evidence that the martian core contains a substantial component of sulfur. In contrast to Mercury, it is still uncertain whether Mars has any magnetic field. A bow shock is observed, but this may be due to an ionosphere, as in the case of Venus. The upper limit for the equatorial surface magnetic field is 6×10^{-4} oersted, about 4000 times weaker than that of the Earth. Although there is no detectable magnetic field at present on Mars, it is likely that a core dynamo operated for the first billion years on Mars, so the older rocks may possess a magnetic signature. The remnant magnetism observed in the SNC meteorites is consistent with the crystallization of these igneous rocks in a field intensity between 0.01 and 0.1 oersted, indicating the possible former presence of a dynamo field, now extinct [101].

The absence of a magnetic field at Venus is usually ascribed either to its slow rotation or to the slightly smaller size of Venus relative to the Earth, which accounts for the probable lack of an inner solid core, the crystallization of which is generally regarded as a crucial factor in the generation of the magnetic field of the Earth [102].

5.8.5. Martian Mantle

In comparison with that of the terrestrial mantle, the martian mantle is thought to be more iron-rich, containing about 16-18% FeO [103]. In this case, the Mg/(Mg + Fe) value for the martian mantle is about 0.75, compared to values of about 0.90 for the much more magnesian-rich terrestrial upper mantle. A significant consequence of the very high Fe and low Si concentrations of such magmas is that the melt has a very high density. The density crossover at which melts become denser than the solids occurs at about 14 kbar or 100 km depth on Mars. Pyroxene and olivine will float in melts at greater depths. Accordingly, magmas produced in such regions may not be able to reach the surface.

Based on Sm-Nd systematics, the following history of martian differentiation has been proposed. An initial mantle-crust separation occurred at about 4.0-4.6 b.y. Crystallization of Nakhlites and Chassigny took place at about 1.2-1.3 b.y., and the Shergottites crystallized at about 0.2 b.y.

5.9. CRUSTAL DEVELOPMENT ON THE TERRESTRIAL PLANETS

There is a basic philosophical problem in dealing with planetary crusts: The surfaces of the 8 planets and the 60 or so satellites all differ from one another. Theories that attempt to provide general principles for planetary crustal evolution tend to founder on the rock of stochastic events. Accordingly, there are difficulties in trying to discover some general patterns of crustal growth in a system in which random events are common. Although there is clear evidence that the Moon was melted and formed a primary crust dominated by plagioclase feldspar, this does not necessarily provide us with a model for crustal development in the early Earth, Venus, or Mars, all of which differ, not only from the lunar example, but from one another.

The solid planets and satellites mostly have relatively thin crusts that differ markedly in composition from their interiors and from primordial solar nebular compositions. Familiarity with our own crust perhaps has obscured how remarkable crusts are, particularly since they concentrate sizable fractions of the planetary budgets of incompatible elements. This is seen most notably in the high surficial abundances of the heat-producing elements (K, U, and Th) in planets and satellites for which we have data.

Although the near-surface concentration of incompatible elements was long ago recognized for the Earth, the lunar samples focused attention on wider aspects of the problem. Pre-Apollo thinking led to the view that the Moon was a primitive undifferentiated object, principally because of its low density [104]. Although this opinion was not universally held [105], the surprising thing was that the samples both from the maria and the lunar highland crust were very highly fractionated. This is in contrast to estimates of primitive solar nebular values, established from the resemblance between the solar photospheric and CI abundances for the nongaseous elements. Indeed, the highland crustal abundances were so enriched in refractory elements that models appeared of crustal formation invoking the late plastering on of a refractory-rich layer [106]; however, these were quickly superseded by magma ocean models [107]

It readily became clear that crusts resulted from internal planetary differentiation, and models invoking late additions of differentiated planetesimals to account for them fell into disfavor. Further exploration revealed that Mercury, Venus, Mars, and many of the larger satellites have surface compositions that differ substantially from any reasonable estimate of their bulk composition; the production of crusts was thus seen to be a general phenomenon in the solar system.

Planetary crusts can arise in three ways: "Primary" crusts can form as a result of planetary differentiation consequent upon planetwide (rather than partial) melting during or shortly following accretion (e.g., lunar highland crust).

"Secondary" crusts arise later in planetary history as a result of partial melting in planetary interiors. These are typically composed of basalt, the primary melt from silicate mantles [108]. Examples of secondary crusts include the lunar maria, the terrestrial oceanic crust, the northern hemisphere of Mars, including the great volcanoes, and the venusian crust. The eucrites, among the meteorites, are probable examples of secondary crustal development on an asteroid. Other possibilities include the water-ice crusts of some of the satellites composed of rock and ice. In many such examples, the distinction between primary crusts produced by accretional melting and secondary crusts formed by partial melting in the satellite interior, or even solid-state resurfacing, must await further study.

"Tertiary" crusts may arise through further melting and differentiation of the mantle-derived material composing the secondary crusts; the continental crust of the Earth may be the sole example of this last type.

5.9.1. Early Terrestrial Crusts

It is frequently supposed, by analogy with the Moon, that the Earth formed an early anorthositic crust. Four reasons make this an unlikely event: (1) The composition of the Moon is probably richer in Ca and Al than that of the terrestrial mantle, leading to the early appearance of plagioclase during crystallization of the lunar magma ocean. (2) Plagioclase is unstable at shallow depths (>40 km) in the Earth and will transform to garnet, thus locking up Ca and Al in a dense phase. In contrast, plagioclase will be stable in the Moon to depths of several hundred kilometers. (3) Plagioclase will float in the bone-dry lunar magma ocean, but will sink in a wet terrestrial basaltic magma. (4) The oldest terrestrial anorthosites are not distinct from and closely resemble younger Archean examples. They must share the same petrogenesis, which makes derivation from a primordial magma ocean of the older examples less likely.

It is sometimes imagined that an early silicic crust is formed "in the beginning." However, studies of the development of the silicic continental crust show that it grew gradually throughout geological time. There is considerable evidence for an episodic crustal growth, with major crust-forming events in the late Precambrian (3000-2500 m.y. ago) and other minor episodes (e.g., at 1800 m.y.) [109]. No rocks older than 3960 m.y. have been identified [110] and are unlikely to be found. The massive pre-3800-m.y. bombardment will have broken up any early crust. The resulting breccias would have been easily removed by erosion (unlike the Moon, where the smashed up highland crust is preserved).

There is evidence from the Sm-Nd isotopic system that some extraction of LREE elements from the mantle occurred before 4000 m.y. ago [111]. The most likely process responsible is formation of an early basaltic crust, the ubiqitious primary partial melt from planetary mantles. This crust, which probably resembled the present basaltic crust of Venus, was removed by a combination of subduction and destruction by the meteoritic bombardment.

5.9.2. Crustal Growth Rates

There are very different growth rates for primary, secondary, and tertiary crusts. The lunar highland crust, forming 10-12% of the Moon was produced within about 100 m.y. Secondary crusts are produced at much slower rates. Thus, the lunar maria, produced by partial melting from the lunar interior over a period exceeding 10^9 years, comprise only 0.1% of lunar mass. The present oceanic crust of the Earth, which is 0.1% of the mass of the planet, formed from the mantle in about 200 m.y. If the present rate is extrapolated back, the total volume of oceanic crust produced is about 2% of the mass of the planet. Over 4000 m.y. of growth of the continental crust of the Earth has resulted in the production of a continental crust that comprises 0.33% of planetary mass. The process is clearly not very efficient.

Growth of primary crusts occurs concomitantly with or shortly following accretion, and is completed on short timescales (10^8 years). Impact-induced melting may blur the distinction between primary and secondary crusts. Growth of secondary and tertiary crusts may extend over the lifetime of the planet. The difficulties in producing a tertiary crust are shown by the small mass of the continental crust relative to that of the Earth, and the long time span taken to produce it [109].

5.9.3. Crustal Bombardments

Because of the continuing sweep-up of large planetesimals, the growth of primary crusts proceeded in a turbulent environment. Most planets and satellites contain ancient battered surfaces. The principal effect, as well shown by the lunar highland samples, is extensive brecciation. The record of such events will be retained on most planets and satellites because of the paucity of effective eroding agents. On the Earth, however, the record is unlikely to be preserved due to the efficiency of terrestrial weathering, eroding, and transporting processes working on the debris resulting from several hundred Mare Orientale-scale impacts.

How long did this bombardment last? The ages of the Imbrium collision and the slightly younger Orientale Basin formation put the terminal stages of these events on the Moon at about 3850 m.y. Lunar accretion was essentially completed by about 4440 m.y. ago, and the later bombardment around 3.9 b.y. was probably a spike or "cataclysm." Since fragments of mare basalt lavas are contained within some of the Apollo 14 breccias [112], some early portions of the secondary lunar crust were caught up in the maelstrom.

Estimates [113] of the number of objects that struck the Moon apparently mostly between 4100 and 3850 m.y. include about 80 basin-forming events (>300 km diameter) and over 10,000 craters with diameters in the range 30-300 km. Conservative estimates for the same interval indicate that over 200 multiring basins (>1000 km diameter) formed on the Earth [114], probably explaining the absence of identifiable rock units older than about 3800 m.y. on this planet. Whether this occurred as a late spike, as on the Moon, is unclear, but the result, destruction of preexisting crust, was probably insensitive to the time interval.

5.9.4. Recycling

Recycling of crusts appears to be rare on a planetary scale. The subducting terrestrial oceanic crust may constitute, like so much else on this planet, a unique example. Crustal growth in the rest of the solar system is essentially an irreversible process. This survey provides no evidence in support of the presence of primary siliceous crusts analogous to those of the terrestrial continental crust, and thus does not favor no-growth recycling models.

Models that involve rifting and separation of an initial world-encircling crust by massive planetary expansion are not supported in any way [115]. The lunar crust has preserved a frozen fossil surface for over 4000 m.y. without any sign of major disruption. The mercurian crust, of similar age, provides evidence of slight contraction, while the small expansion of Ganymede is explicable by the polymorphic transitions of ice.

5.10. TERRESTRIAL PLANETARY ATMOSPHERES AND HYDROSPHERES

The terrestrial planetary atmospheres are all so different that no universal theory can account for them. They have been through a series of events of staggering complexity. A partial list includes the following possibilities: direct capture of nebular gas, addition from planetesimals, mantle outgassing, and absorption or solution in magma oceans, as well as later chemical or isotopic partitioning and hydrodynamic escape of gases and impact-induced degassing and stripping. Impact events, common down to 3.8 b.y. are expected to remove most traces of any primitive atmosphere [116], and the atmospheres of the terrestrial planets are mainly secondary.

TABLE 5.10.1a. Atmospheric gases on Mercury and the Moon.

		Mercury	Moon	
Species	Wavelength (Å)	N_0 (cm^{-3})	N_0 (Day) (cm^{-3})	N_0 (Night) (cm^{-3})
H	1216	23, 230	>10	—
He	584	6.0×10^3	2×10^3	4×10^4
O	1304	4.4×10^4	—	—
Na	5890, 5896	$1.7\text{-}3.8 \times 10^4$	—	—
K	7664, 7699	5×10^2	—	—
Ar	869	$<6.6 \times 10^6$	1.6×10^3	4×10^4

Data from Hunten D. M. et al. (1988) in *Mercury* (F. Vilas et al., eds.), p. 564, Table 1, Univ. of Arizona, Tucson.

TABLE 5.10.1b. Composition of the lunar atmosphere at night.

Concentration	(molecules/cm^3)
H_2	6.5×10^4
^4He	4×10^4
^{20}Ne	8×10^4
^{36}Ar	3×10^3
^{40}Ar	7×10^3
O_2	$<2 \times 10^2$
CO_2	$<3 \times 10^3$

Data from Hoffman J. H. et al. (1973) *Proc. Lunar Sci. Conf. 4th*, p. 2874, Table 2.

Accordingly, they can only contribute indirectly to the main inquiry of this book. The compositions of the present individual planetary atmospheres are first examined, comparisons are attempted, and some constraints on planetary accretion, particularly with respect to the popular concept of "late veneers," are addressed. Readers should be aware that more problems than answers are to be found.

5.10.1. Mercury and Moon

Oxygen is the dominant species in the tenuous atmosphere of Mercury. Atmospheric gas values are given in Table 5.10.1, which also lists data for the Moon, a somewhat comparable case. The presence of Na and K in the tenuous mercurian atmosphere [117] is consistent with a feldspathic surface [118]. Such volatile elements would have been expected to have been lost during any early extensive evaporation episodes, so their presence provides supporting evidence for the massive impact hypothesis [119]. It seems unlikely that the alkali elements are derived from cometary sources, since Na and K ions are also observed up to 1200 km above the surface of the Moon. In the latter case, they are almost certainly derived by sputtering from the feldspathic surface [120].

5.10.2. Venus

Table 5.10.2 gives the atmospheric composition of Venus [121,122]. The most striking feature of the atmosphere of Venus, apart from its mass, is the high abundance of the nonradiogenic noble gases. Relative to the Earth, the venusian atmosphere contains about 80 times as much nonradiogenic argon. This is probably due to a differing impact history, that left Venus with a substantially greater atmosphere [123]. However, if impact erosion is more limited, water is likely to have been retained as well (see below). Although a late veneer of CI composition might add volatiles, the noble gas isotopic signature from such a source does not match the venusian atmospheric data.

5.10.3. Earth

Table 5.10.3 gives atmospheric data for the Earth. The most significant observation in the present context is the evidence for the accretion of the Earth after the nebula had been dissipated. The noble gases (He, Ne, Ar, Kr, and Xe) are strongly depleted in the Earth relative to solar abundances, implying that the gaseous components of the nebula had been dispersed before final accretion of this planet. Most of the evidence for this comes from the abundance and isotopic composition of the noble gases. If the Earth had been immersed in the gas-rich solar nebula during accretion, it is sufficiently massive to have captured a primary atmosphere. Such an atmosphere would have a surface pressure of 10^3 atm and a surface temperature of 4000 K. Absorption of gas from such an atmosphere would result, for example, in a neon budget 10-100 times greater than the present atmospheric content. The low abundance and isotopic composition of the rare gases in the atmosphere do not fit this model. Accordingly, it

TABLE 5.10.2. Venus atmospheric composition.

Assumed Constant with Altitude

Species	Mixing Ratio	
CO_2	$96.5 \pm 0.8\%$	
N_2	$3.5 \pm 0.8\%$	
He	12^{+24}_{-8}	ppm
Ne	7 ± 3	ppm
Ar	70 ± 25	ppm
Kr	0.7 ± 0.35	ppm
	or 0.05 ± 0.025	ppm

Altitude-dependent Minor Constituents

Species	Mixing Ratio (ppm)	Altitude of Measurement (km)
CO	350 to 1400	100
	180	90
	<10	75
	50	64
	30	42
	20	22
H_2O	<1 to 40	cloud top
	100	<55
SO_2	0.05	70
	<10	55
	150	22
H_2S	1	55
	3	<20
HCI	0.4	64
HF	0.005	64
C_2H_6	2	(?)

Isotopic Composition in % Compared with the Terrestrial Atmosphere

Element	Mass Number	Terrestrial Abundance	Venus Abundance
H	1	99.985	
(D)	2	0.015	1.6 ± 0.2
He	3	0.000138	<0.03
	4	99.999862	
C	12	98.90	
	13	1.10	1.12 8u7± 0.02
N	14	99.634	
	15	0.366	0.366 ± 0.075
O	16	99.762	
	17	0.038	
	18	0.200	0.20 ± 0.01
Ne	20	90.51	
	21	0.27	
	22	9.22	7 ± 2
Cl	35	75.77	no significant difference from terrestrial value
	37	24.23	
Ar	36	0.337	44.2
	38	0.063	8.6
	40	99.600	47.2
Kr	78	0.35	
	80	2.25	7
	82	11.6	23
	83	11.5	14
	84	57.0	48
	86	17.3	8

Data from von Zahn U. et al. (1983) in *Venus* (D. M. Hunten et al., eds.), p. 325, Univ. of Arizona, Tucson.

TABLE 5.10.3. Terrestrial atmosphere.

Composition[*]

Gas	%Volume
Nitrogen (N_2)	78.08
Oxygen (O_2)	20.95
Argon (Ar)	0.93
Water vapor (H_2O)	Variable: 0 to few %
Carbon dioxide (CO_2)	0.034

Evolution of Atmospheric Oxygen[‡]

Time	Oxygen Level (% of present level)
2 b.y. ago	1%
1 b.y. ago	5%
670 m.y. ago	7%
550 m.y. ago	10%
400 m.y. ago	100%

Rates of Oxygen Production, Destruction, and Transfer[‡]

Production	
Photosynthesis	10^{16}
Photolysis of water/escape of hydrogen	7×10^9
Destruction	
Respiration and decay	10^{16}
Combustion of fossil fuels	3×10^{14}
Weathering of sedimentary rocks	10^{13}
Reaction with volcanic hydrogen	5×10^{10}
Transfer	
Burial of surface organic matter to sedimentary rocks	10^{13}

Measurements in units of moles O_2/year; the atmosphere contains 3.8×10^{19} moles O_2.

* Data from Levine J. S. (1985) in *The Photochemistry of Atmospheres: Earth, the Other Planets, and Comets,* Academic, New York.
†Data from Cloud P. (1983) *Am. Sci., 249,* 176.
‡Data from Walker J. C. G. (1977) *Evolution of the Atmosphere,* Macmillan, New York.

appears that the nebula had gone by the time the Earth had accreted to about its present size. Large collisions in the final stages of accretion are likely to have removed any primitive atmosphere [116]. The present atmosphere and hydrosphere of the Earth appear to be entirely secondary in origin, and so only indirectly of relevance to this inquiry.

Was this atmosphere slowly evolved from the mantle, was there a sudden early degassing or outgassing event, or was it added as a late veneer? The isotopic composition of the noble gases helium, argon, and xenon have provided crucial evidence on these points. Early extensive degassing is indicated by the argon data. Most of the primitive volatiles were degassed from the mantle in the first half billion years after accretion, before there was significant addition to the mantle of ^{40}Ar from the decay of radiogenic ^{40}K. Xenon data indicate that up to 80% of the degassing occurred within about 50 m.y. following accretion. This early rapid degassing would be consistent with a molten mantle, resulting both from the accretion of large planetesimals and from the formation of the Moon by a massive collision [124,125]. The addition of significant amounts of atmospheric gases by a late veneer of CI composition appears to be ruled out by the differences between the rare gas isotopic signatures of the terrestrial atmosphere and the meteoritic data.

The nature of the prebiotic atmosphere is of much interest with respect to the origin of life. Free oxygen was absent ($>10^{-6}$ atm) and did not become available until about 2.5-2.9 b.y. ago, when photosynthetic bacteria capable of producing oxygen became abundant [126] (see section 4.16). It is not known whether the atmosphere was reducing or nonreducing, but it does not appear to have been strongly reducing. Since this atmosphere was secondary, being derived from outgassing from the mantle, its composition will depend on the oxidation state of the mantle. Reaction with metallic iron, which would produce a strongly reducing atmosphere (H_2, CH_4, and CO) is very unlikely; core formation occurs essentially instantaneously during accretion in terms of the models developed here, and even if outgassing is rapid, it must postdate core formation. This means that the mantle would be much more oxidizing so that in the earliest atmosphere H_2O and CO_2 would predominate over H_2 and CO, with CH_4 virtually absent [127]. Table 5.10.3 gives the present terrestrial atmospheric composition, the history of oxgen evolution, and the sources and sinks of oxygen in the present atmosphere.

TABLE 5.10.4. Composition of the atmosphere of Mars.

Species	Abundance (mole fraction)
CO_2	0.953
N_2	0.027
^{40}Ar	0.016
O_2	0.13%
CO	0.08%
	0.27%
H_2O	(0.03%)
Ne	2.5 ppm
^{36}Ar	0.5 ppm
Kr	0.3 ppm
Xe	0.08 ppm
O_3	(0.03 ppm)
	(0.003 ppm)
Species	**Upper limit (ppm)**
H_2S	<400
C_2H_2, HCN, PH_3, etc.	50
N_2O	18
C_2H_4, CS_2, C_2H_6, etc.	6
CH_4	3.7
N_2O_4	3.3
SF_6, SiF_4, etc.	1.0
HCOOH	0.9
CH_2O	0.7
NO	0.7
COS	0.6
SO_2	0.5
C_3O_2	0.4
NH_3	0.4
NO_2	0.2
HCl	0.1
NO_2	0.1

Data from Lewis J. S. and Prinn R. (1984) *Planets and Their Atmospheres: Origin and Evolution*, Academic, New York.

5.10.4. Early Martian Atmosphere

The composition of the martian atmosphere is shown in Table 5.10.4. Since Mars is more volatile-rich than the Earth, there is no problem with the supply of volatiles. Why then is the atmosphere not thicker? Three mechanisms have been suggested: (1) loss by impacts, (2) hydrodynamic escape, which should result in mass fractionation of the lighter isotopes, and is observed in the present martian noble gas abundances [128], and (3) solar wind interaction with the martian ionosphere is likely to induce plasma velocities that will exceed the escape velocity (5 km/s).

There are two views on the question of an early martian atmosphere. The uniformitarian view is that the present conditions have persisted from the earliest times; the other calls for a wet early Mars [129]. The principal evidence for the latter view is the presence of the ancient dendritic valley networks. If, as on the Earth, these are cut by rainfall, an early martian atmosphere of 5-10 bar of CO_2 is needed to raise the surface temperatures, via a greenhouse effect, to enable water to be the active eroding agent. An alternative view [130] is that the dendritic valleys are cut by headwater sapping, by water released during early intrusion of sills into the crust. It seems less likely that the early high heat flow will provide enough energy to melt water trapped in the early brecciated crust, since the temperature at the surface due to internal heating is very small. Whatever the cause, the dendritic valley networks are early, dating from toward the end of the early great bombardment (Fig. 5.10.1). The valley heads are cirque shaped, about 1 km across, and the interfluves are smooth and uneroded, both features arguing for an origin by headwater sapping rather than by erosion by rainfall.

Fig. 5.10.1. *Valley networks in the cratered terrain in the southern hemisphere of Mars. North is to the lower left and the view is 200 km across. (NASA Viking Orbiter 63A09)*

5.10.5. Late Veneers?

Table 5.10.5 compares the abundances of carbon, nitrogen, and the rare gases in the terrestrial planetary atmospheres, meteorites, and the solar abundances. Table 5.10.6 gives isotope ratio data for the same bodies. These data can be used to place some constraints on the sources of planetary atmospheres. Thus, if a substantial contribution was made by a late veneer of CI-type material, this should be reflected, for example, in the abundance and isotopic composition of xenon. However, since martian xenon is distinct in composition from CI and other meteoritic xenon (as well as from solar xenon), this must severely limit an atmospheric contribution from any late veneer of CI composition [131].

5.10.6. Water in the Nebula

Planetesimals within the inner solar system are not expected to be water rich. Along with the other volatiles, water will be removed from the inner portions of the nebula by early violent solar activity. The

TABLE 5.10.5. Abundances of carbon, nitrogen, and noble gases in the solar system, meteorites, and terrestrial planetary atmospheres.

	^{12}C	^{14}N	^{20}Ne	^{36}Ar	^{84}Kr	^{130}Xe
Solar system	3.90 (-03)	9.42 (-04)	2.24 (-03)	8.97 (-05)	5.84 (-08)	8.07 (-10)
CI carbonaceous chondrites	3.70 (-02)	1.51 (-03)	2.89 (-10)	1.25 (-09)	3.57 (-11)	7.0 (-12)
Planetary atmospheres						
Venus	2.60 (-05)	2.20 (-06)	2.0 (-10)	2.51 (-09)	4.72 (-12)	8.9 (-14)
Earth	1.50 (-05)	1.45 (-06)	1.00 (-11)	3.45 (-11)	1.66 (-12)	1.40 (-14)
Mars	1.11 (-08)	7.30 (-10)	4.38 (-14)	2.16 (-13)	1.76 (-14)	2.08 (-16)

Data from Pepin R. O. (1989) in *Origin and Evolution of Planetary and Satellite Atmospheres* (S. Atreya et al., eds.), p. 294, Univ. of Arizona, Tucson. Numbers in parentheses are powers of 10.

TABLE 5.10.6. Isotopic compositions and elemental ratios of carbon, nitrogen, neon, and argon in the solar system, CI meteorites, and terrestrial planetary atmospheres.

	$\delta^{13}C$	$\delta^{15}N$	$^{14}N/^{12}C$ ($\times 10^{-2}$)	$^{20}Ne/^{22}Ne$	$^{36}Ar/^{38}Ar$	$^{20}Ne/^{36}Ar$ ($\times 10^{-2}$)
Solar system	—	—	24.15 ± 0.20	13.7 ± 0.4	5.6 ± 0.1	2500 ± 300
CI carbonaceous chondrites	-10.3 ± 2.5	42 ± 11	4.1 ± 1.1	8.9 ± 1.3	5.30 ± 0.05	23.1 ± 6.4
Planetary atmospheres						
Venus	-3 ± 18	$\sim 0 \pm 200$	8.5 ± 1.9	11.8 ± 0.7	5.56 ± 0.62	11.6 ± 6.8
Earth	-6.4 ± 1.3	$\equiv 0$	9.7 ± 4.5	9.80 ± 0.08	5.320 ± 0.002	29.0 ± 0.3
Mars	$<+50$	620 ± 160	6.6 ± 2.1	10.1 ± 0.7	4.1 ± 0.2	20.3 ± 6.2

*Units are ‰ deviations from standard compositions for $\delta^{13}C$ and $\delta^{15}N$, atom/atom for isotope ratios, and g/g for element ratios.

Data from Pepin R. O. (1989) in *Origin and Evolution of Planetary and Satellite Atmospheres* (S. Atreya et al., eds.), p. 295, Univ. of Arizona, Tucson.

zoning in the asteroid belt is consistent with water-rich carbonaceous chondrite compositions dominating beyond about 2.5-3 AU in the outer reaches of the belt, but with drier material Sunward. Water ice will condense at about 4-5 AU, consistent with the presence of icy satellites around Jupiter, and is perhaps responsible for the presence of that giant planet. Some water might be retained in the inner nebula in hydrated minerals or in planetesimals that were large enough to survive the early gas and volatile depletion episode.

Although equilibrium condensation models predict a drier Venus, the planetesimal hypothesis predicts only minor changes in composition between the Earth and Venus. It seems unlikely that the Earth and Venus accreted from wet planetesimals [132] or that there is much difference between the water contents of both planets, so the venusian water may be hidden deep in the mantle [133].

5.10.7. Comets as Sources of Water

Current models for the evolution of the nebula predict that the inner region, in which the terrestrial planets accrete, will be depleted in volatiles, notably water. If most of the accreting planetesimals are dry, where do the terrestrial oceans come from? Comets are a possible source. If only 10% of the impacts during the massive bombardment 4.4-3.8 b.y. years ago were cometary, then they could easily deliver an oceanic mass of water [134]. However, cometary impacts at relatively high velocities may remove atmospheres as readily as they deliver them [135], and they appear to be a fickle source for water, volatiles, atmospheres, and life. Only trivial amounts of material on the lunar surface can be ascribed to cometary influx, lending support to this view.

5.10.8. Primitive Oceans on Venus?

Was there a primitive ocean on Venus? Did an early runaway greenhouse boil off the oceans, with water being lost by dissociation in the upper atmosphere [136]? The D/H ratio on Venus is about 120 times the terrestrial value. Such a high value is consistent with the scenario that H is being selectively lost compared to the heavier D from the upper atmosphere of Venus, and would correspond to less than 100 m of water over the planetary surface. However, another scenario is possible. Water may be buried deep in the venusian mantle [133].

All this depends on the question of the initial water content of Venus. If water is continually added to Venus via comets, then the high D/H ratio merely informs us of a steady-state situation, and the high D/H ratio reflects the recent loss of H relative to D without telling us anything about the early amount of water on the planet. The massive collision that reversed the rotation of Venus may have stripped away an early hydrosphere. Alternatively, if Venus was accreted mainly from dry planetesimals, and the water budget was supplied by late accreting comets and planetesimals, no correlation with the water budget on the Earth would be expected.

5.10.9. Early Terrestrial Oceans

The source of water in the Earth remains a problem. As discussed above, little water was available in the zone from which the Earth and the other inner planets were formed, since water would be lost along with the other volatiles in the early heating event. Some water, perhaps present in hydrated minerals in already formed planetesimals, would survive the early intense heating that drove the volatiles out of the inner solar system, and is probably the source of most of the water in the inner planets.

A late accretion of volatile-rich planetesimals is often invoked to account for the volatile (e.g., H_2O) inventory of the Earth. Morgan [100] concludes that an addition of 10^{25} g of CM2 material between 4.5 and 3.8 b.y. would "provide all of the carbon found in the terrestrial crust, oceans and atmosphere . . . of approximately the right isotopic composition . . . and a substantial amount of the earth's water." However, this solution presents its own problems, as noted earlier, and a late cometary influx could be an equally viable source. Water-ice is not expected to be a stable phase in the nebula at distances Sunward of 4-5 AU. This is consistent with the observation that icy satellites are restricted to the region of the giant planets. Carbonaceous chondrites, probably typical of asteroid compositions beyond about 3 AU, contain up to 20 wt% water.

Accretional heating of a water-bearing planet with an ocean the size of that on Earth will cause the oceans to evaporate and the surface temperature to reach 1500 K, melting the surface rocks and forming a magma ocean [137]. Whether the Earth was, however, surrounded by a steam atmosphere is problematical. It depends on the composition of the incoming planetesimals. If these are derived mainly from the inner volatile-depleted nebula, they will be mostly dry. A basic problem that requires resolution is the ultimate source of the Earth's water. Did it come from beyond Mars [138], either in comets [139] or in late-accreting planetesimals, or was it derived from inner solar nebular planetesimals that had both acquired volatiles and grown large enough to escape being swept out of the inner solar sytem by early violent solar activity? The latter model is preferred, since objects coming in from beyond Mars reach high velocities and may erode rather than add material. There are also problems with the isotopic abundances of the rare gases in such models [140].

5.10.10. Water on Mars

A vital question for the composition of Mars is the overall planetary water content. There is no present direct evidence of surface ice, except at the north pole, where in summer the CO_2 seasonal cap evaporates. Surface ice is unstable elsewhere. Ground ice is stable above 40° latitude, from depths of about 1 m down to the base of the permafrost.

The most dramatic evidence for the existence of water on the martian surface is the presence of the large outfow channels. The channels commonly originate in chaotic terrain, which appears to have formed by subsurface removal of material (ice?). There is ample room for water storage in the megaregolith, which resulted from the heavy bombardment. Near the equator, the valley networks provide the principal evidence for near-surface water [141]. Conservative estimates of the water needed to erode the circum-Chryse channels amounts to a depth of 35 m over the whole planet. (The great valley, Valles Marineris, is not primarily of erosional origin, but is largely a structural or tectonic depression; see section 5.6.)

Some features of the outflow channels, such as apparent undulatory floor profiles with amplitudes of 400 m and a wavelength of 50 km, are not consistent with erosion by water, which would have had to flow uphill. Ice has been suggested as an alternative [142]. The northern volcanic plains apparently constitute a veneer of lava flows overlying a volatile-rich older surface. The chief evidence for this is the existence of large channels that run for hundreds of kilometers, particularly around Elysium Mons and Hecate Tholus. These channels have streamlined walls and are complete with teardrop-shaped islands, and appear to start instantly in the midst of the volcanic terrains.

Water lost from low latitudes probably now resides in the polar layered terrains, which are 1-2 km thick in the south and up to 4-6 km thick in the north, both cut by valleys up to 1 km deep. The water from the large catastrophic channel-cutting episodes is probably now in ice-rich deposits in the northern plains. In summary, it appears that water has been lost from low latitudes to depths of perhaps 1 km. If degassing occurred early, then volcanics subsequently erupted are likely to contain relatively low concentrations of water. Since CO_2 may have been expected to have been degassed, possibly 200-500 m of carbonates plus several meters of nitrates may have been folded into the cratered uplands [143].

5.11. THE ASTEROID BELT

> *"The asteroid belt is not a zoo but a wilderness area, where asteroids are preserved in their native habitat" [144],*

This quotation raises a fundamental question. If the asteroid belt is a trapped, esentially random collection of stray bodies (as in a zoo), then it conveys totally different information about the early solar system than if it represents some relic of primordial conditions.

There are 4044 numbered asteroids (as of 1989) with a multitude of smaller bodies. Figure 5.11.1 shows the overall structure, and Figs. 5.11.2 and 5.11.3

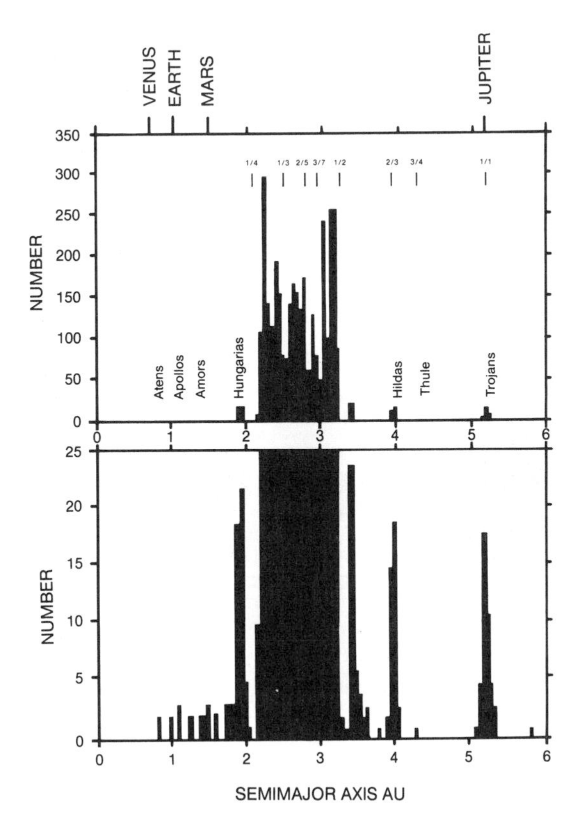

Fig. 5.11.1. *The structure of the asteroid belt as a function of distance from the Sun. The fractions represent the ratios of the orbital periods to that of Jupiter. After Gradie J. C. and Tedesco E. F. (1982) Science, 216, 1405.*

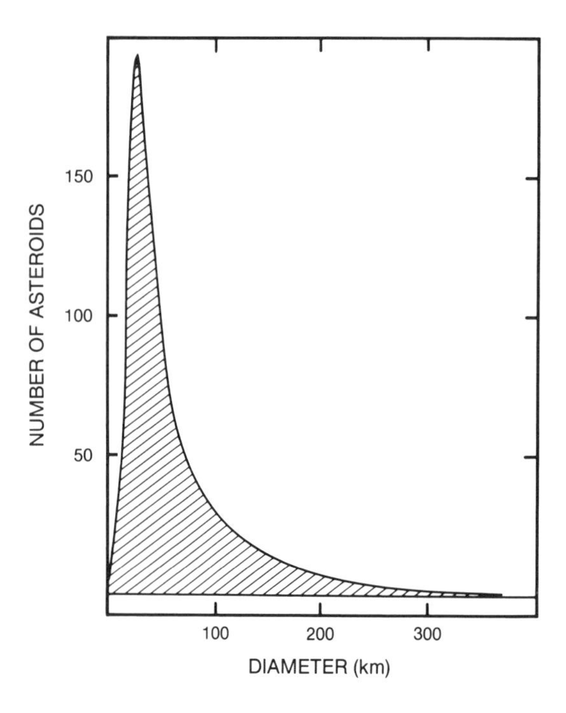

Fig. 5.11.2. *The number of asteroids decreases rapidly with increasing size. Three asteroids (1 Ceres, 2 Pallas, and 4 Vesta) have diameters greater than 500 km. After Veeder G. J., preprint.*

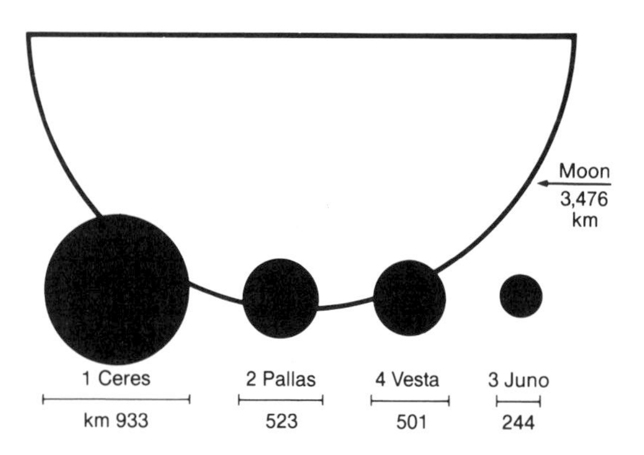

Fig. 5.11.3. *The sizes of the first four numbered asteroids relative to that of the Moon. After Veeder G. J., preprint.*

give information on the size distribution of asteroids, although there are distinctions among the various classes, indicative perhaps of differing strength or collisional histories. The total asteroid mass amounts to only 5% of that of the Moon.

Three main groups are generally recognized: (1) The Near-Earth asteroids, Apollo, Aten, and Amor classes. The Apollos have orbits that cross that of the Earth, the Atens are a subclass of the Apollos, with semimajor axes that are inside that of the Earth, while the Amors are Mars-crossers, with orbits that approach, but do not cross, that of the Earth. (2) The Main Belt, with which are generally included the Hungarias, just on the inside, and the Hildas, which lie just outside the main belt. (3) The Trojans, occupying the L4 and L5 Lagrangian points ahead of and trailing Jupiter in the orbit of that planet, and including a sole martian example.

5.11.1. Kirkwood Gaps

Strong depletions of asteroids occur at the 2/1 (3.28 AU) and 3/1 (2.50 AU) resonance with Jupiter. Other gaps occur at the 3/7, 2/5, and 1/4 ratios of asteroid/jovian orbital periods (Fig. 5.11.1). The ultimate explanation for the Kirkwood Gaps is understood in terms of chaotic behavior. Asteroids occupying orbits that are simple ratios of the jovian orbit may remain in regular orbits for periods of 10^5 yr before suddenly jumping into a chaotic state, with large increases in orbital eccentricity. The resulting orbits may well become Mars- or Earth-crossing, thus providing the terrestrial meteorite flux [146].

The 2/1 and 3/1 Kirkwood Gaps contain few asteroids and contrast sharply with the 3/2 resonance (3.97 AU), which is populated by the Hilda Group, including 153 Hilda and 190 Ismene (80 × 100 km diameter). These differences have always been puzzling in terms of classical dynamics, but are probably due to underlying dynamical differences: "Where the phase space is chaotic there are no asteroids, and where it is quasiperiodic, asteroids are found" [146].

The 4/3 resonance at about 4.2 AU contains 279 Thule, while the 1/1 resonances are occupied by the Trojans at the Lagrangian points ahead of and following the orbit of Jupiter.

5.11.2. Zonation of the Asteroid Belt

Although asteroids vary widely in composition (Table 5.11.1), the most interesting feature is the zonal arrangement (Fig. 5.11.4) [147]. It must be considered fortunate that the asteroid belt straddles that part of the solar system where there is a change-over from the volatile-depleted inner solar system to the volatile-rich outer reaches. The existence of the belt, rather than the presence of a clone of Mars, is of course a consequence of the early formation of Jupiter. If the formation of that gas giant is a

TABLE 5.11.1. Asteroid classes.

Low-Albedo (<0.1) Classes

Class C: Common in the outer part of the main belt, and similar in surface composition to CI and CM chondrites. Subclasses are B, F, and G, with minor spectral or albedo distinctions.
Class D: Rare in the main belt but dominant beyond the 2:1 resonance with Jupiter at 3.25 AU, reddish possibly due to kerogen-like materials. No meteoritic analogues.
Class P: Common near the outer edge of the main belt, probably C-rich. No meteoritic analogues.
Class T: Rare and of unknown composition, possibly highly altered carbonaceous chondrites.
Class K: Possible parents for the CV and CO chondrites

Moderate-Albedo Classes

Class A: A rare type with very reddish spectra, with an olivine infrared absorption feature, possibly similar to brachinites.
Class M: Common in the main belt, probably composed of Fe-Ni metal, analogous to iron meteorites.
Class Q: 1862 Apollo and possibly two other Earth-approaching asteroids are the only examples of this asteroid class, and may be the parents for the ordinary (H, L, and LL) chondrites.
Class R: Unique to 349 Dembowska, which appears to have a surface of olivine, pyroxene, and some metal. The olivine-rich achondrites are possible analogues.
Class S: Very common in the inner parts of the main belt and among Earth-approaching asteroids, with varying proportions of metal, olivine, and pyroxene, possibly parental bodies for the pallasites and some iron meteorites.
Class V: A spectral class unique to 4 Vesta and Amor 3551 1983 RD, showing strong pyroxene features, similar to basaltic achondrites.

High-Albedo (>0.3) Class

Class E: An uncommon type, possibly similar to the enstatite chondrites.

"Primitive" Classes: C, D, J, P, Q; "Metamorphic" Classes: B, G, F, T; "Igneous" Classes: A, E, M, R, S, V, SNC, and lunar.

Adapted from Bell J. F., Chapman C. R., and Gaffey M. J. (personal communication, 1989) and (1989) in *Asteroids II* (R. P. Binzel, ed.), Univ. of Arizona, Tucson.

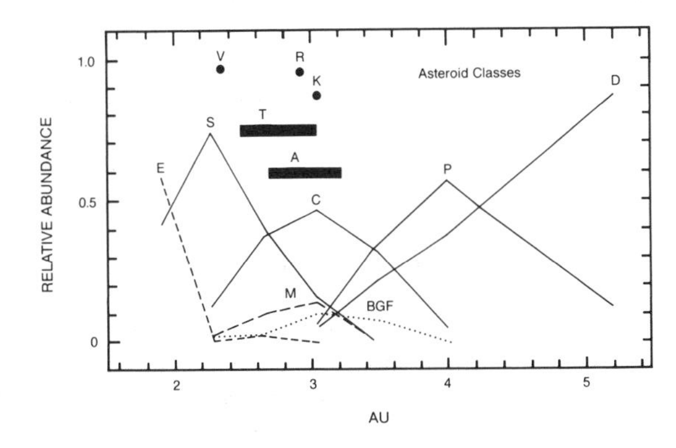

Fig. 5.11.4. *The distribution of taxonomic types in the asteroid belt. After Bell J. F. et al. (1989) in Asteroids II (R. P. Binzel et al., eds.), p. 925, Fig. 1, Univ. of Arizona, Tucson.*

consequence of the pile-up of water ice at the snow line, then such depleted nebular regions as the asteroid belt might be a common feature of planetary systems in general.

Gradie and Tedesco [148] first clearly demonstrated the zonal structure of the asteroid belt (Fig. 5.11.4), a concept reinforced by all later studies. S, C, P, and D types occupy successive rings outward in the belt. Among the rarer types, M types predominate near the middle, while B and F types occur at the outer edge of the belt. The S types can be subdivided into mineralogical subgroups (Ss = silicate rich; Sm = metal rich; Sq = opaque; Sp = pyroxene rich; and So = olivine rich). The Ss asteroids are closest to the inner edge about 2.45 AU, whereas the carbon-rich S types peak at 2.8 AU. The spacecraft Galileo will inspect the S-type asteroid 951 Gaspra in October 1991. One basic question is whether the inner-belt S asteroids are differentiated. Gaffey [149] maintains that all criteria (he lists six) point to their being differentiated assemblages.

The main belt can be divided into three super classes (Fig. 5.11.5) [150]: (1) "igneous" Sunward of 2.7 AU; (2) "metamorphic" around 3.2 AU; and (3) "primitive" outside 3.4 AU.

The broad pattern is that the fractionated types (e.g., S class with Ni/Fe metal, olivine, and pyroxene) predominate in the inner belt, and the low-albedo, "primitive" asteroids (e.g., C class) are the major members of the outer portions of the belt. If this pattern is preserved from the earliest times, as judged here, then it is possible to conclude that all bodies Sunward of 2.0 AU were "igneous," with the proportion declining to zero at 3.5 AU. The implication

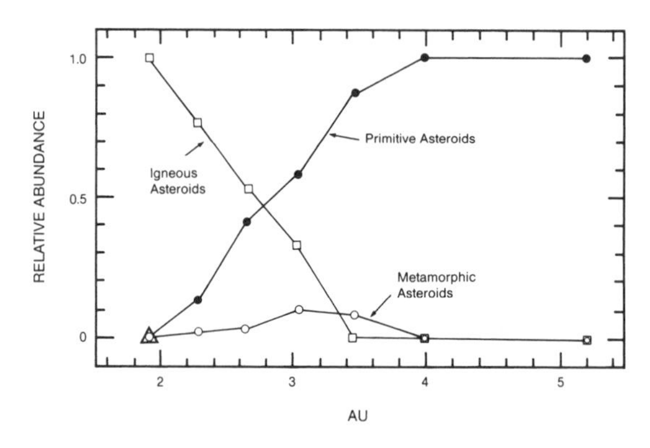

Fig. 5.11.5. *The distribution of asteroid superclasses with increasing distance from the Sun. After Bell J. F. et al. (1989) in Asteroids II (R. P. Binzel et al., eds.), p. 925, Fig. 1, Univ. of Arizona, Tucson.*

for the formation of the terrestrial planets is that they accreted from "igneous" (i.e., differentiated into metal and silicate fractions) planetesimals with little or no input from "primitive" (i.e., CI) material.

The E, S, and M types in the inner belt appear to be igneous with abundant metal on the surfaces of the S asteroids [151]. In the middle belt, C, B, and F types appear to result from thermal metamorphism in the presence of liquid water. In the outer reaches of the belt, the D and P types apparently do not contain hydrated minerals, as is shown by the lack of the strong 3-μm absorption feature in the infrared spectrum in the P and D classes. Water is present as ice; temperatures have not been high enough for the ice to melt and for hydration reactions to occur. The view is taken here that the primordial mineral phases in the nebula were anhydrous; hydrated phases result from subsequent reactions with water following heating sufficient to melt water ice.

5.11.3. 1 Ceres

1 Ceres, which is of interest as the largest asteroid, is an oblate spheroid with a visual albedo of 0.073. Its equatorial radius is 479.6 ± 2.4 km and its polar radius is 453.4 ± 4.5 km. Its effective diameter is 932.6 ± 5.2 km. Ceres has a mean density of 2.3 g/cm^3 [152] and is located at 2.77 AU from the Sun. It belongs to the G subclass of C-type asteroids and, although in hydrostatic equilibrium, apparently has significant topographic relief [153]. It contains about 33% of the mass of the entire asteroid belt. Accordingly, its composition is of interest in that it may be representative of many other asteroids. It shows a strong 3.07-μm absorption band, possibly indicative of the presence of hydroxyl, water ice, or perhaps NH_4-bearing minerals [154]. It is probably closest to CI or CM meteorites. Nine of the 13 most massive asteroids are similar to Ceres.

In most accretion models, Ceres grows quite quickly, from kilometer-sized planetesimals in 10^5 to 10^6 years. Thus, there is no reason for Ceres not to have grown larger unless the belt ran out of material. This is additional circumstantial evidence pointing to an early appearance of massive Jupiter on the planetary scene.

5.11.4. 4 Vesta

4 Vesta has an intact differentiated crust similar to that of basaltic achondrites. Since these differentiated meteorites formed within 20 m.y. of T_0, it is likely that the surface of 4 Vesta has survived from about 10^7 years after T_0. Vesta would be destroyed by collision with an object about 100 km in diameter. Thus, its existence places an upper limit on the frequency of collisions and on vigorous mass loss from the belt.

5.11.5. Asteroid Families

The orbital properties of many asteroids are similar, clustering around preferred values of semimajor axis, inclination, and eccentricity. This originally led Hirayama [155] to erect the concept of asteroid families. These families were generally agreed to be the broken up remnants of precursor asteroids, and so could provide important compositional information; the collisional breakup of a differentiated asteroid could produce several individual asteroids representing different interior zones of the precursor body. Thus, the potential exists to observe and reconstruct the deep interiors of planetesimals. The compositional families are identified by spectral reflectance methods and from albedo measurements. The spectral and albedo data provide information only on the composition of the surfaces of the asteroids. These, however, are generally not heterogeneous [156]. This fact is considered to indicate that the reflective data are not only typical of the surficial coatings, but are representative of the mineralogy at depth. No doubt exceptions occur, but are perhaps rare.

Like many good initial ideas, the family concept has encountered difficulties. It is an excellent example of the sort of problems that can arise due to the enthusiastic adoption of an clarifying concept, particularly when the objects are close to limits of observation. Not unexpectedly, major problems have arisen: e.g., which asteroids belong to which families? Williams [155] and Gradie et al. [155] created more than 100 families to which about 50% of all known asteroids were assigned. Kozai [155] went further. Although he had fewer families than Williams, he grouped about 75% of all asteroids into families. Skeptics soon arose. Carussi and Massaro [155] and Carussi and Valsecchi [155] cast doubt on the reality of most of the families.

Much light has been shed by an 8-color survey for 589 asteroids [157] and a 52-color infrared survey [158]. As a result of these studies, Chapman and coworkers [159] think that only about half of the Williams families, perhaps only 20% of the Kozai families, and 50% of the Carussi and Mezzano families are real. In addition to the four classical families (Nysa, Eos, Koronis, Themis), about which there is little doubt, other real families possibly include: Hungaria, Maria, Eugenie, Lydia, Leto, and Budosa. The Nysa family appears to consist of black objects, that may not be related pieces of a broken-up asteroid. Bell [160] is even more restrictive. He considers only the Themis, Eos, Koronis, and Maria families to be unequivocally established as the result of collisional disruption of large asteroids. The Flora family is probably real, with the spherical Flora being the metallic core of the original parent body [160]. The Phocaea family appears to be an unrelated collection af asteroids clustered together with the same orbital characteristics. The Nysa-Hertha families may be mostly unrelated, with Nysa itself being a foreigner. Thus, many of the smaller "families" are probably dynamical groupings, while interlopers abound among the larger families.

Of the "real families" selected in the genealogical survey by Bell [160], Flora, Koronis, and Maria families are composed of S-type asteroids, Eos of spectral type K asteroids (possibly CO or CV chondrites), and the Themis family of C- and B-type asteroids, possibly metamorphosed carbonaceous chondrites.

In summary, the family concept was a good idea that was pushed beyond reasonable limits. The erection of so many doubtful families is a cautionary tale for classifiers working at the limits of the observational techniques. The families may eventually provide some useful cosmochemical information, but at present it is very difficult to extract. Low-velocity collisions between disparate asteroids may also confuse the record. Trying to unscramble the collisional history of the belt has all the difficulties that beset unmixing studies.

Finally, no evidence has yet been detected for binary asteroids in the asteroid belt proper or dust belts surrounding asteroids [161].

5.11.6. The Apollos, Atens, and Amors

The Apollos are the current suppliers of meteorites to the Earth (section 2.12) and present us with the special hazard of collisional impact with this planet. The Apollos have orbits that cross that of the Earth; the Atens are a subclass of the Apollos with semimajor axes that are inside that of the Earth, while the Amors are Mars-crossers, with orbits that approach, but do not cross, that of the Earth [162]

It is estimated that there are about 1000 Apollos, 100 Atens, and 1000-2000 Amors. Their mean lifetimes are of the order of 10^7-10^8 years, before they either are ejected from the solar system, or collide with a planet. They are a very diverse group of objects, although most of the observed ones are of S type. This may be an observational artifact that has discriminated against finding the darker members [163]. They are smaller than the main-belt asteroids, the largest being 1036 Ganymed, 38.5 km in diameter. This argues that they are collisional fragments. The Earth approachers contain bodies that could be the sources for the ordinary chondrites, parents for which have not yet been identified in the main belt [164]. Curiously, although ordinary chondrites dominate the terrestrial meteorite flux at present, parent bodies have not been identified in the main belt, the inner portions of which are dominated by S asteroids. These do not have ordinary chondrite mineralogy. It is commonly suggested that alteration to the regolith of the S-type asteroids by cosmic radiation might explain this discrepancy; irradiation, however, should darken the surfaces, not produce the reddened surfaces observed.

Dynamical considerations suggest that the inner belt is the home of the H, L, and LL chondrites, but the location of their parent bodies remains uncertain, and the S asteroid-ordinary chondrite controversy remains unresolved (see section 3.13.2). The cautionary tale of the blind men and the elephant is often invoked. Possibly the ordinary chondrites are not common!

Several asteroid classes are not currently delivering meteorites to the Earth. These include classes D, P, and T in the outer belt. One Earth-crossing asteroid of basaltic composition has been identified, and is a possible candidate for the parent body of some of the basaltic achondrites.

Probably 40% of the Apollos, Atens, and Amors are extinct cometary nuclei in which the surface is coated with a thick, dark, nonvolatile lag deposit, which prevents solar heating from releasing any volatile ices left inside. At least one asteroid, 3200

Phaeton, is in the same orbit as a meteor shower (the Gemnids), reinforcing the concept that some "asteroids" may be extinct comets [165].

5.11.7. The Trojans

These distant objects, beyond the main belt, occupy the L4 and L5 Lagrangian points, leading and trailing Jupiter. They have very low albedos, similar to asteroids of the outer belts. Perhaps they are primitive and undifferentiated, but probably they contain some surprises. Two-thirds are class D and the remainder are class P. No meteorites are known that match them. Mineralogically, they are probably a mixture of clays and organic components. Because of both their low relative velocity and the low density of objects, the Trojans undergo only a few percent of the collisions experienced by those of the main belt. These latter have acquired their rotational properties as a result of an extreme history of collisions. Although one might expect that primitive undifferentiated objects with low collisional frequencies would be spherical, in fact they are mostly elongated, possibly with dumbbell shapes. They may represent a more primitive population, with shapes derived from the period of accretion, rather than modified by many collisional events as in the main belt. Possibly they are "well preserved primordial accretion products rather than belt asteroids" [166]. Hence, they might yield useful information about the early solar system, unless they turn out, like the Moon and Comet Halley, to be more complicated objects once they are visited.

5.11.8. Other Asteroid Belts?

One asteroid, 1990 MB, has been recognized trailing Mars at the L5 Lagrangian position, thus becoming the first martian trojan. The L4 and L5 Lagrangian points for Saturn, Uranus, and Neptune are stable, but there is no sign of "Trojan" type asteroids. Why are there no asteroids at these locations? Some orbits between Saturn and Uranus are also stable over periods comparable with the age of the solar system [167]. Chiron, which is either a comet or a left-over icy planetesimal, is in an unstable orbit in this region, but appears to be unique, and is at least three magnitudes brighter than any other object in the 10-AU gap between Saturn and Uranus. One scenario for the occurrence of C-class satellites of the giant planets envisages capture from asteroid belt material scattered by Jupiter. If asteroids were widely scattered from the main belt by Jupiter, some might have been expected to be trapped by Saturn, Uranus, and Neptune, but do not appear to be present.

5.11.9. Mixing in the Belt

The lack of mixing evident among the chondrite groups, which are relatively primitive in composition and very old, is evidence for a strongly zoned belt. Most undifferentiated meteorites are breccias. However, they rarely contain clasts or fragments of other meteorite groups. The low abundance of foreign inclusions in meteoritic breccias implies little lateral mixing during meteorite formation. This indicates that collisional mixing and stirring of the belt must have been relatively minor and the separate chondrite classes accreted within quite narrow nebular zones, perhaps less than 0.1 AU wide.

The rather few genuine asteroid families and the survival of basaltic crusts on such early objects as 4 Vesta indicate that the present belt is not vastly different from that at 4.6 b.y. ago. Large-scale transfer of material through the belt would have destroyed large bodies as well as the zonal structure, although this must have undergone some broadening due to collisional processes.

The zonal structure appears to have been stable for over 4 b.y. The main evidence for this is the rather uniform cratering rate observed on the lunar maria since the close of the great bombardment about 3.8 b.y. ago. If the belt had been unstable, a steady decline in the impact rate, rather than a steady state, would have been observed. In this context "asteroids have nothing to do with the growth of planets" [168] and we can only infer the properties of planetesimals Sunward of the belt.

5.11.10. The Primitive Nature of the Asteroid Belt

There is considerable evidence for a modest primordial asteroid belt [169]. The existence of a preserved basaltic surface on 4 Vesta as noted above seems to indicate little disturbance since its formation. The basalt on 4 Vesta is presumably the same age as that of the basaltic achondrites (4.55 b.y.) and so provides some supporting evidence for the ancient nature of the belt in its present form. The structured nature of the belt makes it difficult to introduce objects or to pass objects through the belt from the outer into the inner solar system, by some kind of "random walk" process; much greater diversity would be expected in the belt if this had happened.

Judging from the meteorite evidence, individual accretion zones for asteroids appear to be quite homogeneous. Very little mixing occurred between the E, O, and C meteorite groups. These contain chondrules with unique oxygen isotope signatures. Mixing among the many chondrite groups is very minor.

The absence of large amounts of 4.5-b.y.-old impact melts in chondrites and the lack of evidence for heavy bombardment within the asteroid belt seem to point to a relatively quiet early accretion history. If the early asteroid belt had been very crowded, dominated by collisions, one might expect that the chondrites would contain much more glass and breccia clasts; instead they are dominated by chondrules. There are no big patches of melted rock, and very little evidence for mixing. Thus, many clues to the accretion process have been preserved. Within ordinary chondrites, for example, variations in composition are usually less than 10%. Thus, there is no evidence, particularly from the lack of impact melts and clasts, to support the hypothesis that chondrules were produced by violent collisions.

It is estimated from the small content of foreign clasts (less than 1%), from the oxygen isotope evidence, and from the general uniformity of compositions within chondrites of one class that "each chondrite class is composed of solids that accreted in a narrow nebular zone <0.1 AU wide, between 2 and 3 AU from the Sun" [170]. The present compositional zones have been widened by subsequent collisions.

5.11.11. Water Ice and Hydrated Minerals

It seems likely that all primitive minerals in the nebula were anhydrous. A probable scenario calls for the accretion of anhydrous silicates, organic components plus water ice [171]. Mild heating will produce aqueous alteration. Such effects have been observed in laboratory simulations at temperatures of 25°C. There is no need to invoke hydrothermal activity; hydrated silicates can form at such low temperatures [172]. There is no correlation between the observations of hydrated silicate (clay minerals, characterized by a 3-μm absorption feature) and asteroid diameter, albedo, or orbital parameters [173]. Although the majority of C-class asteroids have hydrated mineral spectra, some of the largest (300 km diameter) do not. Some of the "volatile-rich" P and D asteroids do not show any sign of the 3-μm absorption band. Probably the hydrated silicates form due to reaction of hydrated minerals with melted ice at low temperatures; the reactions proceed at 25°C [174,175]. In the outer asteroids, the temperatures are probably too cold for the reactions to proceed, and their surfaces reflect a primitive anhydrous mineralogy.

5.11.12. The Origin of the Belt

The existence of the asteroid belt raises many interesting questions relevant to the origin of the solar system. Asteroids accrete from smaller planetesimals or form from collisions, and are not debris from an exploded planet. The total mass is only about 5% of that of the Moon. Why is there such a depletion in mass by about three orders of magnitude at the belt, in comparison with that expected from a simple extrapolation from the planetary masses? Why did a large planetary embryo not grow in the asteroid belt? This is the fundamental problem in the belt since there is no reason to suppose that the region of the solar nebula now occupied by the asteroid belt was initially deficient in material. The early formation of Jupiter starved the belt of material since Jupiter swept up most of that available, and prevented planetesimals from entering the belt area because of gas drag.

Another effect of the early formation of Jupiter was to pump up velocities and eccentricities of the asteroids and so bring accretion to a halt [176], explaining why the present asteroids do not accrete into one body. The general evidence summarized in this section indicates that the asteroid belt has been a relatively quiet place. Growth stopped as the planetesimals reached about 100 km as the belt ran out of material [177]. Although many collisional episodes have occurred since then, this has not resulted in the accretion of a larger object. Among other conclusions, this seems to support planetary growth from narrow feeding zones. If there was much mixing from far and near, one might surmise that we should see a Pluto-sized object at the Titius-Bode rule position, rather than the spread-out zones of differing compositions that we observe in the present belt. The preservation of a basaltic crust on 4 Vesta, remarked on earlier, indicates that Vesta has survived since 4.55 b.y. relatively unscathed. Thus, when Vesta differentiated and produced basaltic lava flows from a silicate mantle, the belt was not greatly different from its present appearance.

The asteroid belt thus contains critical evidence relating to the origin of the solar system. Although it did not contain enough material to form its own planet, this fact sheds much light on the history of the solar system, just as the Moon, a unique object, provides crucial information of the importance of large collisions. It is perhaps ironic that the failure of the asteroid belt to produce a planet is of more value to this inquiry than if we had a somewhat more volatile-rich version of Mars lying between that planet and Jupiter.

In summary, the belt shows strong zoning from "igneous" differentiated asteroids in the inner belt, through "metamorphic" asteroids, affected by liquid water [178], to the more primitive outer-belt asteroids, in which water is present as ice and the minerals are anhydrous. This radial dependence on composition argues that we are looking at the belt in almost its primitive state. The initial mass was only a few times the present mass, following depletion by the early formation of Jupiter. An active early Sun seems indicated as the source of the heating required to differentiate the inner belt. Temperatures nearer the Sun, in the region of the terrestrial planets, must have been higher, a fact consistent with the increased depletion of volatile elements (e.g., lower K/U ratios) in the terrestrial planets compared to the meteorites.

5.12. THE GIANT PLANETS

"I don't think the existence of Jupiter would be predicted if it weren't observed" [179].

These splendid planets, and their associated rings and satellite systems, must excite the admiration of even the most casual of observers. It is interesting that the ancient astronomers, unaware of the true size of Jupiter, nevertheless showed unusual perception in naming it after the chief Roman god. Saturn, recorded from the seventh century B.C., was the outermost planet until the discovery of Uranus in 1781 followed by Neptune in 1846. Reviews are given by Hubbard, Gehrels, and Stevenson [180,181].

5.12.1. Atmospheres

Initially it was believed that the compositions of the atmospheres of Jupiter and Saturn were the same as that of the solar photosphere [182] implying that they were unaltered fragments of the primordial solar nebula. One of the major more recent discoveries is that the atmospheres of Jupiter (Table 5.12.1) and Saturn (Table 5.12.2) show differences from the solar ratio and between themselves in H/He ratios. The solar system He abundance is 27.5 wt%. For Jupiter the ratio is 18 ± 4% and for Saturn is only 6 ± 5%. Uranus is better behaved, with a ratio of 27% within the error limits of the solar ratio. (Table 5.12.3; Fig. 5.12.1). Table 5.12.4 gives ratios of C, N, O, and P to H. There are strong enrichments in C/H ratios,

TABLE 5.12.1. Composition of the atmosphere of Jupiter.

Constituent	Volume Mixing Ratio
H_2	0.89
He	0.11
CH_4	0.00175
C_2H_2	0.02 ppm
C_2H_4	7 ppb
C_2H_6	5 ppm
CH_3C_2H	2.5 ppb
C_6H_6	2 ppb
CH_3D	0.35 ppm
NH_3	180 ppm
PH_3	0.6 ppm
H_2O	1-30 ppm
GeH_4	0.7 ppb
CO	1-10 ppb
HCN	2 ppb

Data from Strobel D. F. (1985) in *Photochemistry of Atmospheres: Earth, the Other Planets and Comets.* (J. S. Levine, ed.), p. 393, Academic, New York.

TABLE 5.12.2. Composition of the saturnian atmosphere.

Constituent	Volume Mixing Ratio
H_2	0.94
He	0.06
CH_4	0.0045
C_2H_2	0.11 ppm
C_2H_6	4.8 ppm
CH_3D	0.23 ppm
PH_3	2 ppm

Data from Strobel D. F. (1985) in *Photochemistry of Atmospheres: Earth, the Other Planets and Comets.* (J. S. Levine, ed.), p. 393, Academic, New York.

TABLE 5.12.3. Helium abundances in the outer atmospheres of the giant planets.

Y (per mass abundance)			
Jupiter	Saturn	Uranus	Protosun
0.18 ± 0.04	0.06 ± 0.05	0.262 ± 0.048	0.274

Data from Gautier D. (1988) *Philos. Trans. R. Soc., A325*, 591.

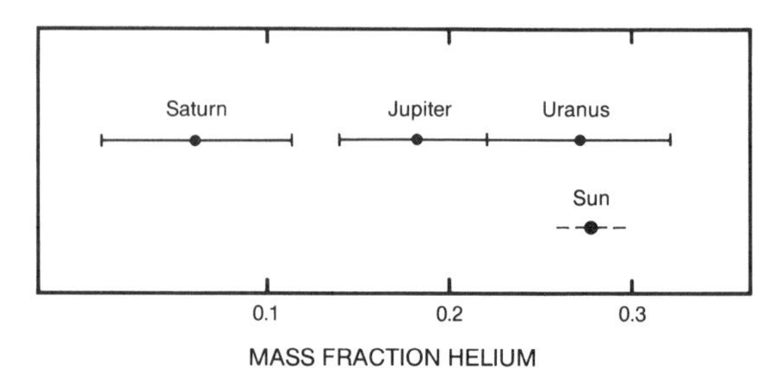

Fig. 5.12.1. *The abundance of He (mass fraction) in the atmospheres of Saturn, Jupiter, and Uranus, compared to that of the Sun, showing the major depletion in the saturnian and jovian atmospheres. After Conrath B. J. et al. (1987) J. Geophys. Res., 92, 15003.*

TABLE 5.12.4. Element ratios, in solar units, in the outer atmospheres of the giant planets.

	Sun	Jupiter / Sun	Saturn / Sun	Uranus / Sun	Neptune / Sun
C/H	4.7×10^{-4}	2.32 ± 0.18	2-6	*ca.* 25	*ca.* 25
N/H	9.8×10^{-5}	*ca.* 2	2-4	$\ll 1$?	$\gg 1$?
O/H	8.3×10^{-4}	1/50 (from H_2O) $\geqq$1/3 (from CO)			
P/H	2.4×10^{-7} (from carbonaceous chondrites)	1 ± 0.3	2.8 ± 1.6		

Data from Gautier D. (1988) *Philos. Trans. R. Soc., A325*, 587.

from 2 at Jupiter to about 25 at Uranus and Neptune. N/H is enriched at Jupiter and Saturn. The CH_4/H_2 ratios for Jupiter and Saturn are 2-5 times solar, and are 20-25 times solar for Uranus and Neptune. Thus, the atmospheric compositions are far removed from anything resembling that of the primitive nebula.

5.12.2. H/He Fractionation

Although the atmospheric ratios of H/He differ from the solar abundances, this does not imply that the bulk planetary values are different. Since processes in the nebula are unlikely to separate H from He, these differences must be due both to differentiation during accretion and to subsequent fractionation processes within the individual planet. Even the lightest gases cannot escape, so the planets must contain the original nebular H/He ratio.

The great depletion of He in the atmosphere of Saturn, and to a lesser extent in that of Jupiter, is ultimately due to the fact that at pressures in excess of 3 Mbar, hydrogen becomes metallic. At low temperatures, He is expected to become immiscible, so that as the planets cool, drops of He will exsolve and fall toward the center of the planet, as the saturation temperature is reached for the He-H^+ mixture. This will lead to a depletion of He relative to H in the outer layers. Temperatures in Jupiter are higher than in Saturn, leading to a less efficient separation of He from H in Jupiter compared to Saturn. Because of its smaller size, He separation thus will begin earlier on Saturn and somewhat later on Jupiter. Uranus and Neptune are too small and their internal pressures thus too low (<2 Mbar) for the 3-Mbar transition from molecular to metallic hydrogen to be reached, so they retain the solar nebular H/He ratio (0.27) in their atmospheres [183] (Fig. 5.12.2).

Figure 5.12.3 shows a possible model for the internal structure of Jupiter and Fig. 5.12.4 shows a possible model for the internal structure of Saturn.

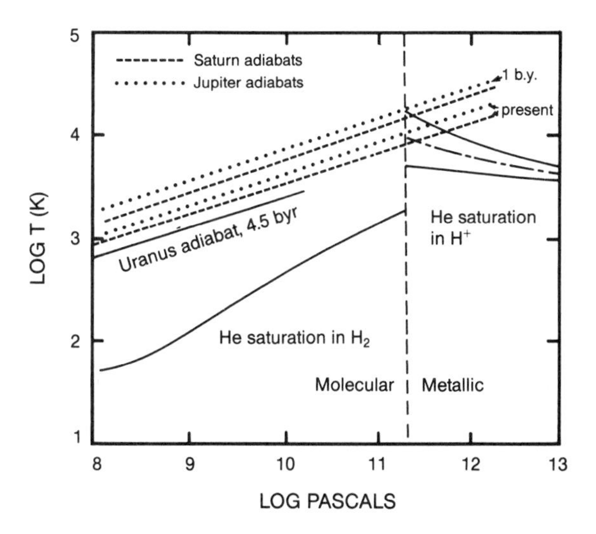

Fig. 5.12.2. *The saturation temperature for He in a 75%H:25%He mixture. The dash-dot curve on the right represents the most plausible solubility of He in metallic H. The uranian and neptunian adiabats are cooler than those of Saturn and Jupiter, and never reach the metallic H transition. In the molecular region all adiabats for the giant planets lie above the saturation line for He in H_2. The saturnian and jovian adiabats extend into the region of metallic H, but initially lie above the He saturation line. As these planets cool, He rains out as the adiabat intercepts the He saturation line in H^+. Saturn cools faster than larger Jupiter, so a larger percentage of He is enriched in the deeper parts of the planet, causing the observed surface depletion. After Gautier D. and Owen T. (1989) in The Origin and Evolution of Planetary and Satellite Atmospheres (S. Atreya et al., eds.), p. 499, Fig. 5, Univ. of Arizona, Tucson.*

5.12.3. Composition

Jupiter and Saturn [184] have radii close to that expected for planets of approximately solar composition and so must be composed primarily of H and

He. However, all the giant planets are enriched relative to the Sun in the elements heavier than H and He. Thus, even Jupiter does not have the primordial solar nebular composition. None of the nuclei of the giant planets captured gas efficiently from the nebula. There is a systematic decrease in the H and He contents with heliocentric distance. This is consistent with decreasing density of the nebula with increasing distance from the Sun, and

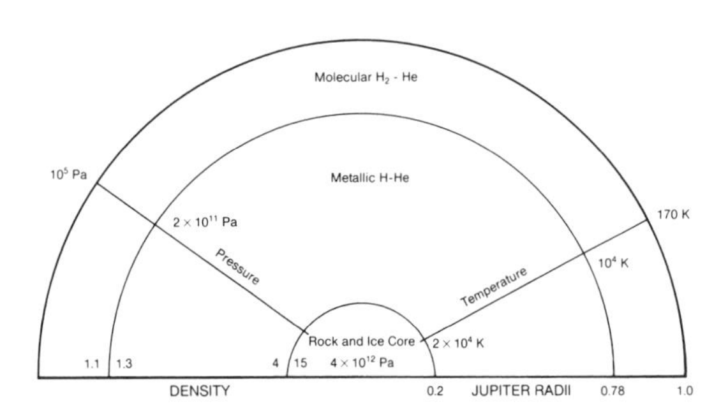

Fig. 5.12.3. *A possible structure for the interior of Jupiter. Pressures are given in Pascals (1 bar = 10^5 Pa). After Stevenson D. J. (1989) in The Formation and Evolution of Planetary Systems (H. A. Weaver and L. Danly, eds.), p. 78, Fig. 1, Cambridge Univ., New York.*

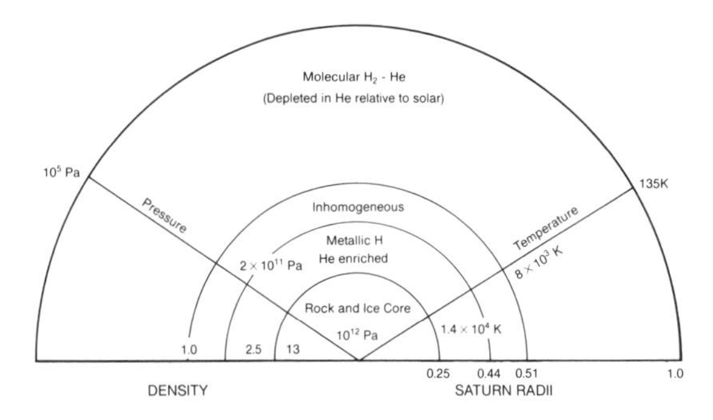

Fig. 5.12.4. *A possible structure for the interior of Saturn. Pressures are given in Pascals (1 bar = 10^5 Pa). After Stevenson D. J. (1989) in The Formation and Evolution of Planetary Systems (H. A. Weaver and L. Danly, eds.), p. 79, Fig. 2, Cambridge Univ., New York.*

with a longer timescale of accretion, so that the gas was being swept out of the nebula as the planets were being assembled (Table 5.12.5).

Deuterium/hydrogen ratios are shown in Fig. 5.12.5. These are significantly depleted in the atmospheres of Jupiter and Saturn, compared with the outer ice giants or Titan, Comet Halley, and the terrestrial oceans, but resemble the solar values. This indicates that the deuterium in these objects was derived from water that had not been in equilibrium with nebular gas.

The gas/(rock + ice) ratio decreases monotonically from Jupiter to Neptune. Jupiter is gas rich, Neptune is ice- and rock-rich. Although the relative proportions of ice and rock increase from Jupiter to Neptune, the absolute amounts are similar within a factor of 2. In contrast, the abundance of gas (H + He) varies by a factor of 100.

The variation of the major CNO compounds with solar distance and estimated temperatures implies that the major CNO species beyond 2 AU are CH_4, NH_3, and H_2O [186,187].

5.12.4. Magnetic Fields

The magnetic fields of Jupiter, Saturn, Uranus, and Neptune vary considerably. The equatorial surface field for Jupiter (4.1 oersted) is an order of magnitude higher than that of Saturn (0.4 oersted) or Uranus (0.23 oersted). The dipole field for Saturn is effectively aligned with the rotation axis, that of Jupiter is offset, like that of the Earth, by about 10°, while the magnetic axis for Uranus lies 60° away from the rotation axis, which, since Uranus is lying on its side, is close to the plane of the ecliptic. The field for Neptune can be modeled by a tilted dipole, inclined at 47° to the rotation axis, and offset from the center by over half the planetary radius. Why these fields for the giant planets are so different is unclear.

5.12.5. Heat Flows

The basic conclusion from the heat flows of the giant planets is that they are consistent with their growth by accretion, so that the planets start hot. The ultimate heat source is thus gravitational energy. In the case of Jupiter, the observed heat flow is explicable on the basis of cooling from an initially hot condition, possibly with some addition from the gravitational energy released from the rainout of helium drops. The same explanation accounts for the heat flow of Saturn, except that the energy from the helium segregation is greater [188].

TABLE 5.12.5. Deuterium abundances in giant planets and in the protosun (10^5D/H).

	Protosun	Jupiter	Saturn	Uranus
from HD	—	3.7-17	4-13	4-10
from CH_3D	—	1.2-5.5	0.7-3	4.5-18
from ^3He (in solar wind)	$2^{+1.5}_{-0.5}$			

Data from Gautier D. (1988) *Philos. Trans. R. Soc., A325,* 590.

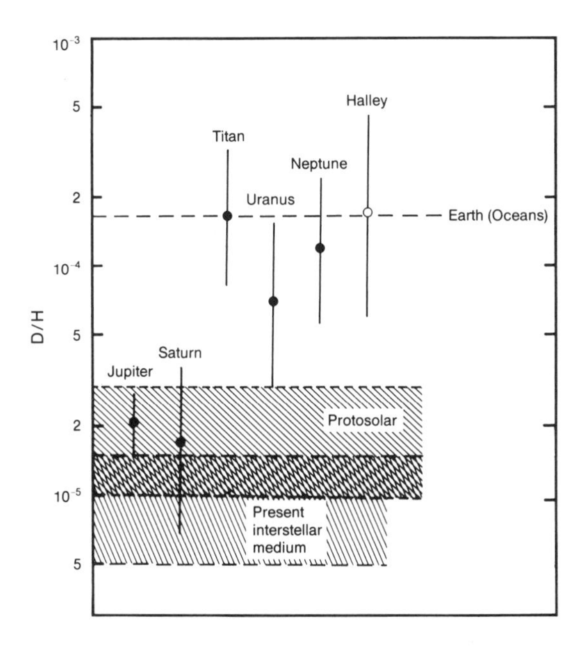

Fig. 5.12.5. *D/H ratios determined in methane in the atmospheres of the giant planets and Titan and in water in Comet Halley, compared with the protosolar, interstellar medium and terrestrial oceanic abundances. After Gautier D. and Owen T. (1989) in The Origin and Evolution of Planetary and Satellite Atmospheres (S. Atreya et al., eds.), p. 504, Fig. 6, Univ. of Arizona, Tucson.*

Uranus and Neptune have distinctly different heat flows, which probably reflects differing evolutionary histories. Neptune emits a modest amount of heat (2×10^{22} erg/s) consistent with a relatively well-mixed interior. Uranus, in contrast, has a very low ($<6 \times 10^{21}$ erg/s) and possibly zero heat flow. These differences are not related to relative distance from the Sun, and present a good example of the sort of detail that needs to be accounted for in theories of the formation of the solar system.

Models that form the giant planets by fragmentation of the nebula predict uniformity. However, the planetesimal hypothesis might explain them along the lines of the following speculative suggestion [189]. This turns on the effects of giant impacts on internal structure during the terminal stages of accretion. The grazing impact by an Earth-sized object that tipped Uranus on its side [190] is unlikely to have promoted efficient mixing deep in the interior of Uranus, leaving a layered internal structure whose compositional differences have trapped heat by hindering convection. The higher heat flux of Neptune, in contrast, is consistent with a heated and well-mixed interior, perhaps the result of a late head-on massive collision [191].

5.12.6. Uranus

Uranus and Neptune are much smaller than would be the case if they were composed mainly of H and He; if this were the case, they would be expected to have radii about twice their present dimensions. Their interiors must be denser, despite the presence of hydrogen in their atmospheres [192]. Figure 5.12.6 shows a possible model for the internal structure of Uranus. Classic models of the interior of Uranus divide the planet into a rocky core, an icy oceanic shell (H_2O, CH_4, and NH_3) and a gaseous (H and He) and icy outer envelope of about 2 Earth masses, in which H and He are enriched over the solar abundances. However, the gravity data suggest that these layers are not clear-cut and that they grade into one another. The surface temperature is about 60 K; a possible central temperature is 7000 K at a pressure of 6 Mbar. The He abundance in the uranian atmosphere is 26.7 ± 4.8%, within the error limits of the solar value [193]. This is consistent with the fact that the pressure in Uranus is not sufficient to reach the metallic hydrogen transition (Fig. 5.12.1), so that this mechanism is not available to fractionate He from H. Of course other processes such as the addition of cometary material might raise the hydrogen-to-helium ratio, so that the present agreement with the solar value might be accidental [194]. Since one

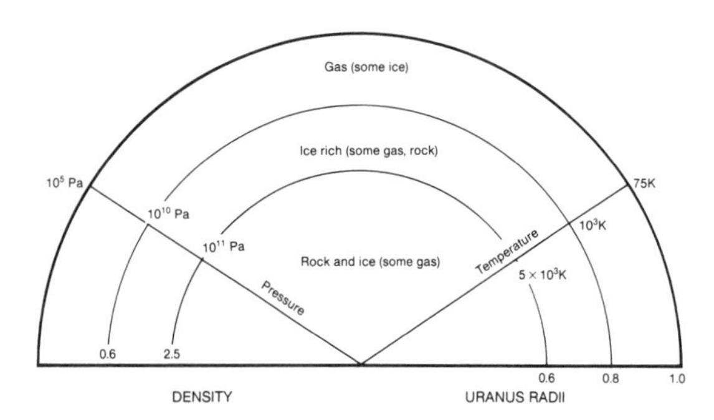

Fig. 5.12.6. *A possible structure for the interior of Uranus. Pressures are given in Pascals (1 bar = 10^5 Pa). After Stevenson D. J. (1989) in The Formation and Evolution of Planetary Systems (H. A. Weaver and L. Danly, eds.), p. 80, Fig. 3, Cambridge Univ., New York.*

current model derives the satellites from an impact-produced disk, possibly Uranus has the same ice/rock ratio as the satellites. However, fractionation, contributions from the impactor, and the derivation of the disk from near the surface of such a large planet all raise many uncertaintites in such a scenario.

5.12.7. Neptune

This ice giant [195] forms the true outer limit of the planetary system. The planet is smaller than Uranus. The oblateness, e, is 0.0209 ± 0.0014 [196]. Since Neptune is more massive (17.2 Earth masses) than Uranus (14.5 Earth masses), it must have a lower content of hydrogen (about 1.1 Earth masses) than Uranus (about 1.6 Earth mass) and a higher content of ice and rock, illustrating yet again the truism that no two planets in the solar system are alike [197].

Neptune is in an extremely circular orbit, with an eccentricity of 0.009. It differs by only 4×10^{-5} from a perfect circle, better, as Bill Kaula has remarked, than most machine shop specifications. This places some stringent limitations on the stability of the solar system. Neptune is only weakly gravitationally bound, and the orbit would easily be changed to one of higher inclination and eccentricity by the passage of a passing star. Calculations indicate that no star with mass greater than 1/10 solar mass has passed through the system; the implication is that the present orbit is primordial, and that no large planets exist beyond Neptune [198].

5.12.8. Origin

Two basically different models exist to explain the existence of the giant planets. The first involves the direct fragmentation of the nebula; the most recent of these models is the giant gaseous protoplanet hypothesis [199]. In this scenario, the giant planets subsequently form cores by infall of the more refractory components. This model suffers from the defect that a core is unlikely to form because of the high solubility of elements, except helium, in hydrogen under megabar pressures.

In the second class of models, ices and silicate accrete to form a core of 10-15 Earth masses, which then accretes gaseous components from the nebula. The critical core mass appears to be about 15 Earth masses, before runaway accretion of H and He can occur from the nebula, but it is possible that much smaller core masses (6 Earth masses) might suffice [200].

The latter general class of hypotheses explains the compositional data; the former would predict that the giant planets should reflect the composition of the nebula directly. The main problem is to grow a core quickly enough to accrete gas before the nebula is dispersed, as discussed earlier. All the giant planets have cores in the range of 10-15 Earth masses. The decrease in total planetary mass with increasing distance from the Sun is explained as resulting from the cores of Saturn, Uranus, and Neptune growing more slowly, so that the gaseous components of the nebula has begun to dissipate by the time that the cores have reached critical mass. This, along with other constraints, indicates that the formation of the giant planets occurs very quickly. It seems highly likely that the growth of Jupiter may well be associated with its position just beyond the "snow line," so that water and other volatiles piled up there as the inner nebula were cleared by early intense solar activity.

Except for Jupiter, with a 3.1° inclination, the axes of rotation of the giant planets are significantly tilted, Saturn by 26.7°, Uranus by 98°, and Neptune by 29°, the probable result of giant collisions, constituting prima facie evidence for the planetesimal hypothesis.

5.13. PLANET X

Are there large planets beyond Neptune? Searches for Planet X have been unsuccessful. No perturbations due to the presence of such a planet have been noted affecting the orbits of Pioneer 10, now over 44 AU distant, far outside the orbit of Neptune (30 AU), or Pioneer 11, at present beyond Uranus at 26 AU [201]. The IRAS survey detected no sign of the postulated Planet X, nor is there any sign of a dark companion star to the Sun. Both of these have been postulated to be responsible for periodic cometary showers, for which there is little firm evidence [202]. In summary, the evidence for Planet X is minimal. Possibly the often-cited perturbations of the orbit of Neptune are due to the Kuiper cometary cloud or to the Inner Oort Cloud.

5.14. THE LONG-TERM STABILITY OF THE SOLAR SYSTEM

It is a widely held belief, reinforced by the geological record, that the solar system has apparently been stable for over 4×10^9 yr. However, definite proof of the stability of the system has not yet been achieved. Project LONGSTOP (long term gravitational stability test of the outer planets) carried out integration studies for the orbits of the outer planets over the past 100 m.y. No instabilities were observed for calculations of orbital behavior over a 10^8-yr period. The Digital Orrery [203] enabled calculations to be extended to 845 m.y. [204,205]. These results have been used to place limits on close encounters of other stars with the solar system [206]. Although the planetary orbits show some quasi-periodic changes, these are small, and there are no secular trends in inclination or eccentricity. Pluto is a probable exception. Otherwise the orbits are stable probably over the age of the system, and do not align nor become more circular with time.

The best parameter for this purpose is the orbit of Neptune, the most weakly bound planet gravitationally. The possibility that this orbit was reduced to its present value from an original much larger inclination and eccentricity by the passage of a passing star is very unlikely. However, the converse is true. The passage of a star would perturb the orbits of the giant planets, increasing both their inclination and eccentricity. The nearly perfectly circular orbit of Neptune indicates that no such event has occurred. Calculations indicate that no star with mass greater than 1/10 solar mass has passed through the system, and no object greater than three Jupiter masses has passed within the orbit of the Earth. Since it appears unlikely that any star has altered the orbits of the giant planets, the orbital characteristics are primordial, and so must be accounted for in theories of the origin of the system.

The present outer boundary of the system is also primordial; no passing star has stripped away planets beyond Neptune. The apparent absence of any distant planets thus indicates that there was never a major planet beyond Neptune [206], a fact that would have disappointed Kant. More recent studies [204] indicate that Pluto is exhibiting chaotic behavior on timescales of the order of 20 m.y. The orbits of the other planets are probably stable over the age of the solar system. In fact, the chief argument, in the absence of definitive proof, for the overall stability of the system, lies in its great age. Geologists and geochemists take comfort from the long-term stability of the terrestrial orbit obvious from the geological record, which indicates great stability of surface temperatures, with evidence for liquid water on the Earth's surface extending back to the limits of the geological record nearly 4 b.y. ago.

Clearly more study of the system is needed, but it should be emphasized that chaotic orbits do not necessarily imply instability, so drastic changes are not anticipated. Another objective is to look for other possible stable orbits, in addition to those of the planets. Many orbits in the asteroid belt, those between Venus and the Earth, and between Saturn and Uranus are stable over the age of the solar system, but most orbits between Jupiter and Saturn and between Uranus and Neptune become unstable (see section 5.11) [207].

NOTES AND REFERENCES

1. McSween H. Y. Jr. (1989) *Am. Sci.,* 77, 146.
2. Kowal C. T. (1989) *Icarus,* 77, 122. Chiron is probably a large comet, but may be a leftover icy planetesimal; the difference is somewhat semantic.
3. Lissauer J. J. (1987) *Icarus,* 69, 249.
4. The definitive work is Vilas F. et al., eds. (1988) *Mercury,* Univ. of Arizona, Tucson, 794 pp. A readable account is given by Strom R. G. (1987) *Mercury, The Elusive Planet,* Smithsonian, Washington, DC, 197 pp. The usefulness of this interesting book is unfortunately diminished by the lack of references to the literature, which makes it difficult to separate fact from opinion.
5. One of the very few theories that attempt to account for the inclination of the spin axis of the Sun is the model in which the Sun captures a filament of material from a passing protostar, as proposed by Dormand J. R. and Woolfson M. M. (1989) *The Origin of the Solar System,* pp. 23, 52, Halstead Press, New York.

6. Anderson J. D. et al. (1987) *Icarus, 71,* 337.
7. Connerney J. E. P. and Ness N. F. (1988) in *Mercury* (F. Vilas et al., eds.), p. 494, Univ. of Arizona, Tucson.
8. Goettel K. A. (1988) in *Mercury* (F. Vilas et al., eds.), p. 617, Univ. of Arizona, Tucson.
9. Tyler A. L. (1988) *Geophys. Res. Lett., 15,* 808.
10. Hunten D. M. et al. (1988) in *Mercury* (F. Vilas et al., eds.), p. 562, Univ. of Arizona, Tucson.
11. For example, Wilhelms D. E. (1976) *Icarus, 28,* 551.
12. For example, Strom R. G. (1984) *NASA SP-469,* p. 13; Spudis P. D. and Guest J. E. (1988) in *Mercury* (F. Vilas et al., eds.), p.152, Univ. of Arizona, Tucson.
13. Scott R. F. (1977) *Earth Sci. Rev., 13,* 379.
14. Solomon S. C. (1976) *Icarus, 28,* 509.
15. Carey S. W. (1983) *Expanding Earth Symposium,* Univ. of Tasmania, Hobart, 423 pp.
16. Goettel K. A. (1988) in *Mercury* (F. Vilas et al., eds.), p. 613, Univ. of Arizona, Tucson; Lewis J. S. (1988) in *Mercury* (F. Vilas et al., eds.), p. 651, Univ. of Arizona, Tucson.
17. Goettel K. A. (1988) in *Mercury* (F. Vilas et al., eds.), p. 616, Univ. of Arizona, Tucson. Other estimates based on theoretical models are given in Table 4.3.2a in BVSP (1981) *Basaltic Volcanism on the Terrestrial Planets,* p. 639, Pergamon, New York, but have only historical interest.
18. Fegley B. Jr. and Cameron A. G. W. (1987) *Earth Planet. Sci. Lett., 82,* 207.
19. Benz W. et al. (1988) *Icarus, 74,* 516; Cameron A. G. W. et al. (1988) in *Mercury* (F. Vilas et al., eds.), p. 692, Univ. of Arizona, Tucson; see also Wetherill G. W. (1988) in *Mercury* (F. Vilas et al., eds.), p. 670, Univ. of Arizona, Tucson.
20. Hunten D. M. et al. (1988) in *Mercury* (F. Vilas et al., eds.), p. 562, Univ. of Arizona, Tucson.
21. Vilas F. (1988) in *Mercury* (F. Vilas et al., eds.), p. 59, Univ. of Arizona, Tucson.
22. Wetherill G. W. (1988) in *Mercury* (F. Vilas et al., eds.), p. 670, Univ. of Arizona, Tucson.
23. Benz W. et al. (1988) *Icarus, 74,* 516.
24. A recent review is given by Kaula W. M. (1990) *Science, 247,* 1191; see also Basilevsky A. T. and Head J. W. (1988) *Annu. Rev. Earth Planet. Sci., 16,* 295; Head J. W. and Crumpler L. S. (1990) *Nature, 346,* 525. Older references are in Hunten D. M. et al., eds. (1983) *Venus,* Univ. of Arizona, Tucson, 1143 pp.; early geophysical results are discussed by Phillips R. J. and Malin M. C. (1983) in *Venus* (D. M. Hunten et al., eds.), p. 159, Univ. of Arizona, Tucson.
25. Surkov Yu. A. (1984) *Proc. Lunar Planet. Sci. Conf. 14th,* in *J. Geophys. Res., 89,* B393.
26. Surkov Yu. A. (1977) *Proc. Lunar Sci. Conf. 8th,* 2665.
27. Surkov Yu. A. et al. (1987) *Proc. Lunar Planet. Sci. Conf. 17th,* in *J. Geophys. Res., 92,* E537.
28. Basilevsky A. T. et al. (1986) *Proc. Lunar Planet. Sci. Conf. 16th,* in *J. Geophys. Res., 91,* D399.
29. Crumpler L. and Head J. W. (1988) *J. Geophys. Res., 93,* 301; Head J. W. and Crumpler L. S. (1989) *Earth Moon Planets, 44,* 219; Head J. W. and Crumpler L. S. (1990) *Nature, 346,* 525.
30. For example, Crumpler L. S. et al. (1986) *Geology, 14,* 1031; Basilevsky A. T. and Head J. W. (1988) *Annu. Rev. Earth Planet. Sci., 16,* 295.
31. Aubele J. C. et al. (1988) in *Lunar and Planetary Science XIX,* p. 21, Lunar and Planetary Institute, Houston; Head J. W. et al. (1991) *Science, 252,* 276.
32. For example, Campbell I. H. and Taylor S. R. (1983) *Geophys. Res. Lett., 10,* 1061.
33. Basilevsky A. T. et al. (1991) in *Lunar and Planetary Science XXII,* p. 57, Lunar and Planetary Institute, Houston.
34. Campbell D. B. et al. (1991) *Science, 251,* 180; Phillips R. J. et al. (1991) *Science, 252,* 288.
35. Parsons B. (1981) *Geophys J., 67,* 437; Reymer A. and Schubert G. (1984) *Tectonics, 3,* 63.
36. Phillips R. J. et al. (1991) *Science, 252,* 651.
37. Solomon S. C. et al. (1991) *Science, 252,* 297.
38. Basilevsky A. T. et al. (1990) *Earth Moon Planets, 50/51,* 3.
39. Head J. W. (1990) *Earth Moon Planets, 50/51,* 25.
40. Saunders R. S. et al. (1991) *Science, 252,* 249; Solomon S. C. and Head J. W. (1991) *Science, 252,* 252.
41. Kaula W. M. (1990) *Science, 247,* 1191.
42. Cameron A. G. W. (1983) *Icarus, 56,* 195; Wetherill G. W. (1990) *Annu. Rev. Earth Planet. Sci., 18,* 205.
43. A comprehensive geophysically based review, with emphasis on the mantle and core, is given by Anderson D. L. (1989) in *Theory of the Earth,* Blackwell, Oxford, 366 pp. The crust is discussed by Taylor S. R. and McLennan S. M. (1985) in *The Continental Crust: Its Composition and Evolution,* Blackwell, Oxford, 312 pp. The immense literature on geological history is splendidly summarized by Cloud P. (1987) *Oasis in Space,* Norton, New York, 508 pp. The plate tectonic revolution is placed in historical context by Marvin U. B. (1973) *Continental Drift: The Evolution of a Concept,* Smithsonian, Washington, DC, 239 pp. A mine of relevant material is to be found in BVSP (1981) *Basaltic Volcanism on the Terrestrial Planets,* Pergamon, New York, 1286 pp. All these sources are replete with references to more detailed studies.
44. Taylor S. R. (1982) *Planetary Science: A Lunar Perspective,* Lunar and Planetary Institute, Houston, 481 pp.; Newsom H. E. and Taylor S. R. (1989) *Nature, 338,* 29.

45. Anderson D. L. (1989) *Theory of the Earth,* Blackwell, Oxford, 366 pp.
46. Walker D. and Agee C. (1989) *Earth Planet. Sci. Lett., 96,* 49.
47. Ringwood A. E. (1989) *Earth Planet. Sci. Lett., 95,* 1; see, however, comments on this paper by Palme H. (1990) *Nature, 343,* 23.
48. A list of theoretical models, now mostly of historical interest, based on such concepts as equilibrium condensation is given in Table 4.3.2.c, p. 641, of BVSP (1981) *Basaltic Volcanism on the Terrestrial Planets,* Pergamon, New York.
49. Wänke H. et al. (1984) in *Archean Geochemistry* (A. Kroner et al., eds.), p.1, Springer-Verlag, Berlin.
50. Stevenson D. J. (1981) *Science, 214,* 611.
51. Kato T. et al. (1987) *Geophys. Res. Lett., 14,* 546; Kato T. et al. (1988) *Earth Planet. Sci. Lett., 89,* 123.
52. Stevenson D. J. (1988) *Nature, 335,* 588.
53. Drake M. J. et al. (1988) *Meteoritics, 23,* 266.
54. Pepin R. O. (1989) in *Origin and Evolution of Planetary and Satellite Atmospheres* (S. Atreya et al., eds.), p. 291, Univ. of Arizona, Tucson.
55. Chyba C. (1990) *Nature, 343,* 129.
56. De Hon R. A. (1988) *LPI Tech. Rpt. 88-05,* p. 52.
57. Carr M. H. (1981) *The Surface of Mars,* Yale Univ., New Haven, 232 pp.
58. Clark B. C. et al. (1982) *J. Geophys. Res., 87,* 10059.
59. Surkov Yu. A. (1989) *Nature, 341,* 597.
60. Taylor S. R. (1989) *Tectonophys., 161,* 147.
61. Christensen P. R. (1986) *J. Geophys. Res., 91,* 3533; Christensen P. R. (1988) *J. Geophys. Res., 93,* 7611.
62. Clark R. et al. (1988) *Eos Trans. AGU,* Dec. 13, p. 1634. The composition of scapolite is 0.75 $Ca_4Al_6Si_6O_{24}CO_3$-0.25 $Na_4Al_3Si_9O_{24}Cl$; Blaney D. L. and McCord T. B. (1989) *Bull. Am. Astron. Soc., 21,* 955.
63. Martin L. (1984) *Icarus, 57,* 317; Zurek R. W. and Haberle R. M. (1989) *LPI Tech. Rpt. 89-01,* p. 40.
64. Smith M. R. et al. (1984) *Proc. Lunar Planet. Sci. Conf. 14th,* in *J. Geophys. Res., 89,* B612; McSween H. Y. (1985) *Rev. Geophys., 23,* 391; Laul J. C. et al. (1986) *Geochim. Cosmochim. Acta, 50,* 909; Treiman A. H. (1986) *Geochim. Cosmochim. Acta, 50,* 1061, 1071.
65. For example, in shergottite EETA 79001.
66. Nakamura N. et al. (1982) *Geochim. Cosmochim. Acta, 46,* 1555.
67. Vickery A. M. and Melosh H. J. (1987) *Science, 247,* 738.
68. Dreibus G. and Wänke H. (1987) *Icarus, 71,* 225.
69. For example, Syria Planum; Tanaka K. L. and Davis P. A. (1988) *J. Geophys. Res., 93,* 14893.
70. Sleep N. H. and Phillips R. J. (1979) *Geophys. Res. Lett., 6,* 803.
71. Phillips R. J. and Malin M. C. (1983) in *Venus* (D. M. Hunten et al., eds.), p. 159, Univ. of Arizona, Tucson.
72. Solomon S. C. and Head J. W. (1982) *J. Geophys. Res., 87,* 9755.
73. Phillips R. J. et al. (1990) *J. Geophys. Res., 95,* 5089; see also Finnerty A. A. et al. (1988) *J. Geophys. Res., 93,* 10225. The radial fracture patterns around Tharsis have often been used to argue for uplift; see, for example, Wise D. U. et al. (1979) *Icarus, 38,* 456.
74. Fanale F. P. et al. (1982) *Icarus, 50,* 381.
75. Ward W. R. et al. (1979) *J. Geophys. Res., 84,* 243.
76. Bills B. G. (1990) *J. Geophys. Res., 95,* 14131.
77. Laskar J. (1989) *Nature, 338,* 237.
78. Bills B. G. and Ferrari A. J. (1978) *J. Geophys. Res., 83,* 3479.
79. Reasenberg R. D. (1977) *J. Geophys. Res., 82,* 369; Kaula W. M. (1979) *Geophys. Res. Lett., 6,* 194.
80. Bills B. G. (1989) *Geophys. Res. Lett., 16,* 385, 1337; Kaula W. M. et al. (1989) *Geophys. Res. Lett., 16,* 1333.
81. A list of these may be found in Table 4.3.2d, p. 642 of BVSP (1981) *Basaltic Volcanism on the Terrestrial Planets,* Pergamon, New York.
82. Laul J. C. et al. (1986) *Geochim. Cosmochim. Acta, 50,* 923, Table 8.
83. Dreibus G. and Wänke H. (1985) *Meteoritics, 20,* 367; Wänke H. et al. (1984) in *Archean Geochemistry* (A. Kroner et al., eds.), p. 1, Springer-Verlag, Berlin.
84. Nakamura N. et al. (1982) *Geochim. Cosmochim. Acta, 46,* 1555; Chen J. H. and Wasserburg G. J. (1986) *Geochim. Cosmochim. Acta, 50,* 955.
85. Tatsumoto M. and Premo W. R. (1988) *LPI Tech Rpt. 88-07,* p. 167.
86. Dreibus G. and Wänke H. (1985) *Meteoritics, 20,* 367.
87. Morgan J. W. and Anders E. (1980) *Proc. Natl. Acad. Sci., 77,* 6973.
88. For example, Campbell I. H. and Taylor S. R. (1983) *Geophys. Res. Lett., 10,* 1061.
89. Surkov Yu. A. (1984) *Proc. Lunar Planet. Sci. Conf. 14th,* in *J. Geophys. Res., 89,* B393.
90. For example, Wänke H. et al. (1984) in *Archean Geochemistry* (A. Kroner et al., eds.), p. 1, Springer-Verlag, Berlin; Ringwood A. E. (1979) *Origin of the Earth and Moon,* Springer-Verlag, Berlin, 295 pp.
91. Chapman C. R. and Gaffey M. (1979) in *Asteroids* (T. Gehrels, ed.) p. 655, Univ. of Arizona, Tucson, list 4 Vesta, 44 Nysa, 64 Angelina, possibly 434 Hungaria, and a few others.

92. Kurat G. et al. (1985) in *Lunar and Planetary Science XVI,* p. 471, Lunar and Planetary Institute, Houston.
93. Kallemeyn G. W. and Wasson J. T. (1981) *Geochim. Cosmochim. Acta, 45,* 1217.
94. Wänke H. et al. (1984) in *Archean Geochemistry* (A. Kroner et al., eds.), p. 1, Springer-Verlag, Berlin.
95. Newsom H. E. and Drake M. J. (1987) in *Lunar and Planetary Science XVIII,* p. 716, Lunar and Planetary Institute, Houston.
96. For example, Ringwood A. E. (1966) *Geochim. Cosmochim. Acta, 30,* 41.
97. Morgan J. W. (1986) *J. Geophys. Res., 91,* 12375.
98. Jones J. H. and Drake M. J. (1986) *Nature, 322,* 221; Brett R. (1984) *Geochim. Cosmochim. Acta, 48,* 1183.
99. See discussion by Benz W. and Cameron A. G. W. (1990) in *Origin of the Earth* (H. E. Newsom and J. H. Jones, eds.), p. 61, Oxford Univ., New York; Newsom H. E. (1990) in *Origin of the Earth* (H. E. Newsom and J. H. Jones, eds.), p. 273, Oxford Univ., New York; Newsom H. E. and Taylor S. R. (1989) *Nature, 338,* 29.
100. Murthy V. R. (1991) *Science, 253,* 303.
101. Collinson D. W. (1986) *Earth Planet. Sci. Lett., 77,* 159; Curtis S. A. and Ness N. F. (1988) *Geophys. Res. Lett., 15,* 737.
102. Fuller M. and Cisowski S. M. (1987) *Geomagnetism 2* (J. A. Jacobs, ed.), p. 307, Academic, New York; see also Stevenson D. J. et al. (1983) *Icarus, 54,* 466, for the model for the Earth's field.
103. Leshin L. A. et al. (1988) in *Lunar and Planetary Science XIX,* p. 677, Lunar and Planetary Institute, Houston.
104. Urey H. C. (1959) *J. Geophys. Res., 64,* 1721.
105. For example, the mare surfaces were correctly identified as lava flows by Baldwin R. B. (1949) *The Face of the Moon,* Univ. of Chicago, 239 pp.
106. For example, Gast P. W. (1972) *Moon, 5,* 121.
107. Taylor S. R. and Jakeš P. (1974) *Proc. Lunar Sci. Conf. 5th,* p. 1287.
108. See BVSP (1981) *Basaltic Volcanism on the Terrestrial Planets,* Pergamon, New York, for an extensive review.
109. Taylor S. R. and McLennan S. M. (1985) *The Continental Crust: Its Composition and Evolution,* Blackwell, Oxford, 312 pp.
110. Bowring S. A. et al. (1989) *Geology, 17,* 971.
111. For example, Hamilton P. J. et al. (1983) *Earth Planet. Sci. Lett. 62,* 263.
112. Taylor L. A. et al. (1983) *Earth Planet. Sci. Lett., 66,* 33.
113. Wilhelms D. E. (1985) in *Lunar and Planetary Science XVI,* p. 904, Lunar and Planetary Institute, Houston.
114. Grieve R. A. F. and Parmentier E. M. (1985) *LPI Tech. Rpt. 85-01,* p. 23.
115. See also negative comments by Taylor S. R. (1983) in *Expanding Earth Symposium,* p. 343, Univ. of Tasmania, Hobart.
116. For example, Cameron A. G. W. (1983) *Icarus, 56,* 195.
117. Potter A. E. and Morgan T. H. (1985) *Science, 229,* 651.
118. Hunten D. M. et al. (1988) in *Mercury* (F. Vilas et al., eds.), p. 561, Univ. of Arizona, Tucson.
119. Fegley B. Jr. and Cameron A. G. W. (1987) *Earth Planet. Sci. Lett., 82,* 207.
120. Potter A. E. and Morgan T. H. (1988) *Science, 241,* 675; Potter A. E. and Morgan T. H. (1988) *Geophys. Res. Lett., 13,* 1515.
121. Donahue T. M. and Pollack J. B. (1983) in *Venus* (D. M. Hunten et al., eds.), p. 1003, Univ. of Arizona, Tucson.
122. Krasnopolsky V. A. (1986) *Photochemistry of the Atmospheres of Mars and Venus,* Springer-Verlag, Berlin, 334 pp.
123. Durham R. and Chamberlain J. W. (1989) *Icarus, 77,* 59.
124. Staudacher T. and Allègre C. (1982) *Earth Planet. Sci. Lett., 60,* 389; Sarda P. et al. (1985) *Earth Planet. Sci. Lett., 72,* 357.
125. Allègre C. J. et al. (1983) *Nature, 303,* 762.
126. Hayes J. M. (1983) in *Earth's Earliest Biosphere* (J. W. Schopf, ed.) p. 291, Princeton Univ.
127. Chang S. et al. (1983) in *Earth's Earliest Biosphere* (J. W. Schopf, ed.) p. 53, Princeton Univ.
128. Hunten D. M. et al. (1987) *Icarus, 69,* 532.
129. Pollack J. B. et al. (1987) *Icarus, 71,* 203.
130. Squyres S. W. et al. (1987) *Icarus, 70,* 385; Wilhelms D. E. and Baldwin R. J. (1988) in *Lunar and Planetary Science XIX,* p. 1270, Lunar and Planetary Institute, Houston.
131. Zahnle K. J. (1990) *Bull. Am. Astron. Soc., 22,* 1073.
132. Kasting J. F. (1988) *Icarus, 74,* 472.
133. Kaula W. M. (1990) *Science, 247,* 1191.
134. Chyba C. F. (1987) *Nature, 330,* Table 1, 632.
135. Melosh H. J. and Vickery A. M. (1989) *Nature, 338,* 487.
136. Kasting J. F. (1988) *Icarus, 74,* 472; Zahnle K. J. et al. (1988) *Icarus, 74,* 62. See comment by Stevenson D. J. (1988) *Nature, 335,* 587.
137. Matsui T. and Abe Y. (1986) *Nature, 319,* 303; Matsui T. and Abe Y. (1986) *Nature, 322,* 526.
138. Wetherill G. W. (1985) *Science, 228,* 877.
139. Chyba C. F. (1987) *Nature, 330,* 632; Chyba C. F. (1990) *Nature, 343,* 129.

140. Pepin R. O. (1989) in *Origin and Evolution of Planetary and Satellite Atmospheres* (S. Atreya et al., eds.), p. 291, Univ. of Arizona, Tucson.
141. Baker V. R. (1982) *The Channels of Mars,* Univ. of Texas, Austin, 198 pp.; Carr M. H. and Clow G. D. (1981) *Icarus, 48,* 91.
142. Lucchitta B. K. (1988) in *Lunar and Planetary Science XIX,* p. 699, Lunar and Planetary Institute, Houston.
143. Carr M. H. (1986) *Icarus, 68,* 187.
144. Scott E. R. D. and Newsom H. E. (1989) *Z. Naturforsch., 44a,* 924.
145. See Lebofsky L. A. et al. (1989) in *Origin and Evolution of Planetary and Satellite Atmospheres* (S. Atreya et al., eds.), p. 192, Univ. of Arizona, Tucson.
146. Wisdom J. (1987) *Icarus, 72,* 273.
147. Chapman C. R. et al. (1989) in *Asteroids II* (R. P. Binzel et al., eds.), p. 386, Univ. of Arizona, Tucson.
148. Gradie J. and Tedesco E. (1982) *Science, 216,* 1405.
149. Gaffey M. J. et al. (1989) in *Asteroids II* (R. P. Binzel et al., eds.), p. 928, Univ. of Arizona, Tucson.
150. Bell J. F. et al. (1988) in *Lunar and Planetary Science XIX,* p. 57, Lunar and Planetary Institute, Houston; Bell J. F. et al. (1989) in *Asteroids II* (R. P. Binzel et al., eds.), p. 921, Univ. of Arizona, Tucson.
151. Gaffey M. J. (1989) in *Lunar and Planetary Science XX,* p. 321, Lunar and Planetary Institute, Houston.
152. Standish E. M. and Hellings R. W. (1989) *Icarus, 80,* 326.
153. Millis R. L. et al. (1987) *Icarus, 72,* 507.
154. See Fig. 11 of Lebofsky L. A. et al. (1989) in *Origin and Evolution of Planetary and Satellite Atmospheres* (S. Atreya et al., eds.) p. 205, Univ. of Arizona, Tucson. The identification of the 3.07-μm absorption feature as due to water ice by these workers has been challenged by King T. V. V. et al. (1990) *Bull. Am. Astron. Soc., 22,* 1123, who consider it to be due to the presence of an NH_4-bearing mineral. Water ice seems more probable.
155. Hirayama K. (1918) *Astron. J., 31,* 185; Williams J. G. (1979) in *Asteroids* (T. Gehrels, ed.), p. 1040, Univ. of Arizona, Tucson; Gradie J. C. et al. (1979) in *Asteroids* (T. Gehrels, ed.), p. 359, Univ. of Arizona, Tucson; Kozai Y. (1979) in *Asteroids* (T. Gehrels, ed.), p. 334, Univ. of Arizona, Tucson; Carussi A. and Massaro E. (1978) *Astron. Astrophys. Supp., 34,* 81; Carussi A. and Valsecchi G. B. (1982) *Astron. Astrophys., 115,* 327.
156. Gaffey M. J. et al. (1989) in *Asteroids II* (R. P. Binzel et al., eds.), p. 98, Univ. of Arizona, Tucson. 4 Vesta, with a density of 3.9 g/cm^3, is an exception; Standish E. M. and Hellings R. W. (1989) *Icarus, 80,* 326.
157. Zellner B. M. et al. (1985) *Icarus, 61,* 355.
158. Bell J. F. et al. (1988) in *Lunar and Planetary Science XIX,* p. 57, Lunar and Planetary Institute, Houston; Bell J. F. (1989) *Icarus, 78,* 426.
159. Chapman C. R. et al. (1989) in *Asteroids II* (R. P. Binzel et al., eds.), p. 386, Univ. of Arizona, Tucson.
160. Bell J. F. (1989) *Icarus, 78,* 426, 436.
161. Gradie J. and Flynn L. (1988) in *Lunar and Planetary Science XIX,* p. 405, Lunar and Planetary Institute, Houston.
162. Shoemaker E. M. et al. (1979) in *Asteroids* (T. Gehrels, ed.), p. 253, Univ. of Arizona, Tucson; McFadden L. A. (1989) in *Asteroids II* (R. P. Binzel et al., eds.), p. 442, Univ. of Arizona, Tucson.
163. Lu J. et al. (1989) *Astron. J., 98,* 1905.
164. McFadden L. A. (1989) in *Asteroids II* (R. P. Binzel et al., eds.), p. 442, Univ. of Arizona, Tucson.
165. Weissman P. R. et al. (1990) in *Asteroids II* (R. P. Binzel et al., eds.), p. 880, Univ. of Arizona, Tucson.
166. Shoemaker E. M. et al. (1989) in *Asteroids II* (R. P. Binzel et al., eds.), p. 487, Univ. of Arizona, Tucson; Hartmann W. K. (1988) in *Lunar and Planetary Science XIX,* p. 453, Lunar and Planetary Institute, Houston; Hartmann W. K. et al. (1988) *Icarus, 73,* 487.
167. Duncan M. et al. (1989) *Icarus, 82,* 402.
168. J. F. Bell, personal communication, 1988.
169. Scott E. R. D. and Newsom H. E. (1989) *Z. Naturforsch., 44a,* 924.
170. Scott E. R. D. (1988) *Meteoritics, 23,* 300; Scott E. R. D. and Taylor G. J. (1987) *Meteoritics, 22,* 497.
171. Lebofsky L. A. et al. (1989) in *Origin and Evolution of Planetary and Satellite Atmospheres* (S. Atreya et al., eds.), p. 192, Univ. of Arizona, Tucson.
172. Zolensky M. E. and McSween H. Y (1988) in *Meteorites and the Early Solar System* (J. F. Kerridge and M. S. Matthews, eds.), p. 114, Univ. of Arizona, Tucson.
173. Jones T. D. et al. (1988) in *Lunar and Planetary Science XIX,* p. 567, Lunar and Planetary Institute, Houston.
174. Zolensky M. E. and McSween H. Y. (1988) in *Meteorites and the Early Solar System* (J. F. Kerridge and M. S. Matthews, eds.), p. 114, Univ. of Arizona, Tucson.
175. Zolensky M. E. et al. (1989) *Icarus, 78,* 411.
176. Weidenschilling S. J. (1988) in *Meteorites and the Early Solar System* (J. F. Kerridge and M. S. Matthews, eds.), p. 348, Univ. of Arizona, Tucson.
177. See comments on Ceres above.
178. Including (ironically) the probable parents of our CI meteorites, long hailed as the type example of unaltered solar nebular material.
179. Wetherill G. W. (1989) in *The Formation and Evolution of Planetary Systems* (H. A. Weaver and L. Danly, eds.), p. 27, Cambridge Univ., New York.

181. Basic references are Gehrels T., ed. (1976) ***Jupiter,*** Univ. of Arizona, Tucson, 1254 pp.; Gehrels T. and Matthews M. S., eds. (1984) ***Saturn,*** Univ. of Arizona, Tucson, 968 pp.
182. Weidenschilling S. J. and Lewis J. S. (1973) ***Icarus, 20,*** 465.
183. For example, Stevenson D. J. (1975) ***Phys. Rev. B., 12,*** 3999; Hubbard W. B. and DeWitt H. E. (1985) ***Astrophys. J., 290,*** 388.
184. Particularly useful papers are Gautier D. (1988) ***Philos. Trans. R. Soc., A325,*** 583, and the review by Gautier D. and Owen T. (1989) in ***Origin and Evolution of Planetary and Satellite Atmospheres*** (S. Atreya et al., eds.) p. 487, Univ. of Arizona, Tucson.
185. Gautier D. (1988) ***Philos. Trans. R. Soc., A325,*** 585.
186. From Lewis J. S. and Prinn R. G. (1980) ***Astrophys. J., 238,*** 357; see also Lunine J. I. and Stevenson D. J. (1985) ***Astrophys. J. Supp., 58,*** 493, for a discussion of clathrates.
187. For information of cloud compositions and structures on the giant planets, see Carlson B. E. et al. (1988) ***J. Atmos. Sci., 45,*** 2066; for magnetic fields, see Ness N. F. et al. (1989) ***Science, 246,*** 1473.
188. Hubbard W. B. (1980) ***Rev. Geophys. Space Phys., 18,*** 1; Hubbard W. B. and Stevenson D. J. (1984) in ***Saturn*** (T. Gehrels, ed.), p. 47, Univ. of Arizona, Tucson.
189. Stevenson D. J. (1989) in ***The Formation and Evolution of Planetary Systems*** (H. A. Weaver and L. Danly, eds.), p. 82, Cambridge Univ., New York.
190. Benz W. et al. (1989) ***Bull. Am. Astron. Soc., 21,*** 916.
191. Stevenson D. J. (1989) in ***The Formation and Evolution of Planetary Systems*** (H. A. Weaver and L. Danly, eds.), p. 82, Cambridge Univ., New York.
192. The pressures of the hydrogen layers within Uranus and Neptune are less than 1 Mbar, for which relevant experimental data exist; Nellis W. J. et al. (1983) ***Phys. Rev., A27,*** 608.
193. Gautier D. and Owen T. (1989) in ***Origin and Evolution of Planetary and Satellite Atmospheres*** (S. K. Atreya et al., eds.), p. 487, Univ. of Arizona, Tucson; Conrath B. J. et al. (1989) in ***Origin and Evolution of Planetary and Satellite Atmospheres*** (S. K. Atreya et al., eds.), p. 538, Univ. of Arizona, Tucson.
194. For example, Pollack J. B. et al. (1986) ***Icarus, 67,*** 409.
195. Stone E. C. and Miner E. D. (1989) ***Science, 246,*** 1417; Smith B. A. (1989) ***Science, 246,*** 1422.
196. Hubbard W. B. et al. (1987) ***Icarus, 72,*** 635.
197. Hubbard W. B. (1989) in ***Origin and Evolution of Planetary and Satellite Atmospheres*** (S. K. Atreya et al., eds.), p. 539, Univ. of Arizona, Tucson.
198. Morris D. E. and O'Neill T. G. (1988) ***Astron. J., 96,*** 1127.
199. de Campli W. M. and Cameron A. G. W. (1979) ***Icarus, 36,*** 367.
200. Hubbard W. B. (1989) in ***Origin and Evolution of Planetary and Satellite Atmospheres*** (S. K. Atreya et al., eds.), p. 539, Univ. of Arizona, Tucson.
201. Anderson J. (1988) ***Planet Report, 8,*** 9.
202. Tremaine S. (1986) in ***The Galaxy and the Solar System*** (R. Smoluchowski et al., eds.), p. 409, Univ. of Arizona, Tucson.
203. Appelgate J. et al. (1985) ***IEEE Trans. Comput., C-34,*** 822.
204. Sussman G. J. and Wisdom J. (1988) ***Science, 241,*** 433.
205. For further information, see Duncombe R. L. et al., eds. (1986) ***The Stability of Planetary Systems,*** Reidel, Dordrecht, 476 pp.
206. Morris D. E. and O'Neill T. G. (1988) ***Astron. J., 96,*** 1127.
207. Duncan M. et al. (1989) ***Icarus, 82,*** 402.

Chapter 6

Rings and Satellites

Illustration, preceding page—*A Voyager view of the inner system of Saturn. The ring thickness is probably less than 50 m, but the diameter of the entire ring system is about the Earth-Moon distance. Note the Cassini Division (see Fig. 6.3.1 for details of the rings). (NASA JPL Voyager P-23068 BW.)*

Rings and Satellites

6

6.1. MINIATURE SOLAR SYSTEMS?

Galileo proposed that the satellites of Jupiter formed a miniature solar system; their rather uniform size and equatorial orbits, coupled with a regular decrease in density from Io to Callisto, encouraged subsequent views that study of the Galilean system would provide insights into the formation of the solar system, serving as a scale model. Indeed, it seems reasonable to have expected at least the satellite systems of the giant planets to exhibit some systematic regularities as a sort of byproduct of planetary formation. Furthermore, while we have only one solar system, three of the giant planets possess regular satellite systems mimicking miniature solar systems, while Neptune has the ruins of one lying nearby, within about five radii of the planet. Accordingly, the satellite systems of the giant planets might be expected to provide some general insights into the origins of planetary systems. The most interesting observation concerning them, however, is that they are all different (Fig. 6.1.1) and "the four giant planets exhibit a startling diversity of satellite systems" [1].

The satellites of Saturn and Uranus differ from those of Jupiter in many ways (Neptune presents a special case). One could ignore the satellites in significantly inclined, eccentric, or retrograde orbits as probable captured objects and hence not necessarily related to the planet to which they have become attached [2]. However, even the regular prograde satellites in equatorial orbits exhibit individual peculiarities of composition, structure, or surface features. It must be concluded that if satellite formation is "an inevitable byproduct of the planetary accumulation process" [3], it does not seem to produce uniform results, even around such relatively similar objects as Jupiter and Saturn. Thus, the compositions of the individual satellites do not provide information on nebular conditions, despite many attempts to infer these, but seem to be dominated by local conditions around each planet. No single model can accommodate the complexities observed, and accordingly we are faced with the same dilemma as with the planets; the absence of a grand unifying theory to accommodate the disparate observations.

6.2. PLANETARY SUBNEBULAE

The regular satellites are generally supposed to form from disks around planets (see section 4.5). This concept raises some interesting questions. How is such material placed into circumplanetary orbits during planetary formation? Is disk formation a common and regular feature of planetary growth? Three modes of formation of such circumplanetary disks have been proposed:

1. They may form directly from the solar nebula as accretion disks during planetary formation, when the angular momentum of the system hinders direct accretion of material to the planet. Collisions between accreting planetesimals may be at appropriate angles and velocities, so that some of the debris may acquire sufficient angular momentum to form a disk [4]. In the case of the giant planets, gas accreting onto the planet may have enough angular momentum to form one [5] (Table 6.2.1). The growth of Jupiter and Saturn, however, is expected to tidally truncate the nebula, so that formation of an accretion disk may be difficult.

An additional process that may also provide a ring of debris around a growing planet is the breakup of

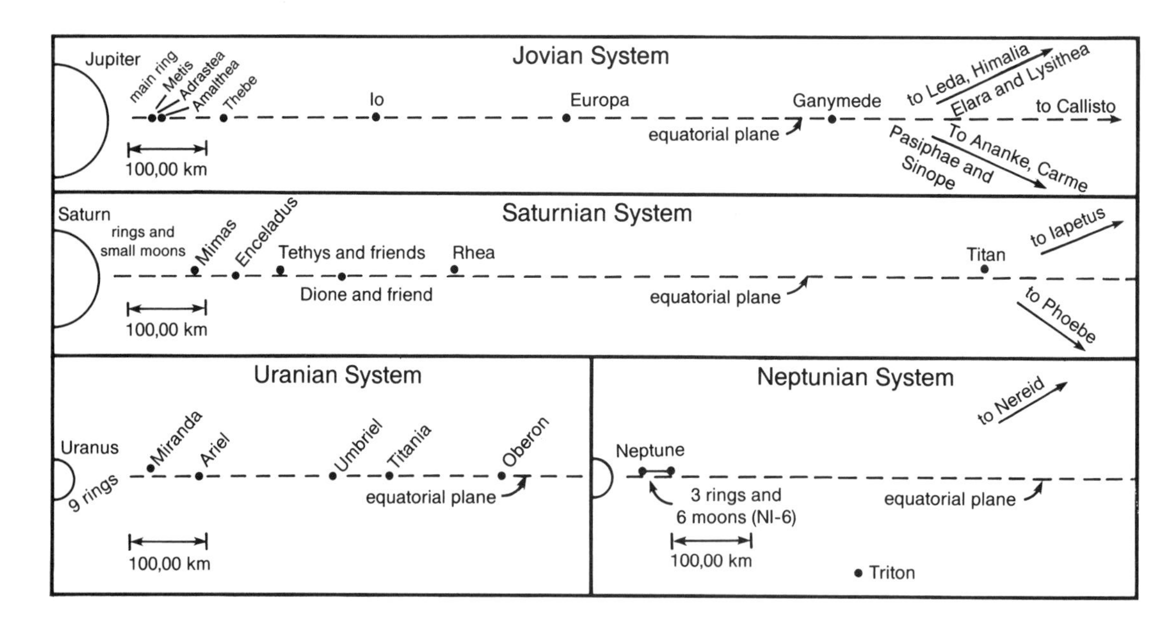

Fig. 6.1.1. *An overall view of the satellite and ring systems of the giant planets. After Elliot J. and Kerr R. (1984) Rings, p. 181, Fig. 10.1, MIT, Cambridge.*

TABLE 6.2.1. Properties of a protosatellite nebula.

Temperature	200 K
Orbit radius	10^{11} cm
Surface density (solids)	3×10^3 g/cm^2
Surface density (gas)	3×10^4 g/cm^2
Gas density	10^{-6} g/cm^3
Cooling/condensation	10^6 yr

Data from Stevenson D. J. (1986) in *Satellites* (J. A. Burns and M. S. Matthews, eds.), p. 65, Table 3, Univ. of Arizona, Tucson.

planetesimals as they come within the Roche limit. In one version, the readily fragmented silicate mantles of differentiated planetesimals form a debris ring; the tougher iron cores are either accreted to the planet or bypass it completely, leaving the silicate rubble in orbit (this was a popular hypothesis for lunar origin since it could produce an iron-poor Moon [6]). Other work has shown that the disruption is unlikely to occur, with the planetesimals either accreting directly to the planet or escaping without fragmentation, and although the planetesimals may be totally molten, they will not be disrupted, even in grazing collisions with growing planets [7]. Since these calculations are dependent on the values used for viscosity, they may not be totally definitive, and some caution is warranted before abandoning the hypothesis of tidal breakup of planetesimals. However, such processes should not be unique, and could reasonably be expected to produce satellites around all the terrestrial planets. These are not observed, reinforcing the notion that the Moon was formed by a unique event.

2. Disks might be spun off due to contraction of the planet and outward transfer of angular momentum. This is a Laplacian-type model. Fission might occur as material is accreted to a growing planet if the angular momentum increases until the system becomes unstable; but if accretion occurs from many bodies as in the planetesimal hypothesis, one would expect the net increase in angular momentum to be minimal. The basic problem is to acquire sufficient excess angular momentum. In addition, fission probably leads to ring formation [8] rather than to production of a separate discrete body. Since fission would thus begin as soon as the stability threshold is crossed, it might prove difficult to split off a large separate fragment unless the system were suddenly spun up. Such an increase could result from a giant impact, so would not really be distinguished from the situation discussed next.

3. Disks could originate through massive collisions. The disks might be derived either from the planet or from the impactor. The best example of the latter process may be the uranian system. The current explanation for the 90° obliquity of Uranus

is that it was caused by collision with an Earth-sized object. During this massive collision, which tilted the planet on its side, the gas blown off by the collision may have condensed to form both the ring system and the regular satellites. These all lie in the equatorial plane, similarly inclined at about 90° to the plane of the ecliptic. They must have formed subsequent to the collisional event, since it seems unlikely that the rings and satellites would reorient themselves through 90° following the large impact. Such scenarios raise some very interesting questions. Satellites originally present would not reorient, but would precess independently about the new equatorial plane. Original rings might, however, collisionally damp down into this plane [9]. Another interesting question is what would happen to a preexisting satellite system in a large collisional event that blows out a disk with a different orientation.

Collisional scenarios are of interest because they form the basis for the most popular model for the formation of the Moon. The collision of a large body with a growing planet has the advantage of providing the appropriate excess angular momentum observed in the Earth-Moon system. These events, in which the successful planet sweeps up the next largest planetesimal in the hierarchical swarm, are likely to be unique; the angles and velocities of such large collisions are unlikely to be duplicated and result in the production of a lunar-type satellite, except under the special circumstances that produced our unique satellite. Such processes seem likely to produce single satellites rather than the multibody systems of Jupiter and Saturn. The special circumstances of the collision required to knock Uranus over may provide a disk for multiple satellite formation, but the amount of material required to account for the uranian rings and satellites is small, placing a constraint on the energetics of the collision [10].

In summary, it is clear that conditions during the formation of the giant planets in a gas-rich nebula are both very different from and more favorable to the production of circumplanetary disks than in the case of the accretion of the terrestrial planets in a gas-free environment. The production of disks around growing giant gaseous planets appears to be a likely consequence of planetary formation. Satellite formation is expected to occur on very short timescales of the order of 10^6 yr, both on account of the gas-rich environment, and also because of the very short lifetimes for such disks [11].

In contrast, the formation of disks from which satellites might form around the inner planets seems to be unlikely, so satellite systems are only expected around gaseous planets. If the terrestrial planets had begun as gaseous protoplanets, this ancestry might have been revealed by the presence of suites of satellites. Only Jupiter and Saturn may have formed subnebulae in the classical sense, since they originated while the gaseous nebula was still present. The satellites of Uranus and Neptune, in contrast, may have formed from blown-out disks, since the gaseous nebula was mostly gone by the time they formed [9].

Temperature and pressures in subnebular disks, produced by whatever process around the giant planets, will be higher than out in the primordial nebula (see also section 6.14). This has some interesting consequences for the compositions of the satellites, which will be commented upon in several places in this chapter. The most significant concerns the role of carbon chemistry. It seems fairly well established that the gaseous component of carbon in the molecular clouds in the galaxy is present as carbon monoxide and that this would have been the dominant component in the primordial solar nebula. In contrast, methane is not uncommon as a constituent of the satellites of the outer planets, and this has raised some interesting controversies [12].

The carbon/oxygen atomic ratio in the primitive nebula, as given by the solar photospheric data, is about 0.42. The remainder of the oxygen not bound to carbon is present as water ice or in silicates, with a rock/ice ratio of about 75/25. Planetesimals forming directly from the nebula can be expected to have such ratios. Pluto and Triton are examples. The satellites of the giant planets generally have lower ice/rock ratios indicating that more water ice as well as methane and ammonia were available during their formation than would be expected in the primitive nebula.

The general explanation is that at the higher pressures temperatures in the subnebulae around the giant planets, some carbon monoxide was converted to methane. The oxygen freed combined with hydrogen to form water ice, thus increasing the ice-to-rock ratio, and lowering the densities of the satellites formed within the disks. The production of ammonia from nitrogen was also accomplished under these conditions, so that methane and ammonia ices and clathrates became available for incorporation into satellites. The residual carbon monoxide and nitrogen were removed along with the other gaseous components during the sweeping out of the nebula. These processes must occur before the dispersal of the gaseous components of the nebula while there is sufficient hydrogen present to react with the carbon monoxide and nitrogen. In summary, "conditions in well-mixed subnebulae around the giant protoplanets . . . provided precisely the chemical thermodynamic and chemical kinetic conditions necesssary to make CH_4 and NH_3 (not CO and N_2) the dominant N and C gases in these subnebulae" [13].

6.3. RING SYSTEMS

"Saturn presents a phenomenon unique in the system of the universe. Two small bodies are always observed on each side of it... Huygens has discovered that they are produced by a large thin ring which surrounds the globe of Saturn and which is everywhere separated from it" [14].

The beauty and strangeness of Saturn's ring system has long excited the admiration of telescope viewers, and it ranks as a favorite astronomical object along with the Moon and the spiral nebulae [15]. Because of our familiarity with rotating flattened systems from the solar system to spiral galaxies, it is interesting to record that spheres rather than disks dominated astronomical thinking from the Greeks to the Copernican Revolution. Once the concept of a nebular disk had been established by Kant and Laplace, the presence of the rings was long thought to be a key to our understanding of planetary formation. Laplace drew the parallel between the formation of the rings and the origin of the solar system. Unlike Kant, who correctly supposed the rings to be composed of many small particles, Laplace, as had Huygens before him, considered them to be solid. Maxwell, and later Jeffreys, confirmed Kant's notion that the rings must be particulate [16].

Expectations of the fundamental significance of rings have continued until very recently, it being felt that such disks of particles, of which the rings of Saturn are the most dramatic example, should serve as an analogue for the solar nebula. Their significance in elucidating such problems has declined, however, with increased understanding. The principal increase in knowledge took place within a brief period from March 1977, when the rings around Uranus were discovered, to the detailed examination of the Neptune ring system by Voyager 2 in August 1989. In the interval, the ring system around Jupiter was discovered in March and July 1979; three spacecraft (Pioneer 2 in September 1979, Voyager 1 in November 1980, and Voyager 2 in August 1981) revealed the extraordinary detail of the saturnian ring system; and Voyager 2 made a close examination of the uranian rings in January, 1986. The resulting flood of material has answered a number of basic questions, but has raised others.

6.3.1. Saturn's Rings

The thickness limit for the main rings of Saturn is about 150-200 m, but the true thickness is probably less, perhaps about 50 m [17] (Fig. 6.3.1). The average particle size is in the range of 3-5 m, although some house- and mountain-sized objects are present as well. The tidal stresses are small, so that there is a sort of quasiequilibrium between disruption and accretion. Fine dust ranging between 0.1 and 10 μm and averaging about 1 μm is also present, particularly in the E, F, and G rings; it probably comprises 100% of the E ring, which, in contrast to the main ring, has a vertical extent of several thousand kilometers. The particles have a bulk density of about 1 g/cm^3, and they appear, from this and reflectance properties, to be made of water ice, but whether this is mixed with or coats silicate grains is uncertain.

It is generally thought that the rings of Saturn are anchored by Mimas; the sharp edge of the outer B ring is in 2/1 resonance with Mimas. This satellite is linked to the more massive Tethys by a 2/1 res-

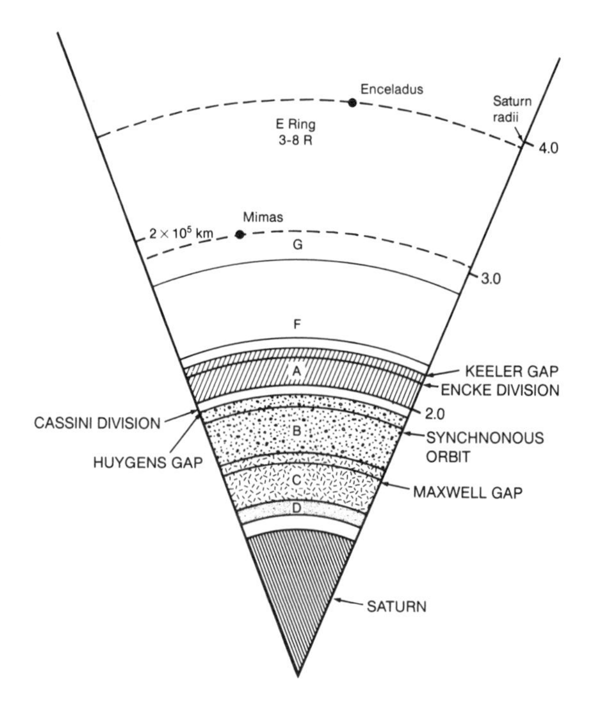

Fig. 6.3.1. *The saturnian ring system and nearby satellites. After Elliot J. and Kerr R. (1984) Rings, p. 193, MIT, Cambridge.*

onance that could help anchor the rings. However, the boundaries of the other major rings, inner A ring edge, B-C boundary, and the inner C ring edge, have no resonances associated with them.

A striking observation is the sharpness of the ring edges. From one of the densest regions of Saturn's rings, the B ring diminishes to nothing over a distance of only 1.3 km. Other rings drop to zero in less than 100 m. This is not readily accounted for by current ring theories. However, in these, the ring particles are treated as smooth, spherical, and of uniform size. It is perhaps not surprising that the real rings refuse to accommodate to this mathematical simplification. Possibly the ring edges are sharpened by the shepherding effects of larger particles or small shepherding satellites [18].

The large (>10^4 km) wedge-shaped "spokes" that appear as transient features of a few hours duration appear to be produced by the impact of small (10 cm-1 m) meteorites impacting at velocities of about 30 km/s. The spokes appear to be formed by showers of tiny grains as a cloud of charged gas produced by the impact moves across the rings.

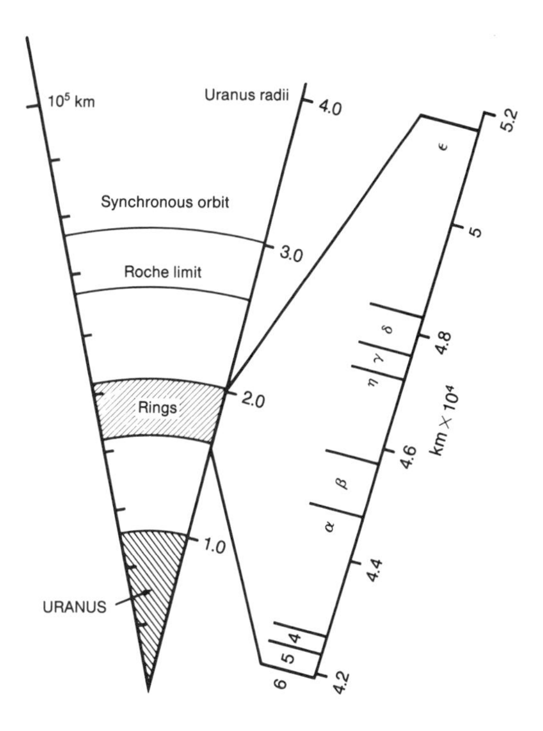

Fig. 6.3.2. *The uranian ring system. After Elliot J. and Nicholson P. D. (1984) in Planetary Rings (R. Brahic and R. Greenberg, eds.), Univ. of Arizona, Tucson.*

6.3.2. Uranian Rings

The nine narrow rings of Uranus (Fig. 6.3.2) all lie within the extended hydrogen atmosphere of the planet. They have widths from about 1 to 12 km, mostly with sharp edges. Ring thicknesses are in the range 7-20 m. They are not as black as commonly painted, but are dark grey, unlike the redder rings of Jupiter and Saturn. The average albedo is 0.04-0.05 [19], similar to the surface reflectance of carbonaceous chondrites [20]. They are very opaque and are joined by broader dusty rings The average ring particles are small (about 1 μm). All the material in the rings and shepherding satellites could be contained within an icy satellite about 75 km in radius [21]. Although some shepherding moonlets are present, they are not massive enough to maintain the present configuration of the rings for extended periods. Accordingly, the rings may be relatively recent, with ages of the order of a few hundred million years, and mechanisms for creating rings by satellite collisions or by disruption of captured comets should be sought [22].

6.3.3. Jovian Ring

The Jupiter ring (Fig. 6.3.3) is very tenuous, apparently made of micrometer-sized particles and accordingly it must be continuously replenished, since the particles will spiral into Jupiter through Poynting-Robertson effects. The ring bears some resemblance to the F and G rings of Saturn. These ethereal Tinker Bell-type rings seem to require a current source of particles. Enceladus may be the source for the E ring of Saturn and hidden sources must exist for the saturnian F and G rings, as well as the jovian ring.

There is, however, so much variety among the rings, as elsewhere in the solar system, that it is difficult to extract any general rules. As Harris [23] has remarked, Jupiter's ring is a shepherd without many sheep. Saturn's rings are mainly sheep, the shepherds having been mostly destroyed by meteoritic influx. The rings of Uranus are dominated by shepherds.

6.3.4. Neptunian Rings

Four rings have been identified [24] (Table 6.3.1). The neptunian rings (Fig. 6.3.4) are striking on account of the three thickened arc segments, respectively 4°, 4°, and 10° in length, in the outer ring. The inner ring is quite broad, about 2500 km wide, with no sharp boundaries, in contrast to the other

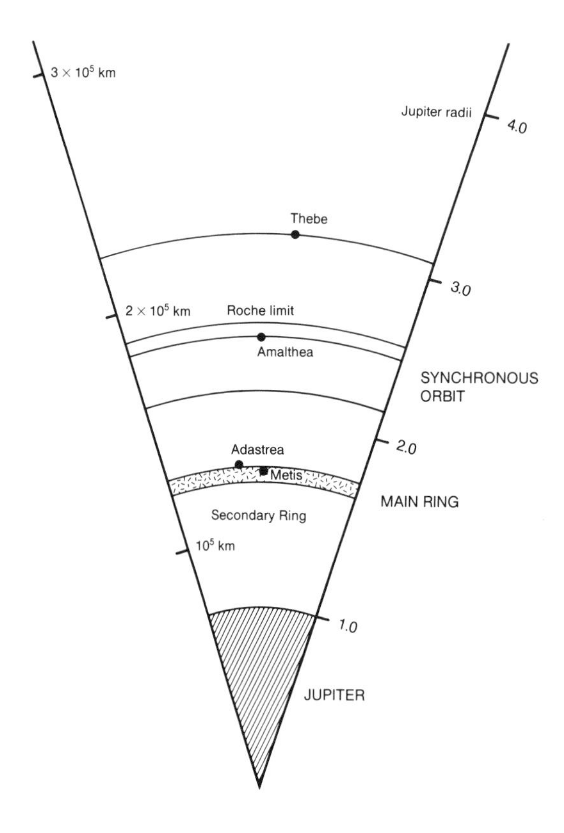

Fig. 6.3.3. *The jovian ring and nearby satellites. After Elliot J. and Kerr R. (1984) Rings, p. 193, MIT, Cambridge.*

rings [21]. The explanation adopted here is that the rings are probably formed from recently disrupted satellites; an unsolved problem is that such material, on current theories, should spread around the ring in a period of a few years rather than remain as clumped segments [25]. The amount of material in the neptunian ring system (Table 6.3.1) is very much smaller than in the uranian ring system, the total area being only about 1% of that of the uranian rings.

The recent discoveries of rings around Uranus, Jupiter, and Neptune naturally raise the question whether the formation of rings is a universal phenomenon during planetary formation. Many models for the formation of the terrestrial planets call for the formation of a circumferential ring of silicate debris. Indeed, this is an essential feature for the double-planet hypotheses of lunar origin. Searches for rings around Mars have, however, have revealed no trace of a ring inside the orbit of Phobos within ±350 km of the equatorial plane [26].

TABLE 6.3.1. Neptune ring data.

Feature	Distance (10^3 km)	Distance (R_N)	Width (km)	Comments
	38.	1.5		Inner extent of 1989N3R?
1989N3R	41.9	1.69	1700	High dust content
	49.	2.0		Outer extent of 1989N3R?
1989N2R	53.2	2.15	*	High dust content
1989N4R (inner)	53.2	2.15		Inner edge of "plateau"
1989N4R (outer)	59.	2.4		Outer edge of "plateau"
1989N1R	62.9	2.54	15	Contains three bright dusty arcs

* 1989N2R is narrow and unresolved in Voyager.

Data from Stone E. C. and Miner E. D. (1989) *Science, 246*, 1419, Table 2.

6.3.5. Ring Origin

"From the beginning, observational resolution seemed to be just short of revealing the essential nature of rings" [27].

Although Jupiter, Saturn, Uranus, and Neptune have ring systems, they are all different and no common theory can account for them [28]. If we do not understand the saturnian system, which has been known for a long time, can we expect to understand the recently discovered jovian, uranian, or neptunian systems [29]?

Two major models seem possible. In one, the rings formed along with the planet, as remnants of primordial accretion disks or from a spun out or knocked out disk more or less coeval with planetary formation. Alternatively they are the debris resulting from the later disruption of a satellite or a large captured comet.

One major problem with the accretion disk scenario is the following: If a growing planet is surrounded by a thick ring, much of the mass of the planet will arrive via ring infall. Accordingly, it will contribute a substantial amount of angular momentum to the planet, and overspin the planet. This is contrary to the observed values [23]. However, there are also problems with the scenario that the rings formed much more recently, representing the disrupted fragments of satellites. How does a large moon form so close to the Roche limit? Even if one existed, there may be no bodies large enough to disrupt it, since all such objects should have been swept up by 4×10^9 yr ago. Thus, if there is no apparent mechanism for forming a ring at a later stage, the rings by default must be ancient [30].

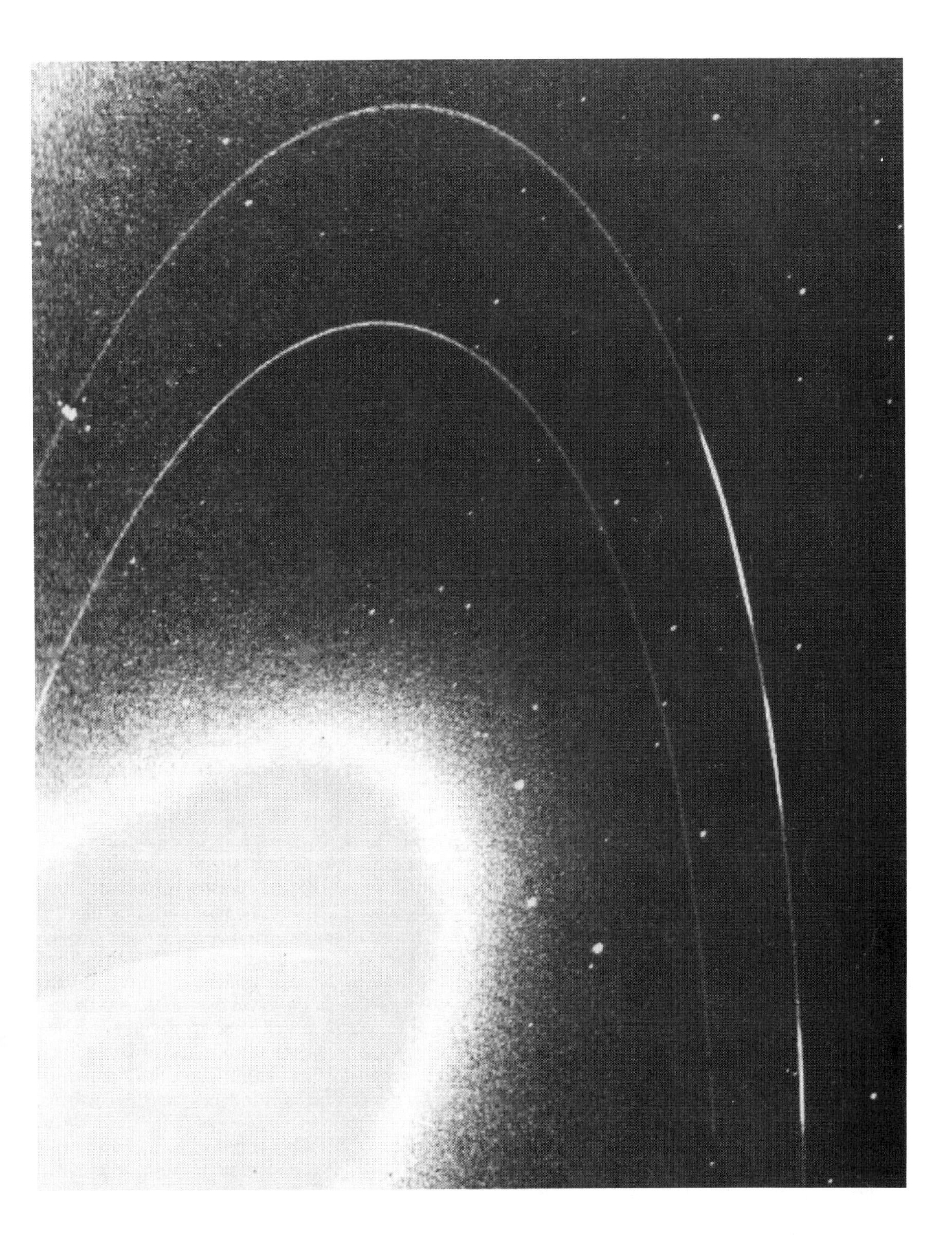

Fig. 6.3.4. *The two main rings of the neptunian ring system backlit by the Sun, from Voyager 2 taken at* 1.1×10^6 *km from Neptune. The rings are at 53 and* 63×10^3 *km from Neptune. The three thickened arc segments in the outer ring are clearly shown. In contrast to the uranian rings, which contain few dust grains, the neptunian rings are bright in this view, due to scattering of light from microscopic dust. (NASA JPL P-34712)*

There is some controversy over this. Thus, by extrapolating the crater density of Iapetus inward toward Saturn, Smith et al. [31] calculate that bodies closer to Saturn have experienced enough collisions to disrupt them several times over. Following satellite disruption, mutual collisions will quickly reduce the fragments to dimensions of about 10 km. This will be followed by "gentle bumping" that is expected to reduce the particle size down to the 5-m average, analogous to rocks in a terrestrial stream bed [32]. Why hasn't the whole belt been reduced to dust in this scenario? Possibly gravitationally bound aggregates keep this from happening. An analogy at a different scale and process is the formation of agglutinates in lunar soils that stabilize the soil particle dimensions at about 60 μm. If the particles became charged, the rings might also possess long-term stability.

A review of the problem of satellite breakup indicates that although current objections have some validity, the question is unresolved [33]. The perception that the rings have limited lifetimes compared to that of the solar system may demand a relatively recent origin. However, other scenarios may be possible. What would be the result of the capture and breakup of a large comet such as Chiron as it came within the Roche limit? Chiron has about the right mass to produce the entire saturnian ring system.

The beautiful rings may thus be disrupted fragments of asteroids, icy planetesimals, or comets, and the more ethereal Tinker Bell-type rings may be real transients. If so, the spectacular rings are a subsidiary event to satellite formation and do not provide us with a key to the formation of the solar system.

In summary, although there is little consensus on time and manner of ring formation, it is the opinion of this reviewer that they are a second-order effect. There are several reasons for this view. The principal one is that it is difficult to maintain the ring structure over the lifetime of the solar system. The thickened arcs in the neptunian system are surely the remnants of disrupted satellites that have not yet spread around the system, although current theory predicts that this should happen within a few years. Since this has not happened, our understanding of ring dynamics needs some revision. More importantly, it probably demonstrates the essentially ephemeral nature of ring systems on solar system timescales. If rings are not produced by recent collisional disruption of satellites, for which considerable arguments can be advanced [32], then the breakup of captured cometary bodies, such as Chiron, is probably responsible. The entire mass of the saturnian ring system is quite small, about equivalent to that of the 195-km-radius satellite, Mimas (famous on account of its large 130-km-diameter crater, Herschel). The close proximity of the rings to large planets means that meteorites and asteroids approach with increased velocities, due to the strong gravitational fields. Small bodies may thus be rather readily disrupted by such encounters. Thus, the rings seem, on solar system timescales, to be relatively ephemeral, and we must judge ourselves fortunate to be present in the solar system at the same time as the rings of Saturn.

6.4. SATELLITE CLASSIFICATION

Various attempts have been made to classify satellites, but like many features of the solar system, they mostly defy being placed into neat pigeonholes. The icy satellites of the giant planets all differ in density and in ice-rock ratios. According to Burns [34] there are three classes of satellites:

1. Regular satellites include almost all the larger satellites: the four Galilean satellites of Jupiter (Io, Europa, Ganymede, and Callisto), the eight classical satellites of Saturn (Mimas, Enceladus, Tethys, Dione, Rhea, Titan, Hyperion, and Iapetus), the five classical satellites of Uranus (Miranda, Ariel, Umbriel, Titania, and Oberon) and the small satellites of Neptune in equatorial orbit (N1 and N2).

2. Irregular satellites have highly inclined, often retrograde and elongate orbits, usually far out from the planet. Members include the prograde irregular cluster of Jupiter (Leda, Himalia, Lysithea, and Elara) at about 160 Jupiter radii; the outer retrograde group (Anake, Carme, Pasiphae, and Sinope) at about 310 Jupiter radii; Phoebe, the outermost satellite of Saturn; and Nereid, the outer satellite of Neptune.

3. Collisional shards are small satellites, most of which have been discovered by the Voyager encounters. They are "tiny, craggy chunks . . . battered and worn down by the ongoing meteoroid flux" [35]. Examples include, in the jovian system, Metis, Adrastea, Amalthea, and Thebe (all embedded in the jovian ring system); and in the saturnian system, Atlas (skirting the outer edge of the main A ring), the F-ring shepherds (Prometheus and Pandora), Janus, Epimetheus, and the Lagrangian satellites of Tethys and Dione (Helene, Telesto, and Calypso). The 10 small inner satellites of Uranus discovered by Voyager in 1985-1986 probably also fall into this category, as well as the recently discovered N3-N6 satellites of Neptune.

Exceptions to the classification include the Moon, Triton, and Charon. All are large, and have undergone tidal evolution. Triton has a retrograde orbit, while Charon may be a collisional fragment from Pluto.

Finally Phobos and Deimos, the tiny satellites of Mars lying in equatorial orbits, are usually considered to be captured, on account of their compositional difference from Mars. However, it has been argued that they formed from collisional debris accumulating in orbit as Mars grew: this might require Mars to have a very primitive anhydrous composition [36]. It has also been flatly stated that "Phobos and Deimos are not captured asteroids. They have been in orbit around Mars ever since... the planets and Sun were formed" [37]. However, there are other views (see section 6.6).

Stevenson [38] detects four regularities among the satellite systems:

1. Jupiter, Saturn, and Uranus possess low-inclination prograde systems of large regular satellites, indicative of formation from an equatorial disk. Although the smaller satellites of Neptune are in equatorial orbit, the arrival of Triton probably disrupted any original satellite system.

2. The regular satellites extend out to 20-50 planet radii, but they do not form a scale model of the solar system (Mercury, the innermost planet, is 60 solar radii distant from the Sun). The mass is a few percent of the parent, roughly analogous to the distribution of mass of the solar system.

3. The smaller satellites tend to be more ice rich (60%) than the larger ones (40% ice; e.g., Ganymede, Callisto, Titan), indicating that the larger satellites may have selectively lost ice or accreted rock.

4. The very volatile ices (CH_4, N_2), appear only on satellites remote from both the Sun and the parent planets, indicative of the importance of radial temperature gradients both in the nebula as a whole and in the immediate vicinity of the planet.

Jupiter, Saturn, and Uranus have 16, 18, and 15 satellites respectively, which include 4, 5, and 5 large ones. In contrast to the Galilean system, the regular satellites of Saturn increase in density outward to Titan and then decrease (if Hyperion, the tumbling satellite, is excluded). Except for Titan, uranian satellites are denser than those of Saturn. Neptune, a look-alike for Uranus, has only one big satellite, Triton, and seven small ones (1989 N1-N6 in equatorial orbit and Nereid). There seem to have been significant differences in satellite formation between Uranus and Neptune [33].

It is clear from this attempt to classify satellites that there are few systematic regularities and that no general theory for satellite formation is obvious. Whether satellites formed from an accretion disk or were seeded from infalling planetesimals is uncertain. The Galilean satellites offer the best case for the former; the regular decrease in density from Io to Callisto argues for a common origin from a circumplanetary subnebula, but this set of four satellites is unique in the solar system.

6.5. REGULAR SATELLITES

"It is observed that there are four small stars round Jupiter, which incessantly accompany it" [39].

The classical example of regular satellites is the jovian system and there is "almost universal agreement that the Galilean satellite system is in some sense 'a miniature solar system' " [40]. The orbits of the satellites are equatorial, coplanar, and nearly circular, the spacing is a variation of the Titius-Bode rule, and there is a regular decrease in density outwards from Io to Callisto. All this provides strong evidence for formation from a protojovian nebula. Such disks will form as an unavoidable consequence in the current models for the formation of Jupiter and Saturn by the collapse of nebular gas onto a 10-20 Earth-mass core.

Some estimates of the mass and composition of the jovian subnebula can be made. The mass of the rocky component in the four satellites is about 2.6×10^{26} g. This is about 6×10^{-3} of the mass of a primordial H- and He-rich nebula. If the protojovian nebula had the same composition as the primitive nebula, it would have a mass of 4.3×10^{28} g or about 0.023 of the mass of Jupiter. Jupiter's mass and luminosity are over three times greater than that of Saturn, so Jupiter has exercised much greater control over its protoplanetary nebula, and over the properties of the regular satellites [41].

The saturnian regular satellite system differs fundamentally from that of the Galilean satellites of Jupiter. This has a significance both outside these systems and the solar system itself. Just as the two gas giants are different from each other, so their satellite systems differ as well. This tells us that no simple sequence of reproducible events has occurred in the solar system. Other planetary systems, if discovered, will be different in detail from our own. What their satellites might look like is only for bold spirits to predict.

The silicate or rocky component of the icy satellites of Jupiter, Saturn, and Uranus, about which we have little compositional information, may be composed of material approaching CI in composition. The rationale for this conclusion is that volatile-refractory-element fractionation appears less likely in the colder outer regions of the solar nebula beyond a "snow line" at about 4 or 5 AU [42]. This assumes that the volatile depletion in the inner regions of the primordial nebula was largely connected with early intense solar activity. This scenario should also be consistent with the composition of comets, but the composition of Halley, if typical of those interesting objects, does not provide much encouragement for this uniformitarian view. The nucleus of Halley has of course been subject to various alteration processes [43], so Earth-approaching comets may not be truly representative of the composition of the primitive outer solar system.

In the absence of other information, the CI composition is probably the only option for models of the "rocky" component [44] although increasing dilution with a carbonaceous or organic component can be expected. If heating and volatile loss occurs during early infall of material into the median plane of the nebula [45], then some volatile-element depletion may extend out beyond the "snow line," and it may thus be premature to assign a CI composition for volatile elements such as potassium.

Although temperatures in the jovian subnebula, from which the Galilean satellites formed, were probably too warm for ammonia hydrates to condense, they may be an important constituent (perhaps 10%) of the outer jovian, as well as of the icy saturnian and uranian satellites. The properties of ammonia hydrates are not well documented, but their major impact is to lower the melting point of water ice. A peritectic mixture of ammonia and water ice contains 33% NH_3 (mass) and has a melting temperature of only 175 K [46]. Radioactive heating in the centers of quite small rock-ice satellites (~500 km radius) will easily reach such temperatures, raising the prospect of ammonia-water "igneous" activity. The density of an ammonia-water liquid at 175 K is 0.946 g/cm^3 [47]. Since typical icy satellites have densities of 1.3 g/cm^3, the ammonia-water liquids can be expected to rise to the surface, either producing patches of mare-basalt-like flooding on small satellites or perhaps completely resurfacing the satellite (analogous to the water-ice surface of Europa). The density of such a surface layer on the uranian satellites at about 60 K would be 0.945 g/cm^3, very close to the liquid density. Arrival of later liquids or mushes at the surface is likely to produce a variety of unfamiliar morphologies, including thick flows. Both intrusive and extrusive "volcanic" features may result, producing puzzling landforms.

6.6. CAPTURED SATELLITES, PHOBOS, AND DEIMOS

The chief significance of captured asteroids in the present discussion is that they are probably prime representatives of planetesimals that survived planetary sweep-up. Their size distribution and composition is accordingly of interest. Candidates include the martian satellites, Phobos and Deimos, and the eight outer satellites of Jupiter. These captured jovian satellites fall into two groups, above and below the equatorial plane of Jupiter respectively (Fig. 6.6.1). Among the saturnian satellites, Phoebe, in retrograde orbit outside the equatorial plane is a likely captured object. Iapetus, has an inclined but circular orbit, and is interpreted as a regular satellite whose orbit has been warped by solar torques [48].

The most interesting case is that of Neptune. Both Triton and Nereid fall far away from the equatorial plane. The orbit of Triton is inclined at a high angle and revolves in the opposite sense to that of Neptune. Triton is almost certainly a captured object. If so, it is the largest object of this class and accordingly of exceptional interest (see section 6.13).

Most of these "distant" satellites are probably stray asteroids or planetesimals that have been captured, while the regular satellites in equatorial orbits originated in subnebulae during the formation of the planet. It has been suggested that many of the captured satellites of Jupiter, Saturn, and Uranus appear to be of spectral class C and so may represent C-class asteroids scattered by Jupiter [49]. If so, it is interesting that few, except perhaps Phobos and Deimos, were captured by the terrestrial planets. Jupiter and Saturn, once formed, act as a barrier between the inner and outer solar system, tossing much material right out of the system.

There is a scarcity or absence of rocky planetesimals, which might have been derived from the terrestrial neighborhood, in the outer solar system. Since they would arrive in the outer reaches of the solar system long after the nebula had dissipated, they would not acquire an icy mantle. Perhaps such objects were never scattered into the outer solar system, or were swept up by the giant planets. The preferred explanation is that there was only very limited mixing throughout the system, although since most of the mass was and is located at Jupiter and beyond, the effect of addition of material from the depleted inner system will be minimal.

These two satellites of Mars are often placed in the category of captured objects, principally on account of their compositional difference from Mars. Although some dynamical evidence indicates that they may be left-over planetesimals from the period of accretion [50], alternative views regard them as captured C-type asteroids because of their low density and albedo. This reviewer is more impressed by the geochemical than by the dynamical evidence:

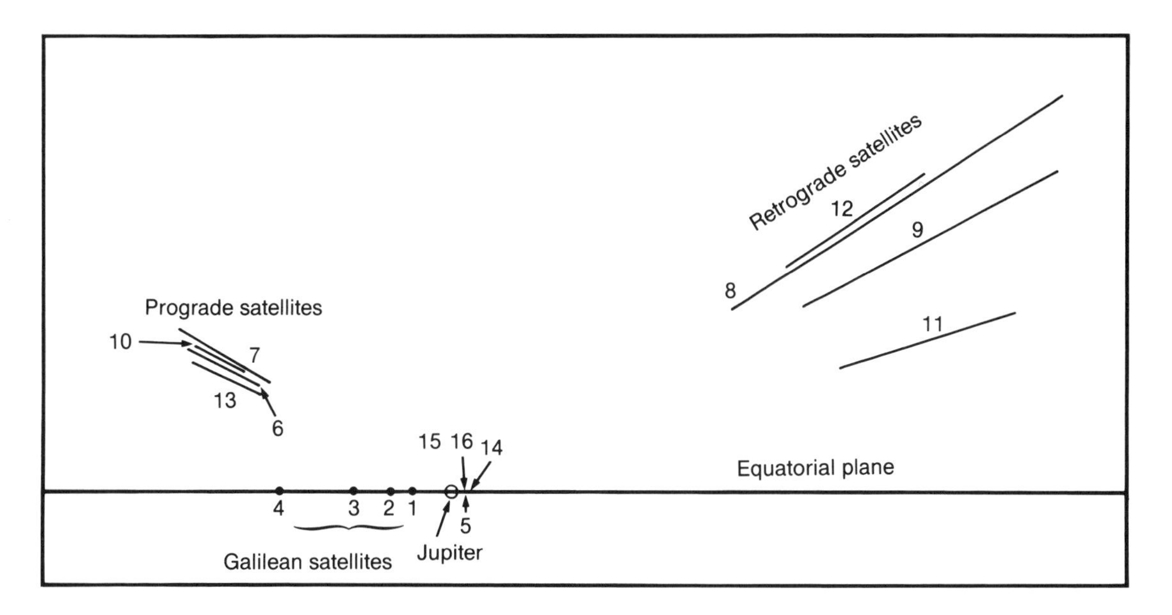

Fig. 6.6.1. *The satellites of Jupiter. The Galilean satellites (1, 2, 3, and 4) and satellites 5, 14, 15, and 16 lie in the equatorial plane. The other two groups in prograde and retrograde orbits with high inclinations probably represent captured objects. After Burns J. A. (1986) in Satellites (J. A. Burns and M. S. Matthews, eds.), p. 18, Fig. 1, Univ. of Arizona, Tucson.*

they are tiny objects presumably quickly put into equatorial orbits by Mars. The density of Phobos is 1.90 ± 0.1 g/cm^3. The albedo is 0.066 ± 0.006, making it similar to the black chondrites, although, once again, there is no single good meteorite match. It is clear that Phobos is not of CI composition [20].

It is a matter of much regret that the long-awaited U.S.S.R. Phobos mission was terminated prematurely [51]. It did, however, manage to shed some light on the problem. The surface of Phobos appears to be heterogeneous and markedly less hydrated than that of Mars, based on the 3-μm hydration absorption spectral feature [52]. These data would indicate a common origin for the satellites, possibly from a CM3 parent body, and distinct from that of the more hydrated planet. Possibly Phobos and Deimos are derived from the outer asteroid belt, where similar dry surface features are seen. Deimos has an orbital period, like the Moon, longer than the day of its parent; Phobos, in contrast, has an orbital period of 7-8 hr, about a third of the martian day. It is slowly spiraling inward and Phobos will impact Mars within the next 40 m.y. [53].

The production of a 200-km-diameter crater on Mars, with the possible arrival of martian meteorites on the Earth, is expected from this rather low-velocity collision, and it is a matter of regret that this spectacular and informative event will occur so far in the future on human timescales [54]. Although the decaying orbit of Phobos is sometimes used to infer that it is a relatively recent capture from the asteroid belt, this does not follow from the dynamics.

6.7. GALILEAN SATELLITES I: IO AND VOLCANISM

"Io is the most active and most bizarre of all known natural satellites" [55].

Most of the satellites of the outer planets have low densities, consistent with their being composed of rock-ice mixtures. Table 6.7.1 lists the satellites of Jupiter. The Galilean satellites of that planet are of special interest since they show a gradation from ice-free Io with its extensive volcanic activity and total lack of impact craters, to frozen Callisto, preserving the most heavily cratered surface in the solar system.

Table 6.7.2 lists properties of Io and Fig. 6.7.1 gives a proposed internal structure for Io [56]. Shortly before the Voyager encounter with Io, Peale et al. [57] predicted that volcanic activity would be widespread on Io due to the dissipation of tidal energy. It is now generally agreed that the heat source is

TABLE 6.7.1. Satellites of Jupiter, in order of increasing distance from the planet.

Satellite	Radius (km)	Density (g/cm^3)	Semimajor Axis (10^3 km)	Eccentricity	Inclination
Metis	20 × ? × 20	—	128	0	0
Adrastrea	12 × 10 × 7	—	129	0	0
Amalthea	135 × 85 × 75	—	181.3	0.003	0.45
Thebe	55 × ? × 45	—	221.9	0.013	0.9
Io*	1815	3.57	421.6	0.004	0.027
Europa*	1569	2.97	670.9	0.0003	0.28
Ganymede*	2631	1.94	1,070	0.001	0.18
Callisto*	2400	1.86	1,880	0.007	0.25
Leda	5	—	11,110	0.147	26.7
Himalia	90	—	11,470	0.158	27.6
Lysithea	10	—	11,710	0.130	29.0
Elara	40	—	11,740	0.207	24.8
Anake	10	—	20,700	0.170	147
Carme	15	—	22,350	0.21	164
Pasiphae	20	—	23,300	0.38	145
Sinope	15	—	23,700	0.28	153

* Galilean satellites.

TABLE 6.7.2. Properties of Io.

Radius	1815 (±5) km
Volume	2.52×10^{10} km^3
Surface area	4.15×10^7 km^2
Mass	8.92×10^{25} g
Relative mass (to Jupiter)	4.703 (±0.006) × 10^{-5}
Density	3.57 g/cm^3
Surface gravity	180 cm/s^2
Escape velocity	2.56 km/s
Orbital escape velocity	≥7.18 km/s (relative to Io surface)
Orbital radius (semimajor axis)	4.216×10^5 km; 5.95 R_J
Orbital period	42.456 hr
Eccentricity	0.0041
Inclination	0.027°
Albedo	0.6
Heliocentric distance	5.20 AU (7.78×10^8 km)
Typical surface temperature	~135 K (subsolar)
Typical hotspot temperature	~300 K
Hotspot heat flow	$>5 \times 10^{13}$ W
Global average heat flow	>1.2 W/m^2
Jovian magnetic field at Io	~2000 γ (or nT)
Age of surface	1×10^6 yr

Data from Nash D. B. et al. (1986) in *Satellites* (J. A. Burns and M. S. Matthews, eds.), p. 632, Table 1, Univ. of Arizona, Tucson; W. B. McKinnon, personal communication, 1990.

tidal, which supplies 2 orders of magnitude more energy than that derived from the radiogenic component [58]. Most of the heat output, which is near the limit obtainable from tidal dissipation, seems to be concentrated in one hemisphere in a few hot spots correlated with dark calderas [59].

Although in popular belief Io is covered with lava flows composed of sulfur, in fact sulfur compounds provide only a surficial coating. Io is characterized by quite rugged relief, with mountains up to 10 km high (on a body about the size of the Moon) and calderas 2 km deep. Only silicates possess the requisite strength to support such structures. Although no silicate spectral features have been observed, a thin ubiquitous coating of sulfur compounds probably masks them, given the nature of the observed volcanic plumes. Sulfur dioxide has been definitely observed, apparently present as frost or an absorbate. The cloud of sodium (and potassium) atoms surrounding Io requires the presence of silicates close to the surface. Various models of intermixed sulfur and silicate lavas have been proposed as forming the surface, lying on a silicate crust and underlain by molten silicates at depths of a few tens of kilometers.

Io has no impact craters greater than the image resolution (600 m). Resurfacing rates due to the volcanic activity are clearly high, perhaps between 10^{-3} and 10^{-1} cm/yr, but may be as high as 10 cm/yr [60]. Many hundreds of volcanic pits and calderas, up to 200 km in diameter, have been mapped (nine active volcanic rents were observed by Voyager I). Some of the mountains may be volcanic constructs, but it is not clear whether there are some tectonic features present. The plains on Io are layered, particularly at high latitudes, and appear to be composed of lava flows.

The spectacular volcanic plumes are 150-550 km in diameter and up to 300 km high. Eight of the plumes were observed by both Voyager missions

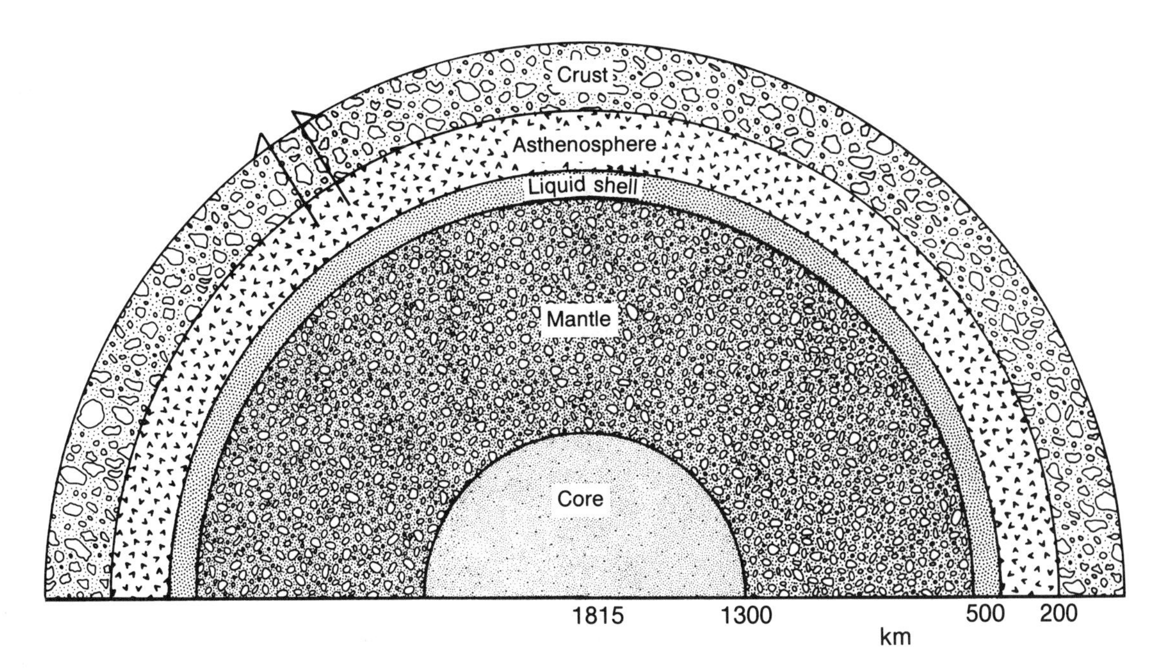

Fig. 6.7.1. *A possible model for the internal structure of Io. After Nash D. B. et al. (1986) in Satellites (J. A. Burns and M. S. Matthews, eds.), p. 660, Fig. 14, Univ. of Arizona, Tucson.*

(four months apart), so the plumes are not ephemeral, but may last for years. A typical plume of this type is Prometheus (3°S, 153°W). Pele, located at 19°S, 257°W, represents a different class of plume, active on a lifetime of a few weeks. The ejection velocity for Pele was about 1 km/s and the plume was 300 km high and 1200 km in diameter [61]. The most probable crustal structure is one of interbedded sulfur and silicate lavas, overlying a silicate crust. The color of the surface has been biased toward the red end of the spectrum on account of the particular set of filters used. The best match for the overall color of Io is "pastel yellow, with orange, green and grey tints" [62,63].

The high density of Io may be due to two causes. Either temperatures were too high in the protojovian nebula so close (six jovian radii) to Jupiter that water ice could not condense, or else the water was lost by impact or heating. The first scenario appears more consistent with the regular decrease in density of the Galilean satellites outward from the planet [64].

6.8. GALILEAN SATELLITES II: GANYMEDE AND PLANETARY EXPANSION

Table 6.8.1 shows that the most distinctive feature of Ganymede is the existence of two different crustal types, each occupying about 50% of the surface area of the satellite. An ancient, dark, densely cratered terrain, which is probably a mixture of ice and meteoritic debris, is fractured and intruded by younger, lighter, grooved terrain, apparently composed of water ice, which appears to have filled wide rift zones in the older crust. The most reasonable interpretation is that the younger, grooved terrain is derived from the interior by later internal melting. The older, more heavily cratered, and darker crust appears to have been split and intruded by the younger event, accompanied by some minor planetary expansion [65].

One explanation is that Ganymede was formed from a mixture of rock and water ice. The internal pressure is sufficient to convert ice to high-pressure polymorphs (e.g., ice VI or VIII, with densities up to 1.67 g/cm^3) (Fig. 6.8.1). Heating in the interior, due to the radioactive elements K, U, and Th in the silicate components, melted the water ice, which then rose to the surface and froze as ice I, leading to a significant density decrease and to minor planetary expansion of a few percent in radius [66]. However, direct limits on the amount of expansion of about 1% can be placed [67], so the explanation based on the thermal expansion of ice is probably only one contributing factor to the strange surface features of Ganymede and the stark contrast they

TABLE 6.8.1. Properties of Ganymede and Callisto.

	Ganymede	Callisto
Radius	2631 ± 10 km	2400 ± 10 km
Mass	$1482.3 \pm 0.5 \times 10^{23}$ g	$1076.6 \pm 0.5 \times 10^{23}$ g
Surface gravity	143 ± 2 cm/s^2	125 ± 4 cm/s^2
Mean density	1.94 g/cm^3	1.86 g/cm^3
Percent silicate by mass	49-59	47-56
Distance from Jupiter	1.070×10^6 km	1.880×10^6 km
	$= 15.1\ R_J$	$= 26.6\ R_J$
from Sun	5.203 AU	5.203 AU
Orbit period	7.155 day	16.689 day
around Sun	11.86 yr	11.86 yr
Eccentricity	0.001 (variable)	0.007
Inclination	0.183°	0.253°
Obliquity (average)	3.08°	3.08°
Visual albedo	0.43	0.19
Solar flux	3.7% × Earth	3.7% × Earth
Subsolar temperature	156 K	168 K
Equatorial subsurface temperature	117 K	126 K

Data from McKinnon W. B. and Parmentier E. M. (1986) in *Satellites* (J. A. Burns and M. S. Matthews, eds.), p. 722, Table 1, Univ. of Arizona, Tucson; W. B. McKinnon, personal communication, 1990.

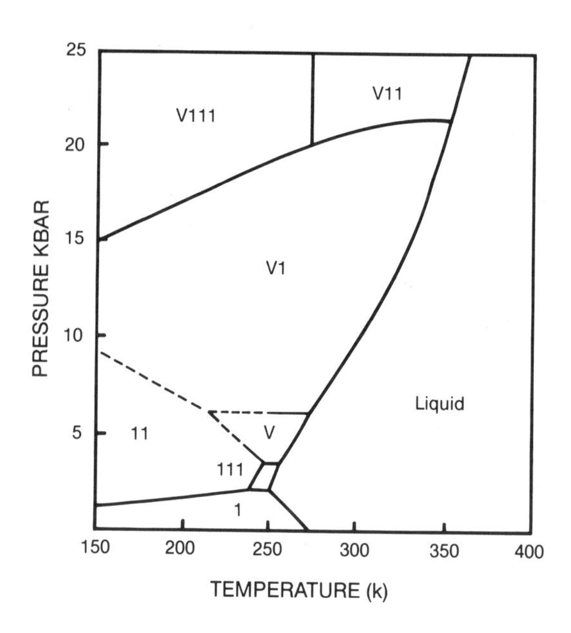

Fig. 6.8.1. *The temperature-pressure phase diagram for water ice. The Roman numerals refer to the various polymorphs in each stability field. Ice VIII has a density of 1.67 g/cm^3, so that a change from ice VIII to ice I will result in a volume expansion.*

make to the heavily cratered apparently primitive surface of Callisto. Other models favor resurfacing by warm ice flows, rather than by liquid water [68].

Ganymede, the largest satellite in the solar system, is thus of extraordinary interest. It displays evidence of slight expansion, which is explicable from the known physical properties of ice. Many problems remain. The age of the older crust from crater counting appears to be in the range 3.7-4.0 b.y. The ages of the bright grooves (sulci) appear to be 3.1-3.7 b.y. It is unclear how Ganymede might form as an undifferentiated object in the first place, or how it would remain unmelted for a period of several hundred million years, which is the apparent difference in ages between the old, dark and the younger, grooved terrain.

Although there have been suggestions that ammonia ice might account for some of the activity on Ganymede, the absence of any internally driven volcanic activity on Callisto, which would be expected to contain significantly more ammonia, reinforces the conclusion that ammonia is not a significant constituent of the Galilean satellites. Ganymede thus presents us with a series of problems that will require new data from the Galileo spacecraft mission to resolve, although "... if past experience is a guide they will be complex ... raising more questions than they resolve" [69]. The great contrast between the two outermost Galilean satellites is analogous to the difference between Venus and the Earth, two apparently similar bodies that differ so greatly in detail. "The difficulty... lies not in proposing an explanation, but in verifying its validity" [69]. This problem is addressed in the next section [70].

6.9. GALILEAN SATELLITES III: EUROPA AND CALLISTO

Although Europa (Table 6.9.1) is mainly rock (r = 1569 km; density = 2.97 g/cm^3), it appears to have a water-ice surface. Whether this crust is primary, due to accretional heating, or a secondary crust derived by later melting of ice within the satellite is not clear, and the difference may be semantic. Europa has only five craters in the 10-30-km-diameter range in the best-imaged region and so must be continually resurfaced, most likely by tidal heating from Jupiter. The most probable model is that moderate tidal heating, together with some radiogenic input, melts the base of the icy crust and causes resurfacing [71].

In sharp contrast to Europa, the outermost Galilean satellite, Callisto (Table 6.8.1), possesses a heavily cratered, apparently primordial crust, unaltered since the termination of the early massive bombardment. Callisto (r = 2400 km; density = 1.86 g/cm^3) is both somewhat smaller and less dense than Ganymede (r = 2638; density = 1.94 g/cm^3).

The ancient nature of the surface indicates that the crust dates from very early in solar system history, so it may represent a primary crust formed during accretion. It appears to have been impacted by a separate population of impactors distinct from that responsible for the heavy cratering in the inner solar system. As the outermost Galilean satellite, it apparently escaped heating by tidal interaction with Jupiter.

TABLE 6.9.1. Properties of Europa.

Radius	1569 ($\pm$ 10) km
Mass	4.87×10^{25} g
Density	2.97 g/cm^3
Orbital semimajor axis	670,900 km
Orbital period	3.551 days
Orbital eccentricity	0.0003
Orbital inclination	0°28′
Percent silicate	97
Age of surface	$<100 \times 10^6$ yr

Data from Malin M. C. and Pieri D. C. (1986) in *Satellites* (J. A. Burns and M. S. Matthews, eds.), p. 690, Table 1, Univ. of Arizona, Tuscon; W. B. McKinnon, personal communication, 1990.

There is still no satisfactory solution to the problem of why Ganymede is evolved, while Callisto appears primitive [70]. Ganymede is both larger and denser than Callisto. Perhaps this difference led to a warmer mantle in Ganymede, leading to melting a few hundred million years after accretion, while the smaller size and lower density of Callisto have inhibited this evolutionary development. Possibly Callisto was accreted slowly enough that it never melted and differentiated. However, it is difficult to accrete Callisto on the basis of current models without melting it and forming an icy crust. The problem is that such a structure is not stable against a second Ganymede-style differentiation, which is not observed [70]. Clearly, Ganymede was on one side of a critical boundary in the Galilean system, and Callisto was on the other, providing yet another example of diversity in this apparently least complex of the satellite systems

6.10. SATURNIAN SATELLITES

"Seven satellites have been observed in motion round this planet, from west to east in orbits nearly circular. The first six move nearly in the plane of the ring. The orbit of the seventh approaches more closely to the plane of the ecliptic" [72].

In contrast to the relatively orderly Galilean system, the saturnian system shows few regularities. The satellites of Saturn (Table 6.10.1) do not resemble the Galilean satellites of Jupiter in detail (Fig. 6.10.1). All except Titan are relatively small. They are less dense than the Galilean satellites, and in contrast to that relatively well-behaved group, do not show any regular variation in density with distance from the planet. Compared to the situation at Jupiter, the subnebula around Saturn was colder, and possibly well mixed, or else homogenization occurred during accretion. There is only one large satellite, Titan. Possibly multiple episodes of satellite formation occurred. In addition to water ice, ammonia and methane clathrates were probably also available for incorporation this far distant from the Sun.

In comparison with the inner solar system, or the jovian satellites, the saturnian satellites appear to have been cratered by a distinct population of objects. Controversy surrounds these arguments for differing cratering populations on the satellites of the giant planets [73]. A basic question is whether the saturnian satellites have been broken up and reassembled many times during their history. Opinion is divided on this issue [33,74]. The surfaces of Mimas, Rhea, and Iapetus do not appear to be saturated with larger craters (d> 64 km), but may be saturated with

TABLE 6.10.1. Satellites of Saturn.

Satellite	Radius (km)	Density g/cm³	Semimajor Axis (Saturn radii)	Eccentricity	Inclination
Atlas	20 × ? × 10	—	2.276	0.002	0.3
Prometheus	70 × 50 × 37	—	2.310	0.004	0.0
Pandora	55 × 45 × 33	—	2.349	0.004	0.1
Janus }	110 × 95 × 80	0.67 ± 0.10			
Epimetheus }	70 × 58 × 50	0.64 ± 0.12	2.51	—	—
Mimas	197 ± 3	1.43 ± 0.2	3.08	0.020	1.5
Enceladus	251 ± 5	1.20 ± 0.5	3.95	0.004	0.0
Tethys	530 ± 10	1.25 ± 0.1	4.88	0.000	1.1
Telesto	15 × 10 × 8	—	4.88	—	—
Calypso	12 × 11 × 11	—	4.88	—	—
Dione	560 ± 5	1.43 ± 0.1	6.26	0.002	0.0
Helene	17 × 16 × 15	—	6.26	0.005	0.2
Rhea	765 ± 5	1.33 ± 0.1	8.73	0.001	0.4
Titan	2575 ± 2	1.881	20.3	0.029	0.3
Hyperion	190 × 145 × 114	—	24.6	0.104	0.4
Iapetus	718 ± 18	1.2 ± 0.1	59	0.028	14.7
Phoebe	110 ± 10	—	215	0.163	150

Data from Morrison D. et al. (1986) in *Satellites* (J. A. Burns and M. S. Matthews, eds.), p. 766, Univ. of Arizona, Tucson; Yoder C. F. et al. (1989) *Astron. J., 98,* 1887.

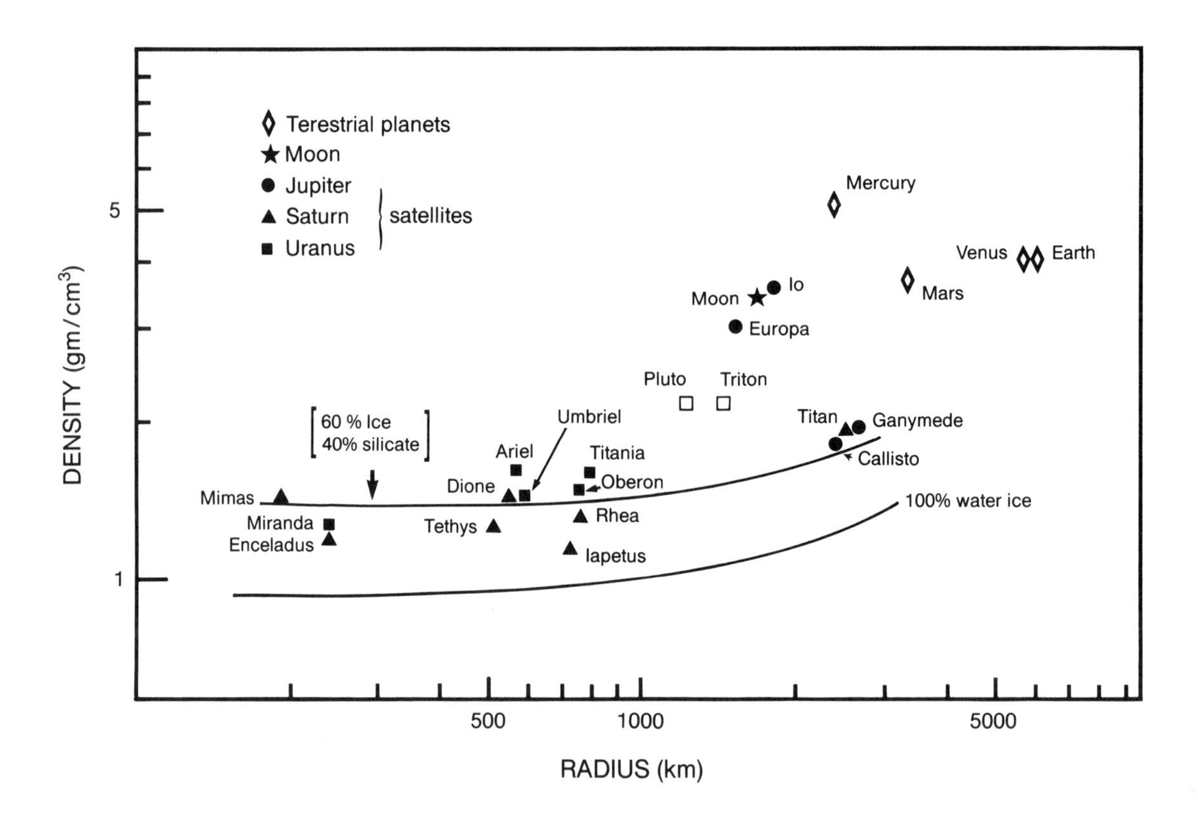

Fig. 6.10.1. *The density of the satellites of the giant planets plotted against radius. Note that the density of the saturnian satellites is lower on average than that of the uranian satellites.*

smaller craters (d < 32 km). The large craters probably represent a production function. One study of crater abundances on the saturnian system suggests that Mimas and the classical satellites are probably original aggregates that were not repeatedly disrupted and reaccreted [30]. If the classical satellites of Saturn are not saturated with large craters, then the surfaces are original and date back to the time of the satellite formation. These views all disagree with the earlier ideas advanced following the initial exploration, in which the satellites were considered to have been broken up and reassembled [75]. In view of the ease with which the surfaces of quite small outer solar system satellites are renewed, the fact that the presently observed crater population is a production function has little real significance for the total history of an individual Moon; it merely records the impact history since the last major resurfacing event, whether by a large impact or by icy volcanism.

I have resisted the great temptation at this stage to include comments on the individual satellites, or to conduct a tour of the fascinating saturnian system [76]. However, some bodies are so bizarre that they cannot escape notice even in a broad survey such as this.

Janus (S10, density = 0.67 ± 0.10 g/cm^3) and Epimetheus (S11, density = 0.64 ± 0.12 g/cm^3) are a strange pair of similar mass, occupying the same orbit, close to the rings. They possess the lowest density of any body in the solar system, lower even than that of water ice (0.92 g/cm^3). The most logical explanation is that they are porous rubble heaps [77].

The case of Hyperion is instructive. This satellite of Saturn, lying between Titan and Iapetus, fifteenth in order outward from the planet, is tumbling chaotically. It is the only satellite at present exhibiting this behavior, although other similar satellites in the past have tumbled [78]. Hyperion is notably irregular, with principal radii of 190 × 145 × 114 km ± 15 km. It is much darker than the other saturnian satellites, and has few large craters. The orbital period is 21 days; its spin rate and axial orientation are changing significantly every few orbits. It possesses a very eccentric orbit (eccentricity ≈ 0.1). These features are consistent with Hyperion being the remnant of a satellite disrupted by a massive collision [79]. Hyperion is unlikely to be slowed by tidal friction [9] and will continue to tumble until the "end of the world" some 5 × 10^9 years hence when the system is overtaken by other events, such as the red giant stage of solar evolution.

6.11. TITAN

This satellite is unique in both the saturnian and the solar system. Titan has a radius of 2575 km and a density of 1.83 g/cm^3, and is the only saturnian satellite comparable in size and density to the Galilean satellites of Jupiter. Its orbit lies between that of Rhea and Hyperion, at 20.3 planetary radii. For comparison, the orbits of Ganymede and Callisto are at 15 and 26.4 planetary radii distant from Jupiter. Titan has a rock/ice ratio of 52/48. The surface temperature is 94 K. This satellite has always excited interest, both on account of its size, a little smaller than Ganymede, and its possession of an atmosphere, with a surface pressure of 1500 ± 20 Mbar, 50% higher than that of the terrestrial atmosphere. No surface details are visible.

The atmosphere of Titan (Table 6.11.1) is dominated by nitrogen, with up to 10% methane, and a possible, but undetected argon component. There is a haze of organic aerosol particles. Methane, ethane (C_2H_6), propane (C_3H_8), HCN, cyanogen (C_2N_2), and cyanoacetylene (HC_3N) have all been detected. Titan may possess an ocean dominated by ethane with some methane and nitrogen, and methane rains appear possible [80]. There is a general problem

TABLE 6.11.1. Composition of the atmosphere of Titan.

Gas		Mole Fraction: Inferred Indirectly	Mole Fraction: Measured Directly
Nitrogen	N_2	0.65-0.98	
Argon(?)	Ar	0-0.25	
Methane	CH_4	0.02-0.10	
Hydrogen	H_2		2×10^{-3}
Carbon monoxide	CO		6×10^{-5}-1.5×10^{-4}
Ethane	C_2H_6		2×10^{-5}
Propane	C_3H_8		4×10^{-6}
Acetylene	C_2H_2		2×10^{-6}
Ethylene	C_2H_4		4×10^{-7}
Hydrogen cyanide	HCN		2×10^{-7}
Methyl acetylene	C_3H_4		3×10^{-8}
Diacetylene	C_4H_2		~10^{-8}-10^{-7}
Cyano-acetylene	HC_3N		~10^{-8}-10^{-7}
Cyanogen	C_2N_2		~10^{-8}-10^{-7}
Carbon dioxide	CO_2		1.5×10^{-9}

Data from Morrison D. et al. (1986) in *Satellites* (J. A. Burns and M. S. Matthews, eds.), p. 790, Table 4, Univ. of Arizona, Tucson.

accounting for the high abundance of methane and for the presence of nitrogen in Titan. Direct capture from the nebula was ruled out by Owen [81] who noted that neon was absent (neon and nitrogen have about the same solar abundance). Prinn and Fegley [82] believed that Titan formed in a methane-rich subnebula that had been spun off from Saturn and in which the conversion of carbon monoxide to methane was complete. Lunine [83] also concludes that the composition and density of Titan are consistent with formation in a dense disk around Saturn.

As discussed in section 6.2, nitrogen and carbon monoxide are the major primitive nebular phases rather than ammonia and methane. Neither these compounds nor clathrates ($NH_3.H_2O$, $CH_4.6H_2O$) are expected as condensates from the primitive nebula, but are likely to form under the higher temperature and pressure conditions in planetary subnebulae [84]. Thus, it seems clear that Titan is not a simple sample of the primordial solar nebula, but has formed in a saturnian subnebula.

6.12. URANIAN SATELLITES

"Dr. Herschel, by means of a very powerful telescope, has discovered six satellites moving round this planet, in orbits almost circular and nearly perpendicular to the plane of the ecliptic" [85].

The uranian satellites have quite high densities, about 1.5 g/cm^3, generally higher than those of the saturnian system (Fig. 6.10.1) indicating that their content of methane ice is low. There is little information on the possible amounts of other icy constituents, that might have lower melting points, and so account for some of the remobilized surface features. The most fascinating observation about the classical uranian satellites (Table 6.12.1) is that some of these quite small icy bodies have been resurfaced [86]. They might have been expected to be small versions of Callisto or Mimas, heavily cratered inert icy lumps, showing no sign of activity since the end of the heavy bombardment. Instead, Miranda, Ariel, and Titania have all been resurfaced, Ariel and Miranda in particular showing surprisingly young surfaces. Ariel and Umbriel present a particularly interesting comparison. They have similar masses and radii, but their surfaces are completely different. Umbriel has the expected primordial surface, like Callisto, unchanged from the end of heavy cratering before about 4 b.y. ago. Ariel, in contrast, has been geologically active.

The question of the supply of heat for the satellites of both Uranus and Saturn has risen because of the surprising diversity of geology on what were expected to be inert bodies like Callisto. Two possible

TABLE 6.12.1. Satellites of Uranus.

*Small Satellites in Order of Distance from the Planet**		
Satellite	Radius (km)	Semimajor Axis (10^3 km)
Cordelia	13 ± 2	49.8
Ophelia	16 ± 2	53.8
Bianca	22 ± 3	59.2
Cressida	33 ± 4	61.8
Desdemona	29 ± 3	62.7
Juliet	42 ± 5	64.4
Portia	55 ± 6	66.1
Rosalind	29 ± 4	69.9
Belinda	34 ± 4	75.3
Puck	77 ± 3	86.0

Classical Satellites, in Order of Distance from Uranus†					
Satellite	Radius (km)	Density (g/cm^3)	Semimajor Axis (10^3 km)	Eccentricity	Inclination
Miranda	242 ± 5	1.26 ± 0.4	129.9	0.0027	4.22°
Ariel	580 ± 5	1.65 ± 0.3	191.0	0.0034	0.31°
Umbriel	595 ± 10	1.44 ± 0.3	266.3	0.0050	0.36°
Titania	805 ± 5	1.59 ± 0.1	435.9	0.0022	0.14°
Oberon	775 ± 10	1.50 ± 0.1	583.5	0.0008	0.10°

*Data from Thomas P. et al. (1989) *Icarus, 81,* 93.
† Data from Tyler G. L. et al. (1986) *Science, 233,* Table 2.

energy sources are tidally induced heating or accretional heating. The latter is clearly insufficient if water ice is a dominant constituent; the required melting temperature of 273 K will not be reached. However, the presence of ammonia changes the situation completely, since the H_2O-NH_3 system has a peritectic point (with 35% NH_3) at 173 K. The density is about that of water ice [87] so that melts will tend to rise, and so be an agent for the observed resurfacing. Accretional temperatures can exceed 173 K for Dione, Rhea, Ariel, Umbriel, Titania, and Oberon, and possibly for Mimas, Enceladus, and Tethys [88]. The difficulty with tidally induced heating in the uranian system, however, is that no orbital resonances, which could induce a temperature rise, melting, and resurfacing, exist at present. Accordingly, it has been a major concern to establish whether such resonances, which could have led to tidal heating, might have existed in the past [89]. Calculations show that Ariel and Umbriel could have existed in a 2/1 orbital resonance that might have induced tidal heating in Ariel (but not in Umbriel). This resonance might eventually have been disrupted by interaction with Titania. For Umbriel, tidal heating is negligible compared to radioactive heating, which has in any event been inadequate to resurface that satellite. For Titania, tidal heating appears inadequate to produce the observed resurfacing [90], but it could induce some melting of ammonia and methane-ice mixtures at depth. Accretional temperatures could reach 276 K at 55 km depth in Titania, for example. Thus, the properties of ammonia-water liquids are probably responsible for some of the observed volcanism on the uranian satellites, and thick semicrystallized flows of this material may erupt through denser silicate-ice crusts, producing unfamiliar landforms. Possibly, more volatile and more readily mobilized compounds exist on Titania [91]. The resurfacing of Ariel and Miranda appears to have taken place via solid-state resurfacing, the first example of such a process in the solar system [92], although it could be partially crystallized mushy $NH_3.2H_2O$ melt [93].

The uranian satellites thus present a fresh set of unknowns. Again, I resist the tempation to conduct a guided tour of the system, but Miranda is too exotic to escape comment.

Miranda has an anomalously large inclination ($i = 4.34°$), about 50 times larger than that of the other uranian satellites, a large eccentricity [94], and yet another extraordinary surface to compete with Io and Triton. There are three coronae, a network of canyons, and a history of cratering and expansion. "The central problem in modelling the thermal histories of the uranian satellites is accounting for Miranda" [95]. The satellite has had such a violent history that heating by tidal interaction appears inadequate to account for it, even if it had been in a 2/1 resonance orbit with Ariel. The presence of adjacent blocks of totally different landscapes makes breakup and reassembly of the satellite an appealing, but perhaps unlikely mechanism. Although some sort of collisional disruption appears to be required, it is not obvious that the present terrain, with relief up to 20 km, would survive catastrophic disruption and reassembly.

6.13. TRITON, NEREID, AND THE OTHER NEPTUNIAN SATELLITES

A massive increase in our understanding, both of Triton and Nereid and of ring systems, resulted from the Voyager II encounter with Neptune in August 1989 (Table 6.13.1 lists data for the neptunian satellites).

6.13.1. Triton

"With its tectonic features, its crisscross of channels, its mushrooms, its wind streaks, its haze, its evidence of condensing volatiles, Triton was beginning to look like Europa, Enceladus, Mars, and Io rolled into one" [96].

Triton has a radius of 1352 ± 3 km, considerably smaller than the Moon (1738 km), but significantly larger than Pluto (1150 km) [97]. It has a nearly circular, but retrograde and highly inclined orbit with an inclination of 21°, the only large satellite with such an orbit. Its mass is 2.141×10^{25} g and

TABLE 6.13.1. The satellites of Neptune in order of increasing distance from the planet.

Satellite	Radius (km)	Semimajor Axis (10^3 km)	Distance R_N
N6 Naiad	27 ± 8	48.0	1.94
N5 Thalassa	40 ± 8	50.0	2.02
N3 Despina	75 ± 15	52.5	2.12
N4 Galatea	90 ± 10	62.0	2.50
N2 Larissa	95 ± 10	73.6	2.97
N1 Proteus	208 ± 8	117.6	4.75
Triton	1352 ± 5	354.8	14.33
Nereid	175 ± 25	5513.4	222.7

N1-N5 are in equatorial orbits with low eccentricities. N6 has an inclination of 4.5°. Triton has a circular orbit, with an inclination of 21° and a density of 2.07. Nereid has an orbit with an eccentricity of 0.756 and inclination of 27.5°. Data from T. Johnson, personal communication, 1989, and W. B. McKinnon, personal communication, 1990.

Fig. 6.13.1. *The Neptune-facing hemisphere of Triton. The sunlit south polar cap probably consists of a slowly evaporating layer of nitrogen ice, deposited during the previous winter. (NASA JPL P-34764)*

its density is 2.075 ± 0.019 g/cm^3, very close to that of Pluto. Its thin atmosphere is mainly composed of nitrogen, with a trace of methane, similar to that of Titan. Neon, argon, and carbon monoxide are all less than 1%. The surface temperature is 38 ± 4 K and the pressure is 16 ± 3 μbar. Liquid nitrogen is not stable on the surface. As the subsolar point moves, nitrogen ice, responsible for the blue color, sublimes and migrates to cooler regions. There are geyser-like plumes in active eruption, probably due to eruptions of nitrogen mixed with dark material, so that resurfacing is occurring as on Io (Fig. 6.13.1). The plume height is 8 km before the plume trails horizontally. The surface relief of about 1 km implies a relatively strong surface material, rigid at temperatures of 40 K. Water ice is the most probable candidate (Fig. 6.13.2).

Triton is almost certainly captured into a retrogade orbit by Neptune, and has demolished any previous regular satellite system in this traumatic episode. The crater density is similar to that of the older lunar maria, indicating that the surface, unlike that of most satellites, is relatively young (about 3 b.y.), but the ages of differing portions of the surface vary by factors of 5. The largest impact crater, Mazomba, is about 27 km in diameter [97].

There is no trace of old heavily cratered terrain or of any early intense bombardment. McKinnon, in a series of papers [98], has pointed out that the record was probably erased by heating during the capture event. Tidal heating was sufficient to cause global melting and differentiation, in which a thin ice shell overlaid a water mantle, floating on top of a silicate core. This may have been capped with a thin layer of rock. Triton is likely to have remained hot for periods exceeding 500 m.y., thus accounting for the absence of old heavily cratered terrain. The contribution from tidal heating will exceed the contribution from radiogenic heating for periods of 1-2 b.y. It is therefore not surprising that Triton shows a comparatively youthful appearance.

The capture of Triton by gas drag into its highly inclined orbit would have to occur before the subnebula was entirely dissipated, so the timing of this event is quite restricted, perhaps to within a million years of the formation of Neptune. However, in order to prevent Triton continuing to spiral into the planet, the nebula needs to have dissipated, so the timing of capture is crucial.

6.13.2. Nereid

Nereid, a somewhat neglected satellite, has an orbit of high eccentricity (0.75) and a high inclination (27.5°) [97]. The distance from Neptune varies from 54 R_N (1.4×10^6 km) to 400 R_N (9.7×10^6 km). The orbital period is 360 days. The radius is 175 ± 25 km and is thus smaller than Proteus (1989 N1) (208 ± 8 km). It is similar in many respects to 2060 Chiron, which most likely is a comet (the distinction between outer solar system icy planetesimals and comets is becoming semantic).

Fig. 6.13.2. *A view of Triton about 500 km across, showing the "lake district." The smooth areas are probably old impact basins that have been extensively modified by flooding and faulting and filled with water ice. (NASA JPL P 34692)*

6.13.3. Inner Satellites

The strange neptunian satellite system seems to be principally the result of the capture of Triton, whose arrival on the scene must have resembled that of a bull entering a china shop. The present satellite system did not exist during the period of Triton's capture and tidal evolution. Any original satellite system was destroyed by mutual collisions on very short timescales, induced by the eccentricities generated by perturbations following the Triton capture. The small inner satellites must have accreted out of the resulting debris. Five of the small satellites probably formed on circular orbits from the resulting equatorial disk of rubble. Naiad (N6) was probably captured into an inclined orbit following a similar origin [99].

6.14. PLUTO AND CHARON

The solar system does not become simpler with increasing distance from the Sun, as Pluto and Charon testify. Pluto is commonly referred to as the ninth planet, but the mass of the Pluto-Charon pair is very small (1.36×10^{25} g), amounting only to 18.5% of the mass of the Moon. The combination of low mass with a highly inclined and eccentric orbit is a major reason for not according Pluto planetary status, although no doubt it will long continue to be referred to as the ninth planet for a combination of traditional and sentimental reasons. It most closely resembles Triton, and both are probably left-over large outer solar system planetesimals that did not accrete to the planets.

Its satellite, Charon, was discovered only very recently [100]. It has a semimajor axis of 19,640 ± 320 km with very low eccentricity [101]. A recent evaluation of the strange case of Pluto and Charon shows that Pluto "... is a new kind of world, intermediate in size between the largest medium-sized icy satellite, Titania (800 km radius) and the smallest icy Galilean satellite, Europa (1569 km)" [102]. Pluto has a high rock/ice ratio (0.68/0.80) in comparison with most other satellites of the outer planets.

Pluto has a radius of 1151 ± 6 km and Charon has a radius of 593 ± 13 km [103]. These values have been obtained due to the fortunate circumstances of mutual eclipses visible from the Earth during the period 1985-1990. Such events occur only twice each 248 years.

6.14.1. Density

The mean density for the Pluto-Charon pair is 2.029 ± 0.032 g/cm^3 [103], very close to that of Triton (2.075 g/cm^3). The value is also close to the densities of Ganymede (1.94 cm/cm^3), Callisto (1.86 g/cm^3), and Titan (1.88 g/cm^3) (Fig. 6.10.1). However, Pluto is so small that its bulk and uncompressed densities differ by less than 1%. The uncompressed densities for Ganymede and Callisto (and Titan) are much lower [104,105] and Pluto has a higher rock/rock + water ice ratio than these satellites. The rock fraction of Pluto, depending on whether it is differentiated or not, ranges by mass from 0.68 to 0.80.

6.14.2. Internal Structure

Pluto is probably differentiated, as indicated by the presence of methane on the surface. The high rock content probably will lead to melting of water ice because of temperature increases from radiogenic heating [106]. Figure 6.14.1 shows a possible model for Pluto, and two alternatives for the internal structure of Charon. Pluto most likely has a hydrated silicate core, surrounded by an icy mantle [107]. The silicates in the core will be hydrated, even if they were anhydrous when accreted, due to interaction with warm water ice in the core before differentiation. Later rises in temperature due to radioactive heating will not reach the 800-1200 K needed to accomplish dehydration.

6.14.3. Atmospheres and Surfaces of Pluto and Charon

Among the many curious features of Pluto and Charon is the difference in the surfaces of two such closely related objects. Pluto has an atmosphere of methane [108,109], with possibly some nitrogen from analogy with Triton. The surface pressure is 10 μbar, and the atmosphere freezes out as it retreats from perihelion; thus, Pluto has an atmosphere that forms and collapses depending on its position in its orbit around the Sun [110,111]. Pluto's atmosphere appears to be extensive, extending to a thickness approaching the radius of the body, definitive evidence having been obtained when Pluto occulted a twelfth-magnitude star [112].

The observed methane surface on Pluto is the best evidence for differentiation; a near-surface reservoir is needed to balance the surface loss of methane over the age of the solar system. In contrast, Charon's surface is probably coated with water ice. The volatile ices appear to be missing, and thus Charon appears to have lost its atmosphere, possibly as a result of collision. The surface composition of Charon may thus provide vital information about the composition of the Pluto-Charon pair, revealing information hidden on Pluto by its shroud of methane.

6.14.4. Orbit

The orbit of Pluto is highly eccentric ($e = 0.250$) and inclined ($i = 17.2°$) in a 2:3 resonance with Neptune, and takes part in at least two other resonances; these ensure that, although the orbits of Pluto and Neptune cross one another, close encounters are avoided. An integration of the orbits of the outer planets for 845 m.y., carried out using the Digital Orrery [113] has found that although the motion of Pluto is apparently chaotic on a 20-m.y. timescale, the level of chaos is low, and it did not evolve out its resonance over the 845-m.y. period studied [114]. Probably Pluto was nudged into resonance with Neptune, with outward transfer of angular momentum. The orbit of Charon about Pluto

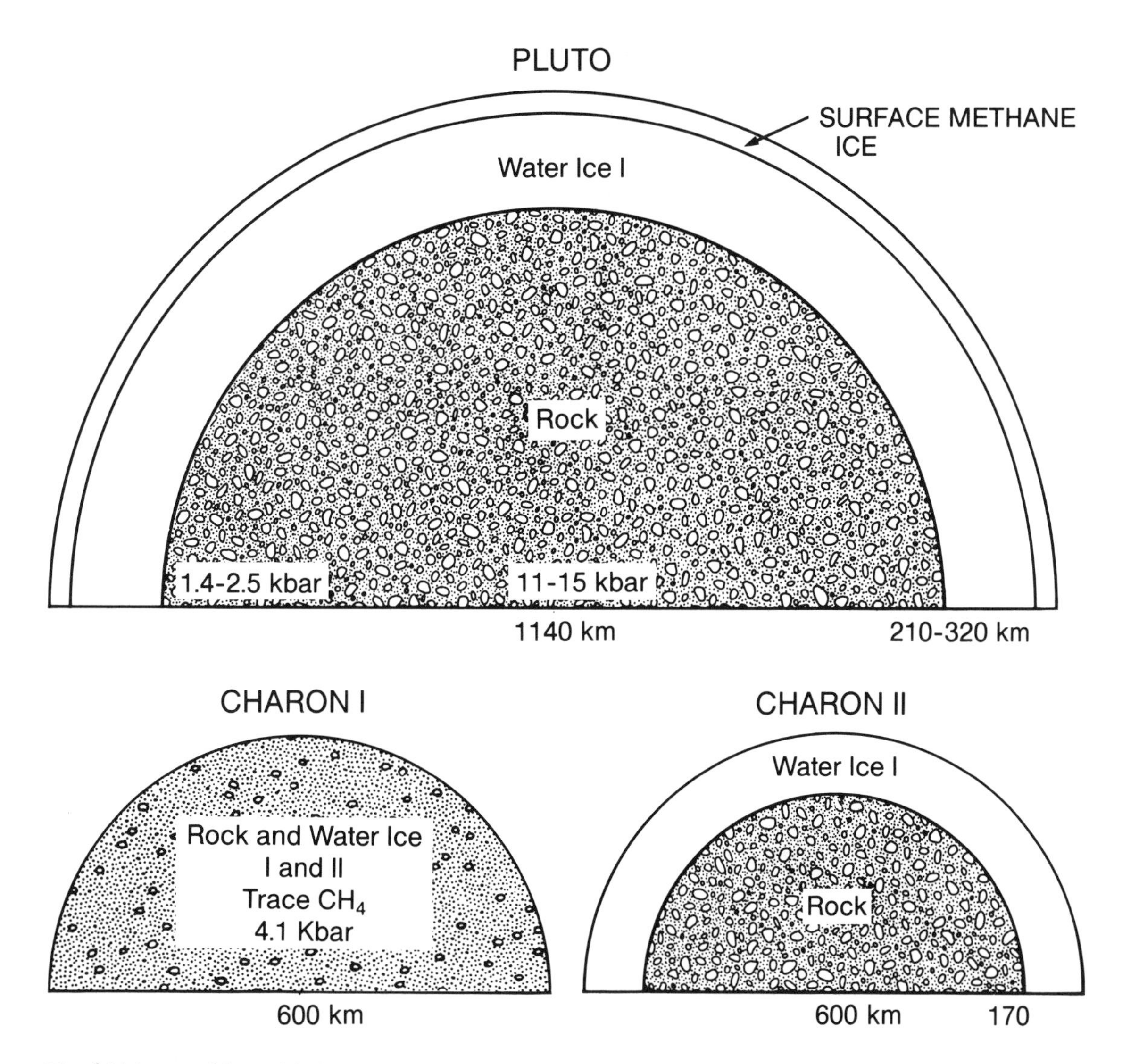

Fig. 6.14.1. *Possible models for the interior structure of Pluto and Charon. After Simonelli D. P. and Reynolds R. T. (1989) Geophys. Res. Lett., 16, 1210, Fig. 1.*

is also bizarre, being inclined at about 90° to the plane of the ecliptic. This is probably the result of a collisional origin for the satellite.

6.14.5. Origin

The chaotic nature of Pluto's orbit makes it difficult to draw firm conclusions about the original location of Pluto and Charon; they might have formed in their current orbit, or in an orbit of lower eccentricity and inclination, and evolved into the present situation through chaotic processes.

Among the many origins that have been proposed for Pluto (and Charon), four may be addressed:

1. An origin in the inner solar system within 4-5 AU (i.e., on the Sunward side of the "snow line") can be excluded on account of the presence of water ice and methane. The difficulties in transporting such an object from the inner to the outer reaches of the solar system seem considerable in view of the zoned structure seen in the asteroid belt, the Jupiter barrier, and the compositional differences among the planets and satellites. If such an interchange were likely and common, one might expect to observe a rather homogeneous solar system.

2. A second possibility, that it formed in the Jupiter-Saturn region, also seems unlikely; most satellites have much lower rock/ice ratios, typically 40/60 for medium-sized saturnian satellites. However, as noted below, a substantial amount of ice might be removed during a massive collision. Even if it formed close to its present location, Pluto is too small and accretional velocities are too low for it to have lost ice by accretional heating [115]. Ganymede, Callisto, and Titan are large enough to have lost some ice in this manner.

3. The notion that it is an escaped satellite of Neptune has been mostly discredited [116]. "It is forbiddingly improbable that Pluto was once a satellite of Neptune that would have required the double chance of something to break it away from Neptune and a quick means of shifting it into the resonance before it collided with Neptune" [117].

4. The most probable explanation is that Pluto represents a large planetesimal formed in the outer solar system with a rock/ice ratio of about 70/30. It is, along with Triton, the rockiest object in the outer solar system. This is consistent with formation in the primitive nebula, rather than in a planetary subnebulae in which the ice/rock ratio is enhanced. Thus, it is really a "big comet" [9] and, along with Triton and Chiron, is a representative of the outer solar system planetesimals.

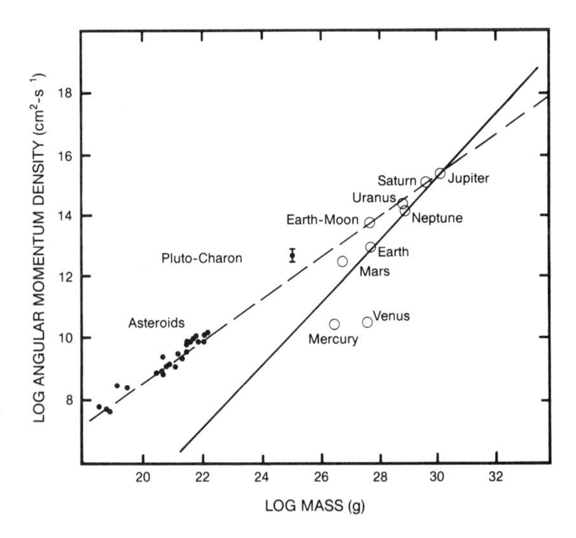

Fig. 6.14.2. ***The angular momentum of bodies in the solar system plotted against mass. Note the high values for the Pluto-Charon and Earth-Moon systems. After Taylor S. R. (1987) Amer. Sci., 75, 476, Fig. 8. Data for the Pluto-Charon pair are from McKinnon W. B. (1989) Astrophys. J. Lett., 344, L41.***

6.14.6. Origin of the Pluto-Charon Pair

Charon is about one-seventh of the mass of the primary, and is in a strange orbit. Among the most compelling evidence for a collisional origin for the pair is the high angular momentum density (Fig. 6.14.2). This is higher than the critical value for a stably rotating single body [118]. The most reasonable explanation is that the high angular momentum of the pair is due to a massive catastrophic collision. The spin axis and orbit of Pluto, as well as the orbit and different surface composition of Charon, are consistent with a collisional origin [118].

Possibly a massive impact might preferentially remove ice relative to silicate, just as on Mercury, where silicate may be removed preferentially to metal by a massive collision. This could account for the high rock/ice ratio of Pluto. A large collision could account for some loss of an icy mantle, provided the collision occurred following differentiation. Jetting during such an event, rather than vaporization, could increase the rock/ice ratio. The idea that it lost ice preferentially due to volatilization during accretion [119] seems to be ruled out because of the small size of Pluto; such effects should be widespread among other satellites, but are not observed.

The relatively low velocities in the outer solar system could account for the reaccretion of Charon and the obliquity of Pluto, while the orbital resonance with Neptune might be explained if the collision occurred near that planet [120]. Although such possibilities must engender caution about using the composition of Pluto to infer the composition of the primitive solar nebula, the similarity in density between Pluto and Triton may indicate that they do represent a fair sample of the early outer nebula.

An alternative scenario for the origin of Pluto and Charon notes that the 7/1 primary/satellite ratio is very high and could be accounted for by binary fission of a rotating molten planetesimal. It seems unlikely, however, that Charon is formed by fission from Pluto because of the chronic problems of overspinning a body, as discussed in section 6.2. There is no obvious way to overspin a body significantly and then have it fly apart [8,121].

Invoking Ockham's razor, the collisional hypothesis solves more problems connected with the Pluto-Charon pair than do the other hypotheses. Pluto and Charon thus exhibit, at the outer edge of the solar system proper, the same properties as the more

Sunward members; a complex evolution dominated by stochastic processes.

6.14.7. Carbon Budget in the Nebula

Pluto and Charon are much rockier, with a rock/ice fraction of about 0.7, compared with lower values (ranging from about 0.47-0.59 depending on the satellite and the model adopted for its interior structure) for the satellites of the outer planets [122]. This is generally interpreted to indicate that they formed within planetary subnebulae, while Pluto and Charon formed out in the nebula, and might thus shed some light on primordial nebular compositions. This view must be tempered by the realization that collisional processes may have selectively removed water ice as noted above (see section 6.2) [123]. The densities of Pluto and Charon do not allow for the presence of much carbon monoxide or methane ice, and they must be dominantly rock and water ice. Although most of the carbon in the nebula was present as gaseous carbon monoxide, carbon monoxide and methane-ices were minor phases in the nebula. Organic material probably constitutes the main solid carbon-containing phase, although carbon monoxide- and methane-clathrate hydrates may have contained a minor portion of the carbon budget [122,124,125].

6.15. THE MOON

"After the Sun, the Moon of all the heavenly bodies is that which interests us the most; its phases afford us a measure of time so remarkable that it has been primitively in use among all people" [126].

The Moon and Mercury (section 5.3) represent special cases even by the standards of the solar system. Mercury is unique due to its high density, which suggests an iron-silicate ratio about twice that of the other inner planets. In contrast, the Moon is of interest because of its bone-dry refractory composition and low density, indicative of a low iron-silicate ratio. Explanations for the peculiar nature of both bodies have a long history and much effort has been expended in attempts to fit one or both into overall schemes of planetary formation, but without conspicuous success. Recent progress in our understanding of the importance of massive impacts during the closing stages of planetary accretion enable us to understand the origins of both bodies, as well as many other features of the solar system (e.g., planetary obliquities, retrograde motion of Venus) [127].

6.15.1. Unique Features

There are many peculiar features about the Moon and the Earth-Moon system that make it unique in the solar system (lunar data are given in Table 6.15.1): (1) The orbit of the Moon about the Earth is neither in the equatorial plane of the Earth, nor in the plane of the ecliptic, but is inclined at 5.1° to the latter. (2) Relative to the Earth, the Moon has the largest mass of any satellite-planet system (1/81.3), except for the Pluto-Charon pair. (3) The Earth-Moon system is unmatched among the inner or terrestrial planets. Neither Venus, close in size to the Earth, nor Mercury have moons, while Phobos and Deimos, the two moons of Mars, are probably tiny captured asteroids with primitive compositions. The satellites of the outer planets are mostly ice-rock mixtures. (4) The Moon has a low density relative to the terrestrial planets, attributable to a low bulk iron content. (5) The U.S. Apollo and U.S.S.R. Luna missions revealed that the Moon has an unusual composition. It is bone dry, depleted in elements such as Tl, Bi, and K, which are volatile below about 1100 K, and also in siderophile elements, and is enriched in refractory elements such as Al and U. (6) The angular momentum of the Earth-Moon system is anomalously high compared to the other planets.

6.15.2. Density

The Moon has a lower bulk density (3.34 g/cm^3) than the Earth (5.54 g/cm^3). However, the lunar density is close to the uncompressed density of the terrestrial mantle, implying that the Moon is composed largely of silicate minerals. Thus, it lacks the massive amounts of metallic iron that are responsible

TABLE 6.15.1. Lunar properties.

Semimajor axis (10^3 km)	384.402
Eccentricity	0.055
Inclination to ecliptic	5°09′
Mean orbital velocity km/s	1.03
Sidereal period of revolution	27.32 days
Aphelion distance (10^6 km)	0.4055
Perihelion distance (10^6 km)	0.3633
Mass (Earth = 1)	0.0123
Equatorial radius (km)	1738
Mean density (g/cm^3)	3.343
Equatorial surface gravity (cm/s^2)	162
Rotation period, synodic	29.53 days
sidereal	27.32 days
Inclination of equator to orbit	1°32′
Surface area ($\times 10^6$ km^2)	37.9
Escape velocity (km/s)	2.37

for the high densities of the inner planets. This similarity in density between the Moon and the Earth's silicate mantle has fueled the speculation, dating back to George Darwin [128], that the Moon fissioned from the Earth's mantle following core formation.

Most of the other satellites in the solar system are mixtures of rock and water ice and have low densities. They formed in the early solar nebula out beyond a "snow line" at about 4-5 AU, where water ice was stable (Io, denser than the Moon, owes its high density to its proximity to Jupiter). The low density of the Moon suggested to Harold Urey that it was a primitive object, since the density matched that of the primitive carbonaceous chondritic (or CI) meteorites.

6.15.3. Crust

The Moon has a thick crust (Table 6.15.2), comprising about 12% of planetary volume, that formed immediately following the accretion of the Moon at about 4.4 b.y. ago. The terrestrial continental crust, in contrast, is relatively much smaller, comprising less than 0.5% of the Earth, and it has grown slowly and episodically throughout geological time. The lunar highland crust is different in composition from that of the interior and contains a large proportion of plagioclase feldspar ($CaAl_2Si_2O_8$) that is responsible for the white color of the lunar highlands (Fig. 6.15.1).

The lunar highland crust is the best-studied example of a primary crust [129]. It is 60-100 km thick on a body whose radius is only 1738 km. Although it is complex in petrographic detail, mainly because of extensive brecciation during the meteoritic bombardment, it is relatively simple geochemically. The generally accepted model proposes that it formed during the crystallization of a deep magma ocean, formed concomitantly during accretion of the Moon. Three major components are identified. A principal

TABLE 6.15.2. The composition of the Moon.

Major Elements

	Bulk Moon	Primitive Mantle (mantle + crust)	Highland Crust	Mantle Following Crust Extraction
SiO_2	43.4	44.4	45.0	44.3
TiO_2	0.3	0.31	0.56	0.28
Al_2O_3	6.0	6.14	24.6	4.1
FeO	10.7	10.9	6.6	11.4
MgO	32.0	32.7	6.8	35.6
CaO	4.5	4.6	15.8	3.36
Na_2O	0.09	0.092	0.45	0.052
K_2O	0.01	0.01	0.075	0.003
Cr_2O_3	0.60	0.61	0.06	0.67
MnO	0.15	0.15	—	—
(Fe, FeS) (core)	2.3	—	—	—
Σ	100.05	99.9	99.7	99.9

Chemical compositions expressed as wt% oxides. A lunar Fe and FeS core comprising 2.3% of the Moon is assumed to be present.

Selected Element Abundances in the Bulk Moon

Li	0.83 ppm	Fe	10.6 wt%	Gd	0.75 ppm
Be	0.18 ppm	Rb	0.28 ppm	Tb	0.14 ppm
B	0.54 ppm	Sr	30 ppm	Dy	0.93 ppm
Na	0.06 wt%	Y	5.1 ppm	Ho	0.21 ppm
Mg	19.3 wt%	Zr	14 ppm	Er	0.61 ppm
Al	3.17 wt%	Nb	1.1 ppm	Tm	0.088 ppm
Si	20.3 wt%	Cs	0.012 ppm	Yb	0.61 ppm
K	83 ppm	Ba	8.8 ppm	Lu	0.093 ppm
Ca	3.22 wt%	La	0.90 ppm	Hf	0.42 ppm
Sc	19 ppm	Ce	2.34 ppm	W	0.74 ppm
Ti	1800 ppm	Pr	0.34 ppm	Th	125 ppb
V	150 ppm	Nd	1.74 ppm	U	33 ppb
Cr	4200 ppm	Sm	0.57 ppm		
Mn	0.12 wt%	Eu	0.21 ppm		

Data from Taylor S. R. (1982) in *Planetary Science: A Lunar Perspective*, Tables 8.1 and 8.4, Lunar and Planetary Institute, Houston.

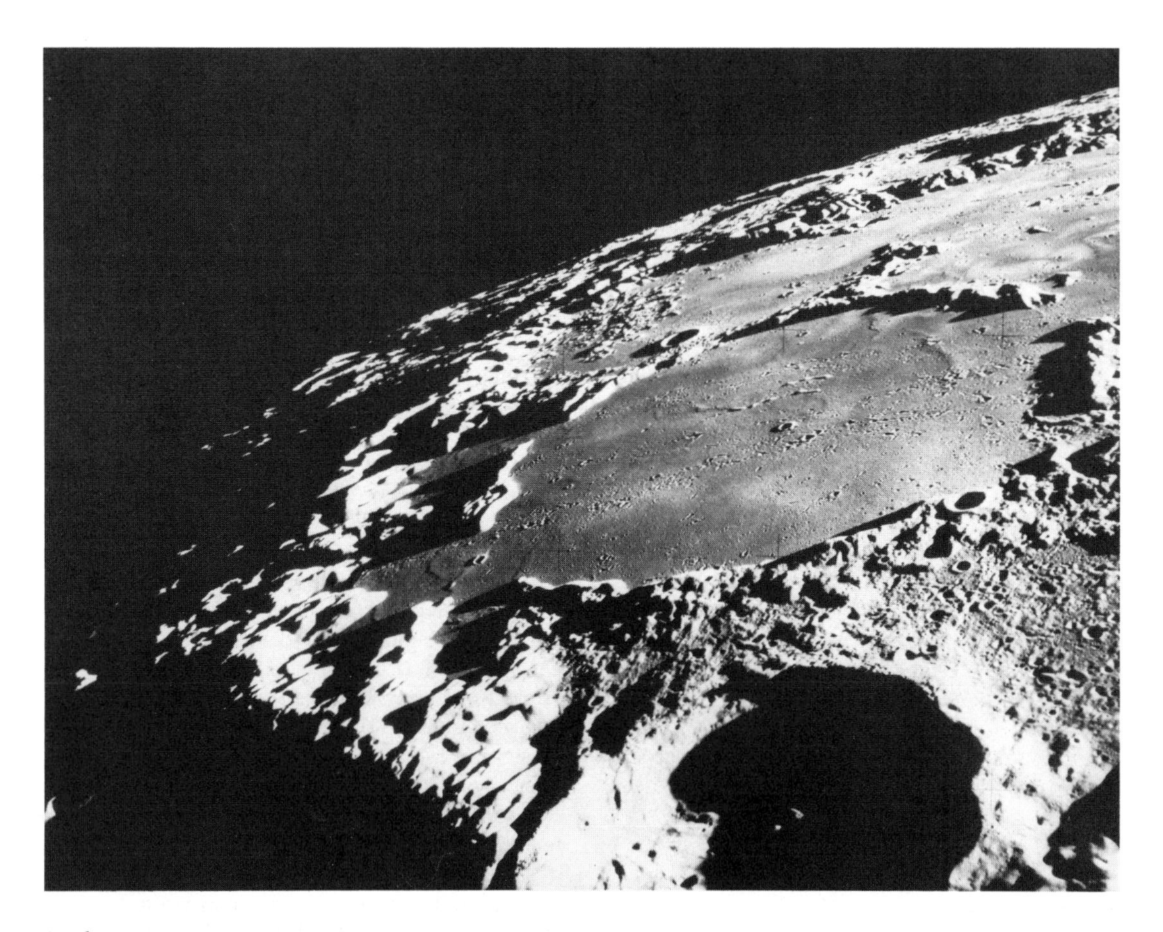

Fig. 6.15.1. *The relationship between the anorthositic lunar highlands and the basaltic lunar maria is well shown in this view of Mare Ingenii on the lunar farside. The large circular crater, filled with mare basalt, is Thomson, 112 km diameter, in the northeast sector of Mare Ingenii, 370 km diameter, centered at 34° S, 164° E. The crater in the right foreground is Zelinsky, 54 km diameter, excavated in the old highland crust. The stratigraphic sequence, from oldest to youngest, is (1) formation of white feldspathic highland crust, (2) excavation of Ingenii Basin, (3) formation of Thomson Crater, (4) formation of Zelinsky Crater, (5) flooding of Ingenii Basin and Thomson Crater with basaltic lava, and (6) production of small impact craters on the smooth mare surface, including a possible chain of secondary craters. (NASA AS15-87-11724)*

unit is composed mainly of plagioclase feldspar (ferroan anorthosite) cumulates formed by flotation in a completely dry magma ocean. The Sm-Nd closure age of 4440 ± 20 m.y. for the ferroan anorthosite 60025 [130] provides a younger limit for plagioclase crystallization and crustal formation. The final stages of solidification of the residual liquid (KREEP) from the magma ocean may have occurred by about 4350 m.y., indicating that about 90 m.y. was apparently required for total crystallization. A third component, the Mg suite, appears to be composed of plutons that intruded the feldspathic crust shortly after its formation. The petrogenesis of the Mg-suite rocks is still unclear, but they were intimately mixed with the feldspathic crust by the meteoritic bombardment. Possibly the Mg suite represents a secondary crustal component produced by impact-induced melting. The enigmatic "KREEP basalts" probably also belong in this category.

The dates for the crystallization of the crust are close to expected ages for the formation of the Moon. Planetary accretion models based on the planetesimal hypothesis require up to 10-100 m.y. from T_0 (4560 m.y.) to complete planetary assembly. Accordingly, the Moon is likely to form as a separate entity perhaps only 10-20 m.y. before the best estimate of the age of 60025. Even allowing for the large uncertainties in all these ages, it is apparent that melting of much of the Moon and development of the thick lunar highland crust proceeded immediately after the formation of the satellite. Such growth rates are essentially instantaneous on a geological timescale.

The observable structure of the lunar crust was dominated by large basin-forming impacts prior to 3800 m.y., which produced the circular mountain arcs, so puzzling to early investigators. A thin veneer (typically 1-2 km thick) of basaltic lavas (the maria), which form the familiar dark features of the "man in the Moon," covers 17% of the lunar surface, mostly on the side facing the Earth. Over 20 distinct types have been erupted, testifying to the zoned mineralogy of the lunar mantle [127]. The eruptions of mare basalt that flooded the impact-produced basins were derived by partial melting of mantle cumulates, and continued down to about 2500 m.y. These cumulates had crystallized from the magma ocean at about 4400 m.y. following extraction of the feldspathic crust (Table 6.15.2).

The highland crust averages 73 km in thickness, ranging from 64 on the nearside, to over 100 on the farside. The abundance of basalts on the visible face, in contrast to their rarity on the farside, is attributable to this difference in crustal thickness, since lavas reach the surface as a consequence of hydrostatic head.

6.15.4. Mantle

The highland crust overlies an upper mantle (see Table 6.15.2). The basaltic lavas that were erupted from this region show that it is composed of zones of varying silicate mineralogy. Following a small drop in seismic velocities between 270 and 500 km depth, the silicate mantle continues to a depth of at least 1000 km, although the interpretation of the seismic data becomes less certain with increasing depth. There is mounting evidence for the existence of a small metallic core, 300-500 km diameter, comprising 2-5% of lunar volume.

One of the more interesting questions turns on the nature of the deep lunar interior. Is it fractionated or does it consist of primitive undifferentiated material? The lunar orange and green glasses are among the most primitive lunar samples available. They possess low μ ($^{238}U/^{204}Pb$) ratios and this is often cited as evidence that they represent samples of the primitive lunar interior. However, they possess other characteristics (e.g., Eu depletion) that indicate that they come from a fractionated source. The general explanation of the low abundance of Eu relative to the other REE in mare basalts derived from the lunar interior is that Eu was depleted by prior crystallization of plagioclase feldspar, which floated to form the thick highland crust [127,131,132,133].

The Moon is depleted in volatile elements relative both to the Earth and to the primordial solar nebula. The terrestrial mantle has a μ ($^{238}U/^{204}Pb$) value of 8. The orange and green glasses have values of 20-30 [134]. The probable μ value for the bulk Moon is about 300. Early cumulates, from which the green and orange glasses are derived, will have μ values as low as 20. In later cumulates, the value will rise, reaching over 500 in the late-stage residual liquids. Feldspars, which accept Pb preferentially to U, will have very low μ values ($\sim$10), so the anorthositic lunar highland crust probably had a low μ value. Mixing in of late-stage differentiates (KREEP) will raise the value dramatically.

Magmas with very high TiO_2 contents (>16 mol%) are too dense to be erupted on the lunar surface. However, such magmas must form a very small proportion of the lunar lavas because of the low intrinsic abundance of Ti. Lavas with 10-13 mol% TiO_2 may come from depths below 400 km, based on buoyancy considerations[135]. This is in contrast to the conclusions of experimental petrology that such magmas come from depths of less than 400 km, and must cast some doubt on the assumptions involved in that technique. Although the volcanic origin for these glasses is generally accepted, the abundance of trace siderophile elements mimics a meteoritic signature and the final resolution of this problem will have to await further lunar missions [136].

6.15.5. Composition

The initial Apollo sample return from the smooth basaltic plains of Mare Tranquillitatis, preceded by hints from the Surveyor remote analyzers, found some unusual chemistry [137]. The extreme depletion of highly volatile elements (e.g., Tl), a lesser depletion of moderately volatile elements (e.g., K), and enrichment of refractory elements (e.g., Zr, U, Al) proved to be general features of lunar samples. Table 6.15.2 gives compositional data for the Moon. The following limits on bulk composition have been established from the Apollo and Luna data [138]:

1. Those elements that are volatile below about 600 K were depleted by factors of about 50 relative to the terrestrial mantle or by factors of about 200 relative to primordial solar nebular abundances. The extreme depletion of the Moon in volatiles is shown by the total absence of water even at parts-per-billion levels, a feature unique among planets (except perhaps Mercury) and their satellites.

2. The moderately volatile elements (including K and Rb), which condense in the range 600-1300 K, are depleted by a factor of about 2 relative to terrestrial abundances or by about 10 relative to primordial solar nebular abundances. The depletions of both sets of volatile elements must have occurred before the accretion of the Moon, since its gravitational field is sufficient to prevent their escape once accreted.

3. Iron and the other siderophile elements, which enter metallic phases, are depleted. Bulk FeO values in the Moon are about 13%, about one-third that of the primordial solar nebula (36% FeO). The trace siderophile elements (W, P, Co, Mo, Ni, Re, and Ir) are depleted in lunar basalts in the order of their metal-silicate partition coefficients. This is consistent with the removal of these elements into a small lunar iron core.

All the above compositional features are well established. The abundances of the refractory lithophile elements are somewhat more controversial. The bulk Moon is probably enriched in refractory elements by a factor of 1.5 compared to the terrestrial mantle, or about 2.5 times the primitive CI abundances. The geochemical arguments have been buttressed by geophysical studies that support a high-alumina Moon. A recent assessment concludes that "only in the case of extreme assumptions can critical aspects of bulk lunar composition be demonstrated to be equivalent to the present-day terrestrial mantle: specifically the Moon has an Mg# that is too low and an alumina abundance that is too high" [139]. Geochemical balance estimates require that at least half the Moon was melted in order to provide sufficient alumina for the observed highland crust [127,138]. Earlier models that restricted the depth of the magma ocean to 200 km [140] have been reevaluated to agree with the deeper geochemical estimates [141]. These refractory elements include Al, Ca, and Ti among the major elements and Th and U among the trace elements. They appear to be enriched in the Moon from the following considerations: (1) The amount of Al in the bulk Moon must be adequate to account for the thick aluminous crust (feldspar contains 36% Al_2O_3). (2) The abundance of Ca and Al in the bulk lunar composition must also be sufficiently high to allow for the early appearance of plagioclase feldspar during crystallization of the magma ocean. (3) The most recent assessments of the lunar mantle seismic velocity data favor high-alumina bulk Moon compositions (>5% Al_2O_3) in contrast to those models that match the terrestrial mantle, which contains about 3.5% [142]. (4) The U and Th content of the bulk Moon must account both for the observed near-surface high concentrations of these elements, and for the measured heat-flow. Most estimates of the U abundance in the bulk Moon exceed 30 ppb, while the terrestrial mantle content, which is tightly constrained by an interlocking set of isotopic and chemical abundance ratios (K/U, K/Rb, Rb/Sr, Sm/Nd), is less than 20 ppb.

6.15.6. Ancient Lunar Magnetic Field

The evidence for a lunar core, although indirect, has become very strong. Among the most surprising results from the Apollo samples was the demonstration of strong natural remnant magnetization, and of surface magnetic fields due to remnant magnetization of the lunar crust. This topic has caused more controversy than any other, despite strong competition from other lunar problems. Some mild degree of consensus is beginning to appear, however [143]. There is now general agreement that there is a case for a lunarwide magnetic field at least between 3.9 and 3.6 b.y. The highland samples are more strongly magnetized than the mare basalts. The maximum strength of the ancient lunar field was probably less than 0.5 oersted. The present terrestrial field strength is about 1 oersted (Fig. 6.15.2). There is a continuing debate whether the field peaks between 3.9 and 3.6 b.y. or whether there is a steady decline from 3.9 to 3.1 b.y. [143]. Since there are few samples with older ages available for measurement, and these are strongly shocked, the question must remain open.

We now turn to the origin of the field. Of the various proposals that have been suggested, most have now been shown to be implausible [144]. Impact-generated fields have become less popular, since the craters are not more magnetized than other areas. Both the solar and the terrestrial fields can be ruled out; there is no evidence for a strong solar field on an interplanetary scale as young as 3.6 b.y., while if the Moon were close enough to the Earth to be affected by the terrestrial field, a very complex interaction, which would not be homogeneous, would result.

The most likely possibility is that the field was generated internally by motions in a fluid lunar iron core. According to Fuller and Cisowski [143] a 400-km core providing a dipole moment about 100 times smaller than the terrestrial dynamo could give a field strength of about 0.5 oersted at the lunar surface. As noted elsewhere, there is strong evidence for the existence of a lunar metallic core. This would form, in current lunar models, at the same time as the Moon; cooling in such a core would be slow on account of the high-temperature origin of the Moon

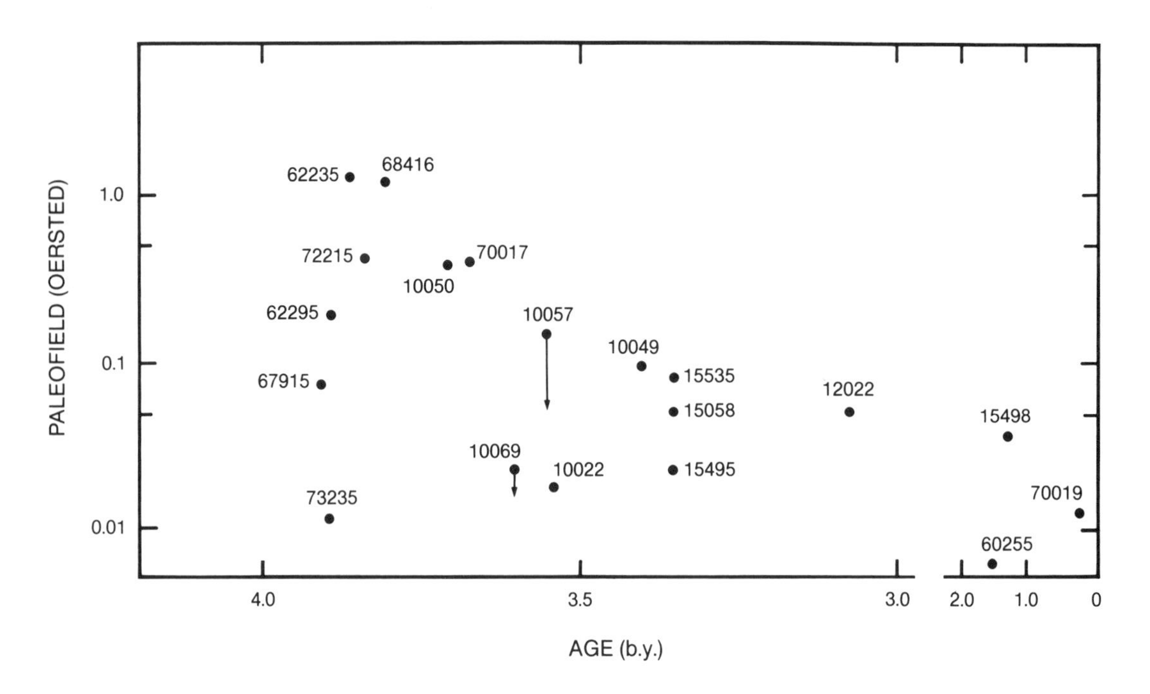

Fig. 6.15.2. *The reliable measurements of paleointensity of lunar samples, indicating a decline in the lunar magnetic paleofield from about 4 to 3 b.y. ago. After Fuller M. and Cisowski S. M. (1987) Geomagnetism 2 (J. A. Jacobs, ed.), p. 420, Fig. 52, Academic, New York.*

and the initial presence of the magma ocean. If the core solidified by about 3 b.y. this could explain the steady decrease in the field from 3.9 b.y. (Fig. 6.15.2).

The field is not necessarily stronger in the period 4.4-3.9 b.y. since the dynamo may not generate a surface field stronger than about 0.5 oersted. Alternatively, if the lunar field were driven, as on the Earth, by freezing of the core, then the apparent lower field strength before 3.9 b.y. (if not an artifact of the sampling) receives a ready explanation, and the time of core freezing is given by the maximum strength of the field between 3.9 and 3.6 b.y.

Lunar swirls (see section 4.10.2) have low albedos due to magnetic shielding of the surface from solar wind ions; accordingly, the soils in those locations have not reached maturity. The strongest magnetic surface fields are found at the antipodes of the youngest impact basins and appear to be due to some kind of shock wave focusing [145].

6.15.7. Atmosphere

The composition of the lunar atmosphere is given in Table 5.10.1. Three comments must suffice. Although all come from old sources, ranging from 1698 to 1854, they scarcely require updating: "The Moon has no air or atmosphere surrounding it as we have . . . [and I] cannot imagine how any plants or animals whose whole nourishment comes from liquid bodies, can thrive in a dry, waterless, parched soil" [146]; ". . . the lunar atmosphere, . . . if any exist, must be of an extreme rarity . . . we may conclude that no terrestrial animal could live or respire at the Moon, and that if it is inhabited, it must be by animals of another nature . . . there is reason to think that all is solid at the surface of the Moon . . . on which some have thought they perceived the effects and even the explosion of volcanoes" [147]; "In the want of water and air, the question as to whether this body is inhabited is no longer equivocal. Its surface resolves itself into a sterile and inhospitable waste, where the lichen which flourishes amidst the frosts and snows of Lapland would quickly wither and die, and no animal with a drop of blood in its veins could exist" [148].

6.16. EVOLUTION OF THE MOON

The broad features of both lunar composition and evolution are well understood and are now known much better than for the Earth. There is very little sign of any features on the lunar surface that can be ascribed to either expansion or compression. This has been used to constrain the intial depth of melting to about 200 km [149]. However, the limits of ±1 km to lunar radius apply only after the end of the massive bombardment at about 3.8 Ae, and a recent reassessment of the problem reveals that "the Moon could have been initially >50% molten (with the remainder relatively close to the solidus) and yet experienced little volume change over the last 3.8 b.y." [141]. This vast mass of molten silicate has been termed the "magma ocean" and a highly energetic and rapid mode of origin for the Moon is required to account for it. The crystallization of the magma ocean is understood in principle [e.g., 127,131,150] (Fig. 6.16.1). Feldspar was an early phase to crystallize. It floated due to the low density of the feldspar crystals and the anhydrous nature of the silicate melt, and formed a thick feldspathic crust, accounting for the depletion of europium in the lunar interior [131,132,133]. Convection during cooling may have swept "rockbergs" of feldspar together, accounting for the variations in crustal thickness. A small lunar iron core about 2-5% by volume formed in the center. This sequestered the siderophile elements from the melted portion of the Moon. The lunar mantle was fully crystallized by about 4.4 b.y. ago, producing a zoned silicate mineralogy, from which the mare basalts were derived much later by partial melting. This cumulate hypothesis for the source region of mare basalts is well established [e.g., 131,150] and is reinforced by fresh evidence [e.g., 151].

As the silicate minerals crystallized, those trace elements that were excluded from their crystal lattices were concentrated in the residual melt fluids. The final stage of magma ocean evolution was the intrusion of this residual liquid into the feldspathic highland crust. The fluid was enriched in elements such as K, REE (rare earth elements), P (from which the acronym KREEP has been coined), Th, U, Zr, Hf, and Nb, and is responsible for the extraordinary near-surface abundance of elements such as K, U, Th, and REE, which may be concentrated by factors of several hundred relative to bulk Moon or primitive nebular values. It pervaded the crust, with which it was intimately mixed by the continuing meteoritic bombardment. The final event in crustal evolution was the intrusion of an Mg-rich suite of rocks, produced either by melting of the already fractionated lunar interior, or perhaps more likely by subcrustal melting induced by the impacts of giant planetesimals.

Two possible alternatives for the initial depth of the magma ocean are a depth of 500 km (about half lunar volume) or total melting of the Moon. In the first case, the deeper parts of the Moon are composed of undifferentiated material of bulk lunar composition; in the second case, they will consist of the earliest phases to crystallize from the magma ocean (Mg-rich olivine and orthopyroxene cumulates). The seismic velocity data do not clearly distinguish between these two models. However, bulk Moon models that contain more than 5% Al_2O_3 provide the best match to the seismic velocity profile, implying that the Moon is enriched in refractory elements relative to both the Earth and primitive solar nebular levels [139,142].

6.17. HYPOTHESES OF LUNAR ORIGIN

"Some partizans [sic] of final causes have imagined that the Moon was given to the Earth to afford it light in the absence of the Sun" [152].

The origin and evolution of the Moon turns out to have been an interesting intellectual exercise. Attempts to explain complex natural phenomena call for a wide variety of skills and the Moon presented a particularly enigmatic example. Preconceived notions and prejudices frequently overrode both data and common sense; there is a close analogy with the question of the origin of tektites, a problem equally efficient at sorting sheep from goats. For this reason, some discussion is given here of earlier views, as a cautionary tale, to illustrate the flaws in scientific philosophy involved, the heroic attempts to force data to fit favored theories, and the nonuniqueness of the procedures used to match lunar and terrestrial compositions [153].

Tidal calculations have often been used to assess the history of the lunar orbit [154], but attempts to determine whether the Moon was once very much closer to the Earth [for example, near the Roche limit (≈18,000 km)], which would place significant constraints on lunar origins, produce nonunique solutions [155]. Work on tidal sequences in South Australia has shown that in the late Precambrian (650 m.y.) the year had 13.1 ± 0.5 months and 400 ± 20 days. The mean lunar distance was 58.4 ± 1.0 Earth radii, so during the Proterozoic, the Moon was only marginally closer to the Earth [156].

The major models for the origin of the Moon can be grouped into five separate categories that include (1) capture from an independent orbit, (2) fission from a rapidly rotating Earth, (3) formation as a double planet, (4) disintegration of incoming planetesimals, and (5) Earth impact by a Mars-sized planetesimal.

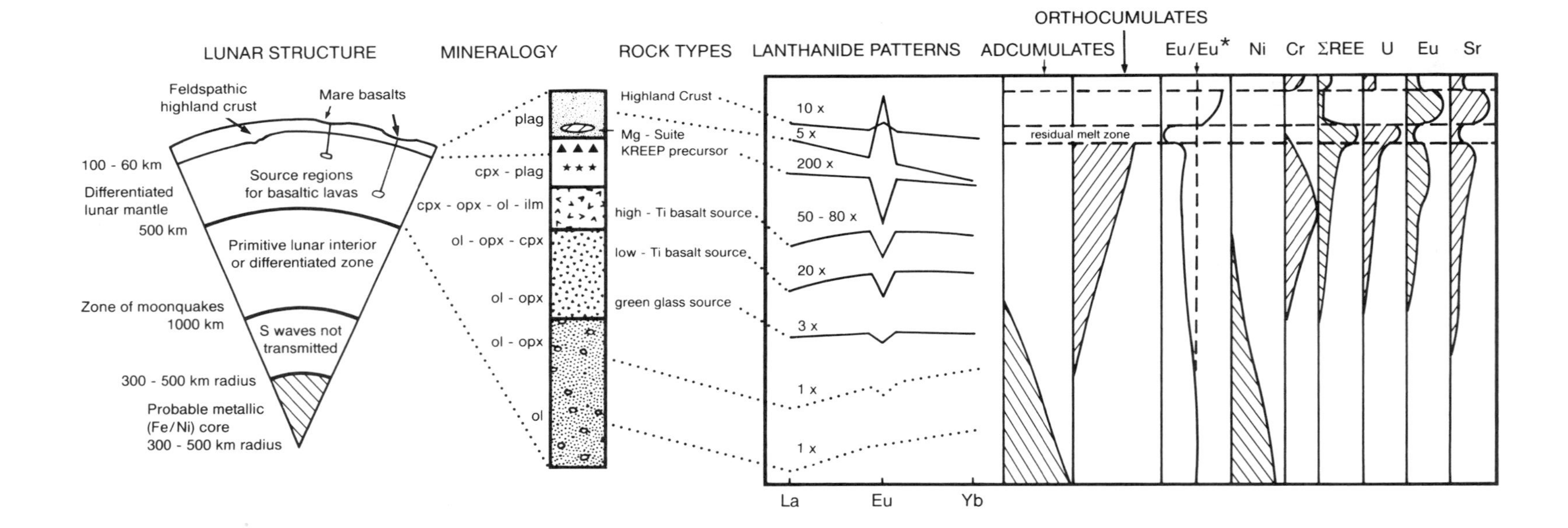

Fig. 6.16.1. *A schematic representation of the crystallization of the lunar magma ocean at about 4.4 b.y. ago (this age is given from the model ages of mare source regions). This idealized section will be modified by some sinking of dense cumulates and convective overturn. Rare earth element patterns of mare source regions and the distribution of some key elements with depth are shown on the right.*

These are not all mutually exclusive and elements of some hypotheses occur in others. The most recent comprehensive survey is by Wood [157]. Hypotheses in which the Earth captures an already formed Moon have generally been abandoned. They have severe inherent dynamical problems and provide no obvious explanation for the exotic lunar geochemistry. Fission hypotheses that derive the material for the Moon from the Earth's mantle encounter two basic difficulties: (1) the angular momentum of the Earth-Moon system, although large, is insufficient by a factor of about 4 to allow for rotational fission [158]; and (2) there is no acceptable mechanism for removing this excess angular momentum following lunar formation. Despite the prediction that the chemistry of the Moon should bear some recognizable signature of the terrestrial mantle, the Moon contains, for example, 50% more iron, and has distinctly different trace siderophile element signatures [e.g., 121,140,159] (Fig. 6.17.1). It also contains lower amounts of volatile elements and higher concentrations of refractory elements. The best estimates for the abundances of the refractory elements (e.g., Al, U) in the Moon exceed those in the terrestrial mantle by 50%. Even though mechanisms for depleting the volatile elements may be incorporated into fission models, the enhancement of Fe and the refractory elements in the Moon cannot be accomplished by such means. The hypothesis thus fails the test of requiring an identifiable chemical signature of the terrestrial mantle in the Moon, despite heroic attempts by geochemists to fit the lunar compositional data to that of the silicate mantle of the Earth [160]. The observed depletion patterns of siderophile elements in the Earth and Moon, relative to chondrites, and plotted as a function of their metal/silicate partition coefficients, are shown in Figs. 5.5.4 and 6.17.1. The Earth and Moon have distinctly different siderophile element patterns. Claims that the siderophile elements in the Moon show a "unique terrestrial signature" have been shown to be erroneous [161]. The similarity in V, Cr, and Mn abundances in the Earth and Moon, sometimes claimed as definitive proof of lunar origin from the terrestrial mantle [162], has been shown to be nonunique (CM, CO, and CV chondrites show the same pattern), and to be due most likely to volatile depletion in precursor planetesimals [163].

Double-planet models that form the Moon and Earth in association possess the twin difficulties of failing to account for the angular momentum of the Earth-Moon system and of readily accounting for the density difference, although Ruskol [164] argued for the formation of a ring of metal-poor silicate debris

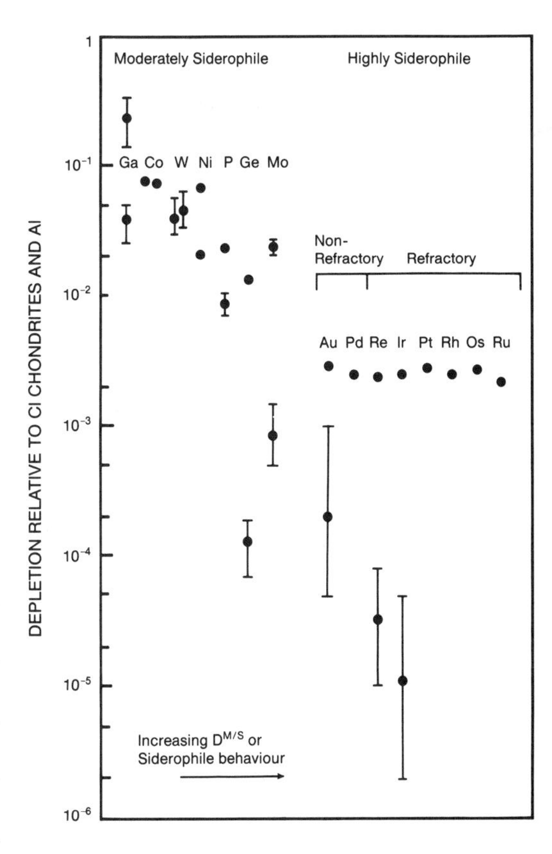

Fig. 6.17.1. *The depletion of the siderophile elements in the Earth and the Moon, relative to CI abundances (normalized to refractory Al), plotted in order of increasing metal/silicate partition coefficients. The geater depletion of Ga and Mo in the Moon is due the fact that they are also volatile elements. Courtesy H. E. Newsom, Univ. of New Mexico.*

from which the Moon could accumulate. They do, however, account for the similarity in oxygen isotopes between Earth and Moon. The eucrite class of basaltic meteorites, formed close to T_0, shows many similarities in composition to lunar basalts and must come from a small Moonlike asteroid (possibly 4 Vesta). However, the difficulties of selectively accreting these rare objects into a circumterrestrial swarm during Earth accretion seem insurmountable.

Several models involve the break-up of differentiated planetesimals as they come within the Roche limit [165]. This process is postulated to result in a circumterrestrial ring of broken-up silicate debris, while the more coherent metallic cores of the either accrete to the Earth or escape. Such scenarios do not solve the angular momentum problem. Moreover, the proposed break-up of the incoming planetesimals may be difficult [155,166]. The previous two hypotheses envisage processes that might be thought to be rather general features of planetary accretion and lead to the presence of Moonlike satellites everywhere. Thus, they fail to account for the unique nature of the Earth-Moon system. Uncommon objects, like our Moon, may require an uncommon origin.

6.18. LARGE IMPACT MODEL FOR THE ORIGIN OF THE MOON

In several respects, the Moon and the Earth-Moon system appear to be unique among the inner or terrestrial planets. Although the dynamical aspects of the problem have been known for a long time, the unusual chemistry was only revealed from the samples returned by the U.S. Apollo and U.S.S.R. Luna missions. The integration of this information into a model for the origin of the Moon to satisfy all the constraints has proven difficult, but something approaching a consensus is now being reached [167].

6.18.1. Angular Momentum

The Earth-Moon system has an anomalously high angular momentum compared to other planets and this has been the rock on which most hypotheses of lunar origin have foundered (Fig. 6.14.2). This excess cannot arise through random small multiple impacts since these must average out. Less than 10% excess angular momentum could arise by such means. However, one very large glancing impact could account for the observed excess [168]. The model proposes that a body larger than Mars (0.14 Earth masses or greater), struck the Earth and a hot disk of material was ejected to form the Moon [169]. This "Mars-sized impactor" hypothesis thus solves the angular momentum problem by definition. The origin of the Moon, as proposed by the giant impactor hypothesis, involves collision with the Earth by the next largest body in the hierarchy (0.1-0.2 Earth masses) and requires prior core formation in the impactor. The Moon is derived from the mantle of the impactor, not that of the Earth, accounting for the significant geochemical differences between the two bodies [170]. The energetic nature of the event provides not only the lunar magma ocean, required by the geochemical evidence, but also induces terrestrial mantle melting [171]. Computer simulations indicate that the material that makes up the Moon comes primarily from the mantle of the impacting body, not from the Earth. The hypothesis thus resolves the geochemical dilemma that the composition of the Moon does not match that of the Earth's mantle by cutting this particular Gordian Knot [172]. An additional advantage accrues since the material that escapes to form the Moon is mostly in liquid or vapor phases accounting for the extreme depletion in volatile elements in the Moon [152, 155,173] Whether this material immediately coalesces into the Moon (within a day or so) or into several smaller lumps that accrete to form the Moon on timescales of 100 years is not clear. The latter scenario would allow for a partially molten Moon, more in accordance with the geochemical evidence that the magma ocean depth was about 500 km (half lunar volume), and for the absence of evidence for extreme temperatures in the protolunar material [174].

Computer simulations on variants of the collisional hypothesis indicate that clumps of molten rock may form and not undergo vaporization [175]. Recondensation can provide not only for the extreme lunar depletion in volatile elements, but also for the enrichment in refractory elements. The large impactor hypothesis also provides for a highly energetic origin for the Moon, and so can account for an initially mainly molten Moon, as required by the geochemical and isotopic evidence for the magma ocean. It appears to be the only current hypothesis that accounts for both the dynamical and geochemical peculiarities of the Moon, and thus warrants further examination.

6.18.2. Large Impactors?

A significant question in the present context is the relative sizes of the planetesimals that, in the current scenario, accreted to form the terrestrial planets. In this scene, large precursor objects were relatively common. Prior to the final sweep-up into the 4 terrestrial planets, Wetherill [176] calculates that 100 objects of lunar mass, 10 with masses exceeding that of Mercury, and several exceeding the mass of Mars should form. He further estimates that perhaps one-third of these objects struck the Earth, which would provide a total of 50-75% of present Earth mass. These bodies may already have been melted and formed metallic cores and silicate mantles, particularly if they were Sunward of the asteroid belt.

As discussed in earlier chapters, other evidence supports the concept of the existence of large precursor bodies during the planetary accretion stage. Giant impacts provide explanations for a variety of other solar system phenomena, such as the retrograde motion of Venus, the absence of a venusian moon, the obliquities of the planets, including the inclination of the axial plane of the Earth to the plane of the ecliptic, and the nonequatorial lunar orbit.

Large bodies capable of producing these effects seem to be required to account for this observational evidence in the solar system.

6.18.3. Chemical Effects

How does the composition of the Moon relate to possible compositions produced during the impact of a Mars- or larger-sized projectile on the Earth? Can limits be placed on the temperatures to which the protolunar material has been subjected? The bulk lunar composition can be separated into groups of elements on the basis of relative volatility [138].

1. The super-refractory elements Zr, Hf, Y, and Sc do not appear to be fractionated relative to the other refractory elements. This contrasts with their behavior in CAI refractory inclusions and in some meteorite minerals (e.g., hibonite) where they are separated from the less refractory elements such as the REE.

2. The refractory elements, which condense at 10^{-3} atm down to temperatures of 1200 K under nebular conditions [177], include an important group of elements such as Ca, Al, and Ti among the major elements, and REE, U, Th, Ba, and Sr among the trace lithophile elements. The siderophile elements Ni, Os, Ir, and Pd that condense in this temperature range are depleted in the lunar mantle due to their siderophile character. Their behavior during the collisional process is difficult to assess. First, they will have undergone extraction into the core of the impactor and the Earth. Second, those that were accreted to the Moon have been sequestered into a lunar core. Among the refractory elements in this temperature range, there is no sign of loss of the more volatile species. These include Ce, Eu, and Yb among the REE (Eu anomalies are of course common in lunar rocks, but are due to crystal-liquid fractionation during internal differentiation processes). Strontium and Ba are the most volatile of the lithophile refractory trace elements, but there is no compelling reason to suppose they are depleted relative to the other refractory elements. It thus appears that all elements with condensation temperatures above 1200 K (at 10^{-3} atm) have condensed in their cosmic abundance proportions. Their enrichment relative to CI abundances is by a factor of 2.5.

3. The moderately volatile elements are depleted in the Moon by rather small factors. Potassium and Rb are depleted by a factor of 2 relative to terrestrial abundances, and by over an order of magnitude relative to CI abundances. This interesting observation illustrates the magnitude of the initial nebular depletion, assuming that the volatile loss in the Earth-precursor planetesimals is typical of the inner nebula. This assumption is supported by the rather uniform K/U ratios for Earth, Venus, and Mars. Manganese is also a moderately volatile element, showing little difference in abundance between Earth and Moon, but is depleted to about 40% of its CI abundance. Clearly these elements were not heavily depleted.

If we assume an impactor with a complement of these elements not grossly different from that of the Earth, then these elements were less depleted by the collisional process than by the initial separation in the nebula at T_0. This conclusion is reinforced by the very low value of the lunar initial $^{87}Sr/^{86}Sr$ ratio (LUNI) noted earlier. Cesium/rubidium ratios are higher in the Moon than in the Earth, contrary to expectations based on relative volatility [178] and may be a feature of the impactor chemistry. Suggestions that Cs was enriched on the Moon by a late addition of CI material [179] clearly fail since many other volatile elements would have been added by such an event [180]. These elements, however, as noted above, are extremely depleted on the Moon. Since trace siderophile elements in the upper mantle of the Earth have relative CI ratios, such a mechanism for adding them to the Earth is often appealed to [181]. However, the attractions of plastering on a late veneer of siderophile and volatile-rich CI material are diminished somewhat by the absence of any evidence for such an enriched layer on the Moon. If such an event occurred on the Earth, either it predated lunar formation or the Moon was remote enough to escape a similar influx. This conclusion adds further complexity to attempts to match lunar and terrestrial mantle siderophile element patterns.

4. The highly volatile elements, including Bi, Tl, Cd, Br, Se, Te, and In, condense below about 800 K (at 10^{-3} atm). They are strongly depleted in the Moon by factors of about 50 relative to terrestrial mantle abundances and by factors of several hundred relative to CI. On the Earth these elements are strongly depleted, by factors of 10-30, relative to CI. Assuming that the impactor had similar levels of depletion for these elements, then a further massive loss of highly volatile elements occurred due to the collisional processes.

5. Noble gases have not been identified as indigenous on the Moon. Those gases that are present in the returned samples are either derived from the solar wind or originate from less volatile radiogenic parents [182,183]. This failure to identify any indigenous lunar noble gases is a first-order observation, implying that there was some additional high-temperature processing of lunar precursor material in addition to that experienced by the terrestrial planets. This implies unique conditions for lunar forma-

tion, distinct from terrestrial planetary accretion models.

In summary, the Moon is heavily depleted in the most volatile elements. It is this feature of lunar geochemistry that makes the Moon unique among satellites and planets. No other satellite, even Io, would appear to be so depleted in volatiles. The Moon has lost the moderately volatile elements by smaller factors, and is enriched in the refractory elements. Since depletion of volatile elements only enriches the residual elements by a few percent, the enrichment of the refractory elements over the CI abundances by a factor exceeding 2 is most readily explained by a condensation mechanism. In the case of the Moon, elements with condensation temperatures in excess of about 1200 K (at 10^{-3} atm) must have recondensed without fractionation.

6.18.4. Impactor Composition

A critical feature of the large impactor hypothesis is that most of the material making the Moon comes from the silicate mantle of the impactor. Thus, its composition is the most significant component for the geochemical constitution of the Moon. The impactor composition is essentially a free parameter, within the constraint that it must have formed in the same part of the solar nebula as the Earth to account for the similarity in oxygen isotopes between Earth and Moon.

Current models assume that the impactor formed a metallic Fe core before the impact occurred, in order to account for the lunar siderophile element abundances and the lunar depletion in iron (13% FeO) relative to the primordial solar nebula volatile-free abundance level (36%). Thus, we envisage that this particular Mars-sized object had already differentiated into a metallic core and silicate mantle before the impact occurred. The abundance of FeO in the mantle of the impactor must, however, have been higher than that in the terrestrial mantle (8% FeO), to account for the higher abundance (13%) of FeO in the bulk Moon. Indeed, since FeO is more volatile than the other major oxides, this value may represent a minimum. The impactor must have been depleted in the volatile element Rb relative to refractory Sr at an earlier stage of nebular history to account for the primitive initial $^{87}Sr/^{86}Sr$ ratios observed in the lunar highland samples [184]. These values (LUNI) are essentially equivalent to those observed in basaltic achondrites (BABI). Since lower ratios are known [185], some precursor history is allowable, but upper limits of perhaps 50 m.y. are placed by these constraints between the initial separation of Rb from Sr (here equated with volatile-element and gas depletion at T_0) and the isolation of the lunar material. Thus, the impactor composition must have shared the depletion in volatile elements that appears to be characteristic of the inner planets.

There is no requirement to postulate an enrichment of refractory elements in the mantle of the impactor over primitive solar nebular values or terrestrial levels, since the lunar enrichment in these elements can be adequately accounted for by the vaporization and recondensation predicted during the impact itself. The only limits we can set at present are that the impactor was higher in iron, and possibly more depleted in Rb relative to Sr than the terrestrial mantle. Much of the impactor, including the core, finishes up being accreted to the Earth's mantle. The effect of this event on terrestrial mantle compositions remains to be evaluated. However, the addition of the impactor mantle, comprising at most 10%, to the mass of the terrestrial mantle seems unlikely to alter the major-element composition significantly, since the overall composition will be generally similar to that of the inner planets. A further consequence is that total melting of the terrestrial mantle occurs.

6.18.5. A Unique Event

Clearly the Moon has a composition that cannot be made by any single-stage process from the material of the primordial solar nebula. The Moon is "bone-dry," being heavily depleted in the most volatile elements. It has lost the moderately volatile elements by smaller factors, is depleted in iron, and is enriched in the refractory elements. These features of lunar geochemistry make the Moon unique among satellites. The compositional differences from that of the primitive solar nebula, the Earth, from Phobos and Deimos, and from the satellites of the outer planets thus call for a distinctive mode of origin.

Unique events are difficult to accommodate in most scientific disciplines. The solar system, however, is not uniform. All nine planets (even such apparent twins as the Earth and Venus) and over 60 satellites are different in detail from one another. All this diversity makes the occurrence of single events more probable in the early stages of solar system history. Consequently, a giant collision with the Earth, at the right angle and velocity to project a jet of hot material from the impactor into circumterrestrial orbit and produce a unique bone-dry refractory satellite with curious orbital properties, becomes a reasonable possibility.

NOTES AND REFERENCES

1. Stevenson D. J. (1989) in *The Formation and Evolution of Planetary Systems* (H. A. Weaver and L. Danly, eds.), p. 82, Cambridge Univ., New York.
2. Regular satellites, if far away from the parent planet, may possess orbits inclined to the equatorial plane due to the effect of solar torques. Thus, Iapetus, 60 saturnian radii from the planet, is probably a regular satellite, even though its orbit has a relatively high inclination; see Ward W. R. (1981) *Icarus, 46,* 97.
3. Stevenson D. J. et al. (1986) in *Satellites* (J. A. Burns and M. S. Matthews, eds.), p. 54, Univ. of Arizona, Tucson.
4. Outside the *Hill sphere,* the spherical region within which the gravitational attraction of the planet is dominant over other forces.
5. See extended discussion by Stevenson D. J. et al. (1986) in *Satellites* (J. A. Burns and M. S. Matthews, eds.), p. 60, Univ. of Arizona, Tucson.
6. The most recent review is by Weidenschilling S. J. et al. (1986) in *Origin of the Moon* (W. K. Hartmann et al., eds.), p. 731, Lunar and Planetary Institute, Houston, following from earlier work by Ruskol El. Y. (1977) *NASA SP-370,* p. 815.
7. Boss A. P. (1986) *Science, 231,* 341; Boss A. P. and Peale S. J. (1986) in *Origin of the Moon* (W. K. Hartmann et al., eds.), p. 609, Lunar and Planetary Institute, Houston; Boss A. P. and Benz W. (1989) *Bull. Am. Astron. Soc., 21,* 915.
8. Durisen R. H. and Scott E. H. (1984) *Icarus, 58,* 153.
9. W. B. McKinnon, personal communication, 1990.
10. Benz W. and Cameron A. G. W. (1989) *Bull. Am. Astron. Soc., 21,* 916.
11. See McKinnon W. B. and Parmentier E. M. (1986) in *Satellites* (J. A. Burns and M. S. Matthews, eds.), p. 720, Univ. of Arizona, Tucson.
12. Prinn R. G. and Fegley B. Jr. (1989) in *Origin and Evolution of Planetary and Satellite Atmospheres* (S. Atreya et al., eds.), p. 78, Univ. of Arizona, Tucson; Fegley B. Jr. and Prinn R. G. (1989) in *The Formation and Evolution of Planetary Systems* (H. A. Weaver and L. Danly, eds.), p. 171, Cambridge Univ., New York; see comments by Stevenson D. J. (1989) in *The Formation and Evolution of Planetary Systems* (H. A. Weaver and L. Danly, eds.), p. 75, Cambridge Univ., New York.
13. Prinn R. G. and Fegley B. Jr. (1989), ibid.
14. Laplace P. S. (1809) *The System of the World, Vol. 1, Book I* (J. Pond, trans.), pp. 88-89, R. Phillips, London.
15. A basic reference is Greenberg R. and Brahic A. (1984) *Planetary Rings,* Univ. of Arizona, Tucson, 784 pp. A readable account is given by Elliot J. and Kerr R. (1984) *Rings: Discoveries from Galileo to Voyager,* MIT, Cambridge, 209 pp.
16. Maxwell J. C. (1890) in *The Scientific Papers of James Clerk Maxwell, Vol. 1* (W. D. Niven, ed.), p. 288, Cambridge Univ., New York; Maxwell J. C. (1983) *Maxwell on Saturn's Rings* (S. G. Brush et al., eds.) MIT, Cambridge, 199 pp.; Jeffreys H. (1947) *Mon. Not. R. Astron. Soc., 107,* 263.
17. Weidenschilling S. J. et al. (1984) in *Rings* (R. Greenberg and A. Brahic, eds.), p. 367, Univ. of Arizona, Tucson.
18. Brophy T. G. and Esposito L. W. (1989) *Icarus, 78,* 181; Borderies N. et al. (1989) *Icarus, 80,* 344.
19. Cuzzi J. N. (1985) *Icarus, 63,* 315.
20. Gaffey M. J. (1976) *J. Geophys. Res., 81,* 905.
21. Smith B. A. et al. (1989) *Science, 246,* 1436.
22. Esposito L. W. and Colwell J. E. (1989) *Nature, 339,* 605.
23. Harris A. W. (1984) in *Rings* (R. Greenberg and A. Brahic, eds.), pp. 656-657, Univ. of Arizona, Tucson.
24. Stone E. C. and Miner E. D. (1989) *Science, 246,* 1417; Smith B. A. (1989) *Science, 246,* 1422; Lane A. L. et al. (1989) *Science, 246,* 1453.
25. Goldreich P. and Tremaine S. (1982) *Annu. Rev. Astron. Astrophys., 20,* 249.
26. Duxbury T. C. and Ocampo A. C. (1988) *Icarus, 76,* 160.
27. Greenberg R. and Brahic A. (1984) in *Rings* (R. Greenberg and A. Brahic, eds.), p. 4, Univ. of Arizona, Tucson.
28. It is generally suggested that, like sheep, ring particles need shepherds to prevent them straying. Sheepdogs are much to be preferred for such tasks, as every New Zealander knows.
29. A perceptive review of the problem is given by Harris A. W. (1984) in *Rings* (R. Greenberg and A. Brahic, eds.), p. 656, Univ. of Arizona, Tucson.
30. Lissauer J. W. et al. (1988) in *Lunar and Planetary Science XIX,* p. 683, Lunar and Planetary Institute, Houston.
31. Smith B. A. et al. (1982) *Science, 215,* 504.
32. This scenario is strongly disputed by Lissauer J. W. (1988) *J. Geophys. Res., 93,* 13776.
33. McKinnon W. B. et al. (1991) in *Uranus,* Univ. of Arizona, Tucson, in press.
34. Burns J. A. (1986) in *Satellites* (J. A. Burns and M. S. Matthews, eds.), p. 16, Univ. of Arizona, Tucson.
35. Burns J. A., ibid., p. 17.
36. Stevenson D. J. (1986) in *Satellites* (J. A. Burns and M. S. Matthews, eds.), p. 83, Univ. of Arizona, Tucson.

37. Everhart E. (1979) in *Asteroids* (T. Gehrels, ed.), p. 288, Univ. of Arizona, Tucson.
38. Stevenson D. J. et al. (1986) in *Satellites* (J. A. Burns and M. S. Matthews, eds.), p. 42, Univ. of Arizona, Tucson.
39. Laplace P. S. (1809) *The System of the World, Vol.1, Book I* (J. Pond, trans.), p. 81, R. Phillips, London.
40. Lunine J. I. and Stevenson D. J. (1982) *Icarus, 52,* 14.
41. Smith B. A. et al. (1981) *Science, 212,* 163.
42. Water ice condenses at 160 K at nebular pressures; B. Fegley Jr., personal communication, 1988.
43. McSween H. Y. and Weissman P. R. (1989) *Geochim. Cosmochim. Acta, 53,* 3263.
44. See for example Mueller S. and McKinnon W. B. (1988) *Icarus, 76,* 437.
45. For example, Wood J. A. and Morfill G. E. (1988) in *Meteorites and the Early Solar System* (J. F. Kerridge and M. S. Matthews, eds.), p. 329, Univ. of Arizona, Tucson.
46. Stevenson D. J. (1982) *Nature, 298,* 142.
47. Croft S. K. et al. (1988) *Icarus, 73,* 279.
48. Ward W. R. (1981) *Icarus, 46,* 97.
49. Hartmann W. K. (1987) *Icarus, 71,* 57.
50. Stevenson D. J. (1986) in *Satellites* (J. A. Burns and M. S. Matthews, eds.), p. 83, Univ. of Arizona, Tucson.
51. Sagdeev R. Z. and Zakharov A. V. (1989) *Nature, 341,* 581.
52. Bibring J.-P. et al. (1989) *Nature, 341,* 591. Deimos shows a similar dry surface from ground-based spectral reflectance data; Bell J. F. et al. (1989) in *Lunar and Planetary Science XX,* p. 58, Lunar and Planetary Institute, Houston.
53. Sinclair A. T. (1989) *Astron. Astrophys., 220,* 321.
54. V. L. Sharpton (personal communication, 1990) has suggested nudging Phobos into collision with Mars, so that we can observe the formation of a ringed basin in "real time."
55. Nash D. B. et al. (1986) in *Satellites* (J. A. Burns and M. S. Matthews, eds.), p. 632, Table 1, and p. 660, Fig. 14, Univ. of Arizona, Tucson.
56. Nash D. B. et al. (1986) in *Satellites* (J. A. Burns and M. S. Matthews, eds.), p. 630, Univ. of Arizona, Tucson; see also McKinnon W. B. (1987) *Rev. Geophys., 25,* 260.
57. Peale S. J. et al. (1979) *Science, 203,* 892.
58. Cassen P. et al. (1982) *Satellites of Jupiter* (D. Morrison, ed.), p. 93, Univ. of Arizona, Tucson.
59. McEwen A. S. et al. (1985) *J. Geophys. Res., 90,* 12345, gives a value of 8×10^{13} W for the thermal flux.
60. Johnson T. V. and Soderblom L. A. (1982) in *Satellites of Jupiter* (D. Morrison, ed.), p. 634, Univ. of Arizona, Tucson.
61. McEwen A. S. and Soderblom L. A. (1983) *Icarus, 55,* 191.
62. Nash D. B. et al. (1986) in *Satellites* (J. A. Burns and M. S. Matthews, eds.), p. 640, Univ. of Arizona, Tucson; see also summary by McKinnon W. B. (1987) *Rev. Geophys., 25,* 260.
63. See Nash D. B. et al. (1986) in *Satellites* (J. A. Burns and M. S. Matthews, eds.), pp. 686-687, Univ. of Arizona, Tucson. Hopefully the Galileo mission will resolve some of these questions. Imaging of Io's surface at 2-km resolution, with some 100-m resolution, will be undertaken.
64. A basic reference is Nash D. B. et al. (1986) in *Satellites* (J. A. Burns and M. S. Matthews, eds.), p. 629, Univ. of Arizona, Tucson.
65. For example, McKinnon W. B. and Parmentier E. M. (1986) in *Satellites* (J. A. Burns and M. S. Matthews, eds.), p. 718, Univ. of Arizona, Tucson; Shoemaker E. M. et al. (1982) in *Satellites of Jupiter* (D. Morrison, ed.), p. 435, Univ. of Arizona, Tucson; Murchie S. L. and Head J. W. (1988) *J. Geophys. Res., 93,* 8795.
66. Squyres S. W. (1980) *Geophys. Res. Lett., 7,* 593
67. McKinnon W. B. (1981) *Proc. Lunar Planet. Sci. 12B,* p. 1585; Golombek M. P. (1982) *Proc. Lunar Planet. Sci. Conf. 13th,* in *J. Geophys. Res., 87,* A77.
68. Kirk R. L. and Stevenson D. J. (1987) *Icarus, 69,* 91.
69. Kirk R. L. and Stevenson D. J., ibid., 126.
70. See discussion by Mueller S. and McKinnon W. B. (1988) *Icarus, 76,* 437.
71. Ojakangas G. W. and Stevenson D. J. (1989) *Icarus, 81,* 220; see also Lunine J. I. and Stevenson D. J. (1982) *Icarus, 52,* 14.
72. Laplace P. S. (1809) *The System of the World, Vol. 1, Book I* (J. Pond, trans.), p. 93, R. Phillips, London.
73. See discussion by McKinnon W. B. et al. (1991) in *Uranus,* Univ. of Arizona, Tucson, in press; see also section 4.13.
74. Lissauer J. W. (1988) *J. Geophys. Res., 93,* 13776.
75. Smith B. A. et al. (1982) *Science, 215,* 504.
76. Detailed accounts may be found in Morrison D. et al. (1986) in *Satellites* (J. A. Burns and M. S. Matthews, eds.), p. 764, Univ. of Arizona, Tucson; McKinnon W. B. (1985) in *Ices in the Solar System* (J. Klinger et al., ed.), p. 829, Reidel, Boston.
77. Yoder C. F. et al. (1989) *Astron. J., 98,* 1875.
78. Wisdom J. (1987) *Icarus, 72,* 241.
79. Farinella P. (1990) *Icarus, 83,* 186.

80. Lunine J. I. et al. (1983) *Science, 222,* 1229; Lunine J. I. (1989) *Icarus, 81,* 1; Lunine J. I. et al. (1989) in *Origin and Evolution of Planetary and Satellite Atmospheres* (S. Atreya et al., eds.), p. 605, Univ. of Arizona, Tucson.
81. Owen T. (1982) *Planet. Space Sci., 30,* 833.
82. Prinn R. G. and Fegley B. Jr. (1989) in *Origin and Evolution of Planetary and Satellite Atmospheres* (S. Atreya et al., eds.), p. 78, Univ. of Arizona, Tucson.
83. Lunine J. I. (1989) *Icarus, 81,* 1,
84. Lunine J. I. and Stevenson D. J. (1985) *Astrophys. J. Suppl. Ser., 58,* 493; Prinn R. G. and Fegley B. Jr. (1989) in *Origin and Evolution of Planetary and Satellite Atmospheres* (S. Atreya et al., eds.), p. 78, Univ. of Arizona, Tucson.
85. Laplace P. S. (1809) *The System of the World, Vol. 1, Book I* (J. Pond, trans.), p. 96, R. Phillips, London.
86. See *Uranus,* Univ. of Arizona, Tucson, in press, especially the discussion by McKinnon et al.
87. Lunine J. I. and Stevenson D. J. (1985) *Astrophys. J. Suppl. Ser., 58,* 493; Croft S. K. et al. (1988) *Icarus, 73,* 279.
88. Squyres S. W. et al. (1988) *J. Geophys. Res., 93,* 8779.
89. Peale S. J. (1988) *Icarus, 74,* 153; Tittemore W. C. and Wisdom J. (1988) *Icarus, 74,* 172.
90. See discussion on accretional heating; Squyres S. W. et al. (1988) *J. Geophys. Res., 93,* 8779.
91. Stevenson D. J. and Lunine J. I. (1986) *Nature, 323,* 46; Squyres S. W. et al. (1983) *Icarus, 53,* 319.
92. Jankowski D. J. and Squyres S. W. (1988) *Science, 241,* 1332.
93. Melosh H. J. and Janes D. M (1989) *Science, 245,* 195.
94. Dermott S. F. et al. (1988) *Icarus, 76,* 295; see also Schenk P. M. (1991) *J. Geophys. Res., 96,* 1887.
95. Croft S. K. (1988) Uranus conference abstracts, Pasadena, 5.10.
96. Cooper H. S. F. (1990) *New Yorker,* June 18, p. 84.
97. Stone E. C. and Miner E. D. (1989) *Science, 246,* 1417; Smith B. A. (1989) *Science, 246,* 1422; Soderblom L. A. et al. (1990) *Science, 250,* 410; Strom R. G. et al. (1990) *Science, 250,* 437.
98. It has long been pointed out by McKinnon and coworkers that tidal heating resulting from the capture of Triton by Neptune would erase any old cratered surfaces; see, for example, McKinnon W. B. (1984) *Nature, 311,* 355; McKinnon W. B. (1988) *Eos Trans. AGU, 69,* 1297; McKinnon W. B. (1989) *Bull. Am. Astron. Soc., 21,* 916; McKinnon W. B. and Benner L. A. M. (1990) in *Lunar and Planetary Science XXI,* p. 777, Lunar and Planetary Institute, Houston; W. B. McKinnon, personal communication, 1990. A similar conclusion has been reached by Ross M. N. and Schubert G. (1990) *Bull. Am. Astron. Soc., 22,* 1127.
99. Banfield D. et al. (1989) *Bull. Am. Astron. Soc., 21,* 911.
100. Christy J. W. and Harrington R. S. (1978) *Astron. J., 83,* 1005.
101. Tholen D. J. and Buie M. W. (1989) *Bull. Am. Astron. Soc., 21,* 982.
102. McKinnon W. B. and Mueller S. (1988) *Nature, 335,* 241.
103. Tholen D. J. and Buie M. W. (1990) *Bull. Am. Astron. Soc., 22,* 1129.
104. 1.53-1.58 g/cm^3; Mueller S. and McKinnon W. B. (1989) *Icarus, 76,* 437.
105. The low density disposes of the old hypothesis that Pluto might be "terrestrial" and an escapee from the inner solar system, as suggested by Whyte A. J. (1980) in *The Planet Pluto,* Pergamon, Oxford, 147 pp.
106. Simonelli D. P. and Reynolds R. T. (1989) *Geophys. Res. Lett., 16,* 1209.
107. Stern S. A. (1989) *Icarus, 81,* 14.
108. Elliot J. L. et al. (1989) *Icarus, 77,* 148.
109. Marcialis R. L. et al. (1987) *Science, 237,* 1349; Trafton L. M. (1989) *Geophys. Res. Lett., 16,* 1213.
110. Hubbard W. B. et al. (1988) *Nature, 226,* 452; Trafton L. M., ibid.
111. Yelle R. V. and Lunine J. I. (1989) *Nature, 339,* 288.
112. Hubbard W. B. et al. (1988) *Nature, 336,* 452.
113. Appelgate J. et al. (1985) *IEEE Trans. Comput., C-34,* 822.
114. Sussman G. J. and Wisdom J. (1988) *Science, 241,* 433.
115. Ahrens T. J. and O'Keefe J. D. (1985) in *Ices in the Solar System* (J. Klinger et al., eds.), p. 631, Reidel, Boston.
116. Drobrovolskis A. R. (1989) *Geophys. Res. Lett., 16,* 1217.
117. W. M. Kaula, personal communication, 1990.
118. McKinnon W. B. (1989) *Astrophys. J., 344,* L41; McKinnon W. B. (1984) *Nature, 311,* 355.
119. Ahrens T. J. and O'Keefe J. D. (1985) in *Ices in the Solar System* (J. Klinger et al., eds.), p. 633, Reidel, Boston.
120. McKinnon W. B. (1989) *Geophys. Res. Lett., 16,* 1237.
121. Peale S. (1991) in *Uranus,* Univ. of Arizona, Tucson, in press.
122. Simonelli D. et al. (1989) *Icarus, 82,* 1.
123. McKinnon W. B. (1989) *Geophys. Res. Lett., 16,* 1238.
124. Prinn R. G. and Fegley B. Jr. (1989) in *Origin and Evolution of Planetary and Satellite Atmospheres* (S. Atreya et al., eds.), p. 78, Univ. of Arizona, Tucson; Fegley B. Jr. and Prinn R. G. (1989) in *The Formation and Evolution of Planetary Systems* (H. A. Weaver and L. Danly, eds.), p. 171, Cambridge Univ., New York.

125. Stevenson D. J. (1989) in *The Formation and Evolution of Planetary Systems* (H. A. Weaver and L. Danly, eds.), p. 75, Cambridge Univ., New York.
126. Laplace P. S. (1809) *The System of the World, Vol.1, Book I* (J. Pond, trans.), p. 40, R. Phillips, London.
127. For general reviews, see Taylor S. R. (1975) *Lunar Science: A Post-Apollo View,* Pergamon, New York, 372 pp.; Taylor S. R. (1982) *Planetary Science: A Lunar Perspective,* Lunar and Planetary Institute, Houston, 481 pp.
128. Darwin G. H. (1879) *Philos. Trans. R. Soc., 170,* 447.
129. Taylor S. R. (1989) *Tectonophysics, 161,* 147.
130. Carlson R. W. and Lugmair G. W. (1988) *Earth Planet. Sci. Lett., 90,* 119.
131. Taylor S. R. and Jakeš P. (1974) *Proc. Lunar Sci. Conf. 5th,* p. 1287.
132. It has recently been suggested that a "... combined crystal chemical-f_{O2} effect... rather than prior removal of plagioclase is responsible for the characteristic lunar mare basalt europium deficiency"; Shearer C. K. and Papike J. J. (1989) *Geochim. Cosmochim. Acta, 53,* 3331.
133. See the negative evaluation of the suggestion of Shearer and Papike [132] by Brophy J. G. and Basu A. (1989) *Proc. Lunar Planet. Sci. Conf. 20th,* p. 25.
134. Tatsumoto M. et al. (1988) in *Lunar and Planetary Science XIX,* p. 1183, Lunar and Planetary Institute, Houston; Premo W. R. et al. (1988) in *Lunar and Planetary Science XIX,* p. 945, Lunar and Planetary Institute, Houston.
135. Delano J. W. (1988) in *Lunar and Planetary Science XIX,* p. 269, Lunar and Planetary Institute, Houston.
136. Morgan J. W. and Wandless G. A. (1979) *Proc. Lunar Planet. Sci. Conf. 10th,* p. 327.
137. Lunar Sample Prelim. Exam. Team (1969) *Science, 165,* 1211.
138. Taylor S. R. (1987) *Geochim. Cosmochim. Acta, 51,* 1297.
139. Mueller S. et al. (1988) *J. Geophys. Res., 93,* 6338.
140. Solomon S. C. and Chaikin J. (1976) *Proc. Lunar Sci. Conf. 7th,* p. 3229.
141. Kirk R. L. and Stevenson D. J. (1989) *J. Geophys. Res., 94,* 12133.
142. Hood L. L. and Jones J. H. (1986) in *Lunar and Planetary Science XVII,* p. 354, Lunar and Planetary Institute, Houston.
143. See review of lunar paleomagnetism by Fuller M. and Cisowski S. M. (1987) in *Geomagnetism 2* (J. A. Jacobs, ed.), Chapter 4, p. 307, Academic, New York.
144. Jacobs J. A., ed. (1987) *Geomagnetism 2,* Academic, New York, 579 pp.
145. Hood L. L. and Williams C. R. (1989) *Proc. Lunar Planet. Sci. Conf. 19th,* p. 99.
146. Huygens C. (1698) *The Celestial Worlds Discovered,* p. 131, T. Childe, London.
147. Laplace P. S. (1809) *The System of the World, Vol. 1, Book I* (J. Pond, trans.), pp. 56-57, R. Phillips, London.
148. Breen J. (1854) *Planetary Worlds,* Robert Hardwicke, London, 250 pp.
149. Solomon S. C. and Chaikin J. (1976) *Proc. Lunar Sci. Conf. 7th,* p. 3229.
150. A review of the magma ocean concept has been given by Warren P. (1985) *Annu. Rev. Earth Planet Sci., 13,* 201, in which he concludes that it remains the most viable hypothesis to explain the geochemical evolution of the Moon.
151. Fujimaki H. and Tatsumoto M. (1984) *Proc. Lunar Planet. Sci. Conf. 14th,* in *J. Geophys. Res., 89,* B445.
152. Laplace P. S. (1809) *The System of the World, Vol. 1, Book IV* (J. Pond, trans.), p. 94, R. Phillips, London.
153. Brush S. G. (1986) in *Origin of the Moon* (W. K. Hartmann et al., eds.), p. 3, Lunar and Planetary Institute, Houston; Brush S. G. (1988) *Space Sci. Rev., 47,* 211; Brush S. G. (1990) *Rev. Mod. Phys., 62,* 43.
154. For example, Hansen K. S. (1982) *Rev. Geophys. Space Phys., 20,* 457.
155. Boss A. P. and Peale S. J. (1986) in *Origin of the Moon* (W. K. Hartmann et al., eds.), p. 59, Lunar and Planetary Institute, Houston.
156. Williams G. E. (1989) *J. Geol. Soc. London, 146,* 97; see also Sonett C. P. et al. (1988) *Nature, 335,* 806.
157. Wood J. A. (1986) in *Origin of the Moon* (W. K. Hartmann et al., eds.), p. 17, Lunar and Planetary Institute, Houston. A history of past endeavors at this intriguing intellectual problem is given by Brush S. G. (1986) in *Origin of the Moon* (W. K. Hartmann et al., eds.), p. 3, Lunar and Planetary Institute, Houston; Brush S. G. (1988) *Space Sci. Rev., 47,* 211; Brush S. G. (1990) *Rev. Mod. Phys., 62,* 43.
158. Durisen R. H. and Gingold R. A. (1986) in *Origin of the Moon* (W. K. Hartmann et al., eds.), p. 487, Lunar and Planetary Institute, Houston.
159. Newsom H. E. and Taylor S. R. (1989) *Nature, 338,* 29.
160. For example, Ringwood A. E. (1986) in *Origin of the Moon* (W. K. Hartmann et al., eds.), p. 673, Lunar and Planetary Institute, Houston.
161. Newsom H. E. (1989) in *Lunar and Planetary Science XX,* p. 784, Lunar and Planetary Institute, Houston.
162. Ringwood A. E. (1986) *Nature, 322,* 323.
163. Drake M. J. et al. (1989) *Geochim. Cosmochim. Acta, 53,* 2101.
164. Ruskol El. Y. (1977) *NASA SP-370,* 815.
165. Wood J. A. and Mitler H. E. (1974) in *Lunar Science V,* p. 851, The Lunar Science Institute, Houston; Smith J. V. (1982) *J. Geol., 43,* 1.

166. Boss A. P. and Benz W. (1989) *Bull. Am. Astron. Soc., 21,* 915.
167. See review by Newsom H. E. and Taylor S. R. (1989) *Nature, 338,* 29.
168. Cameron A. G. W. (1986) in *Origin of the Moon* (W. K. Hartmann et al., eds.), p. 609, Lunar and Planetary Institute, Houston.
169. For example, Benz W. et al. (1989) *Icarus, 89,* 113.
170. For example, Taylor S. R. (1987) *Geochim. Cosmochim. Acta, 51,* 1297; Newsom H. E. and Taylor S. R. (1989) *Nature, 338,* 29.
171. Stevenson D. J. (1987) *Annu. Rev. Earth Planet. Sci., 15,* 271.
172. See Prologue, note 3.
173. Kipp M. E. and Melosh H. J. (1986) in *Origin of the Moon* (W. K. Hartmann et al., eds.), p. 643, Lunar and Planetary Institute, Houston.
174. There are, for example, no coupled Yb and Eu depletions, indicating that the lunar material did not condense from a vapor at temperatures above 1200 K; see below.
175. Cameron A. G. W. and Benz W. (1989) in *Lunar and Planetary Science XX,* p. 137, Lunar and Planetary Institute, Houston.
176. Wetherill G. W. (1986) in *Origin of the Moon* (W. K. Hartmann et al., eds.), p. 519, Lunar and Planetary Institute, Houston.
177. Grossman L. and Larimer J. W. (1974) *Rev. Geophys. Space Phys., 12,* 71.
178. Kreutzberger M. E. et al. (1986) *Geochim. Cosmochim. Acta, 50,* 91.
179. Ringwood A. E. (1986) *Geochim. Cosmochim. Acta, 50,* 1825.
180. For example, Pb, Bi, H_2O, etc.; Jones J. H. and Drake M. J. (1986) *Geochim. Cosmochim. Acta, 50,* 1827.
181. For example, Chou C. L. (1978) *Proc. Lunar Planet. Sci. Conf. 9th,* p. 219.
182. Ozima M. and Podosek F. A. (1983) *Noble Gas Geochemistry,* Cambridge Univ., New York, 367 pp.
183. Swindle T. et al. (1986) in *Origin of the Moon* (W. K. Hartmann et al., eds.), p. 331, Lunar and Planetary Institute, Houston.
184. Nyquist L. E. (1977) *Phys. Chem. Earth, 10,* 103.
185. See section 3.7; $^{87}Sr/^{86}Sr$ for ALL = 0.69881 ± 2.

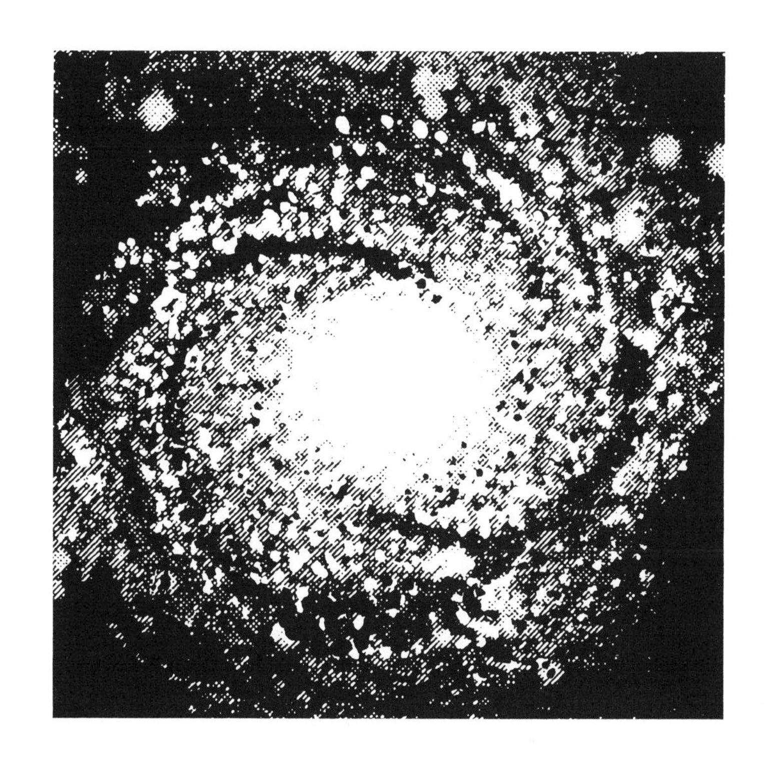

Chapter 7

The New Solar System

The New Solar System

7

7.1. THE END OF CLOCKWORK SOLAR SYSTEMS

"Alphonso, king of Castille (1221–1284 A.D.), was one of the first sovereigns who encouraged the revival of astronomy in Europe. This science can reckon but few such zealous protectors; but he was ill seconded by the astronomers whom he had assembled at a considerable expence [sic], and the tables which they published did not answer to the great cost they had occasioned.... Endowed with a correct judgment, Alphonso was shocked at the confusion of the circles, in which the celestial bodies were supposed to move; he felt that the expedients employed by nature ought to be more simple. 'If the Deity had asked my advice,' said he, 'these things would have been better arranged'" [1].

The problems of the Ptolemaic system had been long understood by acute minds such as Alphonso. The advent of the Copernican system and the success of Newtonian mechanics led to the notion that the solar system was some type of celestial clockwork. Given omnipotent powers, the construction of a well-ordered solar system should not be beyond the powers of a competent clockmaker. Surely God would not have constructed an imperfect system. In the seventeenth and eighteenth centuries, these ideas bore fruit in the construction of mechanical models of the solar system, named orreries after the fourth earl of Orrery, Charles Boyle [2].

These beautiful instruments became very popular. When Louis XV (1710-1774) constructed a new wing at Versailles, an orrery occupied pride of place in the central room, in contrast to the chapel that forms the center of the old wing. This was in keeping with the philosophy of the Age of Enlightenment.

The orderly system embodied in the clockwork orreries influenced thinking about origins of the system until very recently. Theories of a simple hot nebula, cooling down and condensing the planets in an orderly fashion, were constructed until only a few years ago. The concept of a heavenly clockwork, illustrated by the picture of a man peering past a veil of stars to discover a sort of grandfather clock mechanism beyond, typifies this viewpoint. Beyond the clockwork is presumably the clockmaker. Even so distinguished and skeptical a scientist as Kelvin was led to make the following comment: "There may in reality be nothing more of mystery or of difficulty in the automatic progress of the solar system from cold matter diffused through space, to its present manifest order and beauty, lighted and warmed by its brilliant sun, than there is to the winding up of a clock, and letting it go till it stops... a watch spring is much farther beyond our understanding than is a gaseous nebula" [3].

The search for meaning and for regularities and order in the system probably accounts for the popularity of Bode's Rule, surely part of a grand design. But the "rule" is a minor consequence of planetary dynamics, not of fundamental physical significance, nor part of a grand blueprint for constructing planetary systems.

As in so many other fields, ranging from astronomy to genetics, we are reluctantly compelled to realize that we inhabit a system in which chance events play a major role. Thus, the role of collisions has finally been recognized: "The classical celestial mechanics of Newton allowed no collisions... the Sun, moons, and planets do not collide in the normal course of events" [4]. Such notions of the randomness of nature are contrary to the common egocentric philosophies in which man occupies a central role and in which all is designed for his comfort and well-being. The reality is different: "The solar system must be recognized for what it is, just another dynamical system, and as such, the discovery [of] chaotic behaviour... should come as no surprise" [5].

This has been one of the more profound changes in our perception of the world since the construction of Newton's clockwork system could be ascribed to a divine watchmaker.

7.2. THE COLLAPSE OF GRAND UNIFIED THEORIES

"Attempts to find a plausible naturalistic explanation of the origin of the solar system began about 350 years ago... but have not yet been quantitatively successful, making this one of the oldest unsolved problems in modern science" [6].

This book represents a beginning, not an answer, to this question. It discusses the new approach to the problem of the origin of the solar system, now followed by many workers. Rather than constructing grand unified theories, a whole variety of scientific questions that need to be addressed have appeared in each of the preceding six chapters.

A certain amount of confusion has arisen over the question of the existence of other planetary systems. This has been driven by the natural wish to find clones of our system, or of the Earth in particular, complete with intelligent inhabitants. Other planetary systems probably exist, but our experience with our own, with its strong evidence of random or stochastic processes, indicates that the details in other systems can be predicted to be different. Hence, the philosophical question turns not so much on the occurrence of planetary systems, but on whether the detail of our own system is unique.

The division into terrestrial and giant planets, the wide variety of satellites, the existence of that unique satellite of the Earth, the Moon, the asteroid belt, and the many other unusual and astonishing details are as unlikely to be repeated as is the course of the evolution of life on this planet. Thus, the possibility of a clone of our solar system, or of the Earth, is judged to be unlikely.

The great diversity that is observed results from the application of the basic laws of physics and chemistry, but there is no simple recipe from which we can construct the present solar system from first principles, any more than one could predict the existence of elephants from a basic understanding of molecular biology. Such artifacts as the condensation sequence and the attempts to force the anomalous compositions of the Moon and Mercury into the grand plan represent the final attempts to produce a grand unified theory.

Cameron [7] has remarked that, until very recently, the problem of the origin of the solar system has been treated as an intellectual puzzle. Workers took the small number of boundary conditions and attempted to find a complete unified solution. In this endeavor it was customary to give a list of questions that need to be answered by any theory of solar system origin [8].

Usually a dozen or so observations, such as the concentration of mass in the Sun and angular momentum in the planets, the slow rotation of the Sun, the coplanar and prograde rotation of the planets, the Titius-Bode relationship, the number of planets and satellites, the distinction between the terrestrial and giant planets, planetary obliquities, and so forth are listed as important problems to be explained [9].

However, the numerous answers provided over the past 300 years for such questions perhaps explain why there are "... so many theories that all claim to have solved the problem" [10]. Seeking such theories is a false goal, since we cannot identify one dominant process, and only recently has the existence of the solar system been regarded as an "ordinary scientific problem" [7] in which workers investigate individual pieces of the problem. As this book has shown, we have moved into a more realistic and pragmatic view of the solar system in which we live. We are dealing with a system in which many stochastic events occur, with the end result that all planets and satellites are different [11].

In studying the natural world it is difficult to avoid being overwhelmed by detail. Attempting to see the forest for the trees has always been a difficult exercise. The study of chemistry before the discovery of the Periodic Table provides a classic example of a bewildering set of data that was eventually revealed to possess an underlying order. The diversity of organisms in biological systems became understandable as the result of the operation of a single universal law, Darwinian evolution, although the extreme complexity of organisms that have arisen has made any further generalizations difficult. To add further complexity, the course of evolution is driven by chance events, and hence is unpredictable.

In the solar system, a similar diversity results from the application of the basic laws of physics and chemistry. Hence, one is unlikely to find any grand blueprint for constructing such systems from the study of one system; so much of the detail is due to the operation of stochastic processes. Thus, attempts to find some uniform principles, analogous to the Periodic Table or Darwinian evolution, from which one might construct clones of our solar system appear to be on the wrong track.

There is "... a chronic unwillingness to consider the solar system the product of an extremely complicated chain of events, the outcome of the interaction of a very wide range of physical factors. Great as may be the value and attractiveness of the principle of simplicity in the method of exact science, it may very well be that explanation of the solar system demands a very different approach, which in turn could forcefully indicate its excessive rarity if not uniqueness" [12].

Much of our difficulty in trying to understand the solar system stems from the fact that the Earth, with its complex and unique history and obscure cratering record, was not the best place from which to begin. The Moon also turned out to be a unique object. There it is, in plain sight, accessible to naked-eye observation, the closest, but until recently one of the most enigmatic objects in the universe. The controversy over the nature of the lunar dark maria forms a minor but revealing example. Only a few acute observers in pre-Apollo times identified them as basaltic lava floods, since they have few signs of eruptive centers or anything looking like a terrestrial volcano. This is a consequence of the unexpected iron-rich low-viscosity lavas. The Moon contained many other petrological and geochemical surprises. These eventually freed us from the constraining influence of conventional terrestrial experience in the earth sciences [13].

7.3. OUR PRESENT UNDERSTANDING

In the previous chapters, the available evidence has been presented. In this section, I outline, if not a consensus, a reasonable interpretation of the early history of the solar system.

The solar system did not exist for most of the history of the universe. Our collective memory goes back a stupendous 15 or 20 b.y. to the beginning of the observed expansion of the universe. By some 10-15 b.y. after this apparent beginning, the universe had long settled into its present familiar appearance. Galaxies had formed and distributed themselves in patterns of wisps, threads, walls, and knots whose significance we still do not comprehend. Dense molecular clouds formed in the dusty spiral arms of galaxies, breaking into fragments that condensed into stars. Most of these pieces formed binary pairs. Stars were continually being born, evolving, and dying, forming, in the course of these events, the heavier chemical elements and enriching the interstellar medium in them as the stellar material was dispersed.

About 4.6-4.7 b.y. ago, in a universe that would look quite familiar to us, a fragment became detached from a molecular cloud in the spiral arm of one of the 10^{11} galaxies that had come into existence. The fragment was not very large, and it was not spinning very rapidly. This combination of low mass and angular momentum enabled the cloud to condense towards its center, rather than to split into a binary pair. The material was composed of 70% hydrogen, 27% helium, and 2-3% heavier elements. Probably during this chaotic early stage, complex cycles of evaporation, condensation, and melting, repeated many times, produced, like hail, occasional refractory grains that we now observe in meteorites as CAI. This is the first material that we can date at about 4560 m.y. ago.

As mass continued to fall into the center, angular momentum was transferred outward and the remaining material spread out as a rotating dusty disk. This was turbulent and not very well mixed, with substantial variations in the relative abundances of the oxygen isotopes. Dust settled gravitationally toward the midplane of the rotating disk. At this stage, the disk was not symmetrical, but rather resembled a spiral galaxy. Eventually most of the angular momentum was transferred outward, leaving a slowly rotating body in the center, heated, along with the adjacent nebula, through gravitational contraction. Once the central mass reached about 30% of the present solar mass, nuclear reactions started, and a dramatic sequence of events began. The central condensation became convective, and the surface temperature rose to 4000 K. For the first time, the central condensation became visible as a star. As the proto-Sun moved toward the main sequence, bipolar outflows of gas and high-velocity solar winds began to reverse the infall of gas, limiting the size of the Sun. In addition to this violent T Tauri stage, occasional brief, highly luminous events, the FU Orionis stage, added further to the outpouring of energy.

The dust out in the disk was composed of silicates, sulfides, and metal (mostly iron) grains, with some carbonaceous material. Much thermal and magnetic energy was dissipated, mostly in arcing nebular flares high above the midplane. These high-energy events fused dusty silicate balls into chondrules. Mostly, this was a one-shot affair, unlike the repeated cycles of evaporation and condensation that had earlier produced the refractory inclusions. The chondrule factory was, however, much more efficient, processing perhaps half the material now in the chondritic meteorites. Silicates were preferentially melted to form chondrules, while sulfides and metal grains were less efficiently melted, either because they had gravitationally settled preferentially to the midplane, or because they did not stick so well together to form into millimeter-sized balls. Sometimes a refractory inclusion was caught in this maelstrom and remelted to form a chondrule, with only its rare earth element signature surviving as a record of former events.

As all these components sank toward the central plane of the disk, they aggregated into complex mixtures that have survived as the chondritic meteorites. These events occurred very quickly about 4550-4555 m.y. ago. This aggregation of grains into centimeter-, meter-, and kilometer-sized objects remains one of the least understood mechanisms in this opaque topic.

Meanwhile, the early Sun continued its violent career. A combination of high temperatures and strong winds cleared the inner portions of the disk, now identifiable as the solar nebula, of the hydrogen and helium gas. Other gaseous elements, such as the noble gases, were also lost in these episodes from the region in which the terrestrial planets were subsequently to form. There was also, in this part of the nebula, a general depletion of elements volatile below about 1100 K that had not already become incorporated into meter-sized boulders. The general depletion of lead relative to uranium, and of rubidium relative to strontium, provided clocks that can now be read with some difficulty. The inner regions of the asteroid belt were severely heated at this time, causing melting and separation of metal from silicate. In the middle of the belt, mild warming melted the ice in the parent bodies of the carbonaceous chondrites. Sunward of the asteroid belt, the planetesimals were mostly melted and differentiated. Much variation in composition remained, unrelated to distance from the Sun. The outer reaches of the belt, beyond about 3 AU were little affected.

The material swept out from the inner solar nebula piled up at 4 or 5 AU from the Sun, where the temperature was cold enough for water ice to condense. This increase in density had a dramatic effect on the subsequent history of the solar system, for it enabled a rapid accumulation of planetesimals to occur. A rocky and icy body 10-20 times the size of the Earth formed within a million years. It was large enough to cause the gaseous hydrogen and helium, already beginning to be dispersed by the strong solar winds, to collapse by gravitational attraction onto this nucleus. As in most other scenes, the rich become richer, and Jupiter proceeded to collect everything within reach, starving the asteroid belt of material, and ceasing growth only when it had cleaned out the nebula within its tidal reach.

The very early appearance of this giant in the nebula changed all subsequent history. The planetesimals within gravitational reach in the asteroid belt were, if not accreted directly to the massive planet, pumped up into orbits of such inclination and eccentricity that they were unable to collect themselves into a planet. Many of these refugees were either sent to far-distant regions or departed from the system entirely.

Successful though Jupiter was, the nebula was dispersing rapidly, and this planet did not manage to accrete a full solar complement of the gaseous components of the nebula. The situation was worse at Saturn, where the density of the nebula was lower, and it took longer for the core of 10-20 Earth masses of planetesimals to grow large enough to capture the fleeing gas. By the time the ancestors of Uranus and Neptune formed similar cores, the gas was mostly gone, and they accreted mainly ice. Beyond Neptune, the nebula was too diffuse to form planets and the icy components of the nebula formed small planetesimals. Pluto and Triton have survived as examples.

Within the depleted inner regions of the solar nebula, the gas had long since departed. The survivors, a sequence of rocky planetesimals large enough to resist the strong solar winds and heating events that accompanied the violent T Tauri and FU Orionis stages of the early Sun, slowly began to assemble themselves into larger bodies. The accretional energy and, perhaps, other heat sources melted the larger of these ancestors of the terrestrial planets, so that they differentiated into metallic cores and silicate mantles. After 20-50 m.y., four embryos eventually dominated the inner solar system.

As the remaining large bodies were swept up into the planets, dramatic events, whose effects we still observe, occurred. A large body slammed into proto-Mercury, dissipating much of its silicate mantle. Another head-on collision with Venus pushed that planet into a slow backward rotation. A glancing impact between the early Earth and a differentiated body much larger than Mars produced the bone-dry Moon, melted or added heat to an already molten Earth, and affected the tilt of the planet. Mars, already starved by its proximity to Jupiter, suffered a catastrophic bombardment.

The giant planets did not escape. All were knocked about during their final assembly, Uranus, most dramatically, being pushed over through 90°. During these events, satellites settled about the giant planets, some forming from planetary nebulae spun out or knocked out by collisions, and others by capture. Rings of smaller particles formed in more recent times by the collisional disruption of small satellites close to the planets or perhaps from the breakup of icy planetesimals unfortunate enough to come within the embrace of a giant planet.

The residual outer ices in the nebula condensed into bodies mostly a few kilometers in size, and were ejected by gravitational interactions with the outer planets. These icy chunks were tossed out and formed the Kuiper belt in a plane beyond Neptune out to 100 AU or so, and the inner and outer Oort Clouds, the latter comprising a spherical cloud of comets extending out to perhaps 100,000 AU. Stirring of comets by gravitational interactions of these clouds with passing stars placed comets into Sun-crossing orbits. Such processes provided the cometary apparitions that terrified primitive societies by demonstrating that the universe was not fixed and unchanging.

NOTES AND REFERENCES

1. Laplace P. S. (1809) *The System of the World, Vol. 1, Book V* (J. Pond, trans.), p. 293, R. Phillips, London.
2. He commissioned George Graham (d. 1751), renowned as a maker of scientific instruments, to construct an orrery in 1731; the term has been revived for computers dedicated to the calculation of planetary motions (see section 5.14).
3. Thomson Sir W. (Lord Kelvin) (1891) *On the Origin of the Sun's Heat, Popular Lectures and Addresses, Vol. 1,* 2nd edition, pp. 421-422, Macmillan, London.
4. Elliot J. and Kerr R. (1984) in *Rings,* p. 135, MIT, Cambridge.
5. Wisdom J. (1987) *Icarus, 72,* 241.
6. Brush S. G. (1990) *Rev. Mod. Phys., 62,* 43.
7. Cameron A. G. W. (1988) *Annu. Rev. Astron. Astrophys., 26,* 441.
8. See, for example, the list given in Table 1 by Brush S. G. (1990) *Rev. Mod. Phys., 62,* 45.
9. Such lists have some usefulness as a description of the system, but often distract attention from more significant issues. An example from lunar science was the enigma that there are few dark mare basaltic lava flows on the farside of the Moon. This turned out to be a trivial consequence of a thicker farside crust, through which the lavas were unable to rise on account of insufficient hydrostatic head.
10. Haar D. ter and Cameron A. G. W. (1963) in *Origin of the Solar System* (R. Jastrow and A. G. W. Cameron, eds.), p. 34, Academic, New York.
11. If a turning point in thinking about the solar system can be identified, the conference on the *Origin of the Moon,* held in Hawaii in October 1984, may be perceived as the time when the importance of random or stochastic events became widely realized among the community of workers on planetary problems. A similar change at about this period has been noted by the scientific historian Stephen Brush, who identifies the "first space age" as extending from 1956 to 1985; Brush S. G. (1988) *Space Sci. Rev., 47,* 211.
12. Jaki S. L. (1978) *Planets and Planetarians,* p. 248, Halstead/Wiley, New York.
13. An entertaining account of the history of attempts to explain the origin of the Moon is given in Brush S. G. (1988) *Space Sci. Rev., 47,* 211.

Epilogue

The Place of *Homo sapiens* in the Solar System

The creation myths of the world's major religions and other belief systems accept the immensity of space fairly readily as something intuitively obvious from a glance at the night sky. The deep abyss of time, which was the main philosophical consideration from the study of geology, is still not accepted by adherents of some religions. Frazer [1], in *The Golden Bough,* notes that in primitive societies explanations change from magic, through to myth, followed by religion, and then science. He comments, "thus in the acuter minds, magic is gradually superseded by religion, which explains the succession of natural phenomena as regulated by the will, passion or caprice of the spiritual beings like man in kind, though vastly superior to him in power" [1]. Historical attempts to account for the existence of the solar system have followed this pattern.

Theories for the origin of the solar system fall into a sequence. The major turning point can be placed with the work of Laplace, marking the first "modern" attempt to explain the system. Workers before Laplace were not in possession of sufficient factual data to make significant progress. It was first necessary to establish that the Sun, not the Earth, was the center of the system; then it followed that the Earth was just another planet, and not distinct from the rest of the solar system. The Greek philosophers had imagined that the heavenly bodies were made of the shining quintessence, but if the Earth was one of the company, clearly some revision of this cosmology was called for.

One may trace the sequence of thought from Laplace through Darwin, Chamberlin, Moulton, Jeans, and Jeffries to modern attempts. Various major philosophical stumbling blocks have impeded the path. Copernicus dethroned the Earth from a central position in the sixteenth-century view of the universe; more recent discoveries have in turn displaced the Sun from the center of the universe to one star among 10^{11} in a remote corner of one galaxy among 10^{11} galaxies scattered in some presently incomprehensible distribution of wisps, knots, and bubbles. This placing of the Earth and the solar system, itself possibly unique, in a rather distant corner of the universe has implications for the position of *Homo sapiens* in the universe that have not yet been accommodated in most philosophical, mythological, or religious systems.

The next problem was whether planetary systems are normal features, a variant of the common binary star systems, or represent special conditions. Even if single stars are accompanied by Jupiter-sized satellites, this does not imply that other systems like ours have formed—except by chance. The parallels with evolutionary development, with its many random factors, are strong. Perhaps the major question we have to address is not whether planetary systems are common, but whether our present solar system is likely to be duplicated. The outlook for this unfortunately seems poor, despite a frequently expressed human wish to the contrary. It seems clear that many thinkers are driven by the wish to make planetary systems common in the universe, preferably peopled with intelligent beings who can communicate with us once the formidable barriers of time and space are overcome. This cheerful prospect seems unlikely, and *Homo sapiens* is probably alone. Just as the Moon, the inspiration of poets, has turned out to be a unique object and the product of a chance encounter, so our interesting solar system, with its fascinating variety of planets and satellites, a true garden of delights for the observer, is probably one of a kind.

The development of the complex variety of life on this planet is closely linked to random events. The 24-hour period of the rotation of the Earth, partly a consequence of the giant lunar-forming impact, is a critical factor in the evolution of life. A much faster or slower rotation period would have presented evolving organisms with a much more difficult problem. The dynamics that led to the impact of a giant planetesimal at just the appropriate angle and velocity to splash off the Moon, and which removed any initial, probably hostile atmosphere from the Earth, may well account for the benign environment that enabled life to evolve. The subsequent impact 65 m.y. ago that was responsible for the Cretaceous-Tertiary global catastrophe removed the giant reptiles. This set the stage for mammalian evolution to proceed and thus was the precipitating cause of the present dominance of mammals. If the asteroid had missed, it is unlikely that *Homo sapiens* would have evolved at all, and this remarkable planetary system would have continued on its course without developing a species curious enough to reconstruct its history.

Out of 8 planets and 60 satellites, only 1 planet provides a framework for life to develop, which after many a random event, produced *Homo sapiens.* When one combines the many chance events in forming our own system, with those that led to our development, the sober conclusion is that we are alone in the universe. Other planetary systems will differ from the numbers and sizes of planets, just as the geologies of Venus and Mars differ from that of the Earth, and the satellite systems of Jupiter, Saturn, and Uranus differ from one another, so that they might equally well belong to another system entirely.

The philosophical implications of this have scarcely been addressed. The species still retains its highly aggressive instincts, once necessary for survival, but now potentially as much of a danger as the loss of flight was to prove to the Mauritian dodo.

NOTES AND REFERENCES

1. Frazer J. G. (1960) *The Golden Bough: A Study in Magic and Religion,* Macmillan, London, 971 pp.

INDEXES

SUBJECT INDEX

NAME INDEX